Textbook of Cell Biology

Textbook of Cell Biology

Editor: Samantha Granger

CALLISTO REFERENCE

www.callistoreference.com

Callisto Reference,
118-35 Queens Blvd., Suite 400,
Forest Hills, NY 11375, USA

Visit us on the World Wide Web at:
www.callistoreference.com

ISBN: 978-1-63239-921-2 (Hardback)

Cataloging-in-Publication Data

Textbook of cell biology / edited by Samantha Granger.
    p. cm.
Includes bibliographical references and index.
ISBN 978-1-63239-921-2
1. Cytology. 2. Cells. 3. Biology. I. Granger, Samantha.
QH581.2 .T49 2018
571.6--dc23

Table of Contents

Preface

The world is advancing at a fast pace like never before. Therefore, the need is to keep up with the latest developments. This book was an idea that came to fruition when the specialists in the area realized the need to coordinate together and document essential themes in the subject. That's when I was requested to be the editor. Editing this book has been an honour as it brings together diverse authors researching on different streams of the field. The book collates essential materials contributed by veterans in the area which can be utilized by students and researchers alike.

Cell biology discusses the structure and composition cells. DNA damage and replication, cell division and the cell cycle are some of the significant aspects studied under this field. It involves a microscopic as well as molecular study of both prokaryotic and eukaryotic cells. It is an important field which facilitates advancements of related branches like biochemistry, evolution, genetics, nanotechnology, etc. This book explores all the important aspects of cell biology in the present day scenario. It will serve as a valuable source of reference for graduate and post-graduate students.

Each chapter is a sole-standing publication that reflects each author's interpretation. Thus, the book displays a multi-facetted picture of our current understanding of application, resources and aspects of the field. I would like to thank the contributors of this book and my family for their endless support.

Editor

Myxoma Virus Expressing a Fusion Protein of Interleukin-15 (IL15) and IL15 Receptor Alpha Has Enhanced Antitumor Activity

Vesna Tosic[1], Diana L. Thomas[2], David M. Kranz[3], Jia Liu[4], Grant McFadden[4], Joanna L. Shisler[5], Amy L. MacNeill[6], Edward J. Roy[1]*

1 Department of Molecular and Integrative Physiology, University of Illinois at Urbana-Champaign, Urbana, Illinois, United States of America, 2 Neuroscience Program, University of Illinois at Urbana-Champaign, Urbana, Illinois, United States of America, 3 Department of Biochemistry, University of Illinois at Urbana-Champaign, Urbana, Illinois, United States of America, 4 Department of Molecular Genetics and Microbiology, University of Florida, Gainesville, Florida, United States of America, 5 Department of Microbiology, University of Illinois at Urbana-Champaign, Urbana, Illinois, United States of America, 6 Department of Pathobiology at College of Veterinary Medicine, University of Illinois at Urbana-Champaign, Urbana, Illinois, United States of America

Abstract

Myxoma virus, a rabbit poxvirus, can efficiently infect various types of mouse and human cancer cells. It is a strict rabbit-specific pathogen, and is thought to be safe as a therapeutic agent in all non-rabbit hosts tested including mice and humans. Interleukin-15 (IL15) is an immuno-modulatory cytokine with significant potential for stimulating anti-tumor T lymphocytes and NK cells. Co-expression of IL15 with the α subunit of IL15 receptor (IL15Rα) greatly enhances IL15 stability and bioavailability. Therefore, we engineered a new recombinant myxoma virus (vMyx-IL15Rα-tdTr), which expresses an IL15Rα-IL15 fusion protein plus tdTomato red fluorescent reporter protein. Permissive rabbit kidney epithelial (RK-13) cells infected with vMyx-IL15Rα-tdTr expressed and secreted the IL15Rα-IL15 fusion protein. Functional activity was confirmed by demonstrating that the secreted fusion protein stimulated proliferation of cytokine-dependent CTLL-2 cells. Multi-step growth curves showed that murine melanoma (B16-F10 and B16.SIY) cell lines were permissive to vMyx-IL15Rα-tdTr infection. In vivo experiments in RAG1$^{-/-}$ mice showed that subcutaneous B16-F10 tumors treated with vMyx-IL15Rα-tdTr exhibited attenuated tumor growth and a significant survival benefit for the treated group compared to the PBS control and the control viruses (vMyx-IL15-tdTr and vMyx-tdTr). Immunohistological analysis of the subcutaneous tumors showed dramatically increased infiltration of NK cells in vMyx-IL15Rα-tdTr treated tumors compared to the controls. In vivo experiments with immunocompetent C57BL/6 mice revealed a strong infiltrate of both NK cells and CD8$^+$ T cells in response to vMyx-IL15Rα-tdTr, and prolonged survival. We conclude that delivery of IL15Rα-IL15 in a myxoma virus vector stimulates both innate and adaptive components of the immune system.

Editor: Ilya Ulasov, Swedish Medical Center, United States of America

Funding: This study was funded by an anonymous donor arranged by the Illinois Foundation. The funders had no role in study design, data collection and analysis, decision to publish, or preparation of the manuscript.

Competing Interests: The authors have declared that no competing interests exist.

* Email: eroy@illinois.edu

Introduction

The oncolytic potential of many viruses, such as the poxviruses vaccinia virus and myxoma virus, initially suggested that they could be used as cancer therapy, but the efficacy of such viruses as a single agent *in vivo* has been limited [1]. Alternatively, the selectivity of such oncolytic and oncotropic viruses can be used to deliver cytokine genes to cancer cells [2],[3]. One goal of this approach is to shift the immunosuppressive microenvironment found in many solid tumors to an environment that better favors the induction of anti-tumor immune responses.

Myxoma virus is an oncotropic poxvirus that has a particularly attractive safety profile. In the wild, the virus infects only rabbits and other related leporids, and is nonpathogenic in all other nonlagomorph animals tested [4]. Despite its lack of broad pathogenicity other than the rabbit, myxoma virus can replicate in diverse cultured cells from many species, including most human cancer cells, which are particularly permissive for the virus

[5],[6],[7]. It also selectively infects tumors in human xenograft models [8],[9],[10],[11] and primary mouse tumors [12],[13],[11]. It has recently been shown that myxoma virus can discriminate cancerous human myeloid cells from normal CD34$^+$ stem cells, which makes it a potential *ex vivo* purging agent for hematological malignancies [14],[15].

Some oncotropic viruses tested in clinical trials have been modified to express an immunostimulatory cytokine, GM-CSF [16],[17]. Although GM-CSF is a cytokine with potentially favorable anti-tumor activity, it can also stimulate suppressive components of the immune system [18]. Therefore, it is worth exploring other cytokine candidates to be delivered by a tumor-selective viral vector, particularly those that are known to be capable of activating non-responsive or anergic cytotoxic lymphocytes [19].

IL15 is a pro-inflammatory cytokine with significant potential for stimulating T lymphocytes and NK cells against cancer [20].

IL15 expression is tightly regulated at the post-transcriptional level, making IL15 protein largely detectable only in monocytes/macrophages and dendritic cells [21]. Co-expression of IL15 with the α subunit of IL15 receptor (IL15Rα) greatly enhances IL15 stability and function *in vivo* [22], [23],[24]. Since IL-15Rα may be considered a part of the active IL-15 cytokine complex rather than part of the receptor, pre-association of IL-15 with IL-15Rα generates a more potent ligand compared to the cytokine alone [25],[26],[27],[28]. Recombinant myxoma viruses have previously been engineered to express tdTomato red fluorescent protein (vMyx-tdTr) and mouse interleukin-15 (vMyx-IL15-tdTr) [29]. Our previous studies have shown that these myxoma viruses (vMyx-tdTr and vMyx-IL15-tdTr) productively infect cancer cells *in vitro*, but have limited effect on tumor progression of murine melanoma in immune competent mice *in vivo* [30],[31]. In order to deliver the biologically potent form of IL15 with its IL15Rα component *in vivo*, we engineered a new recombinant myxoma virus (vMyx-IL15Rα-tdTr), which expresses IL15Rα-IL15 fusion protein, as well as tdTomato red fluorescent reporter protein.

In this study, we describe the therapeutic effects observed with the new recombinant virus in a mouse model of aggressive melanoma, B16-F10. *In vitro* testing of the virus showed that B16-F10 cells are permissive to the vMyx-IL15Rα-tdTr infection. Secretion of the IL15Rα-IL15 fusion protein was confirmed by ELISA and functional activity of the fusion was assessed by a proliferation assay on IL15-dependent CTLL-2 cells. In *in vivo* experiments, immunohistological analysis of the subcutaneous tumors showed dramatically increased infiltration of NK cells in vMyx-IL15Rα-tdTr treated tumors compared to controls in RAG1$^{-/-}$ mice. In immunocompetent C57BL/6 mice, vMyx-IL15Rα-tdTr increased infiltration by NK cells and CD8$^+$ T cells. RAG1$^{-/-}$ mice with subcutaneous B16-F10 tumors were treated with vMyx-IL15Rα-tdTr, resulting in a significant survival benefit for the treated group compared to the PBS control and the control viruses (vMyx-IL15-tdTr that expresses the native IL15 ligand and control vMyx-tdTr). Treatment of tumor-bearing C57BL/6 mice with vMyx-IL15Rα-tdTr resulted in longer survival than similarly treated RAG1$^{-/-}$ mice. Our results suggest that virally delivered IL15Rα-IL15 drives the recruitment of NK cells and T cells to the site of the tumor and that both the innate and adaptive components of the host immune system play a role in the anti-tumor effect.

Materials and Methods

DNA constructs

pBluescript SK+ plasmid that served as a cloning backbone for the IL15Rα-IL15 – tdTomato expression cassette was obtained from ATCC (Manassas, VA). Whole viral DNA isolated from vMyx-tdTr [29] using the DNeasy Tissue Kit (Qiagen, Valencia, CA) was used as a template for obtaining the PCR fragment containing partial sequences of M135 and M136 genes. HindIII and BamHI cutting sites were introduced in this PCR reaction; primers for all PCR reactions were obtained from Integrated DNA Technologies (Coralville, IA) (Forward primer: 5′- CCA AAG CTT CAC CTG TGT ATG TT -3′, Reverse primer: 5′- CCA GGA TCC ATA ACA CAC AGT TCG G -3′). PCR product from vMyx-tdTr was ligated into pBluescript using the T4 Ligase (Invitrogen, Carlsbad, CA) following sequential digestion with HindIII and BamHI (New England BioLabs, Ipswich, MA). IL15Rα-IL15 fusion protein contains codon-optimized sequence for the murine IL-15Rα sushi domain (amino acids 34–103 of Isoform 1, UniProt accession #Q60819), a linker with the sequence GG(SGG)$_6$ and murine IL-15; it was purchased from

GenScript (Piscataway, NJ) [32]. Poxvirus vvSynE/L promoter, murine Ig κ-chain leader sequence (which directs the protein to the secretory pathway) as well as BspEI and NdeI cutting sites were added and His tag was eliminated from the original IL15Rα-IL15 sequence using forward primer: 5′- CGC AGC TCC GGA AAA AAT TGA AAT TTT ATT TTT TTT TGG AAT ATA AAT AAG ATG GAG ACA GAC ACA CTC CTG CTA TGG GTA CTG CTG CTC TGG GTT CCA GGT TCC ACT GGT GAC ACC ACC TGC CCC CCC CCC GTG -3′ and reverse primer: 5′- TCG CGC CAT ATG TTA TCA GCT GGT GTT GAT GAA CAT CTG CAC G -3′. The resulting IL15Rα-IL15 sequence was cloned into the earlier described pBluescipt construct using BspEI and NdeI (New England BioLabs, Ipswich, MA) and T4 Ligase (Invitrogen, Carlsbad, CA). Finally, tdTomato was cloned out of the plasmid provided by Dr. Brian Freeman (University of Illinois) using primers: forward 5′- GCA GTC GAC ATG GTG AGC AAG G - 3′ and reverse: 5′- CCT GAA TTC TTA CTT GTA CAG CTC G - 3′ and cloned into the existing pBluescript construct using SalI and EcoRI. Resulting plasmid, pBS-IL15Rα-IL15-tdTomatoRed (Figure S1) was used for creating vMyx-IL15Rα-tdTr.

Recombinant viruses

The Lausanne strain of myxoma virus (vMyx-Lau) was used to create recombinant virus expressing tandem dimer Tomato red fluorescent protein (vMyx-tdTr) with or without expression of interleukin-15 (vMyx-IL15-tdTr) by intergenic insertion of the gene cassettes between M135R and M136R of the myxoma virus genome as previously described [29]. Protein expression of IL15 by this recombinant virus is driven by a vaccinia virus late promoter (p11).

The recombinant virus expressing the IL15Rα-IL15 fusion protein (vMyx-IL15Rα-tdTr) was created by homologous recombination in RK-13 cells infected with wild type (WT) vMyx-Lau followed by transfection with the engineered recombination vector pBS-IL15Rα-IL15-tdTomatoRed (Figure S1). The recombination vector contains genes for the IL15Rα-IL15 fusion protein and tdTomato, both under control of the same synthetic vaccinia virus early/late promoter (vvSynE/L promoter). This expression cassette is flanked by M135 and M136 partial gene sequences for the purpose of being transfected into the WT myxoma virus genome between genes M135 and M136. Myxoma virus permissive RK-13 cells were infected with WT vMyx-Lau followed by cationic lipid transfection of the engineered recombination vector pBS with the IL15Rα-IL15-tdTr cassette. After plasmid recombination into the virus, recombinant virus expressing IL15Rα fusion protein was propagated and titrated by focus formation on RK-13 cells. Fluorescent virus foci were harvested, repropagated and titrated on RK-13 cells. This process was repeated three times to isolate a purified virus which contains two exogenous genes; IL-15Rα-IL15 fusion protein and tdTomato. Genomic structure of recombinant virus was confirmed by PCR sequencing.

Cell culture and reagents

Rabbit kidney epithelial (RK-13) cells were a gift from Dr. Richard Moyer (University of Florida, Gainesville, FL; originally from ATCC, Manassas, VA). RK-13 are grown at 37°C, 5% CO2, and 100% humidity in minimum essential medium with Earle's salts (Mediatech, Manassas, VA) supplemented with 2 mM glutamine, 50 U/mL penicillin G, 50 μg/mL streptomycin, 1 mM sodium pyruvate, 0.1 mM nonessential amino acids (MEM-C), and 10% fetal bovine serum (FBS; HyClone, Logan, UT).

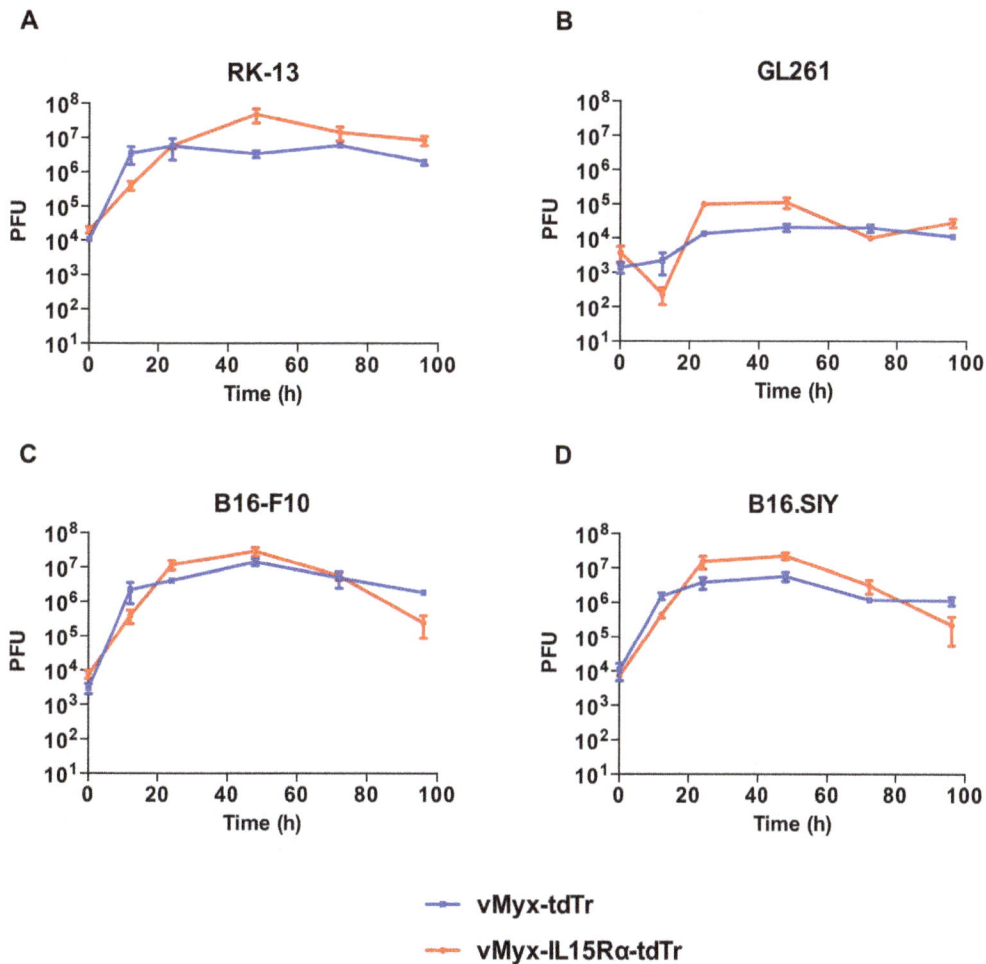

Figure 1. Melanoma and glioma cell lines are permissive to recombinant myxoma virus infection. Cell lines (A) RK-13, (B) GL261, (C) B16-F10, (D) B16.SIY, were infected with either vMyx-tdTr or vMyx-IL15Rα-tdTr at a multiplicity of infection (MOI) of 0.1 to obtain multi-step growth curves. At 0, 12, 24, 48, 72 or 96 h post-infection (p.i.), cells were harvested and lysed, and the viral titer was determined by titration on RK-13 cells. Error bars represent SEM from 3 replicates for each cell line. There was a significant effect of time for each of the cell lines (p<0.001).

The murine melanoma cell line, B16-F10 was purchased from ATCC (Manassas, VA). The murine glioma cell line GL261 was obtained from the National Cancer Institute-Frederick Cancer Research Tumor Repository (Frederick, MD). B16.SIY was derived from B16-F10 cells retrovirally transduced to express green fluorescent protein (GFP) as a fusion protein with SIYRYYGL (SIY) [33,34] and was a gift from Dr. Thomas Gajewski (University of Chicago, Chicago, IL).

Cancer cell lines were cultured in complete Roswell Park Memorial Institute (RPMI) 1640 medium containing 5 mM HEPES, 1.3 mM L-glutamine, 50 µM 2-ME, penicillin, streptomycin and 10% fetal bovine serum (FBS) at 37°C and 5% CO_2.

Cytotoxic T Lymphocyte Line 2 (CTLL-2) cytokine-dependent murine T cell line (ATCC, Manassas, VA) was cultured in complete RPMI 1640 medium additionally supplemented with 10% T-Stim (culture supernatant from rat T cells stimulated with ConA from BD Biosciences, San Jose, CA).

Viral growth curves

Tested cell lines were plated into 6-well cell culture plates (Nunc, Roskilde, Denmark) and grown in MEM-C with 10% FBS until they reached 90–95% confluency. For multi-step growth

curves cells were inoculated with vMyx-IL15Rα-tdTr or vMyx-tdTr diluted in 400 µL MEM-C at a multiplicity of infection (MOI) of 0.1 plaque-forming units (PFU) per cell. Inoculated cells were incubated at 37°C and 5% CO_2 for 1 h, rocking every 15 min. Next, virus was removed, cell monolayers were washed with phosphate-buffered saline (PBS), and MEM-C with 10% FBS was added to each well. Inoculations of each cell line were performed in triplicate. At 0, 4, 8, 12, 24 and 48 h post-inoculation, cells were dislodged by scraping into media. Cells were collected by centrifugation at 500xg for 5 min. Next, each supernatant was removed, and the cellular pellet was resuspended in 0.5 ml PBS and stored at −80°C. Prior to titering the virus, cells were disrupted by three freeze/thaw cycles and sonication in order to release the virus from the cells. Samples were titered in duplicate. Titering was performed by plating 10-fold serial dilutions of the samples in MEM-C onto RK-13 monolayers in 6-well or 24-well culture plates. The inoculated cells were incubated for 1 h at 37°C and 5% CO_2, then the inoculum was removed and an overlay consisting of equal amounts of 1% agarose (Lonza, Rockland, ME) and 2×MEM-C with 20% FBS was added on RK-13 cells. Viral plaques were visualized as small white foci (red foci under fluorescent light) and counted at 6-7 days post-inoculation (dpi).

ELISA analysis of the IL15Rα-IL15 fusion protein

RK-13 cells were plated in 6-well culture plates and upon reaching 90–95% confluency they were inoculated with vMyx-IL15Rα-tdTr or vMyx-tdTr diluted in 400 μL MEM-C at MOI of 5 PFU/cell. After 1 h incubation at 37°C and 5% CO_2, inoculum was replaced with MEM-C with 10% FBS. At different time points post-inoculation, both cell supernatant and cellular extract were collected. Supernatants were centrifuged briefly to remove cellular debris and clarified supernatants were transferred to new tubes and stored at −80C. The remaining cellular monolayer was detached from the well by scraping cells into 1 ml PBS. Cells were collected, pelleted by brief centrifugation (1,300 rpm x 1 min), and cellular pellets were resuspended in Cytoplasmic Extract (CE) buffer supplemented with HALT protease inhibitor cocktail (Thermo Fisher, Rockford, IL). Samples were incubated for 5 min at 4°C and centrifuged at 1,300 rpm for 1 min. Supernatants were moved to new tubes are stored at −80°C.

For IL15Rα-IL15 detection by ELISA, the Mouse IL-15/IL-15R Complex ELISA Ready-SET-Go! kit (eBioscience, San Diego, CA) was used. All samples were 10-fold serially diluted and each dilution was done in duplicate. Each kit included a purified protein standard which was used to establish a standard curve. An ELx800 Absorbance Microplate Reader (BioTek Instruments, Winooski, VT) was used to detect absorbance at 450 nm.

CTLL-2 cell proliferation assay

MTT reagent (3-(4,5-dimethylthiazolyl-2)-2,5-diphenyltetrazo-lium bromide) and detergent were purchased from ATCC (Manassas, VA). Confluent RK-13 cells in 6-well plates were infected with vMyx-IL15Rα-tdTr or vMyx-tdTr at MOI = 5. At 24 h and 48 h p.i., cell-free media was collected and stored at −80°C. CTLL-2 cells were propagated overnight at 37°C, 5% CO_2 in complete RPMI media with no added cytokines. Next, CTLL-2 cells were collected by centrifugation and resuspended at 50,000 cells per well in a 96-well plate in 100 μL complete RPMI containing either 10^{-9}M IL-2, 10^{-9}M TCR-IL15Rα (fusion of the m33 TCR with IL15Rα-IL15, "m33-superfusion" [32]), vMyx-tdTr or vMyx-IL15Rα-tdTr infected cell media. The cells were cultured for 48 h, and then 10 μL MTT was added per well, and the cells were incubated at 37°C, 5% CO_2 for three more hours, and then 100 μL per well detergent was added, and the plate was incubated at room temperature overnight. To estimate CTLL-2 cell proliferation in different conditions, absorbance at 570 nm in each well was read using an EL_x800 universal microplate reader (Bio-Tek Instruments, Winooski, VT).

Animals

C57BL/6 and C57BL/6 RAG1$^{-/-}$ mice originally purchased from The Jackson Laboratory (Bar Harbor, ME, USA) were maintained as colonies and housed in the animal facilities at the University of Illinois. Mice were used in experiments when they were 2–5 months old. All animal studies were approved by the Institutional Animal Care and Use Committee at the University of Illinois Urbana-Champaign (PHS Assurance A3118-01, AAA-LAC, International Accreditation #00766). Anesthesia was used during tumor cell and virus injections, and all efforts were made to minimize suffering.

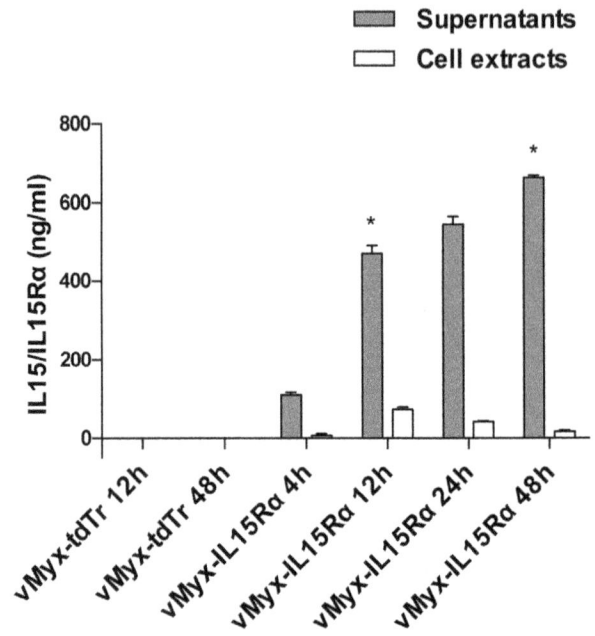

Figure 2. IL15Rα-IL15 fusion protein is present in the supernatants and extracts of cells infected with vMyx-IL15Rα-tdTr. Confluent RK-13 cells in 6-well plates were infected with vMyx-IL15Rα-tdTr or vMyx-tdTr at MOI = 5. At indicated times post-infection, media was collected and cells were scraped, lysed and cytoplasmic extract was harvested. Mean ELISA values with SEM for replicates of the same condition are presented, and the experiment was repeated with similar results. There was a significant increase in IL15Rα-IL15 fusion protein in supernatants of vMyx-IL15Rα-tdTr treated cells compared to vMyx-tdTr treated ones at corresponding timepoints (* - p<0.05).

Subcutaneous tumor establishment and treatment

B16-F10 melanoma cells were harvested and washed twice with Hanks Balanced Salt Solution (HBSS, Cellgro Mediatech Inc., Manassas, VA). Prior to all procedures, mice were anesthetized by isoflurane (Aerrane, Baxter, Deerfield, IL) inhalation using the classic vaporizer unit by E–Z Systems (Palmer, PA). Shaved mice received 1x 10^6 tumor cells in 100 μl HBSS subcutaneously into the right flank. After 7 days, when tumors usually reach a volume of approximately 100 mm^3, mice were assigned to treatment groups and received an intratumoral (i.t.) injection of virus. At this time, tumors were directly injected with 2.6×10^7 PFU of sucrose-pad purified vMyx-IL15α-tdTr, vMyx-IL-15-tdTr or vMyx-tdTr that was in a final volume of 50 μl. A separate set of mice received 50 ul PBS i.t. An additional i.t. inoculation of each virus (2.6×10^7 PFU) occurred 3 days later (day 10 post-implantation). For those tumors that were large, the inoculum was injected into at least three different sites to introduce the virus throughout the mass. Prior to all repeated tumor injections with either virus or PBS, animals were anesthetized by isoflurane inhalation as described earlier. All animals were single housed upon tumor cell implantation and during all subsequent experimental manipulations.

Tumor growth was monitored by measuring tumor length, width and height with a caliper. Tumor volume was calculated as ((length) x (width) x (height))/2. Mice were monitored daily. Mice were humanely euthanized when tumors reached the volume of 3000 mm^3, or showed lethargy or signs of pain, or when reaching 75% baseline body weight. Mice were euthanized by CO_2 asphyxiation followed by cervical dislocation. In some experiments,

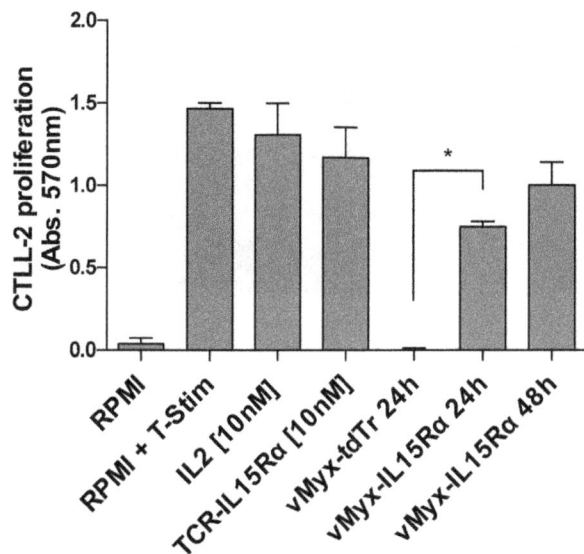

Figure 3. IL15Rα-IL15 fusion protein secreted by virus-infected RK-13 cells is functionally active. Confluent RK-13 cells in 6-well plates were infected with vMyx-IL15Rα-tdTr or vMyx-tdTr at MOI = 5. At 24 h and 48 h p.i., cell media was collected. Cytokine dependent CTLL-2 cells were incubated with un-supplemented media (RPMI), three positive controls (media supplemented with 10% T-Stim (RPMI + T-Stim), 10^{-9}M IL-2, 10^{-9}M TCR-IL15-IL15Rα fusion protein (TCR-IL15Rα)), and vMyx-tdTr or vMyx-IL15Rα-tdTr supernatants. CTLL-2 cell proliferation was analyzed by the MTT assay. Experiment was done in triplicate for each treatment, and mean values with SEM are presented. Positive controls and supernatants from RK-13 cells infected with vMyx-IL15Rα-tdTr (marked by *) showed significant functional IL-15 activity, but supernatant from vMyx-tdTr infected RK-13 cells did not.

samples from some mice were collected 3 days after final virus treatment for histological analysis, and other mice in each treatment group were monitored until they reached a criterion for euthanasia.

Tissue sections and immunostaining

After the mice were euthanized, their subcutaneous tumors were snap-frozen in OCT medium for cryosectioning and immunostaining. Eight μm cryosections were taken. Primary antibodies used for staining were: 4D11 (rat anti-Ly-49G2, BD Pharmingen, San Jose, CA), rabbit anti-CD3 (Abcam, Cambridge, MA), rat anti-CD8 (eBioscience, San Diego, CA), rat anti-CD4 (Abcam, Cambridge, MA). Secondary antibodies used: biotin rabbit anti-rat and biotin goat anti-rabbit (Vector, Burlingame, CA). For immunostaining, slides were fixed in cold 95% ethanol and blocked with Superblock (Thermo Scientific, Rockford IL). Sections were then incubated with a primary antibody in PBS+ 20% glycerol (PBSG) overnight, washed with PBS+0.1% Tween-20 (PBST), and incubated with biotinylated secondary anibody in PBST for 4 h. Slides were washed and incubated with streptavidin-Alexa Fluor 594 (Invitrogen, Carlsbad, CA) or streptavidin-Alexa Fluor 488 (Jackson ImmunoResearch, West Grove, PA) and DAPI (Invitrogen, Carlsbad, CA). Control slides omitting the primary antibody were negative for Alexa Fluor 594 or Alexa Fluor 488. Images were obtained with an Olympus BX-51 microscope at 20x magnification.

Data Analysis

GraphPad Prism software (La Jolla, CA) was used for all statistical analyses and graph presentation. Survival data were

recorded from the time of the tumor cell implantation until euthanasia and were plotted using a Kaplan-Meier curve. Survival treatment groups were compared with a Log-rank (Mantel-Cox) test. Virus growth curves were analyzed by two-way ANOVA, ELISA data were analyzed by t-test at corresponding time points, the bioassay of IL-15 activity was analyzed by one-way ANOVA, and histological cell counts were analyzed by one-way ANOVA with Bonferroni planned comparisons. Significance was considered p<0.05.

Results

Murine melanoma and glioma cell lines are permissive for recombinant myxoma virus infection

Prior to testing the therapeutic capacity of the recombinant IL15Rα-IL15 virus (vMyx-IL15Rα-tdTr), we tested its capacity to infect relevant murine cancer cell lines *in vitro*. Multi-step growth curves of vMyx-IL15Rα-tdTr and the previously characterized vMyx-tdTr control virus [29], showed similar patterns of permissiveness in various cell lines that were tested (Figure 1). Melanoma cell lines (B16-F10 and B16.SIY) were as permissive as the positive control rabbit cell line RK-13. For all three cell lines, infectious viral particles were formed by 12 h post infection, and maximal viral titer was typically obtained at the 48 h time point. vMyx-IL15Rα-tdTr and vMyx-tdTr showed a different growth phenotype in the glioma cell line (GL261) and produced lower viral titers (Figure 1B). Based on the observation that the insertion of the IL15Rα-IL15 gene did not impact the infectivity of the virus, vMyx-IL15Rα-tdTr was considered a useful system to deliver functional IL15Rα-IL15 fusion protein to B16 tumors.

IL15Rα-IL15 fusion protein is expressed and secreted *in vitro* by virus infected cells

To determine if cells infected with vMyx-IL15Rα-tdTr were capable of secreting IL15Rα-IL15 fusion protein, we examined the culture media and cell extracts of infected cells using an ELISA specific for the IL15/IL15R complex. IL15Rα-IL15 was detected in both supernatants and cell extracts of vMyx-IL15Rα-tdTr infected RK-13 cells (MOI = 5) as compared to the control non-cytokine expressing virus vMyx-tdTr (Figure 2). The peak of cell-associated expression of the fusion protein occurred at 12 h post-infection (mean value of 73 ng/ml), while secreted levels peaked at 48 h post-infection (mean value of 663 ng/ml). IL15Rα-IL15 was present in ten-fold higher levels in cellular supernatants versus cell-associated. As would be expected, cells infected with non-cytokine expressing virus (vMyx-tdTr) did not show measurable levels of IL15Rα-IL15. The presence of IL15 and IL15Rα domains was also confirmed by Western blot of supernatants and cell extracts (Figure S2 and Materials and Methods S1). These findings showed that the recombinant virus was transcribed effectively and that the translated IL15Rα-IL15 fusion protein was secreted from infected cells at high levels (over 500 ng/ml).

IL15Rα-IL15 fusion protein secreted by vMyx-IL15Rα-tdTr infected cells is functionally active

Functional activity of the IL15Rα-IL15 fusion protein in the supernatants of virus-infected cells was assayed by its ability to induce proliferation of cytokine-dependent CTLL-2 cells (Figure 3). CTLL-2 is a clone of T cells that requires IL-2 or other growth-promoting cytokines for proliferation [35]. MTT cell proliferation assays showed that CTLL-2 cells cultured in medium supplemented with supernatants of vMyx-IL15Rα-tdTr infected cells proliferated to the similar extent as CTLL-2 cells incubated

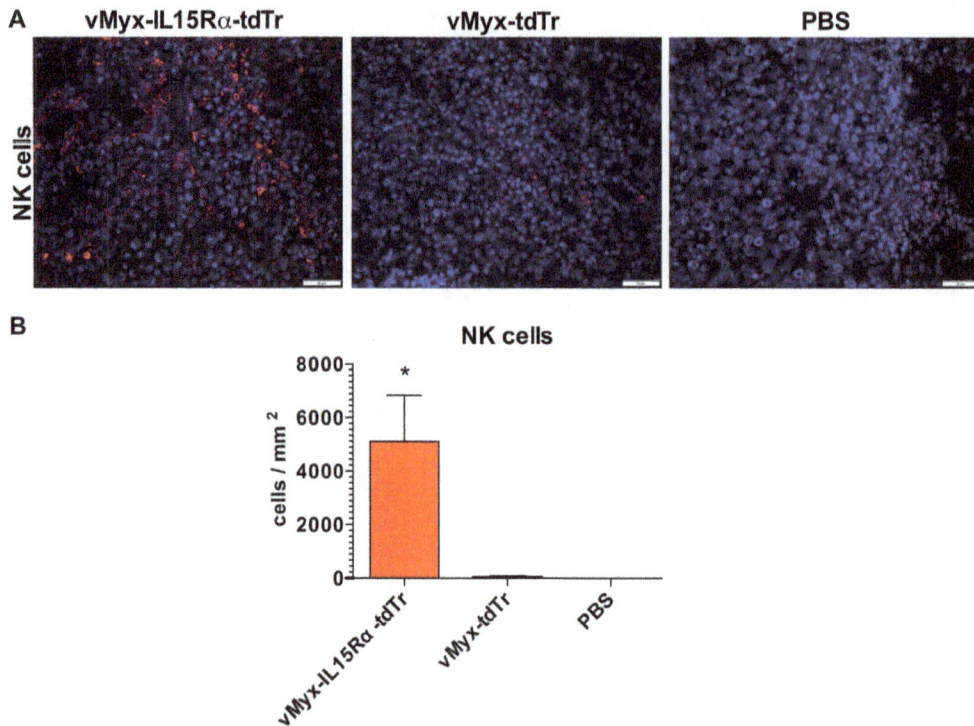

Figure 4. NK cell infiltration of subcutaneous B16-F10 tumors 3 days after intratumoral virus treatment in RAG1$^{-/-}$ mice. RAG1$^{-/-}$ mice (n = 3 per group) were implanted with unilateral subcutaneous B16-F10 tumor cells. The first dose of the virus (2.6×10^7 PFU i.t.) was given on day 7 (when tumors reached approximately 100 mm^3) and the second dose was given on day 10. Treatment groups are: 1. vMyx-IL15Rα-tdTr 2. vMyx-tdTr 3. PBS. Mice were euthanized 3 days after the final virus treatment and tumor sections were analyzed for presence of NK cells by immunostaining for Ly-49G2 (4D11 antibody). Representative tumor sections are shown. (A) Staining for NK cells in tumors. Red – 4D11-positive stain, Blue – DAPI. Scale bar = 50 micrometers. (B) Estimated number of NK cells per square millimeter of a tumor section for each condition. Presented values are mean cell count in tumors from three mice, with SEM. One-way ANOVA showed significant increase in NK cell accumulation in vMyx-IL15Rα-tdTr treated tumors compared to both vMyx-tdTr and PBS treatments (* - p<0.05).

with recombinant IL-2 (10 nM) or a purified fusion protein of IL15Rα-IL15 and a single-chain TCR m33 (10 nM) [32]. CTLL-2 cells cultured with supernatant of the control virus vMyx-tdTr were not stimulated to proliferate.

Treatment with IL15Rα-IL15 fusion protein expressing myxoma virus results in increased presence of NK cells in tumors of RAG1$^{-/-}$ mice

We next tested whether this new recombinant virus would affect cellular immune responses *in vivo*. Because NK cells are responsive to IL15 [36], we investigated whether treatment with vMyx-IL15Rα-tdTr was associated with the presence of NK cells in subcutaneous tumors of RAG1$^{-/-}$ mice, which have NK cells but no T or B cells. Accordingly, RAG1$^{-/-}$ mice bearing established subcutaneous B16-F10 tumors were injected intratumorally (i.t.) with vMyx-IL15Rα-tdTr, vMyx-tdTr or PBS on days 7 and 10 post tumor cell injection. Tumor sections of mice treated with the virus expressing IL15Rα-IL15 fusion protein showed dramatic and significant increase in numbers of infiltrating NK cells, compared to vMyx-tdTr and PBS treated tumor (Figure 4). This evidence suggests a role of NK cells as a component of the host immune system that may contribute to an anti-tumor effect.

vMyx-IL15Rα-tdTr treatment enhances both NK cell and T cell recruitment to subcutaneous tumors in C57BL/6 mice

To determine the effects of the virus in fully immunocompetent animals, we repeated the experiment using C57BL/6 mice. C57BL/6 mice with subcutaneous B16-F10 tumors were injected intratumorally with vMyx-IL15Rα-tdTr, vMyx-tdTr or PBS on days 7 and 10 post tumor cell injection. Similar to the effect observed in RAG1$^{-/-}$ mice, C57BL/6 mice treated with vMyx-IL15Rα-tdTr also had significant intra-tumor infiltration of NK cells, compared to both tdTomato expressing virus and PBS treatment (Figure 5A, B). In addition, T cell infiltration mirrored that of NK cells (Figure 5C, D). Analysis of T cell subsets in this response revealed that most tumor infiltrating T cells were CD8$^+$, although CD4$^+$ cells were also elevated in vMyx-IL15Rα-tdTr treated tumors compared to controls (Figure 6).

Mice bearing subcutaneous melanoma tumors live longer when treated with IL15Rα-IL15 fusion protein-expressing virus compared to control viruses

For survival experiments, mice with established B16-F10 s.c. tumors were treated the same way as described for histological analysis (intratumoral virus treatment on days 7 and 10 post tumor cell injection) and were monitored for survival. For RAG1$^{-/-}$ mice, treatment groups were vMyx-IL15Rα-tdTr, vMyx-IL15-tdTr, vMyx-tdTr and PBS. Without any treatment (PBS), B16-F10 grows as an exceptionally aggressive tumor, with a median survival

A

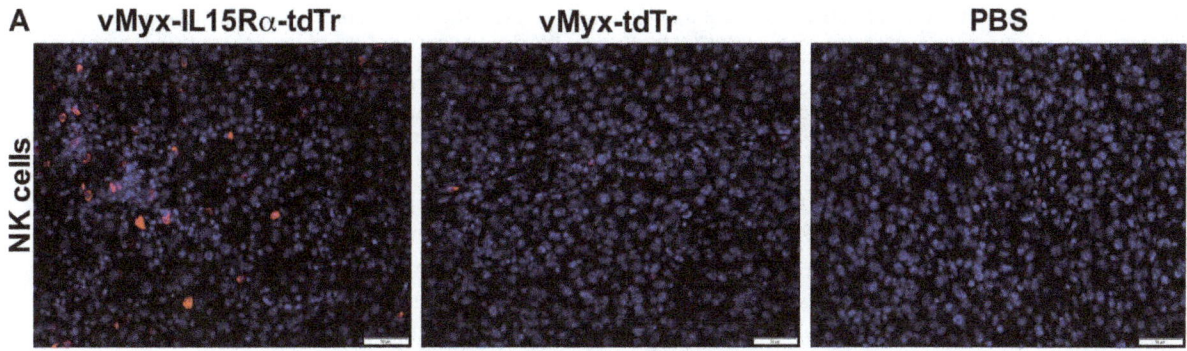

B

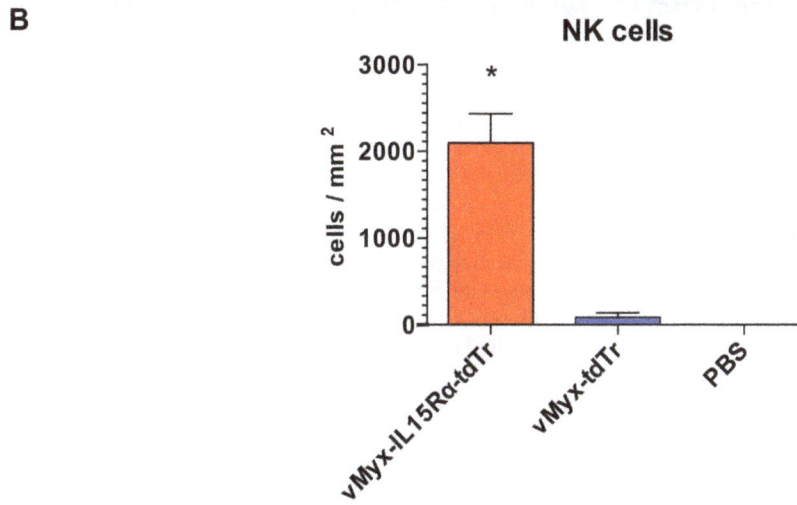

C

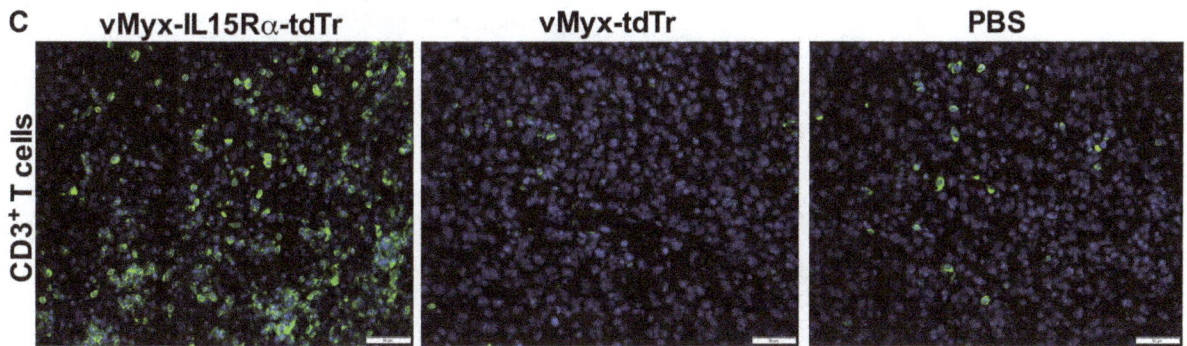

D

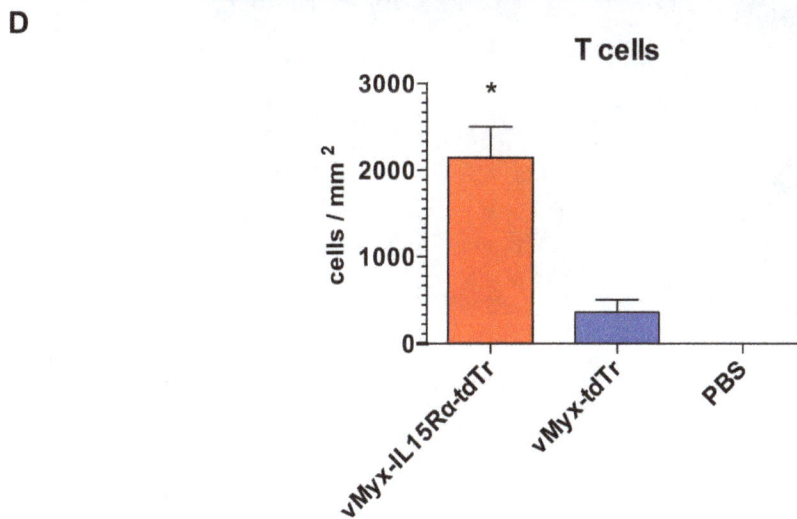

Figure 5. NK and T cell infiltration of subcutaneous B16-F10 tumors 3 days after intratumoral virus treatment in C57BL/6 mice. C57BL/6 mice (n = 3 per group) were implanted with unilateral subcutaneous B16-F10 tumor cells. The first dose of the virus (2.6×10^7 PFU i.t.) was given on day 7 (when tumors reached approximately 100 mm^3) and the second dose was given on day 10. Treatment groups are: 1. vMyx-IL15Rα-tdTr, 2. vMyx-tdTr, 3. PBS. Mice were euthanized 3 days after the final virus treatment and tumor sections were analyzed for presence of NK cells and T cells by immunostaining for Ly-49G2 (4D11 antibody) and CD3, respectively. Representative tumor sections are shown. (A) Staining for NK cells in tumors. Red – 4D11-positive stain, Blue – DAPI. Scale bar = 50 micrometers. (B) Estimated number of NK cells per square millimeter of a tumor section for each condition, mean values and SEM from 3 mice per group. One-way ANOVA showed significant increase in NK cell accumulation in vMyx-IL15Rα-tdTr treated tumors compared to both vMyx-tdTr and PBS treatments (* - p <0.05). (C) Staining for T cells in tumors. Green – CD3-positive cells, Blue – DAPI. Scale bar = 50 micrometers. (C) Estimated number of T cells per square millimeter of a tumor section for each condition, mean values and SEM from 3 mice per group. One-way ANOVA showed significant increase in T cell accumulation in vMyx-IL15Rα-tdTr treated tumors compared to both vMyx-tdTr and PBS treatments (* - p<0.05).

of 17 days. A small survival benefit was observed in the tdTomato-only expressing virus group, similar to values obtained in a slightly different experimental setting [30]. Addition of the IL15 alone to the virus construct did not result in any improvement above this survival in RAG1$^{-/-}$ mice. However, addition of the IL15Rα-IL15 fusion protein improved therapeutic efficacy of myxoma virus compared to the other virus controls, including myxoma virus that expressed only the native IL15 domain (Figure 7). vMyx-IL15Rα-tdTr treatment resulted in tumor stabilization in the majority of animals until day 20, while mice given other treatments were succumbing to tumors at this point (Figure 7A). For C57BL/6 mice, treatment groups were vMyx-IL15Rα-tdTr, vMyx-tdTr and PBS. The anti-tumor effect of vMyx-IL15Rα-tdTr in immuno-competent animals showed the same pattern as in RAG1$^{-/-}$ mice, with overall longer median survival in corresponding groups (Figure 8). The effect on both strains is especially notable given that the time of treatment was when the tumors were already established and at the start of their aggressive growth phase.

Discussion

IL15 has been proposed as a useful cytokine for immunotherapy for cancer, and the complexing of IL15 with its receptor alpha component has been shown to enhance its biological activity. We therefore modified a viral system to deliver the fusion protein of

IL15Rα-IL15, employing a myxoma virus vector with a strong safety profile. We confirmed that the vMyx-IL15Rα-tdTr virus has the same ability to infect melanoma cells as the previously characterized vMyx-tdTr control virus, and that it secretes biologically active IL15Rα-IL15.

IL15Rα-IL15 could potentially be delivered to tumors by a variety of means. For example, Bessard et al. delivered an IL15Rα-IL15 fusion protein by three systemic injections, prolonging the survival with a B16-F10 model for 7 days [28], and Dubois et al. injected IL15 preassociated with IL15RαFc, repeated as many as nine times, prolonging survival of B16-F10 bearing mice for 5 days [24]. In comparison, in the present study two injections of vMyx-IL15Rα-tdTr resulted in a prolongation of survival of 12 days in RAG1$^{-/-}$ mice and 20 days in C57BL/6 mice. Delivery by a viral vector results in secretion of virally encoded proteins peaking at 48–72 h and persisting for up to a week [31], so most likely fewer treatments would be needed to maintain the presence of the cytokine in the tumor environment.

In the survival experiments in RAG1$^{-/-}$ mice we compared effects of vMyx-IL15Rα-tdTr with vMyx-IL15-tdTr [29] (virus expressing IL15 but without the IL15Rα fusion component) as well as non-cytokine expressing vMyx-tdTr. Based on recent literature [23],[37],[38],[39],[40], adding the IL15Rα significantly improves IL15 effects compared to the cytokine itself. This was confirmed in our experimental setting: IL15-only expressing virus,

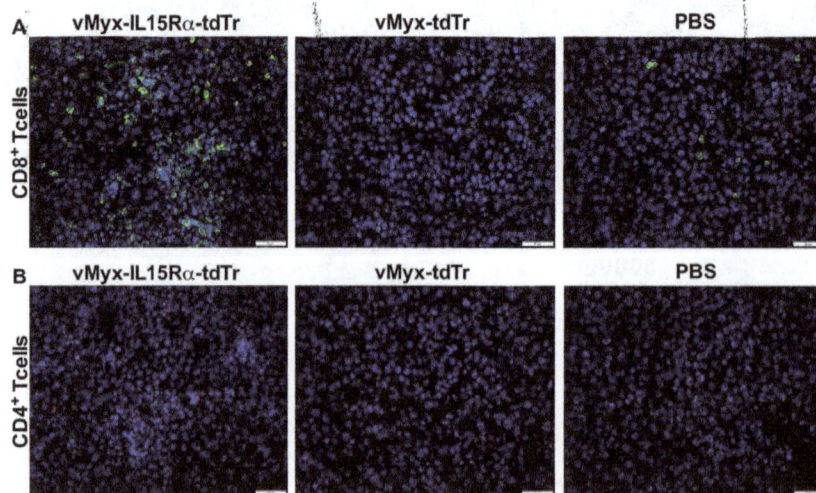

Figure 6. Analysis of subsets of T cells infiltrating subcutaneous B16-F10 tumors 3 days after intratumoral virus treatment in C57BL/6 mice. C57BL/6 mice (n = 3 per group) were implanted with unilateral subcutaneous B16-F10 tumor cells. The first dose of the virus (2.6×10^7 PFU i.t.) was given on day 7 (when tumors reached approximately 100 mm^3) and the second dose was given on day 10. Treatment groups are: 1. vMyx-IL15Rα-tdTr, 2. vMyx-tdTr, 3. PBS. Mice were euthanized 3 days after the final virus treatment and tumor sections were analyzed for presence of T cells by immunostaining for CD4 and CD8 markers. Representative tumor sections are shown. (A) Staining for CD8$^+$ T cells in tumors. Green – CD8-positive cells, Blue – DAPI. Scale bar = 50 micrometers. (B) Staining for CD4$^+$ T cells in tumors. Red – CD4-positive cells, Blue – DAPI. Scale bar = 50 micrometers.

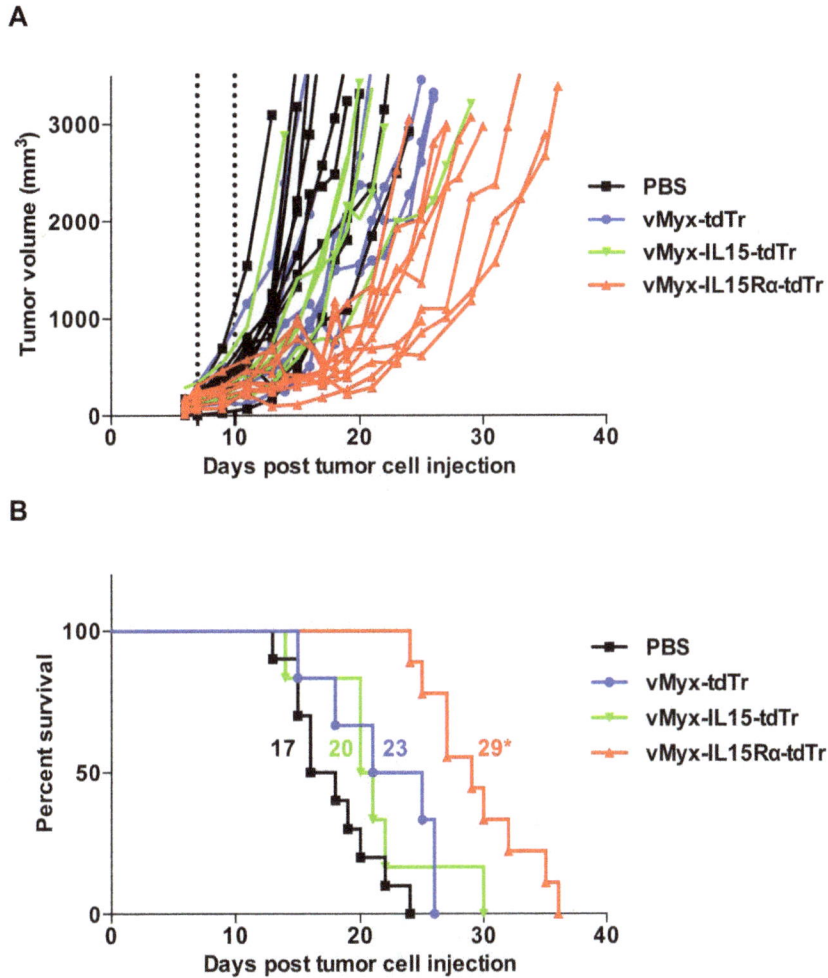

Figure 7. Prolonged survival of RAG1$^{-/-}$ mice bearing subcutaneous B16-F10 tumors treated with vMyx-IL15Rα-tdTr intratumorally. RAG1$^{-/-}$ mice (10 mice in vMyx-IL15Rα-tdTr and PBS and 6 mice in vMyx-IL15-tdTr and vMyx-tdTr treatment groups) were implanted with subcutaneous B16-F10 tumor cells. 7 days later, when tumors reached approximately 100 mm^3, virus was inoculated intratumorally (i.t.) with 2.6×10^7 PFU vMyx-IL15Rα-tdTr, vMyx-IL15-tdTr, vMyx-tdTr or PBS. Mice received a second i.t. inoculation of 2.6×10^7 PFU of each virus on day 10. (A) Growth of individual tumors. Dashed lines designate time of virus treatment, and growth of tumors was measured every 2 days. (B) Kaplan–Meier survival curve of the same experimental subjects. Numbers next to corresponding survival curves designate median survival time (days). (* - p<0.05 for vMyx-IL15Rα-tdTr treated group compared to other vMyx, as well as PBS treatment).

consistent with published data [30], showed therapeutic effect against murine melanoma tumors in the RAG1-knockout background indistinguishable from vMyx-tdTr. Hence, for most of our other studies we compared the novel recombinant virus with the variant that was closer to wild type, expressing only the fluorescent protein.

Both NK cells and CD8$^+$ T cells responded to vMyx-IL15Rα-tdTr. In RAG1$^{-/-}$ mice, histological analysis revealed robust NK cell accumulation in the tumors of the treated animals. Previous studies have done depletion of NK cells prior to treatment to show that NK cells contribute to the anti-tumor effect of IL15/IL15Rα [28],[41]. In some models, the effect of IL15Rα-IL15 is more dependent on CD8$^+$ T cells [27]. In immunocompetent C57BL/6 mice, both NK cells and CD8$^+$ T cells heavily infiltrated the tumors following vMyx-IL15Rα-tdTr treatment. Consistent with the idea that both cell types play a role in the effects of vMyx-IL15Rα-tdTr, treated C57BL/6 mice survived longer than treated RAG1$^{-/-}$ mice (43 days versus 29 days, p<0.05). In our previous report of the effects of vMyx-IL15-tdTr without the receptor-α

[30], we observed a significant increase in the number of CD3$^+$ cells but no significant increase in survival. Quantitation of cell numbers was conducted differently in the two studies, but the density of infiltration by CD3$^+$ cells following vMyx-IL15Rα-tdTr treatment appears greater than following vMyx-IL15-tdTr treatment, consistent with numerous reports of greater biological activity of IL15 when it is combined with its receptor-α subunit.

Elpek et al. observed that sustained activation of NK cells by IL15/IL15Rα treatment (5 injections over 2 weeks) can lead to functional exhaustion of effector functions of NK cells [42]. Viral delivery by myxoma virus produces IL15Rα-IL15 secretion that is intermediate between a rapidly cleared systemic injection and chronic exposure. Future studies could determine an optimal interval for repetitive treatments to minimize NK cell exhaustion. In order to assess the effector function of infiltrating NK cells and T cells, we attempted to dual-stain tumor sections with anti-interferon-γ and markers for NK cells or T cells. However, consistent with a report by Van der Loos [43] that interferon-γ is

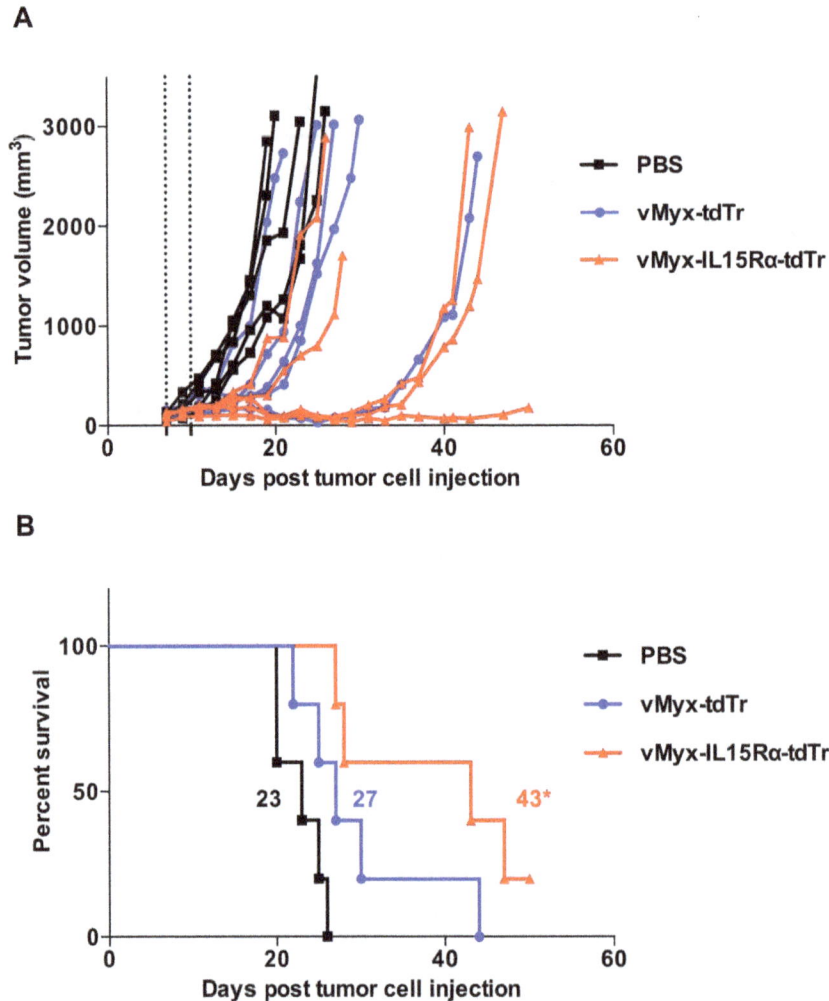

Figure 8. Prolonged survival of C57BL/6 mice bearing subcutaneous B16-F10 tumors treated with vMyx-IL15Rα-tdTr intratumorally. C57BL/6 mice (5 mice per group) were implanted with subcutaneous B16-F10 tumor cells. 7 days later, when tumors reached approximately 100 mm³, virus was inoculated intratumorally (i.t.) with 2.6×10^7 PFU vMyx-IL15Rα-tdTr, vMyx-tdTr or PBS. Mice received a second i.t. inoculation of 2.6×10^7 PFU of each virus on day 10. (A) Growth of individual tumors. Dashed lines designate time of virus treatment, and tumor size was measured every 2 days. (B) Kaplan–Meier survival curve of the same experimental subjects. Numbers next to corresponding survival curves designate median survival time (days). (* - $p < 0.05$ for vMyx-IL15Rα-tdTr treated group compared to PBS).

lost from sectioned tissue, we did not observe interferon-γ staining in the tumor, lymph nodes, or spleen.

In addition to the delivery of IL15Rα-IL15, the myxoma construct itself may contribute to an enhanced immune response. Previously we demonstrated the feasibility of combining adoptive T cell therapy with concurrent administration of an oncolytic virus [31]. There are at least three potential mechanisms by which myxoma virus could kill susceptible tumor cells: First, virus can directly kill tumor cells by viral oncolysis; second, local production of anti-tumor cytokines caused by viral infection can lead to recruitment and activation of immune cells that better recognize and kill tumor cells; third, killed cancer cells can be a more potent source of cross-presented tumor peptides by tumor stroma to further enhance the acquired anti-tumor immune response [44]. Manipulation of tumor microenvironment is an important strategy to improve adoptive T cell therapy and eliminate occurrence of antigen loss variants (ALV), cells that lose the T cell reactive epitope and eventually lead to tumor outgrowth [31]. We hypothesized that delivery of a highly functional and potent

IL15Rα-IL15 cytokine, especially in the context of viral infection, would provide a necessary boost to immune cells in driving their functional anti-tumor activities. Potential combination therapy along with the immunomodulating activities of anti-PD1/PDL1 antibodies, might provide an even more robust initial response and elimination of ALVs [45],[46],[27].

In summary, the use of delivery systems such as vMyx-IL15Rα-tdTr, and related genetically modified viruses, has the potential to improve clinical outcomes of cancer therapy.

Supporting Information

Figure S1 Recombinant plasmid for modifying WT myxoma virus and generating vMyx-IL15Rα-tdTr. Plasmid pBS-IL15Rα-IL15-tdTomatoRed (6267bp) is based on the pBluescript backbone on which M135 and M136 partial viral gene sequences are flanking genes for IL15Rα-IL15 fusion protein and tdTomato red fluorescent protein, both under control of vvSynE/L viral promoters. This expression cassette is flanked by partial viral gene sequences for the purpose of being transfected into the WT vMyx-Lau virus genome between genes M135 and M136.

Figure S2 Western blot showing presence of IL15Rα-IL15 fusion protein in the supernatants of cells infected with vMyx-IL15Rα-tdTr. Confluent RK-13 cells in 6-well plates were infected with vMyx-IL15Rα-tdTr or vMyx-tdTr at MOI = 5. Cell media was collected and cells were scraped, lysed and cytoplasmic extract was harvested at 48 h post-infection. Membranes blotted with supernatants and cell extracts of virus infected cells were stained for IL15 (left panel) or IL15Rα (right panel). Experiment was repeated five times with similar results. (SNT – supernatant, CE – cell extract).

Acknowledgments

Dr. Jennifer Stone (Dr. David Kranz Laboratory, Biochemistry Department, University of Illinois at Urbana-Champaign) provided us with the fusion protein of the m33 TCR with IL15Rα-IL15 and advice and assistance on IL15Rα-IL15 construct design, Western blot, MTT and other assays. We thank Dr. Thomas Gajewski for his gift of B16.SIY. We would also like to thank Caroline Johnson, Melanie Studzinski, and Sydney Sherman for help with survival experiments.

Author Contributions

Conceived and designed the experiments: VT DLT JL ALM ER JLS. Performed the experiments: VT ALM. Analyzed the data: VT ALM ER. Contributed reagents/materials/analysis tools: JL GM DMK ALM ER JLS. Wrote the paper: VT ER.

References

1. Russell SJ, Peng K-W, Bell JC (2012) Oncolytic virotherapy. Nat Biotechnol 30: 658–670. doi:10.1038/nbt.2287.

2. Stephenson KB, Barra NG, Davies E, Ashkar AA, Lichty BD (2011) Expressing human interleukin-15 from oncolytic vesicular stomatitis virus improves survival in a murine metastatic colon adenocarcinoma model through the enhancement of anti-tumor immunity. Cancer Gene Ther 19: 238–246. doi:10.1038/cgt.2011.81.

3. Lun X, Chan J, Zhou H, Sun B, Kelly JJP, et al. (2010) Efficacy and safety/toxicity study of recombinant vaccinia virus JX-594 in two immunocompetent animal models of glioma. Mol Ther J Am Soc Gene Ther 18: 1927–1936. doi:10.1038/mt.2010.183.

4. Chan WM, Rahman MM, McFadden G (2013) Oncolytic myxoma virus: the path to clinic. Vaccine 31: 4252–4258. doi:10.1016/j.vaccine.2013.05.056.

5. Sypula J., Wang F., Ma Y., Bell J., & McFadden G. (2004) Myxoma virus tropism in human tumor cells. Gene Ther Mol Biol 8 103–114.

6. Barrett JW, Alston LR, Wang F, Stanford MM, Gilbert P-A, et al. (2007) Identification of host range mutants of myxoma virus with altered oncolytic potential in human glioma cells. J Neurovirol 13: 549–560. doi:10.1080/13550280701591526.

7. Wang G, Barrett JW, Stanford M, Werden SJ, Johnston JB, et al. (2006) Infection of human cancer cells with myxoma virus requires Akt activation via interaction with a viral ankyrin-repeat host range factor. Proc Natl Acad Sci U S A 103: 4640–4645. doi:10.1073/pnas.0509341103.

8. Lun X, Yang W, Alain T, Shi Z-Q, Muzik H, et al. (2005) Myxoma Virus Is a Novel Oncolytic Virus with Significant Antitumor Activity against Experimental Human Gliomas. Cancer Res 65: 9982–9990. doi:10.1158/0008-5472.CAN-05-1201.

9. Lun XQ, Zhou H, Alain T, Sun B, Wang L, et al. (2007) Targeting human medulloblastoma: oncolytic virotherapy with myxoma virus is enhanced by rapamycin. Cancer Res 67: 8818–8827. doi:10.1158/0008-5472.CAN-07-1214.

10. Wu Y, Lun X, Zhou H, Wang L, Sun B, et al. (2008) Oncolytic efficacy of recombinant vesicular stomatitis virus and myxoma virus in experimental models of rhabdoid tumors. Clin Cancer Res Off J Am Assoc Cancer Res 14: 1218–1227. doi:10.1158/1078-0432.CCR-07-1330.

11. Wennier ST, Liu J, Li S, Rahman MM, Mona M, et al. (2012) Myxoma Virus Sensitizes Cancer Cells to Gemcitabine and Is an Effective Oncolytic Virotherapeutic in Models of Disseminated Pancreatic Cancer. Mol Ther 20: 759–768. doi:10.1038/mt.2011.293.

12. Stanford MM, Shaban M, Barrett JW, Werden SJ, Gilbert P-A, et al. (2008) Myxoma virus oncolysis of primary and metastatic B16F10 mouse tumors in vivo. Mol Ther J Am Soc Gene Ther 16: 52–59. doi:10.1038/sj.mt.6300348.

13. Lun X, Alain T, Zemp FJ, Zhou H, Rahman MM, et al. (2010) Myxoma Virus Virotherapy for Glioma in Immunocompetent Animal Models: Optimizing Administration Routes and Synergy with Rapamycin. Cancer Res 70: 598–608. doi:10.1158/0008-5472.CAN-09-1510.

14. Bartee E, Chan WM, Moreb JS, Cogle CR, McFadden G (2012) Selective purging of human multiple myeloma cells from autologous stem cell transplantation grafts using oncolytic myxoma virus. Biol Blood Marrow Transplant J Am Soc Blood Marrow Transplant 18: 1540–1551. doi:10.1016/j.bbmt.2012.04.004.

15. Rahman MM, Madlambayan GJ, Cogle CR, McFadden G (2010) Oncolytic viral purging of leukemic hematopoietic stem and progenitor cells with Myxoma virus. Cytokine Growth Factor Rev 21: 169–175. doi:10.1016/j.cytogfr.2010.02.010.

16. Senzer NN, Kaufman HL, Amatruda T, Nemunaitis M, Reid T, et al. (2009) Phase II clinical trial of a granulocyte-macrophage colony-stimulating factor-encoding, second-generation oncolytic herpesvirus in patients with unresectable metastatic melanoma. J Clin Oncol Off J Am Soc Clin Oncol 27: 5763–5771. doi:10.1200/JCO.2009.24.3675.

17. Heo J, Reid T, Ruo L, Breitbach CJ, Rose S, et al. (2013) Randomized dose-finding clinical trial of oncolytic immunotherapeutic vaccinia JX-594 in liver cancer. Nat Med. doi:10.1038/nm.3089.

18. Parmiani G, Castelli C, Pilla L, Santinami M, Colombo MP, et al. (2007) Opposite immune functions of GM-CSF administered as vaccine adjuvant in cancer patients. Ann Oncol Off J Eur Soc Med Oncol ESMO 18: 226–232. doi:10.1093/annonc/mdl158.

19. Cawood R, Hills T, Wong SL, Alamoudi AA, Beadle S, et al. (2012) Recombinant viral vaccines for cancer. Trends Mol Med 18: 564–574. doi:10.1016/j.molmed.2012.07.007.

20. Cheever MA (2008) Twelve immunotherapy drugs that could cure cancers. Immunol Rev 222: 357–368. doi:10.1111/j.1600-065X.2008.00604.x.

21. Steel JC, Waldmann TA, Morris JC (2012) Interleukin-15 biology and its therapeutic implications in cancer. Trends Pharmacol Sci 33: 35–41. doi:10.1016/j.tips.2011.09.004.

22. Dubois S, Mariner J, Waldmann TA, Tagaya Y (2002) IL-15Rα Recycles and Presents IL-15 In trans to Neighboring Cells. Immunity 17: 537–547. doi:10.1016/S1074-7613(02)00429-6.

23. Stoklasek TA, Schluns KS, Lefrançois L (2006) Combined IL-15/IL-15Rα Immunotherapy Maximizes IL-15 Activity In Vivo. J Immunol 177: 6072–6080.

24. Dubois S, Patel HJ, Zhang M, Waldmann TA, Müller JR (2008) Preassociation of IL-15 with IL-15Rα-IgG1-Fc Enhances Its Activity on Proliferation of NK and CD8+/CD44high T Cells and Its Antitumor Action. J Immunol 180: 2099–2106.

25. Rubinstein MP, Kovar M, Purton JF, Cho J-H, Boyman O, et al. (2006) Converting IL-15 to a superagonist by binding to soluble IL-15R{alpha}. Proc Natl Acad Sci U S A 103: 9166–9171. doi:10.1073/pnas.0600240103.

26. Jakobisiak M, Golab J, Lasek W (2011) Interleukin 15 as a promising candidate for tumor immunotherapy. Cytokine Growth Factor Rev 22: 99–108. doi:10.1016/j.cytogfr.2011.04.001.

27. Epardaud M, Elpek KG, Rubinstein MP, Yonekura A, Bellemare-Pelletier A, et al. (2008) Interleukin-15/interleukin-15R alpha complexes promote destruction of established tumors by reviving tumor-resident CD8+ T cells. Cancer Res 68: 2972–2983. doi:10.1158/0008-5472.CAN-08-0045.

28. Bessard A, Solé V, Bouchaud G, Quéméner A, Jacques Y (2009) High antitumor activity of RLI, an interleukin-15 (IL-15)-IL-15 receptor alpha fusion protein, in metastatic melanoma and colorectal cancer. Mol Cancer Ther 8: 2736–2745. doi:10.1158/1535-7163.MCT-09-0275.

29. Liu J, Wennier S, Reinhard M, Roy E, MacNeill A, et al. (2009) Myxoma virus expressing interleukin-15 fails to cause lethal myxomatosis in European rabbits. J Virol 83: 5933–5938. doi:10.1128/JVI.00204-09.

30. MacNeill AL, Doty RA, Liu J, McFadden G, Roy EJ (2013) Histological evaluation of intratumoral myxoma virus treatment in an immunocompetent mouse model of melanoma. Oncolytic Virotherapy: 1. doi:10.2147/OV.S37971.

31. Thomas DL, Doty R, Tosic V, Liu J, Kranz DM, et al. (2011) Myxoma virus combined with rapamycin treatment enhances adoptive T cell therapy for murine melanoma brain tumors. Cancer Immunol Immunother CII 60: 1461–1472. doi:10.1007/s00262-011-1045-z.

32. Stone JD, Chervin AS, Schreiber H, Kranz DM (2012) Design and characterization of a protein superagonist of IL-15 fused with IL-15Rα and a high-affinity T cell receptor. Biotechnol Prog 28: 1588–1597. doi:10.1002/btpr.1631.

33. Spiotto MT, Yu P, Rowley DA, Nishimura MI, Meredith SC, et al. (2002) Increasing tumor antigen expression overcomes "ignorance" to solid tumors via crosspresentation by bone marrow-derived stromal cells. Immunity 17: 737–747.

34. Blank C, Brown I, Peterson AC, Spiotto M, Iwai Y, et al. (2004) PD-L1/B7H-1 Inhibits the Effector Phase of Tumor Rejection by T Cell Receptor (TCR) Transgenic CD8+ T Cells. Cancer Res 64: 1140–1145. doi:10.1158/0008-5472.CAN-03-3259.

35. Kaspar M, Trachsel E, Neri D (2007) The Antibody-Mediated Targeted Delivery of Interleukin-15 and GM-CSF to the Tumor Neovasculature Inhibits Tumor Growth and Metastasis. Cancer Res 67: 4940–4948. doi:10.1158/0008-5472.CAN-07-0283.

36. Marçais A, Viel S, Grau M, Henry T, Marvel J, et al. (2013) Regulation of Mouse NK Cell Development and Function by Cytokines. Front Immunol 4: 450. doi:10.3389/fimmu.2013.00450.

37. Castillo EF, Schluns KS (2012) Regulating the immune system via IL-15 transpresentation. Cytokine 59: 479–490. doi:10.1016/j.cyto.2012.06.017.

38. Kermer V, Baum V, Hornig N, Kontermann RE, Müller D (2012) An Antibody Fusion Protein for Cancer Immunotherapy Mimicking IL-15 trans-Presentation at the Tumor Site. Mol Cancer Ther 11: 1279–1288. doi:10.1158/1535-7163.MCT-12-0019.

39. Bouchaud G, Garrigue-Antar L, Solé V, Quéméner A, Boublik Y, et al. (2008) The exon-3-encoded domain of IL-15ralpha contributes to IL-15 high-affinity binding and is crucial for the IL-15 antagonistic effect of soluble IL-15Ralpha. J Mol Biol 382: 1–12. doi:10.1016/j.jmb.2008.07.019.

40. Liu RB, Engels B, Arina A, Schreiber K, Hyjek E, et al. (2012) Densely Granulated Murine NK Cells Eradicate Large Solid Tumors. Cancer Res 72: 1964–1974. doi:10.1158/0008-5472.CAN-11-3208.

41. Rowley J, Monie A, Hung C-F, Wu T-C (2008) Inhibition of Tumor Growth by NK1.1+ Cells and CD8+ T Cells Activated by IL-15 through Receptor β/Common γ Signaling in trans. J Immunol 181: 8237–8247.

42. Elpek KG, Rubinstein MP, Bellemare-Pelletier A, Goldrath AW, Turley SJ (2010) Mature natural killer cells with phenotypic and functional alterations accumulate upon sustained stimulation with IL-15/IL-15Rα complexes. Proc Natl Acad Sci 107: 21647–21652. doi:10.1073/pnas.1012128107.

43. Van der Loos CM, Houtkamp MA, de Boer OJ, Teeling P, van der Wal AC, et al. (2001) Immunohistochemical detection of interferon-gamma: fake or fact? J Histochem Cytochem Off J Histochem Soc 49: 699–710.

44. Zhang B, Bowerman NA, Salama JK, Schmidt H, Spiotto MT, et al. (2007) Induced sensitization of tumor stroma leads to eradication of established cancer by T cells. J Exp Med 204: 49–55. doi:10.1084/jem.20062056.

45. Pardoll DM (2012) The blockade of immune checkpoints in cancer immunotherapy. Nat Rev Cancer 12: 252–264. doi:10.1038/nrc3239.

46. Xu W, Jones M, Liu B, Zhu X, Johnson CB, et al. (2013) Efficacy and Mechanism-of-Action of a Novel Superagonist Interleukin-15: Interleukin-15 Receptor αSu/Fc Fusion Complex in Syngeneic Murine Models of Multiple Myeloma. Cancer Res 73: 3075–3086. doi:10.1158/0008-5472.CAN-12-2357.

2

In-Depth Transcriptome Analysis of the Red Swamp Crayfish *Procambarus clarkii*

Huaishun Shen[1,2]*, Yacheng Hu[1,2], Yuanchao Ma[1,2], Xin Zhou[1]*, Zenghong Xu[1], Yan Shui[1], Chunyan Li[3], Peng Xu[3], Xiaowen Sun[3]

1 Key Laboratory of Freshwater Fisheries and Germplasm Resources Utilization, Ministry of Agriculture, Freshwater Fisheries Research Center, Chinese Academy of Fishery Sciences, Wuxi, China, **2** Wuxi Fisheries College, Nanjing Agricultural University, Nanjing, China, **3** The Center for Applied Aquatic Genomics, Chinese Academy of Fishery Sciences, Beijing, China

Abstract

The red swamp crayfish *Procambarus clarkii* is a highly adaptable, tolerant, and fecund freshwater crayfish that inhabits a wide range of aquatic environments. It is an important crustacean model organism that is used in many research fields, including animal behavior, environmental stress and toxicity, and studies of viral infection. Despite its widespread use, knowledge of the crayfish genome is very limited and insufficient for meaningful research. This is the use of next-generation sequencing techniques to analyze the crayfish transcriptome. A total of 324.97 million raw reads of 100 base pairs were generated, and a total of 88,463 transcripts were assembled *de novo* using Trinity software, producing 55,278 non-redundant transcripts. Comparison of digital gene expression between four different tissues revealed differentially expressed genes, in which more overexpressed genes were found in the hepatopancreas than in other tissues, and more underexpressed genes were found in the testis and the ovary than in other tissues. Gene ontology (GO) and KEGG enrichment analysis of differentially expressed genes revealed that metabolite- and immune-related pathway genes were enriched in the hepatopancreas, and DNA replication-related pathway genes were enriched in the ovary and the testis, which is consistent with the important role of the hepatopancreas in metabolism, immunity, and the stress response, and with that of the ovary and the testis in reproduction. It was also found that 14 vitellogenin transcripts were highly expressed specifically in the hepatopancreas, and 6 transcripts were highly expressed specifically in the ovary, but no vitellogenin transcripts were highly expressed in both the hepatopancreas and the ovary. These results provide new insight into the role of vitellogenin in crustaceans. In addition, 243,764 SNP sites and 43,205 microsatellite sequences were identified in the sequencing data. We believe that our results provide an important genome resource for the crayfish.

Editor: Pikul Jiravanichpaisal, Fish Vet Group, Thailand

Funding: This work is supported by the Science & Technology Pillar Program of Jiangsu Province (BE2013316) (http://www.jskjjh.gov.cn/13kjskj2/) and the Natural Science Foundation of Jiangsu Province (BK2012534) (http://www.jskjjh.gov.cn/13kjskj2/). The funders had no role in study design, data collection and analysis, decision to publish, or preparation of the manuscript.

Competing Interests: The authors have declared that no competing interests exist.

* Email: shenhs@ffrc.cn (HS); zhoux@ffrc.cn (XZ)

Introduction

The red swamp crayfish *Procambarus clarkii* is a freshwater crayfish species that is native to parts of Mexico and the United States [1], but is also commonly found outside its natural range in Asia, Africa, Europe, and elsewhere in the Americas, where it is often considered to be an invasive pest [2]. *P. clarkii* was introduced to China from Japan in the 1930s [3]. Crayfish farming began in in the 18th century in Louisiana in the USA, where the species was cultivated in rice fields. Crayfish have been farmed extensively in China since the 1990s, and China is now the world's leading crayfish producer [4].

P. clarkii is a highly adaptable, tolerant, and fecund freshwater crayfish that can inhabit a wide range of aquatic environments, including those with moderate salinity, low oxygen levels, extreme temperatures, and pollution [2,5]. Because of these characteristics, in addition to its economic role, the crayfish has become an important crustacean model organism in research on viral infection [6–10], animal behavior [11–17], and environmental stress and toxicity [18–23].

Despite great interest in this organism, knowledge of the crayfish genome is very limited, and gene discovery has been performed on a relatively small scale. Only 330 expressed sequences (EST) and 547 nucleotide sequences have been deposited in GenBank (accessed on Jul 29, 2014) for the crayfish, which is fewer than the close relative *Pacifastacus leniusculus* (1063 EST and 1100 nucleotide sequences) and far less fewer than other economically important crustaceans, such as the freshwater prawn *Macrobrachium nipponense*, the giant freshwater prawn *Macrobrachium rosenbergii*, the pacific white shrimp *Litopenaeus vannamei*, and others. Furthermore, only a very few genetic markers have been discovered for *P. clarkii* [3,24–26].

The traditional methodology to explore expressed sequence tags (ESTs) involves construction of a cDNA library followed by Sanger sequencing, which is time-consuming and inefficient. Normally, the numbers of ESTs generated using this method is no more than ten thousand [27]. In recent years, next-generation sequencing technologies from companies such as 454 Life Sciences, Illumina, and Applied Biosystems (SOLiD sequencing) have been widely used to explore genomic information in model and non-model organisms. In comparison to traditional Sanger sequencing technology, next-generation sequencing technologies are superior in many aspects, and in general they are able to provide enormous amounts of sequence data with a greater breadth and depth of information, in shorter times and at a significantly lower cost [28–30]. The expressed sequences generated using next-generation sequencing technologies are often on the order of thousands or hundreds of thousands of sequences, which are ten-fold or one-hundred-fold greater than the number identified by traditional technologies.

Crustaceans studied using next-generation sequencing technologies include the giant freshwater prawn *Macrobrachium rosenbergii* [31], the orient river prawn *Macrobrachium nipponense* [32], the Chinese mitten crab *Eriocheir sinensis* [33–36], the pacific white shrimp *Litopenaeus vannamei* [37–40], the Chinese shrimp *Fenneropenaeus chinensis* [41], the pandalid shrimp *Pandalus latirostris* [42], and the crab *Portunus trituberculatus* [43–44]. These data have significantly enriched our genetic and genomic knowledge of crustaceans.

In this study, hi-seq sequencing technology was used to sequence the transcriptomes of 4 major organs in the crayfish: hepatopancreas, muscle, ovary, and testis. This data was used to generate expressed sequence data, simple sequence repeat markers, and SNP markers that represent a resource for trait mapping, as well as differential organ gene expression profiles, to better understand the functions of the studied organs in the crayfish. We believe that the data obtained from this study represent an import resource for crayfish research into gene function, molecular events associated with breeding, and other areas.

Material and Methods

Ethics statement

This study was approved by the Animal Care and Use Committee of the Center for Applied Aquatic Genomics at Chinese Academy of Fishery Sciences.

Animal collection

P. clarkii weighing approximately 10–20 g were collected from a crayfish farm in Xuyi, Jiangsu Province, China. Collected crayfish were cultured in water tanks with adequate aeration at 20°C and a natural photoperiod, and were fed with a commercial crayfish diet once per day. Four tissue types (hepatopancreas, muscle, ovary, and testis) were collected, and each group of tissues contained samples from approximately ten crayfish. The tissue samples were frozen immediately in liquid nitrogen, and stored at −80°C.

RNA isolation and Illumina sequencing

Total RNA from various tissues was isolated using the RNeasy Plus Mini Kit (Qiagen, Valencia, CA, USA) according to the manufacturer's protocol, and treated with RNase-free DNase I (Qiagen) to remove genomic DNA. RNA integrity was evaluated by 1.5% agarose gel electrophoresis. RNA concentrations were measured and purity was determined using a NanoDrop ND-1000 spectrophotometer (NanoDrop Technologies, Wilmington, DE, USA).

RNA-seq library preparation and sequencing was carried out by the Genomic Analysis Lab of The Institute of Genetics and Developmental Biology of the Chinese Academy of Sciences (Beijing, China). Approximately 5 μg of DNase-treated total RNA was used to construct a cDNA library following the protocols of the Illumina TruSeq RNA Sample Preparation Kit (Illumina, San Diego, CA 92122, USA). The cDNA libraries were amplified by PCR and contained TruSeq indexes 1–4 within the adaptors. Amplified libraries yielded approximately 500 ng of cDNA with an average length of approximately 270 base pairs (bp). Finally, the libraries were sequenced on an Illumina HiSeq 2000 instrument with 100 bp paired-end (PE) reads.

De novo assembly and transcriptome analysis

Raw reads, which were generated by the Illumina/Solexa sequencer, were first trimmed by removing adapter sequences. Low quality reads (quality scores less than 20) were trimmed and short length reads (<10 bp) were removed [45–46]. The resulting high-quality reads were used in subsequent assembly. The Crayfish transcriptome was *de novo* assembled using Trinity software (vision 2013.02.25) with the default parameters [47]. In brief, three steps were performed. First, data was processed by Inchworm, in which the high-quality reads were combined to form longer fragments called contigs. Second, data was processed by Chrysalis, in which sequences were obtained by connecting contigs in such a manner that they could not be extended on either end, which resulted in de Bruijn graphs. Finally, de Bruijn graphs were further treated by Butterfly to obtain transcripts.

Transcriptome annotation and gene ontology analysis

All transcripts were compared with the NCBI non-redundant (nr) protein database, GO database, COG database, and KEGG database for functional annotation using BLAST software with an e-value cutoff of 1e-5 [48]. Functional annotation was performed with gene ontology (GO) terms (www.geneontology.org) that were analyzed using Blast2go software (http://www.blast2go.com/b2ghome) [49]. The COG and KEGG pathway annotations were performed using Blastall software against the COG and KEGG databases [48].

Differentially expressed genes

To obtain expression levels for every transcript in different tissues, cleaned reads were first mapped to all transcripts using Bowtie software [50–51], then the FPKM (fragments per kilobase of exon per million fragments mapped) value of every transcript was obtained using RSEM (RNASeq by expectation maximization, http://deweylab.biostat.wisc.edu/rsem/) software. Differentially expressed genes were identified using edgeR (empirical analysis of digital gene expression data in R, http://www.bioconductor.org/packages/release/bioc/html/edgeR.html) software [52–53]. For this analysis step, the filtering threshold was set as an FDR (false discovery rate) <0.5.

RT-PCR amplification of transcripts

To validate the assembly of the crayfish transcriptome, 20 selected transcripts were used for expression analysis by RT-PCR. Total RNA was prepared from the four tissues (hepatopancreas, muscle, ovary, and testis) of crayfish using Trizol reagent in accordance with the manufacturer's instructions. Total RNA was treated with RQ1 RNase-free DNase (Promega, Madison, WI, USA) to avoid genomic DNA contamination, and then reverse transcribed using M-MLV reverse transcriptase (Promega, USA) according to the manufacturer's instructions. The synthesized cDNA was used as a template for PCR. The following PCR program was used: denaturation at 94°C for 3 min, 35 amplification cycles of 94°C for 30 s, 58°C for 30 s, and 72°C for 30 s, and a final extension at 72°C for 10 min. The PCR products were determined by 1.5% agarose gel electrophoresis using DNA markers. The expression of the 18S RNA gene of *Procambrus clarkii* (accession number: EU920952.1) was selected as reference gene, using the primer pair *Pc18S*-F (5'-ATCACGTCTCTGACCGCAAG-3') and *Pc18S*-R (5'-GACACTTGAAAGATGCGGCG-3').

SNP and microsatellite sequence identification

The raw reads were exported in FASTQ format to allow them to be imported into software for SNP calling. SAMtools (http://samtools.sourceforge.net/) and VarScan (http://varscan.sourceforge.net/) software were applied to align reads to the reference transcriptome and to detect SNPs [54–55]. For this analysis step, the filtering threshold was set as a quality score no less than 20.

Msatcommander software was used to identify microsatellites from assembled contigs, as well as for primer design [56]. The mononucleotide repeats were ignored by modifying the configuration file. The repeat thresholds for di-, tri-, tetra-, penta-, and hexa-nucleotide motifs were set as 8, 5, 5, 5, and 5, respectively. Only microsatellite sequences with flanking sequences longer than 50 bp on both sides were collected for future marker development.

Results and Discussion

Illumina sequencing from crayfish tissues

Illumina-based RNA-sequencing (RNA-Seq) was performed with samples of four tissue types from crayfish. A total of 324.97 million paired-end reads were generated with a read length of 100 bp, of which 102.46 million reads were from the hepatopancreas, 83.51 million reads were from muscle, 84.94 million reads were from the ovary, and 36.06 million reads were from the testis. All raw sequence data were deposited in the NCBI Sequence Read Archive (SRA) under accession code SRP044128. After trimming of low quality reads and short reads, a total of 306.73 million high-quality sequences (94.73%) were obtained (Table S1 in File S1), and these sequences were used for further analysis.

De novo assembly of the transcriptome

At present, *P. clarkii* has no reference genome sequence, therefore a *de novo* assembly strategy was utilized, in which the crayfish transcriptome was *de novo* assembled by Trinity software (version 2013.02.25) using the default parameters. *De novo* assembly of 306.73 million high-quality sequences generated a total of 88,463 transcripts that ranged from 351 to 34,708 bp in length, with an average length of 1655.49 bp (Table 1). The length distribution of transcripts is shown in Figure 1. Most of the transcripts (23.71%) were 401–600 bp in length, 11.51% ranged from 601–800 bp, and 10.90% ranged from 1–400 bp. These 88,463 transcripts yielded a total of 50,219 non-redundant transcripts because of alternative splicing; therefore, it was possible to match two or more transcripts to one gene. Our sequence data provided a large number of transcripts as compared to publicly available data from the Genbank database, which represent a convenient source of information for future full length cDNA cloning and gene function research in *P. clarkii*. Prior to this study, only 330 EST and 547 nucleotide sequences were listed in the GenBank database. Our study supplied 50,219 non-redundant transcripts, which has significantly enriched our knowledge of the *P. clarkii* genome and will facilitate further study of the functions of *P. clarkii* genes.

Functional annotation

Protein coding sequences of transcripts were predicted using a tool supplied by Trinity software (http://trinityrnaseq.sourceforge.net/analysis/extract_proteins_from_trinity_transcripts.html). Of the 88,463 transcripts, 42,905 were found to contain open reading frames (ORFs), with an average protein coding length of 552.78 bp and a mean nucleotide length of 2551.34 bp. These isosequences likely represent genes that play essential roles in *P. clarkii* biological processes.

88,463 transcripts were compared with the NCBI non-redundant (nr) protein database, GO database, COG database, and KEGG database for functional annotation using BlastX with an e-value cutoff of 1e-5 (Table S2 in File S1). A total of 31,763 transcripts (35.91% of all transcripts) had significant hits in at least one of these databases, which corresponded to 11,222 genes (21.63% of all genes). A total of 30,779 transcripts (96.90% of all annotated transcripts) had significant hits in the nr protein database, which corresponded to 10,862 genes (96.79% of all annotated genes). The gene names of top BLAST hits were assigned to each transcript with significant hits, and 3110 transcripts from *P. clarkii* were best matched with genes from *Daphnia pulex*, 2675 transcripts were best matched with genes from *Tribolium castaneum*, and 1837 transcripts were best matched with genes from *Pediculus humanus* (Figure S1). *Daphnia pulex* is a primitive water flea, *Tribolium castaneum* is a type of beetle, and *Pediculus humanus* is a louse species that infests humans. Thus, the genes from *P. clarkii* were most similar to those known from crustaceans and insects, and the distribution of significant BLAST hits over different organisms reflects the phylogenetic relationship between *P. clarkii* and other species.

GO analysis was conducted on annotated transcripts using blast2go software. A total of 15,457 transcripts, corresponding to 5890 genes, were assigned at least one GO term for biological processes, molecular functions, and cellular components, and the output of the GO annotations was plotted (Figure 2). Terms from the molecular function term group made up the majority of significant terms (12,842 transcripts, 88.08%), followed by the biological process group (10,241 transcripts, 66.25%) and the cellular component group (7,406 transcripts, 47.91%). For biological processes, genes involved in cellular processes (GO: 0009987, 8,148 transcripts), metabolic processes (GO: 0008152, 6,622 transcripts), and single-organism process (GO: 0044699, 5,623 transcripts) were highly represented. For molecular functions, binding (GO: 0005488, 8,313 transcripts) and catalytic activity (GO: 0003824, 6,781 transcripts) were the most represented GO terms. For cellular components, cells (GO: 0005623, 5,778 transcripts), cell part (GO: 0044464, 5,778 transcripts), and organelles (GO: 0043226, 3,632 transcripts) were the most represented terms. There were 9 identified terms that contained fewer than 10 transcripts.

Transcripts were also compared with the COG database, and 6,034 transcripts (2,386 genes) were matched to database entries. These transcripts were classified into 25 functional categories (Figure 3), among which the largest group (2031 transcripts) was signal transduction mechanisms, followed by general function prediction only (1718 transcripts), transcription (914 transcripts), posttranslational modification, protein turnover, and chaperones (782 transcripts). Genes matched to the nuclear structure category (17 transcripts) represented the smallest group.

In addition, a KEGG pathway analysis was performed on all assembled transcripts as an alternative approach for functional categorization and annotation. A total of 14,596 transcripts, corresponding to 5414 genes, were categorized into functional groups, in which the metabolism group was the most well

Table 1. Summary of Illumina Hiseq2000 assembly and analysis of *P. clarkii* transcriptomic sequences.

Type	Number
Total genes	50219
Total transcripts	88463
Total residues	146449732
Average length	1655.49
Largest transcript	34708
Smallest transcript	351

represented, with 7821 transcripts, followed by the human disease group (7597 transcripts), organismal systems group (7179 transcripts), environmental information processing group (3460 transcripts), cellular processes group (3420 transcripts), and genetic information processing group (2481 transcripts) (Table S3 in File S1). Each functional group was made up of genes from different KEGG pathways. In addition, the number of transcripts in each KEGG pathway was counted, and the most abundant 20 KEGG pathways are shown in Figure 4. In brief, 2016 transcripts were categorized into metabolic pathways, followed by pathways for biosynthesis of secondary metabolites (552 transcripts), cancer (435 transcripts), focal adhesion (431 transcripts), and endocytosis (419 transcripts).

Differential analysis of gene expression profiles between tissues

The expression levels of whole transcripts in the hepatopancreas, the testis, the ovary, and muscle were evaluated (Table S4 and S5 in File S1). Transcriptomic analysis of these tissues showed that

more genes were overexpressed in the hepatopancreas as compared with genes expressed in the other three tissues, while more genes were underexpressed in the testis and the ovary compared with genes expressed in the hepatopancreas and muscle (Figure 5). Interestingly, although more genes were overexpressed in the hepatopancreas, fewer genes were expressed in the hepatopancreas (32999 genes) than in the muscle (37873 genes), the ovary (45083 genes), and the testis (39503 genes). This result may be due to the crucial role of the hepatopancreas in growth, which resulted in genes for metabolism being actively and highly expressed, while the ovary and the testis are reproductive organs, and thus more functional molecules are needed in reserve, but are not highly expressed.

Enriched pathways in the hepatopancreas. Metabolism is the basic physiological process that sustains living organisms, and it includes multiple reactions, such as the synthesis of digestive enzymes, secretion, digestion, nutrient absorption, excretion, lipid and glycogen storage, and mobilization [31]. In crustaceans, the hepatopancreas is the major metabolic organ. KEGG pathway enrichment analysis showed that regulation of amino acids, carbohydrates, lipids, and glycan metabolism were significantly enriched in the hepatopancreas compared to the other three tissues (Table 2). For example, 59 transcripts were identified in the fatty acid metabolism pathway (ko00071), and 26 of these transcripts (21 genes) were found to be significantly overexpressed in the hepatopancreas compared to muscle, including ACOX1 (acyl-CoA oxidase [EC:1.3.3.6]), ACADS (butyryl-CoA dehydrogenase [EC:1.3.8.1]), ALDH7A1 (aldehyde dehydrogenase family 7 member A1 [EC:1.2.1.3, 1.2.1.8, 1.2.1.31]), and ACAA2 (acetyl-CoA acyltransferase 2 [EC:2.3.1.16]), and only 7 transcripts (3 genes) were found to be significantly underexpressed. It was also found that the expression levels of genes in the fatty acid metabolism pathway in the hepatopancreas were significantly higher than those in the ovary and the testis, indicating that active

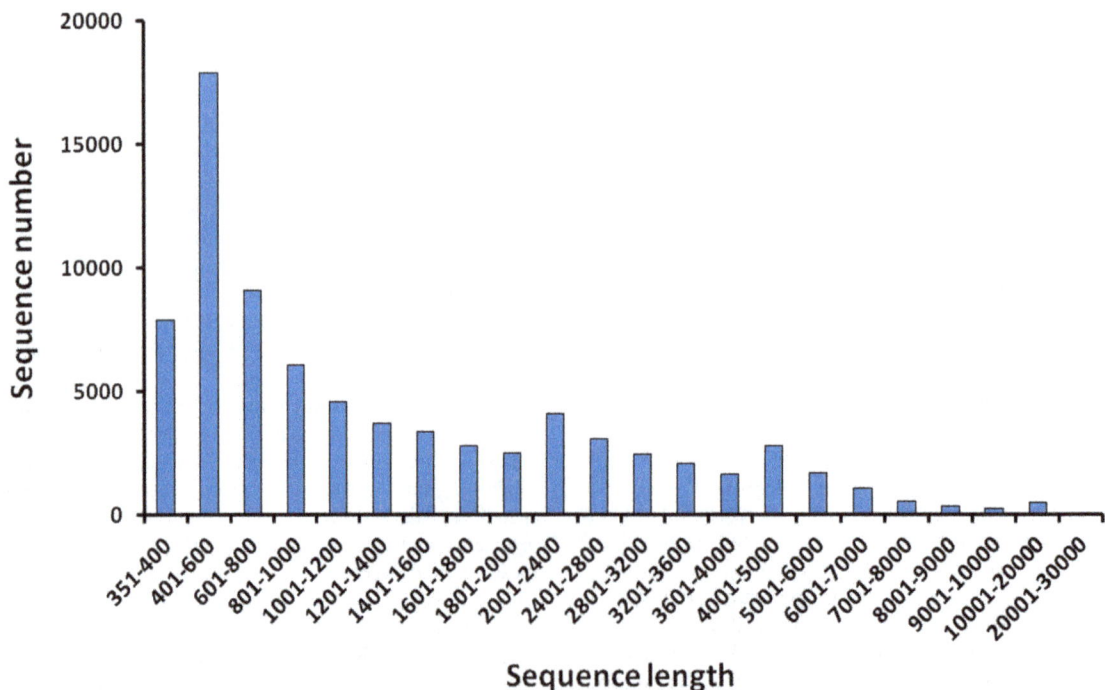

Figure 1. Length distribution of assembled transcripts of *P. clarkii*.

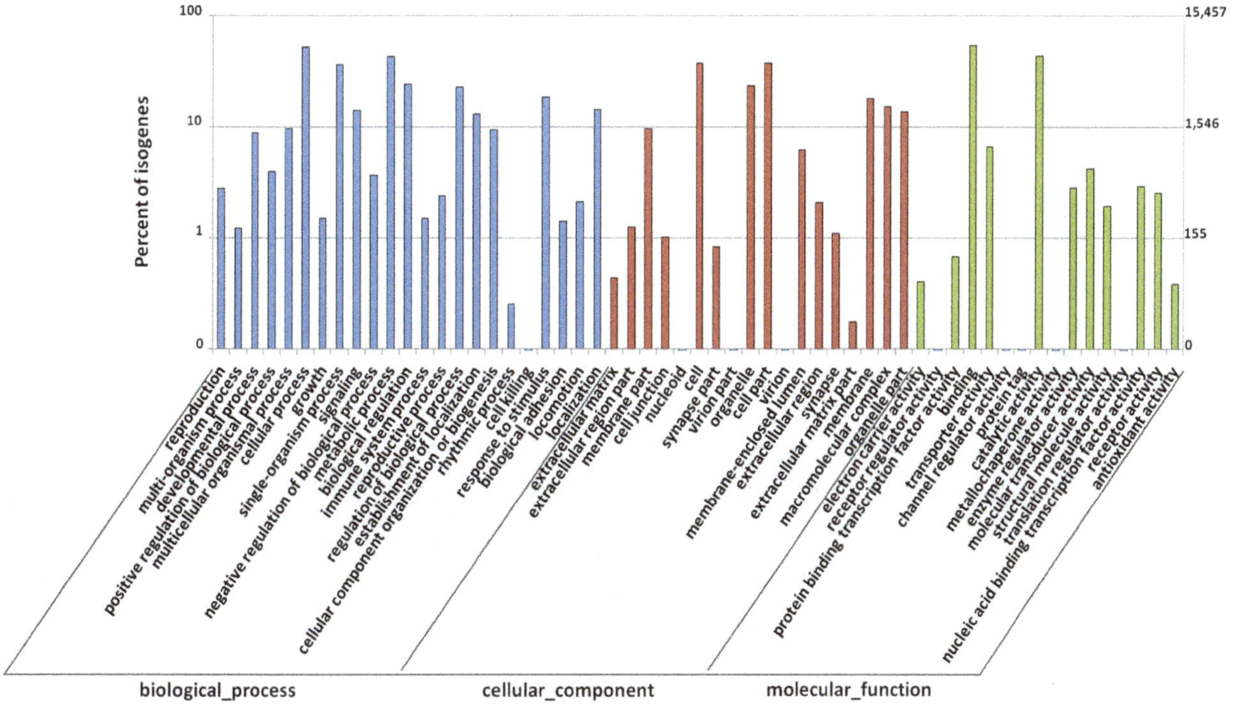

Figure 2. Gene ontology (GO) classification of transcripts of *P. clarkii.* GO terms were processed by Blast2Go and categorized at level 2 under three main categories (biological process, cellular component, and molecular function).

fatty acid metabolism takes place in the hepatopancreas of *P. clarkii* (Table S6 in File S1).

Enriched pathways in xenobiotic metabolism, heavy metal and oxidative stress, and the innate immune system were also found in

the hepatopancreas. The crayfish is well-known for its ability to survive in polluted environments, including water polluted by heavy metals, pesticides, and other chemicals, and also to tolerate hypoxia. The hepatopancreas is the primary site for accumulation

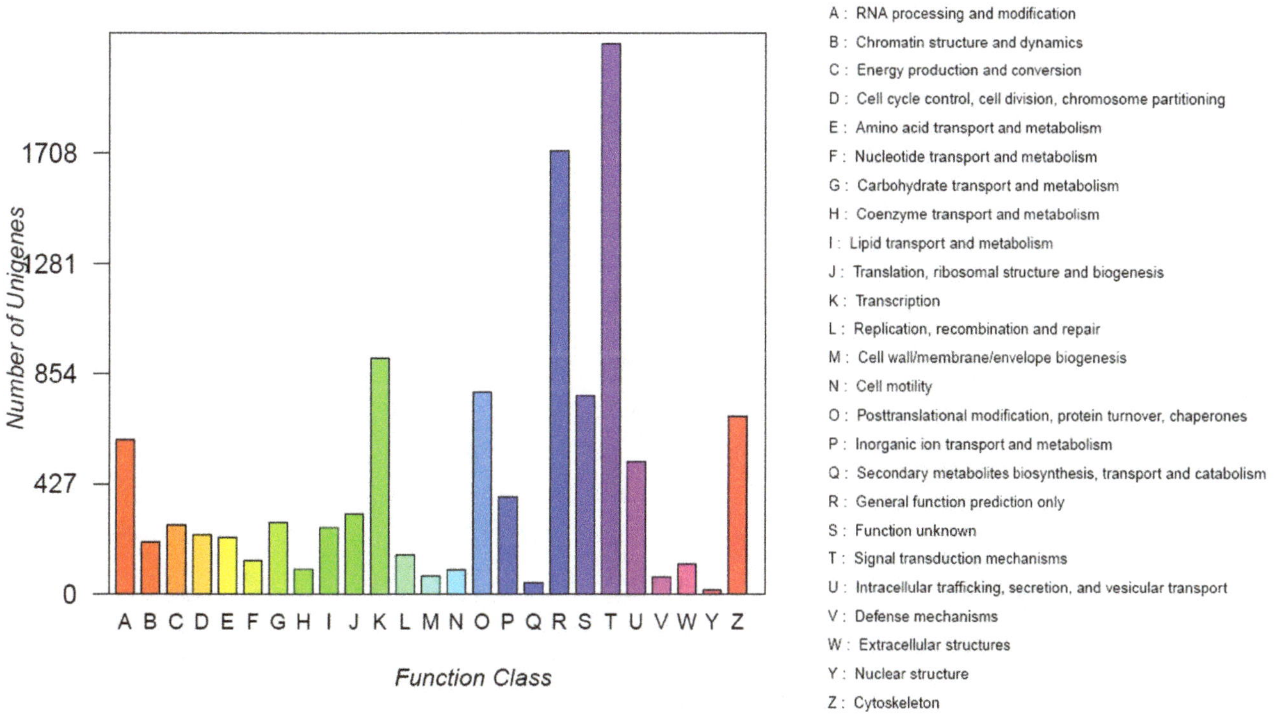

A : RNA processing and modification
B : Chromatin structure and dynamics
C : Energy production and conversion
D : Cell cycle control, cell division, chromosome partitioning
E : Amino acid transport and metabolism
F : Nucleotide transport and metabolism
G : Carbohydrate transport and metabolism
H : Coenzyme transport and metabolism
I : Lipid transport and metabolism
J : Translation, ribosomal structure and biogenesis
K : Transcription
L : Replication, recombination and repair
M : Cell wall/membrane/envelope biogenesis
N : Cell motility
O : Posttranslational modification, protein turnover, chaperones
P : Inorganic ion transport and metabolism
Q : Secondary metabolites biosynthesis, transport and catabolism
R : General function prediction only
S : Function unknown
T : Signal transduction mechanisms
U : Intracellular trafficking, secretion, and vesicular transport
V : Defense mechanisms
W : Extracellular structures
Y : Nuclear structure
Z : Cytoskeleton

Figure 3. Cluster of orthologous groups (COG) classification of putative proteins.

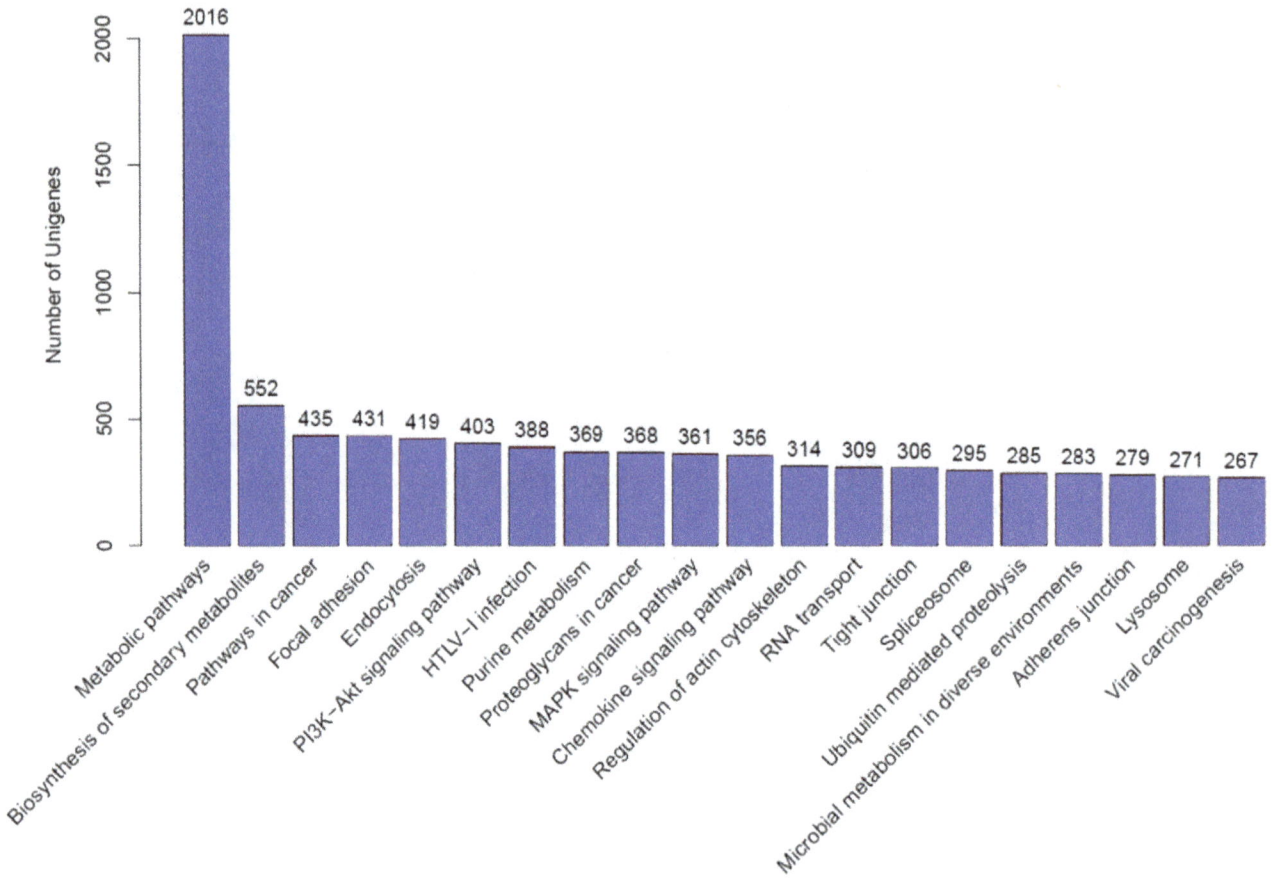

Figure 4. KEGG Classification of the genes. 14596 transcripts were assigned to 311 KEGG pathways. The top 20 most abundant KEGG pathways are shown.

and detoxification of xenobiotic pollutants in lysosomes. The R cells in the hepatopancreas perform biotransformation using enzymes, such as cytochrome p450, to sequester and detoxify xenobiotic pollutants [31]. Compared to the other tissues, several pathways were significantly enriched in the hepatopancreas: "lysosome", "peroxisome", "metabolism of xenobiotics by cytochrome p450", and "drug metabolism - cytochrome P450" (Table 2). Peroxisomes are essential organelles that play key roles

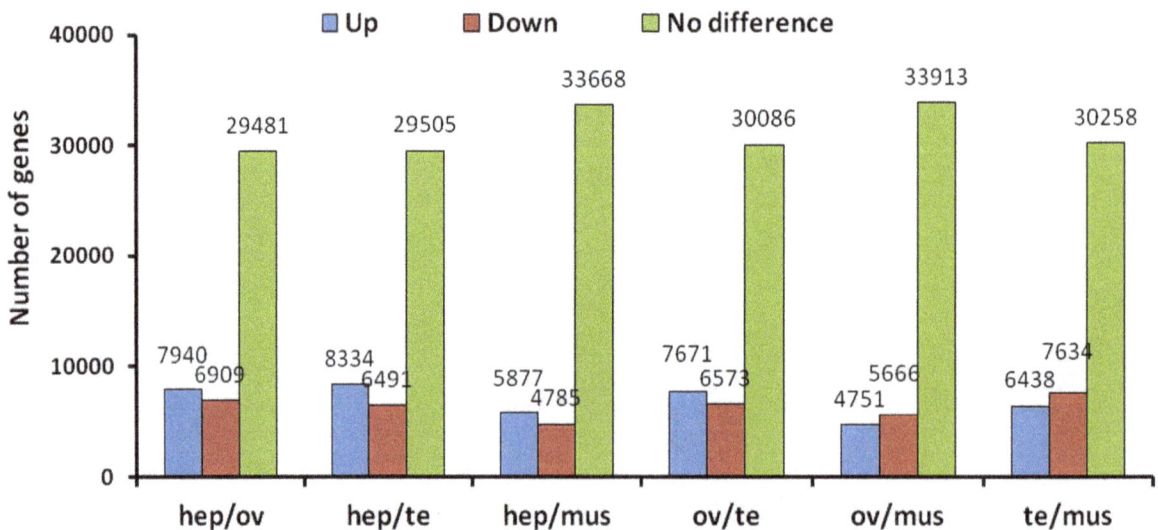

Figure 5. Differentially expressed genes analysis of four different crayfish tissues. hep: hepatopancreas; mu: muscle; ov: ovary; te: testis.

Table 2. KEGG pathways enriched in 4 crayfish tissues with Bonferroni-corrected p-values.

tissue 1	tissue 2	#Term	Sample number	Background number	Corrected P-Value
hepato-pancreas	muscle	Lysosome	148	272	7.27E-07
		Galactose metabolism	45	66	1.96E-05
		Peroxisome	79	138	5.42E-05
		Regulation of actin cytoskeleton	169	342	5.45E-05
		Rheumatoid arthritis	39	58	0.000139
		Methane metabolism	34	49	0.000183
		PPAR signaling pathway	49	79	0.000186
		Bacterial invasion of epithelial cells	90	169	0.000266
		Starch and sucrose metabolism	64	112	0.000267
		Focal adhesion	184	389	0.000278
		Arachidonic acid metabolism	61	107	0.000422
		Basal cell carcinoma	38	60	0.000693
		Sphingolipid metabolism	56	98	0.000766
		Glutathione metabolism	58	103	0.000904
		Glycolysis/Gluconeogenesis	56	100	0.001221
		Other glycan degradation	42	71	0.001747
		Retinol metabolism	22	31	0.001909
		Metabolism of xenobiotics by cytochrome P450	37	61	0.00197
		Alanine, aspartate and glutamate metabolism	32	51	0.002217
		Pentose phosphate pathway	35	58	0.003069
		Type II diabetes mellitus	30	48	0.003427
		Shigellosis	66	128	0.005031
		Phagosome	100	208	0.005674
		Fatty acid metabolism	34	58	0.006587
		Viral myocarditis	81	165	0.007891
		Drug metabolism - other enzymes	35	61	0.008885
		Nitrogen metabolism	22	34	0.009292
		Endometrial cancer	53	102	0.012052
		Pyruvate metabolism	39	71	0.013601
		Pentose and glucuronate interconversions	24	39	0.014534
		Amino sugar and nucleotide sugar metabolism	61	124	0.026912
		Caffeine metabolism	9	11	0.028532
		Antigen processing and presentation	33	61	0.037375
		Proximal tubule bicarbonate reclamation	21	35	0.038357
		Cytokine-cytokine receptor interaction	16	25	0.044251
		Drug metabolism - cytochrome P450	30	55	0.044251
hepato-pancreas	ovary	Ribosome	85	91	0
		Lysosome	188	272	4.53E-08
		Drug metabolism - cytochrome P450	49	55	2.00E-07
		Metabolism of xenobiotics by cytochrome P450	51	61	7.90E-06
		Oxidative phosphorylation	96	135	5.70E-05
		Glutathione metabolism	76	103	6.91E-05
		Rheumatoid arthritis	44	58	0.004021
		Sphingolipid metabolism	68	98	0.004644
		Retinol metabolism	26	31	0.004991
		Lysine degradation	92	140	0.006405
		Pentose and glucuronate interconversions	31	39	0.006589
		Fatty acid metabolism	42	58	0.017066
		Ascorbate and aldarate metabolism	18	21	0.023131
		Cell cycle	116	188	0.023826

Table 2. Cont.

tissue 1	tissue 2	#Term	Sample number	Background number	Corrected P-Value
		Arginine and proline metabolism	59	88	0.023826
		Arachidonic acid metabolism	70	107	0.023826
		Parkinson's disease	82	129	0.031582
		PPAR signaling pathway	53	79	0.036184
		Melanoma	41	59	0.039756
hepato-pancreas	testis	Lysosome	163	272	4.82E-07
		Peroxisome	91	138	2.02E-06
		Amino sugar and nucleotide sugar metabolism	83	124	2.14E-06
		Fatty acid metabolism	44	58	1.18E-05
		Rheumatoid arthritis	43	58	4.29E-05
		Starch and sucrose metabolism	72	112	7.17E-05
		Oxidative phosphorylation	84	135	7.17E-05
		Metabolism of xenobiotics by cytochrome P450	44	61	7.95E-05
		Drug metabolism - cytochrome P450	40	55	0.000155
		Glutathione metabolism	65	103	0.0004
		Drug metabolism - other enzymes	42	61	0.000658
		Retinol metabolism	24	31	0.002129
		Galactose metabolism	43	66	0.003549
		Pentose and glucuronate interconversions	28	39	0.004508
		Valine, leucine and isoleucine degradation	41	63	0.004508
		Sphingolipid metabolism	59	98	0.004508
		Aminobenzoate degradation	16	19	0.004508
		Butanoate metabolism	30	43	0.004508
		Other glycan degradation	45	71	0.004508
		beta-Alanine metabolism	26	36	0.004797
		Alanine, aspartate and glutamate metabolism	34	51	0.006232
		Arginine and proline metabolism	53	88	0.007429
		DNA replication	42	67	0.008497
		PPAR signaling pathway	48	79	0.009379
		Arachidonic acid metabolism	62	107	0.009379
		Histidine metabolism	19	25	0.009451
		Cysteine and methionine metabolism	36	56	0.009451
		Two-component system	16	20	0.009557
		Nitrogen metabolism	24	34	0.010345
		Parkinson's disease	72	129	0.013287
		Antigen processing and presentation	38	61	0.014212
		Tyrosine metabolism	28	43	0.023795
		Propanoate metabolism	29	45	0.024472
		Glycosaminoglycan degradation	30	47	0.024998
		Pyrimidine metabolism	89	168	0.026852
		Glycine, serine and threonine metabolism	44	75	0.029221
		Synthesis and degradation of ketone bodies	15	20	0.032987
		Glycolysis/Gluconeogenesis	56	100	0.032987
		Tryptophan metabolism	32	52	0.035613
		Meiosis - yeast	63	115	0.035613
		Nucleotide excision repair	43	74	0.036095
		Collecting duct acid secretion	17	24	0.041595
		Pyruvate metabolism	41	71	0.048078
ovary	muscle	Ribosome	81	91	0

Table 2. Cont.

tissue 1	tissue 2	#Term	Sample number	Background number	Corrected P-Value
		Protein processing in endoplasmic reticulum	134	240	0.001548
		Type II diabetes mellitus	35	48	0.001548
		Glycolysis/Gluconeogenesis	63	100	0.001843
		DNA replication	43	67	0.014867
		Colorectal cancer	61	102	0.014867
		Insulin signaling pathway	123	230	0.019284
		Cell cycle	102	188	0.024154
		Focal adhesion	196	389	0.024154
		Mismatch repair	32	49	0.027405
		Homologous recombination	38	61	0.02987
		Fanconi anemia pathway	60	105	0.033469
		Basal cell carcinoma	37	60	0.036003
		Thyroid cancer	14	18	0.04045
		Melanoma	36	59	0.044224
		Prostate cancer	51	89	0.044446
testis	muscle	RNA polymerase	47	69	0.000652
		Focal adhesion	199	389	0.002043
		Glycolysis/Gluconeogenesis	61	100	0.002043
		Carbon fixation in photosynthetic organisms	29	42	0.007085
		Bladder cancer	19	25	0.011212
		Basal cell carcinoma	38	60	0.011212
		Cytosolic DNA-sensing pathway	30	45	0.011555
		Pyrimidine metabolism	90	168	0.018107
		Colorectal cancer	58	102	0.021629
		Regulation of actin cytoskeleton	168	342	0.03891
ovary	testis	Ribosome	80	91	0

in redox signaling and lipid homeostasis [57]. Here, 146 transcripts were identified in the peroxisome pathway (ko04146), of which 71 transcripts were expressed at significantly higher levels in the hepatopancreas than in muscle tissue, including SOD2 (superoxide dismutase, Fe-Mn family [EC:1.15.1.1]), CAT (catalase [EC:1.11.1.6]), DDO (D-aspartate oxidase [EC:1.4.3.1]), DAO (D-amino-acid oxidase [EC:1.4.3.3]), SCP2 (sterol carrier protein 2 [EC:2.3.1.176]), and PIPOX (sarcosine oxidase/L-pipecolate oxidase [EC:1.5.3.7, 1.5.3.1]) (Figure S2, Table S7 in File S1). Only 11 transcripts were expressed at significantly lower levels in the hepatopancreas than in muscle tissue.

Methyl farnesoate and ecdysteroids are important hormones in crustaceans. Methyl farnesoate, which is synthesized in the mandibular organ (MO), is an insect juvenile hormone homologue that is believed to act as a juvenile hormone in crustaceans [58]. Juvenile hormones are involved in many biological processes, including development and reproduction. The major function of ecdysteroids is to control molting, but they are also involved in reproduction [59]. Here, genes in the insect hormone biosynthesis pathway were identified from the *P. clarkii* transcriptome. Among these genes, genes in the juvenile hormone synthesis pathway were significantly overexpressed in the hepatopancreas compared to the other three tissues (Table S8 in File S1). In particular, seven of nine transcripts that encoded a CYP15A1 (cytochrome P450, family 15, subfamily A, polypeptide 1) homologue were highly expressed in the hepatopancreas, in which the expression levels of

these transcripts were more than 100-fold greater than in the other tissues. These results indicate that the hepatopancreas is an important site for genes that are responsible for the synthesis of juvenile hormone. In contrast, genes in the ecdysteroid synthesis pathways did not show the same trend, and the differences in expression of these genes were not significant between the examined tissues.

Enriched pathways in the ovary and the testis. The ovary and the testis are the major reproductive organs, in which the processes of oogenesis, sperm genesis, DNA replication, and meiosis occur frequently. As expected, KEGG pathway analysis showed that the pathways of "DNA replication", "cell cycle", "mismatch repair", "homologous recombination", were significantly enriched in the ovary compared to muscle and the hepatopancreas. The pathways of "DNA replication", "pyrimidine metabolism", "meiosis-yeast", and "Nucleotide excision repair" were significantly enriched in the testis compared to muscle and the hepatopancreas. For example, 51 transcripts were identified in the DNA replication pathway (ko03030), of which 32 transcripts (26 genes) were expressed at significantly greater levels in the ovary than in the hepatopancreas, and of which only 2 genes were expressed at significantly lower levels in the ovary than in the hepatopancreas. Analysis showed that 26 transcripts (22 genes) were expressed at significantly higher levels in the testis than in the hepatopancreas, and only 6 transcripts (4 genes) were expressed at significantly lower levels in the testis than in the hepatopancreas (Table S9 in File S1).

In oviparous animals, vitellin is the major yolk protein that provides nutrition during embryonic development. The precursor of vitellin is vitellogenin (Vg). It is believed that extraovarian Vg is synthesized in the hepatopancreas and secreted in the hemolymph, where it is sequestered into developing oocytes by the Vg receptor (VgR) through receptor-mediated endocytosis [60]. It has been reported that multiple genes encode vitellogenin in various crustaceans, such as the shrimp *Metapenaeus ensis*, the freshwater water flea *Daphnia magna*, and the banana shrimp *Penaeus merguiensis* [61–63]. Here 29 transcripts (20 genes) were determined to encode vitellogenin, of which 14 transcripts were highly expressed specifically in the hepatopancreas, 6 transcripts were highly expressed specifically in the ovary, and no transcript was found to be highly expressed in the testis or in muscle. Indeed, vitellogenin was extremely difficult to detect in muscle (Table S10 in File S1), suggesting that vitellogenin is synthesized in the hepatopancreas and the ovary of *P. clarkii*. This result is consistent with previous reports in the Chinese mitten-handed crab *Eriocheir sinensis*, the tiger shrimp *Penaeus monodon,* the blue crab *Callinectes sapidus*, the freshwater crayfish *Cherax quadricarinatus*, the freshwater prawn *Macrobrachium rosenbergii*, the green mud crab *Scylla paramamosain*, and other species of shrimp and crab [64–68]. Interestingly, no transcript among the 20 transcripts determined to encode vitellogenin was found to be highly expressed in both the hepatopancreas and the ovary. Thus, expression of the identified vitellogenin transcripts was tissue-specific, including 14 transcripts that were hepatopancreas-specific and 6 transcripts that were ovary-specific. It has been reported that *MeVg1*, one of two vitellogenin genes in the shrimp *Metapenaeus ensis*, is expressed only in the ovary and the hepatopancreas, while the other vitellogenin gene, *MeVg2*, is expressed exclusively in the hepatopancreas [62,69]. These results provide new insight into the expression of vitellogenin genes in the

hepatopancreas and the ovary, and provide the basis for future studies on the manner in which vitellogenin genes collaboratively perform their specific functions at different developmental stages in the ovary.

The vitellogenin receptor is located in the cell membrane of oocytes and mediates vitellogenin absorption by oocytes through receptor-mediated endocytosis (RME) [70]. Unlike vitellogenin in *P. clarkii*, which was highly expressed in the hepatopancreas and in the ovary, the vitellogenin receptor gene was highly expressed in the ovary only, which is consistent with its ovary-specific expression pattern in the shrimp *Penaeus monodon* and in the freshwater prawn *Macrobrachium rosenbergii* (Table S10 in File S1) [60,64].

RT-PCR Assays. To validate the assembled transcripts and their expression profiles in the 4 collected tissue types, 20 transcript sequences were selected for RT-PCR (reverse transcription polymerase chain reaction) amplification. Their putative gene names, primer sequences, and expected PCR product sizes are shown in Table S11 in File S1. All 20 primer pairs gave amplification products of the expected sizes (Figure 6). For G09 and G13, in addition to the expected PCR products, larger PCR unexpected products were also found in the ovary. Analysis of the FPKM levels of these 20 selected transcript sequences showed that 8 sequences (G01–G08) were specifically expressed in the hepatopancreas, 3 sequences (G09–G11) were specifically expressed in muscle, 3 sequences (G12–G14) were specifically expressed in the ovary, 1 sequence (G15) was specifically expressed in the testis, and the other 5 sequences (G16–G20) were highly expressed in 3 or 4 tissue types (Table S11 in File S1). RT-PCR analysis showed that, with the exception of sequence G11, which was indicated by FPKM analysis to be specifically expressed in muscle, but was indicated by RT-PCR to be highly expressed in both muscle and testis, the expression modes of the other 19 sequences in the 4 tissue types were consistent with their FPKM levels (Figure 6). The evaluation and validation of the assembled

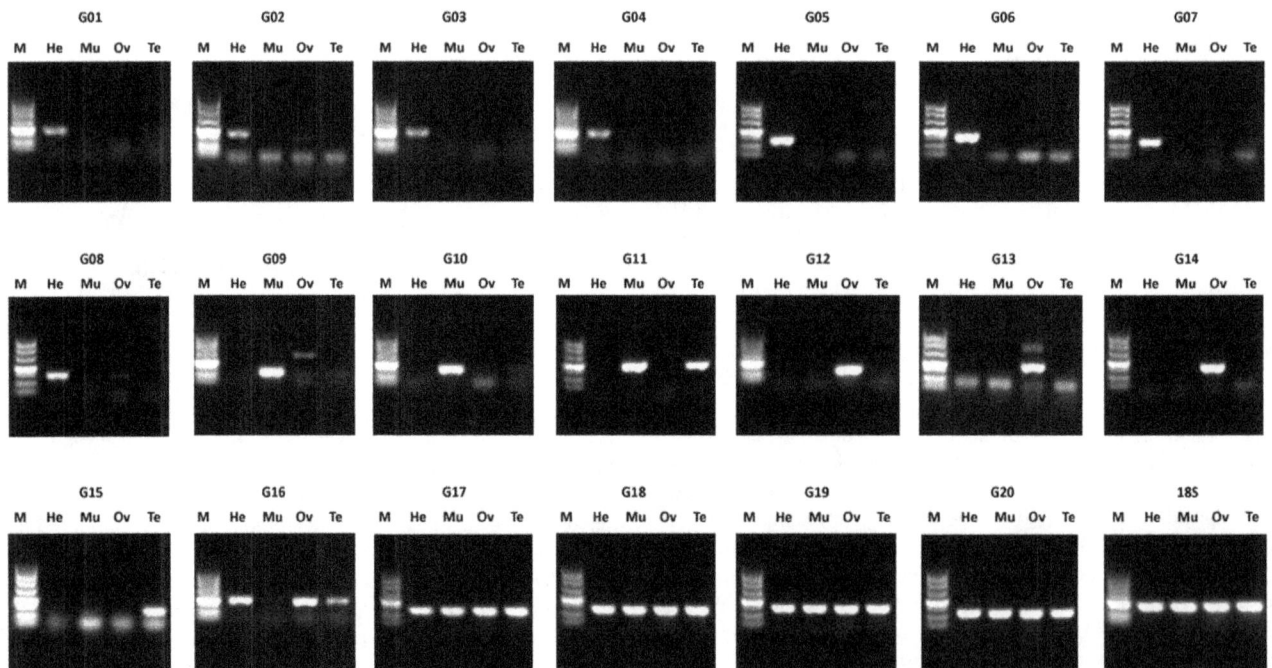

Figure 6. RT-PCR amplification and agarose gel (1.5%) electrophoresis of 20 transcripts. G01–G20: names of transcripts, represented transcript_ID given in Table S10 in File S1; 18S: 18S rRNA transcript; He: hepatopancreas; Mu: muscle; Ov: ovary; Te: testis; M: DNA marker.

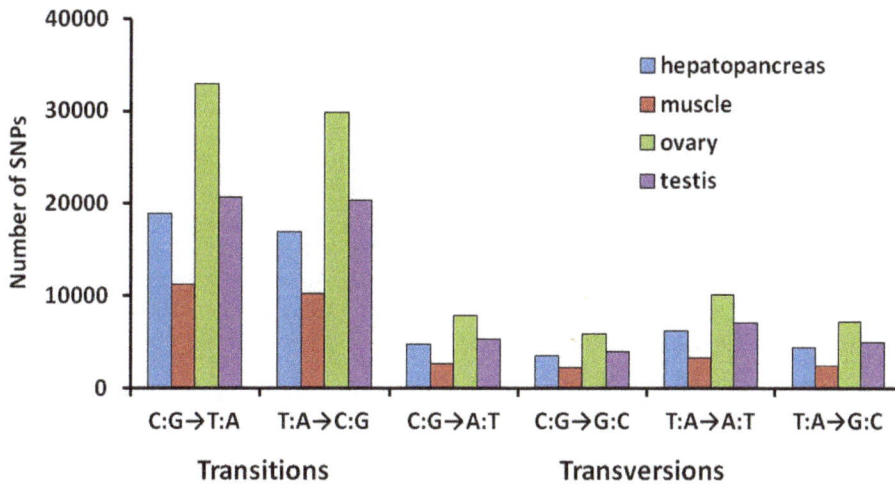

Figure 7. Classification of SNPs identified in the crayfish transcriptome.

transcripts verified the high accuracy of Illumina paired-end sequencing and de novo assembly, and thus indicated that our study could be useful for further research into gene function.

SNP identification

Single-nucleotide polymorphisms (SNPs) are the most common type of variation in the genome. SNPs were identified by alignments of multiple sequences used for contig assembly. After excluding those that had a base mutation frequency of less than 1%, a total of 243,764 SNPs were obtained (Figure 7). The proportions of transition substitutions were 34.44% for C:G→T:A and 31.74% for T:A→C:G, compared with smaller proportions of transversion for C:G→A:T (8.49%), C:G→G:C (6.42%),

T:A→A:T (11.05%) and T:A→G:C (7.86%). The total transition:transversion ratio was 1.96:1. Differences in base structure and the numbers of hydrogen bonds between different bases resulted in a large proportion of transition type SNPs and a small proportion of transversion type SNPs. The ovary had the most SNPs (94023 SNPs), followed by the testis (62601 SNPs), hepatopancreas (54855 SNPs), and muscle (32285 SNPs). Statistics for identified SNPs in the crayfish transcriptome are shown in Figure 7.

Microsatellite sequence identification

Microsatellite sequences, or simple sequence repeats (SSRs), are polymorphic loci present in genomic DNA that consist of repeated

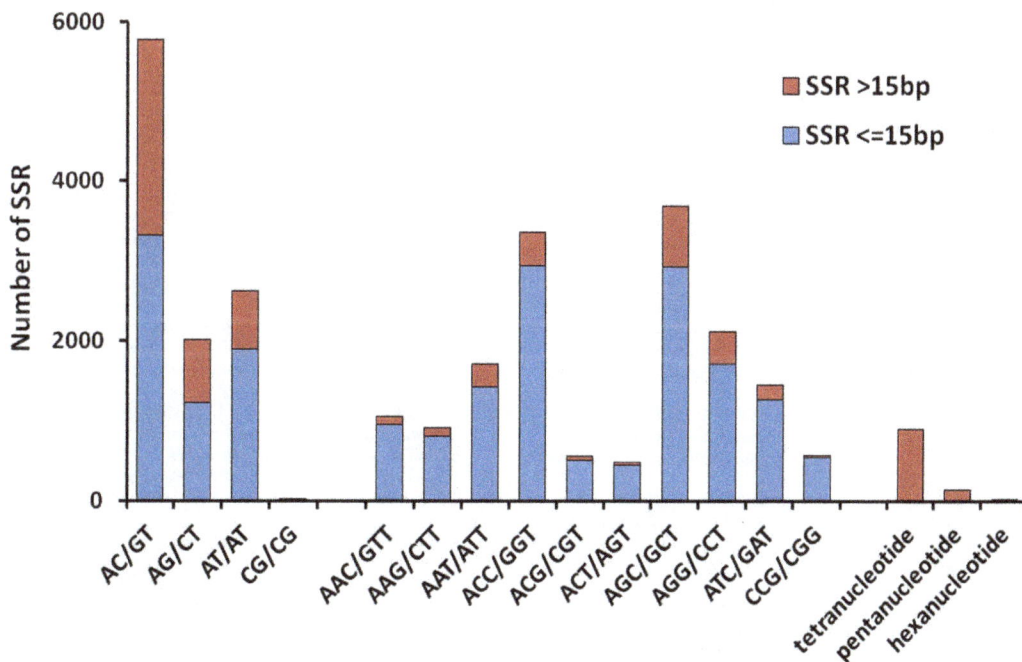

Figure 8. Distribution of simple sequence repeat (SSR) nucleotide classes among different nucleotide types found in the transcriptome of *P. clarkii*.

core sequences of 2–6 base pairs in length [71]. A total of 27,451 SSRs were initially identified from 29,534 transcripts, including 36.92% trinucleotide repeats, 24.14% di-nucleotide repeats, and 2.48% tetra/penta/hexanucleotide repeats (Figure 8). In addition, a total of 4775 SSRs (27.12%) were found that were more than 15 base pairs in length. Among the tri-nucleotide repeat motifs, (AGC/GCT)n (3,693 SSRs, 23.16%) and (ACC/GGT)n (3368 SSRs, 22.13%) were the most common types, and appeared significantly more than the other types of tri-nucleotide repeat motifs (Figure 8). After removing the microsatellites that lacked sufficient flanking sequences for primer design, 16953 unique sequences with microsatellites possessed sufficient flanking sequences on both sides of the microsatellites to allow the design of primers for genotyping.

Conclusions

This is the report on the transcriptome of *P. clarkii* using *de novo* assembly techniques with next-generation sequencing. We identified 50,219 non-redundant transcripts that will provide the basis for future studies on crayfish gene function. We also explored gene expression patterns in four different tissues from *P. clarkii*, and a number of candidate novel genes were identified that may be involved in important physiological processes and are worthy of further investigation. In addition, a large number of predicted SNPs and SSRs were reported that provide a basis for further genetic analysis and crayfish breeding.

Supporting Information

Figure S1 The hit species distribution based on BLASTx.

Figure S2 Transcripts identified as differentially expressed between muscle tissue and the hepatopancreas in the peroxisome pathway (ko04146). Genes for which expression levels in muscle were higher than those in the hepatopancreas are shown with a red frame, and genes for which expression levels in muscle were lower than those in the hepatopancreas are shown with a green frame.

File S1 Tables S1–S11. Table S1, Statistics for *P. clarkii* sequencing data. Table S2, Summary of BLASTX search results for the *P. clarkii* transcriptome. Table S3, Number of transcripts (genes) for each KEGG functional group. Table S4, Annotation and FPKM values for transcripts with ORFs in 4 tissue types. Table S5, Annotation and FPKM values of transcripts without ORFs in 4 tissue types. Table S6, Transcripts identified in the fatty acid metabolism pathway (ko00071) and differential expression analysis between the hepatopancreas and muscle tissue. Table S7, Transcripts identified in the peroxisome pathway (ko04146) and differential expression analysis between the hepatopancreas and muscle tissue. Table S8, Transcripts identified in the insect hormone biosynthesis pathway (ko00981). Table S9, Transcripts identified in the DNA replication pathway (ko03030) and differential expression analysis between the ovary and the hepatopancreas and between the testis and the hepatopancreas. Table S10, Transcripts encoding vitellogenin and the vitellogenin receptor identified from the *P. clarkii* transcriptome. Table S11, Information on the 20 transcript sequences selected for RT-PCR.

Acknowledgments

We thank Borun Beijing Innovation Technology Co., Ltd., for the revised English. We acknowledge scientists from Shanghai Majorbio Bio-pharm Biotechnology Co., Ltd. for their kind help with the bioinformatics analysis.

Author Contributions

Conceived and designed the experiments: HS. Performed the experiments: HS YH. Analyzed the data: HS YM. Contributed reagents/materials/analysis tools: XZ ZX YS CL PX XS. Wrote the paper: HS.

References

1. Banci KRS, Viera NFT, Marinho PS, Calixto PdO, Marques OAV (2013) Predation of Rhinella ornata (Anura, Bufonidae) by the alien crayfish (Crustacea, Astacidae) *Procambarus clarkii* (Girard, 1852) in São Paulo, Brazil. Herpetology Notes 6: 339–341.

2. Gherardi F (2006) Crayfish invading Europe: the case study of *Procambarus clarkii*. Marine and Freshwater Behaviour and Physiology 39: 175–191.

3. Yue GH, Wang GL, Zhu BQ, Wang CM, Zhu ZY, et al. (2008) Discovery of four natural clones in a crayfish species *Procambarus clarkii*. Int J Biol Sci 4: 279–282.

4. Wang W, Gu W, Ding Z, Ren Y, Chen J, et al. (2005) A novel Spiroplasma pathogen causing systemic infection in the crayfish *Procambarus clarkii* (Crustacea: Decapod), in China. FEMS Microbiol Lett 249: 131–137.

5. Cruz MJ, Rebelo R (2007) Colonization of freshwater habitats by an introduced crayfish, *Procambarus clarkii*, in Southwest Iberian Peninsula. Hydrobiologia 575: 191–201.

6. Chen AJ, Gao L, Wang XW, Zhao XF, Wang JX (2013) SUMO-conjugating enzyme E2 UBC9 mediates viral immediate-early protein SUMOylation in crayfish to facilitate reproduction of white spot syndrome virus. J Virol 87: 636–647.

7. Lin LJ, Chen YJ, Chang YS, Lee CY (2013) Neuroendocrine responses of a crustacean host to viral infection: effects of infection of white spot syndrome virus on the expression and release of crustacean hyperglycemic hormone in the crayfish *Procambarus clarkii*. Comp Biochem Physiol A Mol Integr Physiol 164: 327–332.

8. Du HH, Hou CL, Wu XG, Xie RH, Wang YZ (2013) Antigenic and immunogenic properties of truncated VP28 protein of white spot syndrome virus in *Procambarus clarkii*. Fish Shellfish Immunol 34: 332–338.

9. Wu XG, Xiong HT, Wang YZ, Du HH (2012) Evidence for cell apoptosis suppressing white spot syndrome virus replication in *Procambarus clarkii* at high temperature. Dis Aquat Organ 102: 13–21.

10. El-Din AH, Varjabedian KG, Abdel-Gaber RA, Mohamed MM (2013) Antiviral immunity in the red swamp crayfish, *Procambarus clarkii*: hemocyte production, proliferation and apoptosis. J Egypt Soc Parasitol 43: 71–86.

11. Tattersall GJ, Luebbert JP, LePine OK, Ormerod KG, Mercier AJ (2012) Thermal games in crayfish depend on establishment of social hierarchies. J Exp Biol 215: 1892–1904.

12. Buscaino G, Filiciotto F, Buffa G, Di Stefano V, Maccarrone V, et al. (2012) The underwater acoustic activities of the red swamp crayfish *Procambarus clarkii*. J Acoust Soc Am 132: 1792–1798.

13. Celi M, Filiciotto F, Parrinello D, Buscaino G, Damiano MA, et al. (2013) Physiological and agonistic behavioural response of *Procambarus clarkii* to an acoustic stimulus. J Exp Biol 216: 709–718.

14. Tomina Y, Kibayashi A, Yoshii T, Takahata M (2013) Chronic electromyographic analysis of circadian locomotor activity in crayfish. Behav Brain Res 249: 90–103.

15. Tierney AJ, Andrews K, Happer KR, White MK (2013) Dear enemies and nasty neighbors in crayfish: Effects of social status and sex on responses to familiar and unfamiliar conspecifics. Behav Processes.

16. Ameyaw-Akumfi C, Hazlett BA (1975) Sex recognition in the crayfish *Procambarus clarkii*. Science 190: 1225–1226.

17. Araki M, Hasegawa T, Komatsuda S, Nagayama T (2013) Social status-dependent modulation of LG-flip habituation in the crayfish. J Exp Biol 216: 681–686.

18. Leung TS, Naqvi SM, Naqvi NZ (1980) Paraquat toxicity to Louisiana crayfish (*Procambarus clarkii*). Bull Environ Contam Toxicol 25: 465–469.

19. Barbee GC, McClain WR, Lanka SK, Stout MJ (2010) Acute toxicity of chlorantraniliprole to non-target crayfish (*Procambarus clarkii*) associated with rice-crayfish cropping systems. Pest Manag Sci 66: 996–1001.

20. Al Kaddissi S, Legeay A, Elia AC, Gonzalez P, Camilleri V, et al. (2012) Effects of uranium on crayfish *Procambarus clarkii* mitochondria and antioxidants responses after chronic exposure: what have we learned? Ecotoxicol Environ Saf 78: 218–224.

21. Al Kaddissi S, Frelon S, Elia AC, Legeay A, Gonzalez P, et al. (2012) Are antioxidant and transcriptional responses useful for discriminating between chemo- and radiotoxicity of uranium in the crayfish *Procambarus clarkii*? Ecotoxicol Environ Saf 80: 266–272.

22. Bonvillain CP, Rutherford DA, Kelso WE, Green CC (2012) Physiological biomarkers for hypoxic stress in red swamp crayfish *Procambarus clarkii* from field and laboratory experiments. Comp Biochem Physiol A Mol Integr Physiol 163: 15–21.

23. Tan SH, Yuan ZD, Liu YF, Yang YN (2012) [Effects of Cd2+ on antioxidant system in hepatopancreas of *Procambarus clarkii*]. Ying Yong Sheng Tai Xue Bao 23: 2595–2601.

24. Belfiore NM, May B (2000) Variable microsatellite loci in red swamp crayfish, *Procambarus clarkii*, and their characterization in other crayfish taxa. Mol Ecol 9: 2231–2234.

25. Yue GH, Li JL, Wang CM, Xia JH, Wang GL, et al. (2010) High prevalence of multiple paternity in the invasive crayfish species, *Procambarus clarkii*. Int J Biol Sci 6: 107–115.

26. Li Y, Guo X, Cao X, Deng W, Luo W, et al. (2012) Population genetic structure and post-establishment dispersal patterns of the red swamp crayfish *Procambarus clarkii* in China. PLoS One 7: e40652.

27. Wu P, Qi D, Chen L, Zhang H, Zhang X, et al. (2009) Gene discovery from an ovary cDNA library of oriental river prawn *Macrobrachium nipponense* by ESTs annotation. Comp Biochem Physiol Part D Genomics Proteomics 4: 111–120.

28. Margulies M, Egholm M, Altman WE, Attiya S, Bader JS, et al. (2005) Genome sequencing in microfabricated high-density picolitre reactors. Nature 437: 376–380.

29. Huse SM, Huber JA, Morrison HG, Sogin ML, Welch DM (2007) Accuracy and quality of massively parallel DNA pyrosequencing. Genome Biol 8: R143.

30. Novaes E, Drost DR, Farmerie WG, Pappas GJ Jr., Grattapaglia D, et al. (2008) High-throughput gene and SNP discovery in Eucalyptus grandis, an uncharacterized genome. BMC Genomics 9: 312.

31. Mohd-Shamsudin MI, Kang Y, Lili Z, Tan TT, Kwong QB, et al. (2013) In-depth tanscriptomic analysis on giant freshwater prawns. PLoS One 8: e60839.

32. Ma K, Qiu G, Feng J, Li J (2012) Transcriptome analysis of the oriental river prawn, *Macrobrachium nipponense* using 454 pyrosequencing for discovery of genes and markers. PLoS One 7: e39727.

33. Li X, Cui Z, Liu Y, Song C, Shi G (2013) Transcriptome Analysis and Discovery of Genes Involved in Immune Pathways from Hepatopancreas of Microbial Challenged Mitten Crab *Eriocheir sinensis*. PLoS One 8: e68233.

34. He L, Wang Q, Jin X, Wang Y, Chen L, et al. (2012) Transcriptome profiling of testis during sexual maturation stages in *Eriocheir sinensis* using Illumina sequencing. PLoS One 7: e33735.

35. Lemgruber Rde S, Marshall NA, Ghelfi A, Fagundes DB, Val AL (2013) Functional categorization of transcriptome in the species *Symphysodon aequifasciatus* Pellegrin 1904 (Perciformes: Cichlidae) exposed to benzo[a]pyrene and phenanthrene. PLoS One 8: e81083.

36. Li E, Wang S, Li C, Wang X, Chen K, et al. (2014) Transcriptome sequencing revealed the genes and pathways involved in salinity stress of Chinese mitten crab, *Eriocheir sinensis*. Physiol Genomics 46: 177–190.

37. Zeng D, Chen X, Xie D, Zhao Y, Yang C, et al. (2013) Transcriptome analysis of Pacific white shrimp (*Litopenaeus vannamei*) hepatopancreas in response to Taura syndrome Virus (TSV) experimental infection. PLoS One 8: e57515.

38. Chen X, Zeng D, Xie D, Zhao Y, Yang C, et al. (2013) Transcriptome Analysis of *Litopenaeus vannamei* in Response to White Spot Syndrome Virus Infection. PLoS One 8: e73218.

39. Sookruksawong S, Sun F, Liu Z, Tassanakajon A (2013) RNA-Seq analysis reveals genes associated with resistance to Taura syndrome virus (TSV) in the Pacific white shrimp *Litopenaeus vannamei*. Dev Comp Immunol 41: 523–533.

40. Li C, Weng S, Chen Y, Yu X, Lu L, et al. (2012) Analysis of *Litopenaeus vannamei* transcriptome using the next-generation DNA sequencing technique. PLoS One 7: e47442.

41. Li S, Zhang X, Sun Z, Li F, Xiang J (2013) Transcriptome analysis on Chinese shrimp *Fenneropenaeus chinensis* during WSSV acute infection. PLoS One 8: e58627.

42. Kawahara-Miki R, Wada K, Azuma N, Chiba S (2011) Expression profiling without genome sequence information in a non-model species, Pandalid shrimp (*Pandalus latirostris*), by next-generation sequencing. PLoS One 6: e26043.

43. Wang W, Wu X, Liu Z, Zheng H, Cheng Y (2014) Insights into hepatopancreatic functions for nutrition metabolism and ovarian development in the crab *Portunus trituberculatus*: gene discovery in the comparative transcriptome of different hepatopancreas stages. PLoS One 9: e84921.

44. Lv J, Liu P, Gao B, Wang Y, Wang Z, et al. (2014) Transcriptome Analysis of the *Portunus trituberculatus*: De Novo Assembly, Growth-Related Gene Identification and Marker Discovery. PLoS One 9: e94055.

45. Cock PJ, Fields CJ, Goto N, Heuer ML, Rice PM (2010) The Sanger FASTQ file format for sequences with quality scores, and the Solexa/Illumina FASTQ variants. Nucleic Acids Res 38: 1767–1771.

46. Erlich Y, Mitra PP, delaBastide M, McCombie WR, Hannon GJ (2008) Alta-Cyclic: a self-optimizing base caller for next-generation sequencing. Nat Methods 5: 679–682.

47. Grabherr MG, Haas BJ, Yassour M, Levin JZ, Thompson DA, et al. (2011) Full-length transcriptome assembly from RNA-Seq data without a reference genome. Nat Biotechnol 29: 644–652.

48. Camacho C, Coulouris G, Avagyan V, Ma N, Papadopoulos J, et al. (2009) BLAST+: architecture and applications. BMC Bioinformatics 10: 421.

49. Conesa A, Gotz S, Garcia-Gomez JM, Terol J, Talon M, et al. (2005) Blast2GO: a universal tool for annotation, visualization and analysis in functional genomics research. Bioinformatics 21: 3674–3676.

50. Langmead B, Trapnell C, Pop M, Salzberg SL (2009) Ultrafast and memory-efficient alignment of short DNA sequences to the human genome. Genome Biol 10: R25.

51. Langmead B, Salzberg SL (2012) Fast gapped-read alignment with Bowtie 2. Nat Methods 9: 357–359.

52. Reiner A, Yekutieli D, Benjamini Y (2003) Identifying differentially expressed genes using false discovery rate controlling procedures. Bioinformatics 19: 368–375.

53. Robinson MD, Smyth GK (2007) Moderated statistical tests for assessing differences in tag abundance. Bioinformatics 23: 2881–2887.

54. Li H (2011) A statistical framework for SNP calling, mutation discovery, association mapping and population genetical parameter estimation from sequencing data. Bioinformatics 27: 2987–2993.

55. Koboldt DC, Chen K, Wylie T, Larson DE, McLellan MD, et al. (2009) VarScan: variant detection in massively parallel sequencing of individual and pooled samples. Bioinformatics 25: 2283–2285.

56. Faircloth BC (2008) msatcommander: detection of microsatellite repeat arrays and automated, locus-specific primer design. Mol Ecol Resour 8: 92–94.

57. Rottensteiner H, Theodoulou FL (2006) The ins and outs of peroxisomes: co-ordination of membrane transport and peroxisomal metabolism. Biochim Biophys Acta 1763: 1527–1540.

58. Nagaraju GPC (2007) Is methyl farnesoate a crustacean hormone? Aquaculture 272: 39–54.

59. Nagaraju GP (2011) Reproductive regulators in decapod crustaceans: an overview. J Exp Biol 214: 3–16.

60. Tiu SH, Benzie J, Chan SM (2008) From hepatopancreas to ovary: molecular characterization of a shrimp vitellogenin receptor involved in the processing of vitellogenin. Biol Reprod 79: 66–74.

61. Tokishita S, Kato Y, Kobayashi T, Nakamura S, Ohta T, et al. (2006) Organization and repression by juvenile hormone of a vitellogenin gene cluster in the crustacean, *Daphnia magna*. Biochem Biophys Res Commun 345: 362–370.

62. Tsang WS, Quackenbush LS, Chow BK, Tiu SH, He JG, et al. (2003) Organization of the shrimp vitellogenin gene: evidence of multiple genes and tissue specific expression by the ovary and hepatopancreas. Gene 303: 99–109.

63. Phiriyangkul P, Puengyam P, Jakobsen IB, Utarabhand P (2007) Dynamics of vitellogenin mRNA expression during vitellogenesis in the banana shrimp Penaeus (*Fenneropenaeus merguiensis*) using real-time PCR. Mol Reprod Dev 74: 1198–1207.

64. Revathi P, Iyapparaj P, Munuswamy N, Krishnan M (2012) Vitellogenesis during the ovarian development in freshwater female prawn *Macrobrachium rosenbergii* (De Man). International Journal of Aquatic Science 3: 13–27.

65. Jia X, Chen Y, Zou Z, Lin P, Wang Y, et al. (2013) Characterization and expression profile of Vitellogenin gene from *Scylla paramamosain*. Gene 520: 119–130.

66. Ferre LE, Medesani DA, Garcia CF, Grodzielski M, Rodriguez EM (2012) Vitellogenin levels in hemolymph, ovary and hepatopancreas of the freshwater crayfish *Cherax quadricarinatus* (Decapoda: Parastacidae) during the reproductive cycle. Rev Biol Trop 60: 253–261.

67. Zmora N, Trant J, Chan SM, Chung JS (2007) Vitellogenin and its messenger RNA during ovarian development in the female blue crab, *Callinectes sapidus*: gene expression, synthesis, transport, and cleavage. Biol Reprod 77: 138–146.

68. Li K, Chen L, Zhou Z, Li E, Zhao X, et al. (2006) The site of vitellogenin synthesis in Chinese mitten-handed crab *Eriocheir sinensis*. Comp Biochem Physiol B Biochem Mol Biol 143: 453–458.

69. Tiu SH, Hui JH, He JG, Tobe SS, Chan SM (2006) Characterization of vitellogenin in the shrimp *Metapenaeus ensis*: expression studies and hormonal regulation of MeVg1 transcription in vitro. Mol Reprod Dev 73: 424–436.

70. Roth Z, Khalaila I (2012) Identification and characterization of the vitellogenin receptor in *Macrobrachium rosenbergii* and its expression during vitellogenesis. Mol Reprod Dev 79: 478–487.

71. Queller DC, Strassmann JE, Hughes CR (1993) Microsatellites and kinship. Trends Ecol Evol 8: 285–288.

3

Effect of Antigen Shedding on Targeted Delivery of Immunotoxins in Solid Tumors from a Mathematical Model

Youngshang Pak[1,2]*, Ira Pastan[2], Robert J. Kreitman[2], Byungkook Lee[2]*

1 Department of Chemistry and Institute of Functional Materials, Pusan National University, Busan, Republic of Korea, 2 Laboratory of Molecular Biology, National Cancer Institute, National Institutes of Health, Bethesda, Maryland, United States of America

Abstract

Most cancer-specific antigens used as targets of antibody-drug conjugates and immunotoxins are shed from the cell surface (Zhang & Pastan (2008) Clin. Cancer Res. 14: 7981-7986), although at widely varying rates and by different mechanisms (Dello Sbarba & Rovida (2002) Biol. Chem. 383: 69–83). Why many cancer-specific antigens are shed and how the shedding affects delivery efficiency of antibody-based protein drugs are poorly understood questions at present. Before a detailed numerical study, it was assumed that antigen shedding would reduce the efficacy of antibody-drug conjugates and immunotoxins. However, our previous study using a comprehensive mathematical model showed that antigen shedding can significantly improve the efficacy of the mesothelin-binding immunotoxin, SS1P (anti-mesothelin-Fv-PE38), and suggested that receptor shedding can be a general mechanism for enhancing the effect of inter-cellular signaling molecules. Here, we improved this model and applied it to both SS1P and another recombinant immunotoxin, LMB-2, which targets CD25. We show that the effect of antigen shedding is influenced by a number of factors including the number of antigen molecules on the cell surface and the endocytosis rate. The high shedding rate of mesothelin is beneficial for SS1P, for which the antigen is large in number and endocytosed rapidly. On the other hand, the slow shedding of CD25 is beneficial for LMB-2, for which the antigen is small in number and endocytosed slowly.

Editor: Andrew J. Yates, University of Glasgow, United Kingdom

Funding: This work was supported in part by an appointment to the ORISE Research Participation Program at NCI. This program is administered by the Oak Ridge Institute for Science and Education through an interagency agreement between the US department of Energy and the National Cancer Institute. This research was supported by the Intramural Research Program of the NIH, National Cancer Institute, Center for Cancer Research. The funders had no role in study design, data collection and analysis, decision to publish, or preparation of the manuscript.

Competing Interests: The authors have declared that no competing interests exist.

* Email: bk@nih.gov (BL); ypak@pusan.ac.kr (YP)

Introduction

Recombinant immunotoxins (immunotoxins for short) and antibody-drug conjugates (ADCs) are protein and chemical toxins, respectively, that are conjugated to an antibody or antibody fragment. As promising anti-cancer agents, immunotoxins and antibody-drug conjugates are designed to kill only the cancer cells by binding to specific target antigens expressed on tumor cell surface [1–3]. In this report, we consider the delivery efficiency of immunotoxins only, although a similar consideration will apply for the antibody-drug conjugates as well.

The potency of an immunotoxin can be reduced by incomplete penetration through the solid tumor tissue [4]. The immunotoxin delivery process to solid tumors consists of a series of kinetic events. Upon injection into the blood stream, immunotoxin molecules permeate through the blood vessel wall into the extra-cellular space of the tumor, diffuse in the extra-cellular space, and bind to and dissociate from surface antigen molecules on tumor cells (Fig. 1). The surface bound immunotoxin molecules are internalized by receptor-mediated endocytosis and processed. The processed toxin molecules are then translocated into the cytosol where they begin the process that leads to cell death. Understanding these long and complex kinetic events for the immuno-

toxin delivery process is of practical importance for designing a better delivery strategy to improve the efficacy of immunotoxins.

In preclinical tests, different doses of immunotoxin are injected into the blood stream of tumor-bearing mice and subsequently tumor volume changes are monitored with time. These dose-dependent anti-tumor activity data, along with *in vitro* binding and cytotoxicity data, give a valuable insight into the factors that determine the effectiveness of immunotoxins. We previously reported on a mathematical model [5], which consisted of a system of partial differential equations and incorporated many key experimentally determined biological variables. It tracks the concentration of immunotoxin in all parts of the tumor tissue at all times. This model was applied to SS1P (anti-mesothelin-Fv-PE38), which is an immunotoxin targeted to mesothelin, a protein antigen highly expressed in several cancers [6]. SS1P is an excellent benchmark system for the development of our mathematical model because extensive pharmacokinetic parameters are available for this system from careful preclinical studies [7,8]. The model reproduced experimental tumor response data upon injection of different amounts of SS1P. It also exhibited the well-known binding site barrier effect [4,9], which refers to the hindrance to penetration of the antibody-based toxin into the

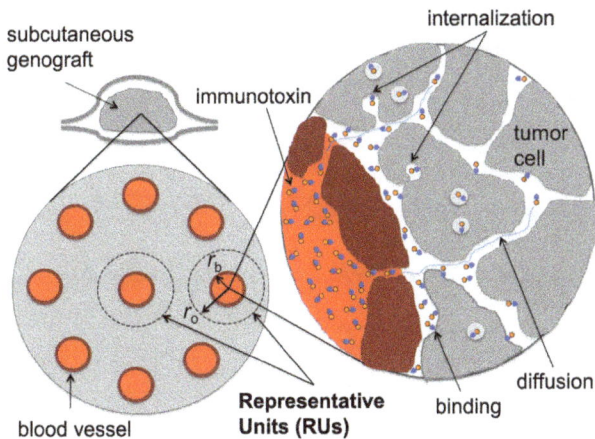

Figure 1. The vascular tumor model. It consists of a collection of representative units (RUs), each a cylinder of radius r_0 centered on a cylindrical blood vessel of radius r_b, which serves as the source of the administered immunotoxin. The values of r_b and r_0 used were 11 and 50 µm, respectively. Expanded view: The immunotoxin molecules, each represented as a small orange circle with a blue arrowhead, leave from blood vessel by permeating through the blood vessel wall into the extra-cellular space of the tumor tissue, diffuse in this space, bind to surface antigen molecules, and then become internalized by endocytosis.

tumor tissue by excessive binding to antigens on the cells nearby the blood vessel.

The most interesting finding was related to the effect of antigen shedding. Most cancer-specific antigens are shed into the extracellular space and into the blood [10,11]. Prior to this study, the antigen shedding was expected to hinder the immunotoxin delivery [11] since the shed antigen can act as a decoy for the immunotoxins and since the immunotoxin-antigen complex formed on the cell surface can be shed before being internalized. However, the model unexpectedly showed that receptor shedding enhanced the delivery efficacy of SS1P in solid tumors [5]. This was because shed antigen molecules in the extra-cellular space became a reservoir and carrier of immunotoxins and promoted a more uniform immunotoxin distribution in the tumor tissue by circumventing the binding site barrier effect. Based on this new finding, several suggestions were made to improve the effectiveness of immunotoxins in treating solid tumors.

Here we present a new, improved model. The modifications made are described in the Methods section and in more detail in Text S1. They are mostly technical and, with a couple of exceptions, result in small, although clearly discernable, improvements in the fit between the calculated and experimentally measured changes in volume of human tumors growing in mice with time after immunotoxin administration. Briefly, the two important exceptions are: (1) Back permeation into blood is allowed for all species including the shed antigen-immunotoxin complex. Previous model allowed back permeation of only the shed antigen species. This change results in a marked reduction of the beneficial effect of shed antigen. And (2) the permeation rate constant is reduced when the tumor size becomes large. This is not so much to reflect that the permeability of the capillaries in tumor varies with tumor size, although it may, but to effectively mimic the condition of very aberrant and relatively poorer vascularization for larger tumors [12,13]. In the previous model, the permeation rate constant was kept fixed, independent of tumor size. This change is needed in order to reproduce the experimental

observation that the shed antigen level in the extracellular space of solid tumors is larger for larger tumors [8].

We applied the new model to SS1P and another immunotoxin, LMB-2. LMB-2 (anti-Tac-Fv-PE38) is an anti-CD25 immunotoxin targeting CD25 displayed on cells of various hematological malignancies [14]. LMB-2 and SS1P share the same toxin fragment of Pseudomonas exotoxin A (PE). The preclinical efficacy data for SS1P were obtained using mice that bear the mesothelin expressing A431/H9 cell derived tumors. Similar data exist for LMB-2 using mice that bear ATAC-4 cell derived solid tumors [15]. A431/H9 and ATAC-4 cells are both A431 cells (human epidermoid carcinoma cell line), the former transfected with mesothelin, and the latter with CD25, expressing vectors. Thus, the two systems are very similar. But they also have important differences. For example, the number of antigen molecules on the surface is measured to be $\sim 10^6$ per cell in the case of the mesothelin expressing A431/H9 cells [7] whereas it is only about 2×10^5 for the CD25 expressing ATAC-4 cells [16]. Both antigens are membrane-bound, mesothelin by a GPI anchor [17] and CD25 by a trans-membrane peptide segment near the C-terminus [18]. Both are shed from the cell surface apparently by proteolytic cleavage from the membrane-bound C-terminal region of the polypeptide chain [19,20]. But the shedding rate is estimated to be about 20% per hour for mesothelin and only 0.2% per hour for CD25, according to our analysis (see Methods) of the experimental data [7,21]. Such similarities and differences, as well as the availability of extensive data, make these two systems highly valuable for testing and refining our mathematical model. The model includes some 31 parameters (Table 1), 18 of which have the same values for both systems (Table 2) while 13 others have different values (Table 3).

SS1P and LMB-2 have comparable affinities to their respective antigens; K_D values are 1.2 nM for SS1P (unpublished data) and 1.4 nM for LMB-2 [22]. However, the potencies of the two immunotoxins may well differ since there are more mesothelin molecules on the surface, which are also shed more. Experimental data that directly and clearly compare the efficacies of the two immunotoxins do not exist. However, data from independent experiments suggest that LMB-2 is equally, or somewhat more, effective than SS1P: injection of three 160 µg/kg doses of LMB-2 to ATAC-4 tumor bearing mice [15] appeared to produce more tumor growth suppression than injection of three 200 µg/kg doses of SS1P to A431/H9 tumor bearing mice [7].

Our new model reproduces tumor volume changes for both of these systems. The model shows that shedding can reduce, as well as enhance, antitumor activity depending on the number of antigen molecules on the cell surface: It predicts that shedding is beneficial for SS1P but reduces the efficacy of LMB-2. The fact that antigen shedding can both enhance and retard the efficacy of immunotoxins raises the possibility of a new mechanism by which receptor shedding can regulate signaling in normal tissues [23,24].

Results

Shed antigen levels

Figs. 2A and 2B show the calculated shed antigen levels in the extra-cellular space and blood, respectively, for the mesothelin-expressing A431/H9 tumors without the administration of SS1P. The calculated values reproduce the characteristic feature of the experimental data by Zhang et al. [8] that the shed antigen concentrations in the extra-cellular space and blood plasma both increase with tumor size. The tumor size-dependence of the shed antigen level is produced by the implementation of a new permeability function that reduces the permeability as the tumor

Table 1. Symbols, definitions, and units of the model parameters. (ECS: extra-cellular space).

Parameters	Unit	Definition
Mw (immunotoxin)	kDa	Molecular weight of immunotoxin
Mw (shed antigen)	kDa	Molecular weight of shed-antigen
R^*	number/cell	Number of total surface receptors per cell
ρ^*	cells/cm^3	Tumor cell density in EVS
φ_e		Volume fraction of the tumor extracellular space per tumor extravascular volume
r_b	μm	Blood vessel radius
r_o	μm	Outer radius of RU
t_α (immunotoxin)	min.	Half-life of immunotoxin in the blood
t_α (shed antigen)	min.	Half-life of shed-antigen in the blood
P^f_{eff} (low)	cm/s	Lower value of forward permeability (to ECS)
P^f_{eff} (high)	cm/s	Higher value of forward permeability
V_c	mm^3	Volume at which the permeability is the average of the low and high values.
a	mm^{-3}	Slope of the sigmoidal function for permeability at V_c
D_{eff}	cm^2/s	Diffusion constant of immunotoxin in ECS
χ_{efT}	hr^{-1}	Degradation rate constant of immunotoxin in ECS
χ_{efR}	hr^{-1}	Degradation rate constant of free antigen in ECS
χ_{ecR}	hr^{-1}	Degradation rate constant of complexed antigen in ECS
χ_{ce}	hr^{-1}	Endosomal degradation rate constant of immunotoxin
χ_{cc}	hr^{-1}	Cytosolic degradation rate constant of immunotoxin
k_d	hr^{-1}	Immunotoxin-receptor dissociation rate constant
k_a	hr^{-1} nM^{-1}	Immunotoxin-receptor association rate constant
$K_D \equiv k_d/k_a$	nM	Binding affinity of immunotoxins to receptors
k_s	hr^{-1}	Receptor shedding rate constant
k_e	hr^{-1}	Endocytosis rate constant
k_t	hr^{-1}	Immunotoxin translocation rate constant
$\Omega \equiv k_t/(\chi_{ce}+k_t)$		Fraction translocated from endosome
k^{max}_{cat}	hr^{-1}	Maximum intoxication rate constant
T_0		Number of cytosolic toxin molecules per type 1 cell at which the intoxication rate is half of k^{max}_{cat}
Γ_0	hr^{-1}	Tumor growth rate constant at small volume
α	mm^{-3}	Tumor growth rate decay parameter with tumor volume
Δ	hr^{-1}	Cell death rate constant
X	Hr^{-1}	Dead cell clearance rate constant

volume increases according to Eqs. (2) and (3) in Methods. The previous model used a constant permeability, which led to a flat extra-cellular level with tumor volume.

This tumor cell line expresses a large number of mesothelin molecules (10^6/cell). The shedding rate constant determined from the best fit of the calculated extra-cellular and blood concentrations of shed antigen to the experimental data is 0.20 hr^{-1} (Table 3). This shedding rate is smaller than the previous value of 0.4 hr^{-1} [5], which was based on *in vitro* experimental data.

Figs. 2C and 2D show the shed antigen (CD25) concentrations for the ATAC-4 tumor. Unlike the case of mesothelin, the shed CD25 concentrations in the extra-cellular space and blood are orders of magnitude smaller and the data points are scattered [15]. Nevertheless, these data provide useful information on the shedding rate: Assuming the same permeability as for the A431/H9 tumor (Table 2), the experimental shed antigen levels can be obtained only when the shedding rate is much smaller than that of

mesothelin. Numerically, the shedding rate constant of 0.002 hr^{-1} gives the best fit for CD25.

Tumor volume responses

Fig. 3A shows the predicted tumor volume responses upon three SS1P injections with a dose of 200 μg/kg (62 nM) every other day. All calculated volumes fall within the variation of experimental data points. The fit is better than in the previous work on the same system [5], particularly for early and late tumor volumes. The better fit for the early times occurs mainly upon the use of tumor size dependent vascular permeability. The better fit for the late tumor volume data is due to the use of the Gompertzian cell growth model (see Eq. (4) in Methods).

Fig. 3B shows the experimental and simulated tumor volume profiles upon administration of LMB-2 to the ATAC-4 tumor bearing mice. Here, the LMB-2 with a dose of 160 μg/kg (51 nM) was injected three times every other day. The calculated volumes

Table 2. Parameters that were unchanged between SS1P and LMB-2.

Parameters	SS1P LMB-2	Comment/Reference
ρ^* (cells/cm^3)	0.5×10^9	[25]
φ_e	0.1	
r_b (μm)	11.0	
r_o (μm)	50.0	Assume
P^f_{eff} (low) (cm/s)	7.1×10^{-7}	
P^f_{eff} (high)(cm/s)	8.0×10^{-6}	
V_c (mm^3)	250.0	
a (mm^{-3})	0.04	
D_{eff} (cm^2/s)	2.5×10^{-8}	[33]
χ_{efT} (hr^{-1})	0.1	
χ_{efR} (hr^{-1})	0.1	
χ_{ecR} (hr^{-1})	0.1	
χ_{ce} (hr^{-1})	0.021	[33]
k_t (hr^{-1})	0.69	[5]
k_{cat}^{max} (hr^{-1})	0.252	
T_0	400	
Δ (hr^{-1})	0.17	
X (hr^{-1})	0.23	

compare well with the recent experimental data of Singh et al. [15].

Antigen shedding improves the efficacy of SS1P on the A431/H9 tumor

The shed antigen is found in the blood as well as in the extra-cellular space [7,11]. In this work, we assume that both the shed antigen and the shed antigen-immunotoxin complex that forms in the extra-cellular space back-permeate to the blood. We investigated the effect of the back-permeation on the tumor volume response upon the administration of SS1P. Fig. 4A shows the simulated tumor volume profiles for the non-shedding case (red line) and the shedding case with (black line) and without (blue line) back-permeation of the shed antigen-immunotoxin complex. The latter case corresponds to our previous model wherein the back-permeation rate for the antigen-immunotoxin complex was set to zero. In the current model, the back-permeation is allowed for the complex with the permeability set to be the same as that for the shed antigen and the same as the forward (from blood to the tumor extra-cellular space) permeability (see Methods). The Figure shows that antigen shedding improves the efficacy of SS1P in both models. However, the improvement is much less when back-permeation is allowed (current model) than when it is not allowed

Table 3. Parameters that were changed for LMB-2.

Parameters	SS1P	Comment/Reference	LMB-2	Comment/Reference
Mw (immunotoxin) (kDa)	64	[7]	63	[14]
Mw (shed antigen) (kDa)	42	[7]	45	[34]
R^*	1.0×10^6	[7]	2.0×10^5	[16]
t_α (immunotoxin) (min)	24	[35]	13	
t_α (shed antigen) (min)	3	[5]	5	
k_d (hr^{-1})	0.61	u	0.71	Assume
K_D (nM)	1.2	u	1.4	[22]
k_s (hr^{-1})	0.2		0.0020	
k_e (hr^{-1})	0.22		0.08	
Ω	0.065		0.012	
Γ_0 (hr^{-1})	0.0301		0.0170	
α (mm^{-3})	0.00127		0.00197	
χ_{ce} (hr^{-1})	9.9		56.8	

u unpublished data.

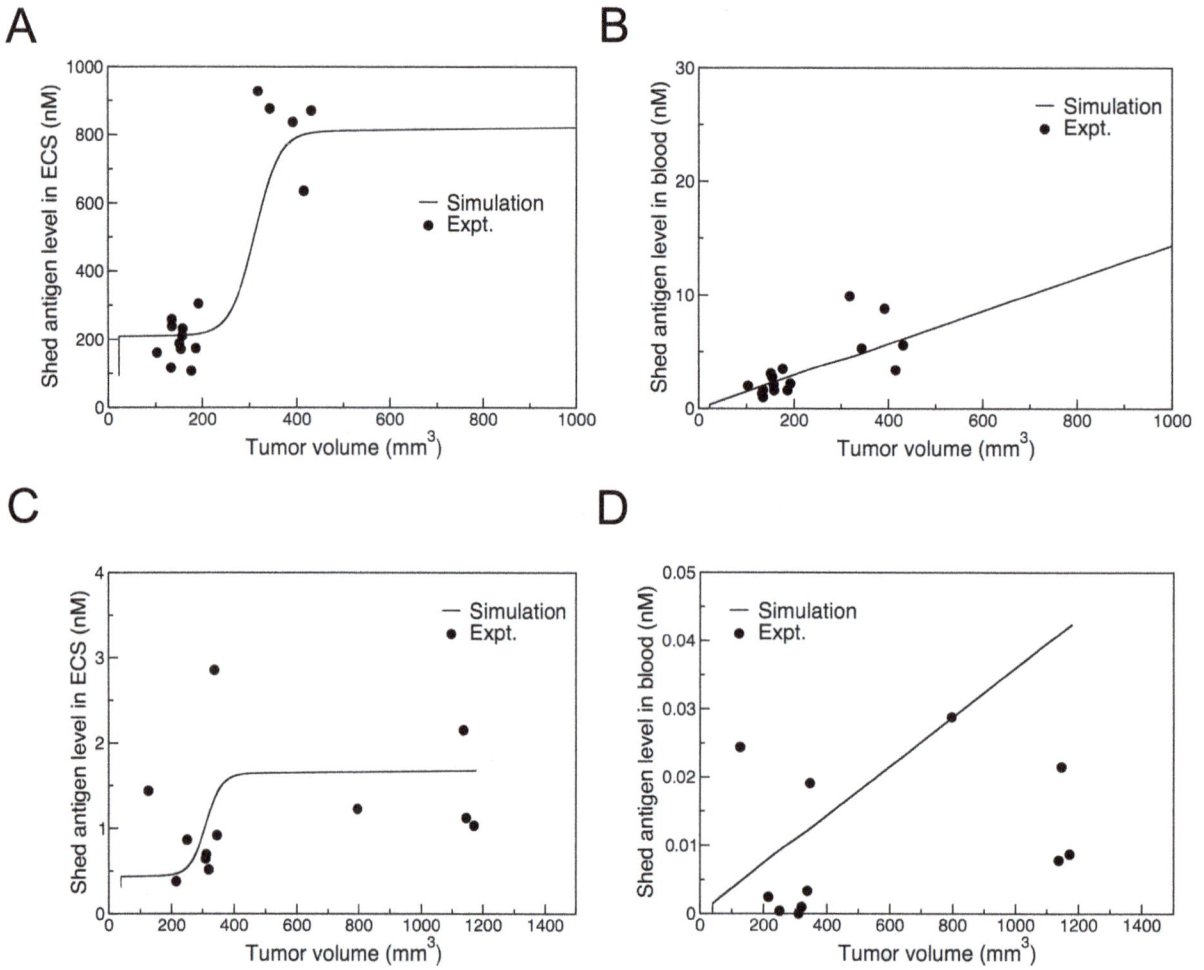

Figure 2. Tumor volume dependence of the measured (symbols) and computed (line) shed antigen levels in the extra-cellular space (A, C) and blood (B, D) for A431/H9 (A, B) and ATAC-4 cells (C, D). The experimental data for mesothelin on A431/H9 cell (panels A and B) are from Zhang et al. [7] and those for CD25 on ATAC-4 cell (panels C and D) are from Singh et al. [15].

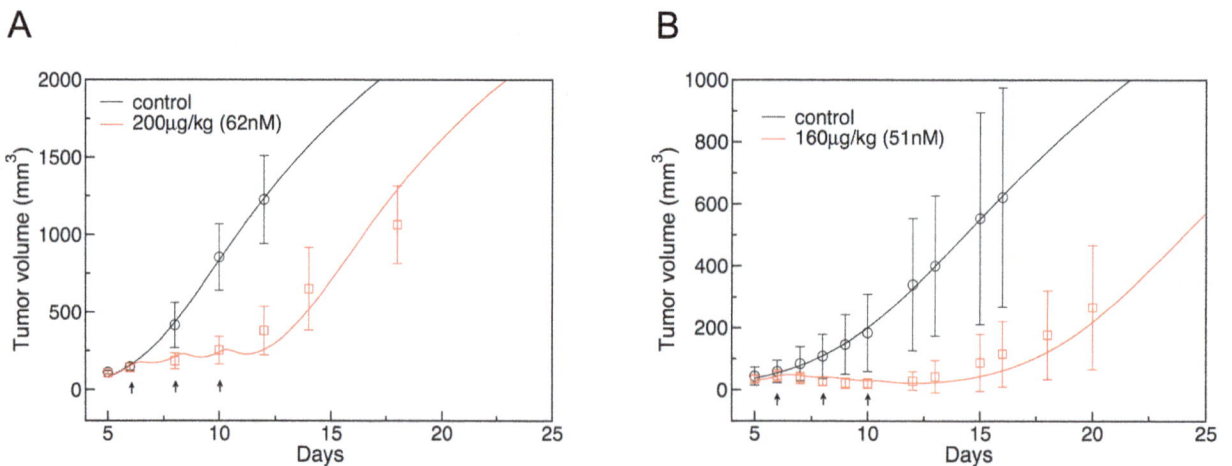

Figure 3. Measured (symbols with error bars) and computed (line) tumor volumes with time of A431/H9 (A) and ATAC-4 (B) cell tumors growing in nude mice. The experimental data are from Zhang et al. [7] and Singh et al. [15]. In each panel, the black line and symbols are for control with no immunotoxin and the red line and symbols are for the case when immunotoxin was given three times, at times indicated by the black arrows. The dose level was 200 µg/kg (62 nM) for SS1P (panel A) and 160 µg/kg (51 nM) for LMB-2 (panel B).

A

B

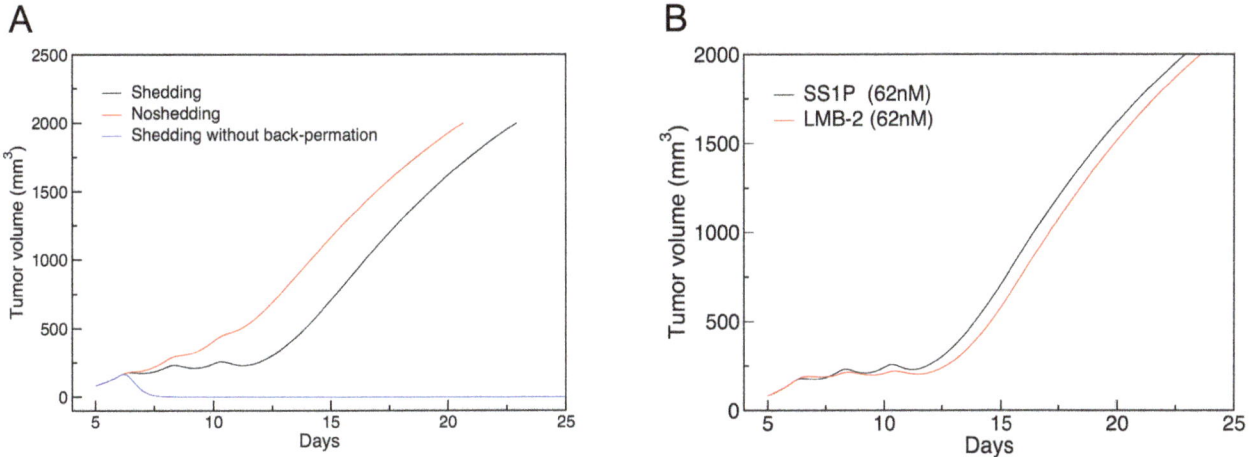

Figure 4. Computed tumor volume profiles using altered model parameters. (A) A431/H9 cell tumor with 3×62 nM injections of SS1P for three cases: No shedding (red line) and shedding with (black line; current model) and without (blue line; previous model) the back-permeation of shed antigen-immunotoxin complex to the blood vessel. **(B)** Comparison of the A431/H9 cell tumor with SS1P injection (black line) and ATAC-4 cell tumor with LMB-2 injection (red line) using the same dose (62 nM), initial volume (110 mm³), and tumor growth rate ($\Gamma_0 = 0.0301 \text{ hr}^{-1}$, $\alpha = 0.00127 \text{ mm}^{-3}$).

(previous model). Thus, the beneficial effect of antigen shedding reported in the previous study may have been over-estimated (see Discussion).

SS1P and LMB-2 have a similar predicted potency

Having demonstrated the robust fitting capability with our refined models, next we compared the *in vivo* potency of these two immunotoxins. In order to make the comparison simple, we made the initial conditions for tumor volume simulations the same by keeping their initial tumor volume to 110 mm³ for both tumors and the dose to 62 nM for both immunotoxins. As seen from the tumor growth profiles without toxins (the control curves in Fig. 3), the experimental tumor growth rates for the A431/H9 and ATAC-4 cells are quite different, although both are derived from the same cell line. For the efficacy test, any bias resulting from two different tumor cell growth rates needs to be eliminated. For the purpose of comparison, we therefore assigned the same tumor growth rate to both cells by changing the values of the Γ_0 and α parameters (see Table 1) of ATAC-4 cells to those of the A431/H9 cells. Other parameters for the LMB-2 case were kept as given in Tables 2 and 3. The parameters for the SS1P case were not changed. Then, we recalculated the tumor volume profiles for both immunotoxin cases. The resulting tumor volume profiles for the two cases are remarkably similar (Fig. 4B), indicating that these two immunotoxins are equally effective.

The effect of antigen shedding depends on the number of antigen molecules on the cell surface and the endocytosis rate

Although the models predict a similar anti-tumor activity for SS1P and LMB-2, there are large differences in some of the parameter values. As summarized in Table 3, in comparison with SS1P targeting mesothelin, there are a smaller number of target antigen molecules ($R^* = 2 \times 10^5$) for LMB-2, with a negligible shedding rate ($k_s = 0.002 \text{ hr}^{-1}$) and which undergoes a slower endocytosis ($k_e = 0.08 \text{ hr}^{-1}$ vs. 0.22 hr^{-1} for mesothelin). Also, the fraction of toxin translocated into the cytosol (Ω) was lower for LMB-2, which indicates that more toxin molecules are degraded

in the endosome for LMB-2 than for SS1P. (See Table 1 for the names of the parameters.)

In an attempt to explain the similar efficacy of LMB-2 and SS1P and to understand these differences in parameter values between the two cases, we calculated the tumor volume profile for SS1P for R^* values ranging from 10^3 to 10^7 for both shedding ($k_s = 0.2 \text{ hr}^{-1}$) and non-shedding ($k_s = 0.0$) cases, each using two different endocytosis rate constants. (All other parameters remain the same as those listed in Tables 2 and 3 for SS1P). The results are shown in Fig. 5, where the simulated tumor volumes 2 days after the third injection (day 12) of 62 nM (panel A) or 124 nM (panel B) are plotted at different R^* values for both shedding and non-shedding cases of SS1P. The Figures show that there exists an optimal range of R^* values where the immunotoxins are maximally effective. The optimal range varies depending on the dose applied: $10^4 < R^* < 10^5$ when 62 nM is given (Fig. 5A) and $10^4 < R^* < 2 \times 10^5$ when 128 nM is given (Fig. 5B). For both shedding and non-shedding cases, immunotoxin becomes less effective as R^* increases beyond these ranges, presumably because the binding site barrier increases.

One can also note that, for R^* values of half a million or more, the receptor shedding is clearly beneficial for the immunotoxin delivery and its beneficial effect becomes more pronounced with increasing R^*. (Compare the solid black and red lines in Figs. 5.) For smaller values of R^*, however, shedding in fact reduces the effectiveness of immunotoxin. Thus, the beneficial effect of antigen shedding which we reported in the earlier study is more likely limited to the case of a large number of target binding sites on the cell when the binding site barrier effect is high.

Endocytosis rate also matters. Figs. 5 show that a smaller k_e value makes immunotoxins more effective for the non-shedding case (black broken line), but less effective for the shedding case (red broken line). This finding can be explained along the following line of reasoning. In the non-shedding case, binding site barrier effect is the controlling factor as the tumor cells near the capillary act as a sink by absorbing a large amount of immunotoxin, preventing penetration and uniform distribution of immunotoxin into tumor tissues. A lower endocytosis rate helps in this case by reducing the immunotoxin loss to the sink thereby making more immunotoxin available for deeper penetration into the tumor tissue. In the

A

B

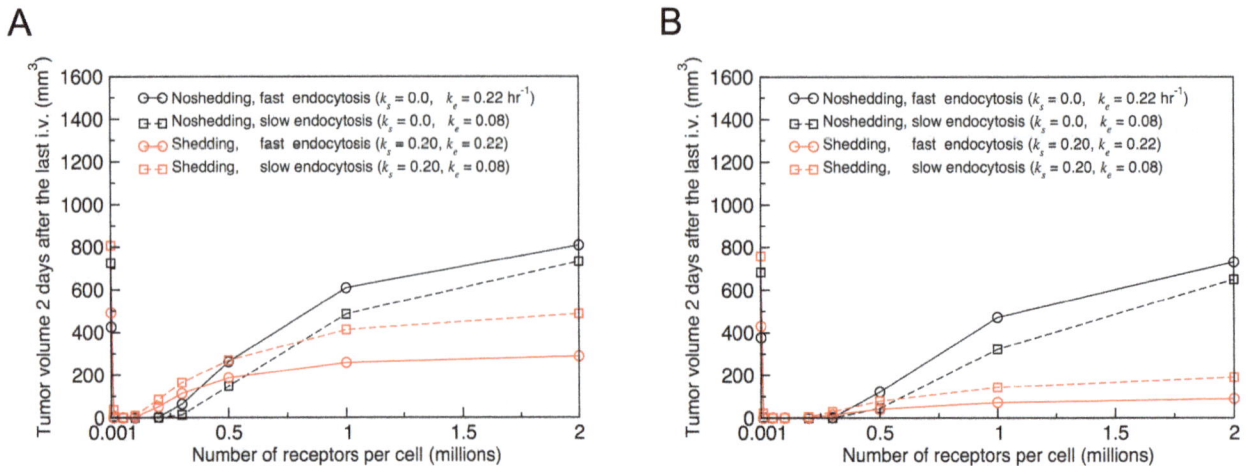

Figure 5. Predicted tumor volumes at two days after the last injection versus number of surface antigen molecules per cell ($R*$) for the A431/H9 cell tumor with 3×62 nM (A) or 3×124 nM (B) injections of SS1P for the shedding and no shedding cases with two different endocytosis rate constants (k_e=0.08 and 0.22 hr^{-1}).

shedding case, binding site barrier effect is not as important since the presence of free immunotoxin-receptor complex effectively circumvents the barrier. A larger k_e value helps in this case because the antigen shedding competes with the endocytosis process to retard the internalization of immunotoxins.

Comparing Figs. 5A and 5B, we also note that increasing the dose level has a large effect when antigen is shed (red lines) but has relatively little effect when there is no shedding (black lines). This latter feature is the expected behavior when the binding site barrier is the limiting factor.

The model predicts faster translocation rate for SS1P than LMB-2

Table 3 shows that the value of the Ω parameter is smaller for the LMB-2 than for the SS1P case, which implies that LMB-2 is degraded more in endosome (or less translocated to the cytosol) than SS1P. This is an expected result because, in terms of delivery efficiency, LMB-2 with a smaller receptor number ($R* = 2 \times 10^5$) and a negligible shedding (or non-shedding; $k_s = 0.0$) should be better than SS1P with a larger receptor number ($R* = 1 \times 10^6$) and a substantial shedding ($k_s = 0.2$ hr^{-1}) (Fig. 5). Therefore, within our model, the only way the efficacies of SS1P and LMB-2 become similar when the natural growth rates of the different cell lines are made the same (Fig. 4B) is that LMB-2 is degraded more in endosome (or less translocated to the cytosol). This conclusion has some experimental support: the measured IC50 value for LMB-2 on ATAC-4 cells is 0.4 pM [25] whereas that for SS1P against A431/H9 cells is 0.2 pM (unpublished data). Although definitive conclusions cannot be made from these IC50 values, which depend sensitively upon details of the experimental conditions, they are consistent with the possibility that LMB-2 is less toxic on ATAC-4 cells than SS1P is on A431/H9 cells. We assume that this is possible if CD25 and mesothelin end up in different endosomal compartments upon endocytosis.

Discussion

In our earlier study [5], we presented a mathematical model, which tracks immunotoxin concentrations in different compartments of a solid tumor tissue. For the first time, this model incorporated effects of antigen shedding into the detailed kinetic events that immunotoxin molecules encounter during their transit from the blood to tumor cells. We showed that this model reproduced dose-dependent antitumor activity of SS1P on the A431/H9 cell-derived tumor growing in mice. The model predicted that antigen shedding improved delivery efficiency of the immunotoxin in solid tumors.

In the present study, we improved the model by incorporating several new features and applied the improved model on two systems, the immunotoxin LMB-2 targeting CD25 of the ATAC-4 cell tumor as well as the original SS1P targeting mesothelin of the A431/H9 cell tumor.

One of these improvements is the introduction of a tumor volume dependent vascular permeability to better fit the experimental shed antigen level. In the absence of lymphatic drainage, the concentration of shed antigen in the extra-cellular space increases by shedding and decreases mainly by back-permeation. The shedding rate depends on the number of antigen molecules per cell and the number of cells per unit volume of the extra-cellular space. When immunotoxin is not given, both of these are constants, independent of the tumor size in our model. The back-permeation rate depends on the blood vessel surface area per volume of the extra-cellular space, which is also constant. Therefore, the steady state extra-cellular space concentration of shed antigen cannot depend on the tumor size in our model. However, the recent experiments on the mesothelin-expressing tumor [8] showed that, in the absence of immunotoxin, shed antigen level in the extra-cellular space increased with tumor size. This implies that the tumor capillary does not grow proportionally with the tumor volume as required by our model. Within the framework of our model, a simple way to produce a similar effect is to make the blood vessel less permeable as the tumor size increases. Clearly, this new scheme led to much better fits to the experimental shed antigen level (Fig. 2).

Another major modification is to allow all species, including immunotoxin and the shed antigen-immunotoxin complexes, to back-permeate from the extra-cellular space to the blood. In the previous model, we assumed that only the shed antigen could back-permeate. In a normal tissue, materials extravasate into the interstitial space mainly by the blood-tissue hydrostatic pressure differential [26] and are cleared from there by the lymphatic system; there should be little back permeation into the blood stream. In the tumor tissue, back permeation should increase both because there is no lymphatic drainage and because of the hydrostatic pressure that builds up in the interior of the tumor tissue [13]. The rate of back-permeation is not known, but presumably it approaches, but probably not exceeds, the forward permeation rate. In this study, we determined the back-permeation rate constant for the shed antigen by assuming that it is the same for both mesothelin and CD-25 and by using their blood and extra-cellular space concentrations in the absence of immunotoxin (see Methods and Text S1). We then assumed that the forward as well as the back-permeation rate constants for all molecular species were the same as the back-permeation rate constant determined for the free shed antigen. This probably over-estimates the back-permeation of the shed antigen-immunotoxin complex relative to that of the free shed antigen, since the complex molecule is larger in size, although permeability is not too sensitive to the molecular size in the 70 kD range [12,27], and relative to the forward permeation of immunotoxin, in contrast to the under-estimation made in the previous model by dis-allowing the back-permeation altogether. Under this condition of relatively high back-permeation of the shed antigen-immunotoxin complex, the benefit of shedding is greatly reduced but still present in the case of SS1P (Fig. 4A).

This study also shows the value of considering more than one system. Whereas shedding is clearly beneficial for SS1P (compare red vs. black lines at R^* value of 1 million in Fig. 5), it would reduce the efficacy of LMB-2 (compare red vs. black lines at R^* value of 0.2 million in Fig. 5). When the number of target antigen molecules is large, as in the case of mesothelin-expressing A431/H9 cells, the free immunotoxin penetration is severely retarded by a significant binding site barrier and shedding is beneficial as the shed antigen acts as a protective carrier of immunotoxin, thus circumventing the binding site barrier effect. This finding confirms the previous result on the A431/H9/SS1P [5]. On the other hand, when the number of target antigen molecules is small, as in the case of CD25-expressing ATAC-4 cells, the binding site barrier is lower and the transport of the free immunotoxin in the extra-cellular space is less hindered. In this situation, the negative effects of antigen shedding become dominant: The shed antigen acts as a decoy to decrease the free immunotoxin level in the extra-cellular space and the antigen-immunotoxin complex can be shed, thus reducing the rate of productive entry of immunotoxin into the cell by endocytosis. This is in line with the other view that was suggested before [7,11]. Therefore, our expanded perspective on the antigen shedding is that the shed antigen can act as either a promoter or a suppressor of immunotoxin deliveries depending on the number of target binding sites per cell.

In the previous work [5], we suggested that binding site barrier effect must exist also for natural extra-cellular signaling molecules operating in normal tissues and that receptor shedding could be a natural strategy to improve the range and duration of signaling molecules. The findings in this study now suggest that the effect of shedding can be either positive or negative depending on the number of receptors and that receptor shedding can be used to control the efficiency of signaling. However, the point at which the effect switches from being negative to positive will generally not be the same as that found in this study with the immunotoxins on cancer tissues. The switching point will depend on a number of factors including the endocytosis rate (see below), number of ligand molecules required to cause an effect in each cell, and the concentration of ligands.

The rate of entry of immunotoxin molecules into the cell is governed by several competing processes: the association with and dissociation from the surface-bound antigen, antigen shedding,

and endocytosis. The two immunotoxins have a comparable association and dissociation rates. But the endocytosis rates are very different (Table 3). We find that a larger endocytosis rate does not always ensure an increased efficacy of immunotoxins. Our model shows (Fig. 5) that a smaller k_e value is more effective for the non-shedding case where the binding site barrier is the main obstacle whereas a larger k_e value is more effective for the shedding case. Also, except for the case of extremely small number of antigen molecules (<1000), in which the internalization of immunotoxin is severely restricted, a smaller number of binding sites is always better for both shedding and non-shedding cases. This last observation is supported by other models (see, for example, [28]) and serves as a warning that many cancer antigens, which are identified because of their overexpression in cancer cells, may not be the ideal targets for immunotherapy [29].

Since the number of binding sites, the shedding rate, and the endocytosis rate all favor LMB-2 over SS1P, LMB-2 should reach more cancer cells. However, the models that best fit the experimental data indicate that LMB-2 and SS1P have remarkably similar potencies (Fig. 4B). Within our model, these facts imply that the translocation efficiency (Ω) in endosome must be reduced significantly for LMB-2 (Table 3). Thus, rather surprisingly, our delivery simulation model suggests that CD25 and mesothelin go to different endosomal compartments where the cargo that they carried in is degraded or translocated into the cytosol at different rates. Obviously, any prediction made using a highly simplified mathematical model needs to be verified experimentally, but the fact that such a prediction can be made at all shows the usefulness of a comprehensive model.

Materials and Methods

Model system

A tumor is modeled as a collection of m identical representative units (RUs, Fig. 1). Each RU is a cylinder of 50 μm radius and has a cylindrical blood vessel at its center. The blood vessel is the source of immunotoxin. As the tumor grows or shrinks, the total number of RUs increases or decreases and the tumor volume is simply given by m times the volume of an RU.

The model consists mainly of two sets of partial differential equations (see Text S1 for detailed equations). One set of equations governs immunotoxin concentrations in the blood, in the extra-cellular space (ECS), and in the three compartments of the tumor cell: surface, endosome, and cytosol (Fig. 6). The other set of differential equations describes density changes of three different types of tumor cells (Fig. 7): un-intoxicated (type 1), intoxicated (type 2) and dead (type 3) cells. The un-intoxicated cells divide and shed their surface antigens into the ECS. They become intoxicated by the action of immunotoxin. Protein synthesis is arrested in intoxicated cells, which do not produce new antigen molecules and do not divide, but endocytosis, intracellular trafficking, and surface antigen shedding are still presumed to go on until the cells die. Only the antigen shedding and the on and off reactions of free immunotoxins with the surface antigens are still presumed to go on in the dead cells, which nonetheless occupy volume in the tumor mass until physically cleared. Tumor cells in each RU can move out of RU as the tumor cells increase in number by cell division or into RU as space is created when intoxicated cells die and get cleared. A simple flux consideration of the cell flow in and out of the RU gives the governing equation for tumor volume profile with time (Eq. 8 in Text S1).

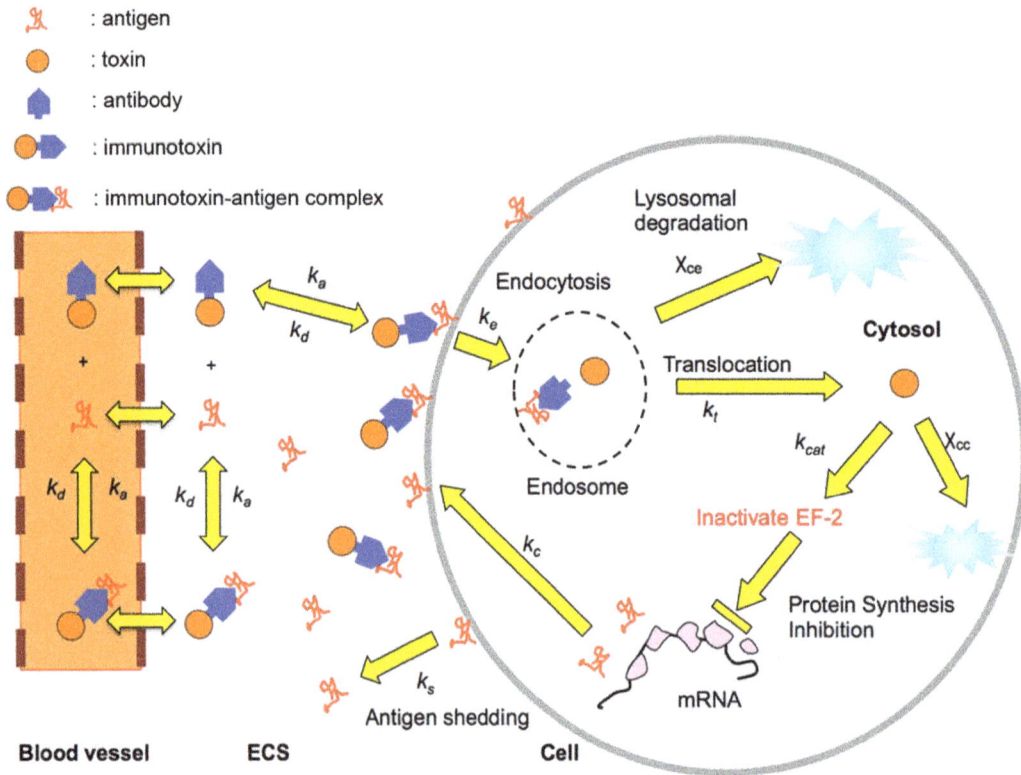

Figure 6. Kinetic events involved in immunotoxin-antigen binding and intracellular trafficking. Each yellow arrow indicates a kinetic step of the model. The tumor cell sheds the surface antigen and the surface complexed antigen at a certain rate. The immunotoxin molecule exiting from the capillary diffuses in the extra-cellular space and binds to either the surface antigen or shed antigen by the association reaction between the antigen and immunotoxin. The surface-bound immunotoxin is internalized by the receptor-mediated endocytosis and mostly inactivated in the endosomal stage. The surviving toxin translocates to the cytosol, where the toxin inhibits protein synthesis and eventually causes cell death. In non-intoxicated cells, the antigen is replenished by fresh protein synthesis.

Modifications and improvements

The basic concept and equations of the present model are the same as those of the earlier model [5], (See Text S1 for details) except for the modifications summarized below.

(i) The back-permeation into the blood vessel is allowed for all molecular species in ECS, including the shed antigen-immunotoxin complex, which was not allowed to back-permeate in the previous model. Specifically, for any relevant quantity Q_i in ECS, the boundary condition at the blood vessel wall (r_b) was given by allowing both forward and backward permeations across the blood vessel boundary:

$$P_i^f \cdot Q_i(blood) - P_i^b \cdot Q_i(r_b) = -\phi_e D_i \frac{\partial Q_i}{\partial r}\Big|_{r_b}, \qquad (1)$$

where P_i^f and P_i^b are the forward (from blood to ECS) and backward (from ECS to blood) permeabilities of molecular species i (in unit of cm/s), D_i is the diffusion constant of species i in ECS, and φ_e is the volume ratio of the ECS to the RU.

(ii) The permeability of shed antigens and other relevant molecular species traveling across the blood vessel wall decreases when tumor volume becomes large. This makes the shed antigen level to increase with tumor volume (see Fig. 2) in line with the recent experimental observation [8] that the concentration of shed antigen is higher both in the ECS and in the blood when tumor size is large. In the previous model, the permeability was assumed to be constant, which resulted in a constant level of shed antigen

concentration in the ECS, independent of the tumor size. In the current model, we introduce a permeability that varies from low P_{low} to high P_{high} values with tumor volume V according to

$$P(V) = P_{high} \cdot [1 - S(V)] + P_{low} \cdot S(V), \qquad (2)$$

where the sigmoidal switch function $S(V)$ is given by

$$S(V) = \frac{1}{1 + \exp[-a \cdot (V - V_c)]}, \qquad (3)$$

where a (slope) and V_c (center) are adjustable parameters. As explained in the Introduction and Discussion sections, this is an artificial way by which the model effectively accounts for the fact that the vasculature in large tumors is very aberrant and there exist pockets of poor blood supply.

(iii) The tumor cell growth rate constant used in the previous model has been replaced with a growth rate function $\Gamma(t)$, which decreases as the tumor volume increases,

$$\Gamma(t) = \Gamma_0 \cdot \exp[-\alpha \cdot V(t)], \qquad (4)$$

where $V(t)$ is the total tumor volume at time t, and Γ_0 and α are adjustable parameters. This is like Gompertz's model [30] except that it uses time for the exponential damping. We use tumor volume instead of time because our tumor volume can shrink

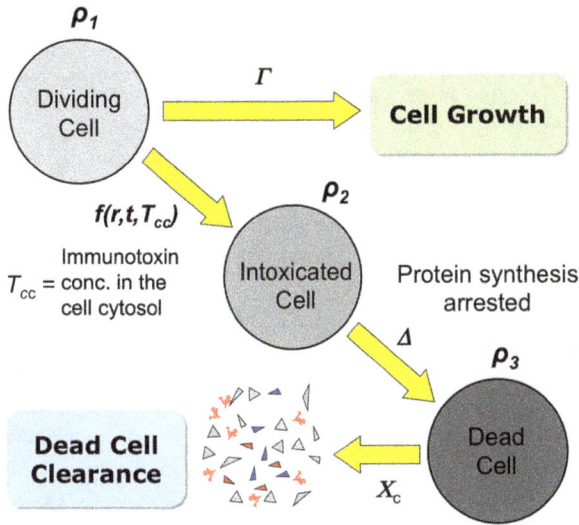

Figure 7. Kinetics of tumor cell population changes. The model defines three different tumor cell types: un-intoxicated, intoxicated, and dead cells, with densities ρ_1, ρ_2, and ρ_3, respectively. The un-intoxicated cells proliferate, but the intoxicated and dead cells do not. The conversion from un-intoxicated to dead cells is irreversible. The number of the surface antigen molecules on an un-intoxicated cell remains constant, but it decays on an intoxicated cell through endocytosis and shedding without supply of newly synthesized internal antigen. Both protein synthesis and endocytosis have stopped in the dead cell, but the shedding and the on and off reaction between the surface antigen and immunotoxin still go on. The dead cells are removed from the tissue, presumably by macrophages, at a certain rate.

(upon administration of immunotoxin) as well as grow with time. Use of Eq. (4) produces a noticeably better fit to the experimental tumor volume data at large tumor volumes both when immunotoxin is not given and long time after immunotoxin is given.

(iv) The cell intoxication follows the Michaelis-Menten type kinetics [31]. The cell intoxication rate is governed by the number of toxin molecules accumulated in the cell cytosol per tumor cell (T_1^{CC}). In the previous work, the use of a cell intoxication rate function of $f(r,t) = k_{cat} \cdot T_1^{CC}$ overestimated the rate at large T_1^{CC} and also produced a numerical instability when the type 1 cell density became small. In the current work, we use the new intoxication rate function

$$f(r,t) = \frac{k_{cat}^{\max} \cdot T_1^{CC}(r,t)}{T_0 + T_1^{CC}(r,t)} \qquad (5)$$

where $k_{cat}^{\max}$, the maximum intoxication rate, and T_0, the number of toxin molecules per cell at which the reaction rate is half of $k_{cat}^{\max}$, are adjustable parameters.

(v) The RU size has been expanded from 38 to 50 μm, so that the inter-capillary distance is 100 μm. The larger distance is more consistent with the median distance between cells to the blood vessel of, for example, 53–63 μm reported for the squamous cell carcinoma xenografts [32]. Also, cylindrical, rather than spherical, geometry is used for RU, with the cylinder radius set to 50 μm; the model is independent of the cylinder height, which never enters into the calculation. The blood vessel is also cylindrical. When a typical blood volume fraction of 5% is applied, the blood vessel radius (r_b) becomes 11 μm. The cylindrical RU provides more

layers of cells surrounding the blood vessel than a comparable RU with spherical geometry, but the model is rather insensitive to the geometry used: The quality of fit between the calculated and experimental tumor volume profiles is quite similar for SS1P whether cylindrical or spherical RU is used, but is slightly better with the cylindrical RU for LMB-2 (data not shown).

(vi) Finally, the cell density ρ^* has been reduced to half of the value used before. (The value of ρ^* given in the the earlier work (Table S1 in ref. [5]) has a typographical error – the actual value used in the model was 1.0×10^9 cells/cm^3 [25], not 1.9×10^9 cells/cm^3.) This is to effectively take into account the presence of non-tumor cells, e.g., stromal cells, macrophages, and other normal cells, which we assume to make up half of all the cells in the tumor tissue. Therefore, the cell density parameter in the current model refers to the three types (un-intoxicated, intoxicated, and dead) of tumor cells only. We kept the extra-cellular volume fraction φ_e the same as before.

All the equations describing our model system are given in Text S1.

Model parameters

All the parameters employed in the present work are listed in Table 1. Fig. S1 gives definitions of additional parameters used in the detailed model described in Text S1. Many parameter values in this model were adopted from the previous work [33] or published [7,14,15] or unpublished *in vivo* or *in vitro* experimental data. Other parameters were determined by the best fit to experimental data. For example, the endocytosis rate constant (k_e) of SS1P was determined by fitting the *in vitro* dye-labeled SS1P internalization data on the A431/H9 tumor cell [7]. For LMB-2, the k_e value was obtained by fitting *in vitro* internalization data of ^{111}In-labeled LMB-2 into the ATAC-4 tumor cell [21]. The tumor growth rate parameters (Γ_0, α) were determined by fitting to the experimental *in vivo* tumor growth curve (without immunotoxin injection). The shedding rate constant (k_s), vascular permeability parameters (P_{low}, P_{high}, V_c, and a), and the decay rate constant of the shed free antigen in the ECS and blood (χ_{efR} and χ_{bfR}) were obtained by fitting to the *in vivo* experimental shed antigen data in the ECS and blood [7,8,15]. Since we could not find reliable information on the permeation rates in the literature, we assumed that the forward and backward permeabilities of all molecular species were the same and equal to the back-permeability of shed antigens determined as above. Other remaining parameter values were determined by direct fits to experimental *in vivo* tumor volume profiles upon the administration of immunotoxins. Detailed procedures to obtain the model parameter values are given in Text S1. All the parameter values from our best fit with the current model are given in the Tables 2 and 3. Since the current model is modified in several different ways, the present parameter values for SS1P are not the same as before [5]. The model is variably sensitive to the model parameters. For a rather extensive sensitivity analysis of model parameters, see Chen et al. [33]. This analysis was done using an early model, but we expect the sensitivity to remain substantially the same for the new model, which is descended from and basically the same as the earlier model.

Supporting Information

Figure S1 Parameters involved in the kinetics of receptor recycling. (Similar to Figure S2 of ref. 5 but redrawn.) The surface receptors (R_s) are placed on the cell surface (yellow ribbon) with a rate constant k_c and depleted by endocytosis (rate constant k_e) and shedding (rate constant k_s). The endocytosed receptors are mostly degraded (rate constant k_{deg}). The remainder combines with the newly synthesized (at rate G) to form the internal receptor pool (R_i) which is recycled to the surface (rate constant k_c). For the type 1 cell, a steady state condition is imposed, such that the number of the surface receptors (R_s) and the internal receptors (R_i) per cell are kept constant. Furthermore, all endocytosed receptors are degraded and that all receptors that are presented on the surface are newly synthesized.

Author Contributions

Conceived and designed the experiments: BL. Performed the experiments: YP BL. Analyzed the data: YP BL IP. Contributed reagents/materials/analysis tools: IP RJK YP. Wrote the paper: YP BL IP RJK. Supplied the experimental data: IP RJK. Wrote the software: YP.

References

1. Adams GP, Weiner LM (2005) Monoclonal antibody therapy of cancer. Nature biotechnology 23: 1147–1157.
2. Dosio F, Stella B, Cerioni S, Gastaldi D, Arpicco S (2014) Advances in anticancer antibody-drug conjugates and immunotoxins. Recent Pat Anticancer Drug Discov 9: 35–65.
3. Pastan I, Hassan R, Fitzgerald DJ, Kreitman RJ (2006) Immunotoxin therapy of cancer. Nature reviews Cancer 6: 559–565.
4. Thurber GM, Schmidt MM, Wittrup KD (2008) Antibody tumor penetration: transport opposed by systemic and antigen-mediated clearance. Advanced drug delivery reviews 60: 1421–1434.
5. Pak Y, Zhang Y, Pastan I, Lee B (2012) Antigen shedding may improve efficiencies for delivery of antibody-based anticancer agents in solid tumors. Cancer research 72: 3143–3152.
6. Hassan R, Ho M (2008) Mesothelin targeted cancer immunotherapy. European journal of cancer 44: 46–53.
7. Zhang Y, Hansen JK, Xiang L, Kawa S, Onda M, et al. (2010) A flow cytometry method to quantitate internalized immunotoxins shows that taxol synergistically increases cellular immunotoxins uptake. Cancer research 70: 1082–1089.
8. Zhang Y, Xiang L, Hassan R, Pastan I (2007) Immunotoxin and Taxol synergy results from a decrease in shed mesothelin levels in the extracellular space of tumors. Proc Natl Acad Sci U S A 104: 17099–17104.
9. Fujimori K, Covell DG, Fletcher JE, Weinstein JN (1990) A modeling analysis of monoclonal antibody percolation through tumors: a binding-site barrier. Journal of nuclear medicine: official publication, Society of Nuclear Medicine 31: 1191–1198.
10. Kulasingam V, Diamandis EP (2008) Strategies for discovering novel cancer biomarkers through utilization of emerging technologies. Nat Clin Pract Oncol 5: 588–599.
11. Zhang Y, Pastan I (2008) High shed antigen levels within tumors: an additional barrier to immunoconjugate therapy. Clinical cancer research: an official journal of the American Association for Cancer Research 14: 7981–7986.
12. Dreher MR, Liu W, Michelich CR, Dewhirst MW, Yuan F, et al. (2006) Tumor vascular permeability, accumulation, and penetration of macromolecular drug carriers. J Natl Cancer Inst 98: 335–344.
13. Jain RK (2001) Delivery of molecular and cellular medicine to solid tumors. Advanced drug delivery reviews 46: 149–168.
14. Tsutsumi Y, Onda M, Nagata S, Lee B, Kreitman RJ, et al. (2000) Site-specific chemical modification with polyethylene glycol of recombinant immunotoxin anti-Tac(Fv)-PE38 (LMB-2) improves antitumor activity and reduces animal toxicity and immunogenicity. Proceedings of the National Academy of Sciences of the United States of America 97: 8548–8553.
15. Singh R, Zhang Y, Pastan I, Kreitman RJ (2012) Synergistic antitumor activity of anti-CD25 recombinant immunotoxin LMB-2 with chemotherapy. Clinical cancer research: an official journal of the American Association for Cancer Research 18: 152–160.
16. Kreitman RJ, Bailon P, Chaudhary VK, FitzGerald DJ, Pastan I (1994) Recombinant immunotoxins containing anti-Tac(Fv) and derivatives of Pseudomonas exotoxin produce complete regression in mice of an interleukin-2 receptor-expressing human carcinoma. Blood 83: 426–434.
17. Hassan R, Bera T, Pastan I (2004) Mesothelin: a new target for immunotherapy. Clin Cancer Res 10: 3937–3942.
18. Greene WC, Leonard WJ, Depper JM, Nelson DL, Waldmann TA (1986) The human interleukin-2 receptor: normal and abnormal expression in T cells and in leukemias induced by the human T-lymphotropic retroviruses. Ann Intern Med 105: 560–572.
19. Junghans RP, Waldmann TA (1996) Metabolism of Tac (IL2Ralpha): physiology of cell surface shedding and renal catabolism, and suppression of catabolism by antibody binding. J Exp Med 183: 1587–1602.
20. Zhang Y, Chertov O, Zhang J, Hassan R, Pastan I (2011) Cytotoxic activity of immunotoxin SS1P is modulated by TACE-dependent mesothelin shedding. Cancer research 71: 5915–5922.
21. Kobayashi H, Kao CH, Kreitman RJ, Le N, Kim MK, et al. (2000) Pharmacokinetics of 111In- and 125I-labeled antiTac single-chain Fv recombinant immunotoxin. Journal of nuclear medicine: official publication, Society of Nuclear Medicine 41: 755–762.
22. Reiter Y, Brinkmann U, Lee B, Pastan I (1996) Engineering antibody Fv fragments for cancer detection and therapy: disulfide-stabilized Fv fragments. Nat Biotechnol 14: 1239–1245.
23. Arribas J, Borroto A (2002) Protein ectodomain shedding. Chem Rev 102: 4627–4638.
24. Dello Sbarba P, Rovida E (2002) Transmodulation of cell surface regulatory molecules via ectodomain shedding. Biol Chem 383: 69–83.
25. Kreitman RJ, Pastan I (1998) Accumulation of a recombinant immunotoxin in a tumor in vivo: fewer than 1000 molecules per cell are sufficient for complete responses. Cancer research 58: 968–975.
26. Rippe B, Haraldsson B (1985) Solvent drag component of unidirectional albumin out-flux. Microvascular research 30: 246–248.
27. Schmidt MM, Wittrup KD (2009) A modeling analysis of the effects of molecular size and binding affinity on tumor targeting. Mol Cancer Ther 8: 2861–2871.
28. Ackerman ME, Pawlowski D, Wittrup KD (2008) Effect of antigen turnover rate and expression level on antibody penetration into tumor spheroids. Mol Cancer Ther 7: 2233–2240.
29. Cheever MA, Allison JP, Ferris AS, Finn OJ, Hastings BM, et al. (2009) The prioritization of cancer antigens: a national cancer institute pilot project for the acceleration of translational research. Clinical cancer research: an official journal of the American Association for Cancer Research 15: 5323–5337.
30. Laird AK (1964) Dynamics of Tumor Growth. Br J Cancer 13: 490–502.
31. Michaelis L, Menten ML (1913) The kenetics of the inversion effect. Biochemische Zeitschrift 49: 333–369.
32. Lauk S, Zietman A, Skates S, Fabian R, Suit HD (1989) Comparative morphometric study of tumor vasculature in human squamous cell carcinomas and their xenotransplants in athymic nude mice. Cancer Res 49: 4557–4561.
33. Chen KC, Kim JH, Li XM, Lee B (2008) Modeling recombinant immunotoxin efficacies in solid tumors. Annals of Biomedical Engineering 36: 486–512.
34. Honda M, Kitamura K, Takeshita T, Sugamura K, Tokunaga T (1990) Identification of a soluble IL-2 receptor beta-chain from human lymphoid cell line cells. Journal of immunology 145: 4131–4135.
35. Onda M, Nagata S, Tsutsumi Y, Vincent JJ, Wang Q, et al. (2001) Lowering the isoelectric point of the Fv portion of recombinant immunotoxins leads to decreased nonspecific animal toxicity without affecting antitumor activity. Cancer research 61: 5070–5077.

4

Acylation of Glucagon-Like Peptide-2: Interaction with Lipid Membranes and *In Vitro* Intestinal Permeability

Sofie Trier[1,2], Lars Linderoth[2], Simon Bjerregaard[2], Thomas Lars Andresen[1], Ulrik Lytt Rahbek[2]*

1 Dept. of Micro- and Nanotechnology, Center for Nanomedicine and Theranostics, Technical University of Denmark, Kgs. Lyngby, Denmark, 2 Diabetes Research Unit, Novo Nordisk, Maaloev, Denmark

Abstract

Background: Acylation of peptide drugs with fatty acid chains has proven beneficial for prolonging systemic circulation as well as increasing enzymatic stability without disrupting biological potency. Acylation has furthermore been shown to increase interactions with the lipid membranes of mammalian cells. The extent to which such interactions hinder or benefit delivery of acylated peptide drugs across cellular barriers such as the intestinal epithelia is currently unknown. The present study investigates the effect of acylating peptide drugs from a drug delivery perspective.

Purpose: We hypothesize that the membrane interaction is an important parameter for intestinal translocation, which may be used to optimize the acylation chain length for intestinal permeation. This work aims to characterize acylated analogues of the intestinotrophic Glucagon-like peptide-2 by systematically increasing acyl chain length, in order to elucidate its influence on membrane interaction and intestinal cell translocation *in vitro*.

Results: Peptide self-association and binding to both model lipid and cell membranes was found to increase gradually with acyl chain length, whereas translocation across Caco-2 cells depended non-linearly on chain length. Short and medium acyl chains increased translocation compared to the native peptide, but long chain acylation displayed no improvement in translocation. Co-administration of a paracellular absorption enhancer was found to increase translocation irrespective of acyl chain length, whereas a transcellular enhancer displayed increased synergy with the long chain acylation.

Conclusions: These results show that membrane interactions play a prominent role during intestinal translocation of an acylated peptide. Acylation benefits permeation for shorter and medium chains due to increased membrane interactions, however, for longer chains insertion in the membrane becomes dominant and hinders translocation, i.e. the peptides get 'stuck' in the cell membrane. Applying a transcellular absorption enhancer increases the dynamics of membrane insertion and detachment by fluidizing the membrane, thus facilitating its effects primarily on membrane associated peptides.

Editor: Miguel A. R. B. Castanho, Faculdade de Medicina da Universidade de Lisboa, Portugal

Funding: ST was partially funded by The Danish Agency for Science, Technology and Innovation, and TLA was funded by the Technical University of Denmark. The funder Novo Nordisk A/S provided support in the form of salaries for authors ST, LL, SB and ULR, but did not have any additional role in the study design, data collection and analysis, decision to publish, or preparation of the manuscript. The specific roles of these authors are articulated in the 'author contributions' section.

Competing Interests: ST, LL, SB and ULR are employed by Novo Nordisk A/S, and LL, SB and ULR are shareholders in Novo Nordisk A/S.

* Email: ulyr@novonordisk.com

Introduction

Acylation of peptides with fatty acids is a naturally occurring post-translational modification, which has inspired alteration of therapeutic peptides for drug delivery. Acylation prolongs the systemic circulation half-life of otherwise rapidly cleared peptide drugs, through increased enzymatic stability [1–3] and binding to - and piggy-backing on - serum albumin [4]. An additional effect of acylation is increased peptide self-association and aggregation, which has been employed to ensure prolonged release of peptide drugs following subcutaneous injection [5]. Acylation can be performed without disrupting the peptide's biological potency [6], and has been employed for a multitude of therapeutic peptides

[2,4,7,8], including several marketed drugs (e.g. insulin and Glucagon-like peptide-1).

The increased enzymatic stability of acylated peptides is particularly beneficial for oral administration, due to the highly metabolic environment in the stomach and intestine [9]. Another requirement for oral drug delivery is adequate absorption through the intestinal epithelial barrier, which is a major challenge for large, hydrophillic peptide drugs [10]. A widely used method for predicting oral absorption *in vivo* is *in vitro* quantification of translocation across monolayers of the human colon cancer cell line (Caco-2), which has been shown to correlate well with oral bioavailability [11,12]. Acylation has previously been shown to increase intestinal permeability of peptide drugs [6,7,13], but detailed investigations of systematic acyl variations are lacking, which would benefit rational new designs of peptide drugs. The *in vitro* intestinal translocation studies can be further supplemented by measurements of peptide binding to model lipid membranes [14–16] in order to investigate the influence of membrane binding of acylated peptides on cellular membrane translocation.

Glucagon-like peptide-2 (GLP-2) is a 33 amino acid peptide, which is secreted from the human intestine following nutrient intake [17,18]. Therapeutically, GLP-2 stimulates intestinal growth and is employed in the treatment of inflammatory bowel diseases (e.g. Crohn's disease) and short bowel syndrome (e.g. following intestinal surgery) [19,20]. The plasma half-life of GLP-2 in humans is limited to a few minutes [21] due to extensive renal clearance and rapid enzymatic degradation by dipeptidyl peptidase-4 [21,22]. Furthermore, GLP-2 is presently administered as subcutaneous injections, which compromises patient comfort and compliance, in particular for chronic diseases like Crohn's. It would be highly beneficial to enable oral administration, and the combined effects of prolonged circulation time, improved enzymatic stability and intestinal permeability may render acylated GLP-2 a suitable candidate for oral drug delivery. Currently, however, there are no reports on the intestinal permeability or oral drug delivery potential of acylated GLP-2.

In the present study we synthesized and characterized acylated analogues of GLP-2, with systematically increasing acyl chain length, in order to investigate the effect of the acyl chain on membrane interaction and *in vitro* intestinal permeability. This was achieved by combining investigations of the interaction with lipid membranes and translocation across an intestinal cell model, as outlined in fig. 1.

We hypothesize that the acylation chain length can be optimized for translocation across the intestinal barrier, i.e. a moderate interaction with the lipid cell membrane is beneficial for translocation, whereas a stronger interaction may impair translocation. Acylation is expected to confer membrane affinity to GLP-2, as the native peptide is not membrane active. In this regard, GLP-2 was employed as a model peptide, however, the results may be applicable for development of a rational acylation strategy for other peptide drugs.

Absorption enhancers are often employed to increase oral peptide absorption, which makes it interesting to investigate how these affect the translocation of acylated peptides [23]. In the present study we included two enhancers with different enhancing mechanism, in order to investigate the effect of the enhancing mechanism. Ethylene glycol-bis(β-aminoethyl ether)-N, N, N, N-tetraacetic acid (EGTA) is a paracellular enhancer which increases transport between the cells by opening of the tight junctions [24], and sodium dodecyl sulfate (SDS) is a transcellular enhancer which increases transport through the cells at low concentrations, predominantly by fluidizing the cell membrane [25].

We hypothesize that the effect of paracellular enhancers will not be influenced by acylation, whereas the effect of transcellular enhancers that directly interact with the cell membrane may depend on the peptide-membrane interaction, through altered membrane affinity and/or dynamics of membrane insertion.

Materials and Methods

Materials

Resin and natural amino acids were purchased at Novabiochem (Germany). c8, c12 and c16 carboxylic acids, Fmoc-beta-Alanine and native GLP-2 were provided by Novo Nordisk A/S. Palmitoyloleoylglycerophosphocholine (POPC) was purchased from Avanti Polar Lipids (USA). HEPES, Ovalbumin (OVA, from chicken egg white) and other standard chemicals were purchased from Sigma-Aldrich (Denmark). DMEM medium, l-glutamine and penicillin/streptomycin was purchased from Lonza (Switzerland). HBSS buffer, fetal bovine serum (FBS), nonessential amino acid and other standard cell culture products were purchased from Gibco (Denmark). Radioactively labeled [^{3}H]mannitol, scintillation fluid (Microscint-40), luciferase substrate (SteadyLite) and 96-well plates for luciferase assay (CulturPlate, black) were purchased from PerkinElmer (USA). 12 well Transwell plates for Caco-2 cell monolayers (polycarbonate, 12 mm, pore size 0.4 μM) were purchased from Corning Costar Corp. (USA). GLP-2R BHK cells were provided by Novo Nordisk (the cloning was previously described by Thulesen et al. [26] and Sams et al. [27]) and Caco-2 cells (HTB-37) were purchased from ATCC.

Peptide synthesis

The peptides were synthesized by automated Fmoc based SPPS, using a preloaded Fmoc-Asp(OtBu)-Wang polystyrene LL resin in 0.25 mmol scale on a CEM Liberty microwave peptide synthesizer (CEM Corporation, NC) using standard protocols, with a modified coupling temperature of 50°C [28]. Fmoc deprotection was carried out in 5% piperidine and 0.05 M HOBt in NMP.

The acylation was conjugated to the lysine side chain by incorporation of lysine as Lys(Mtt), which allowed chemical modification using the standard coupling procedures stated above, following Mtt removal. The Mtt group was removed by washing the resin with DCM and suspending the resin in neat (undiluted) hexafluoroisopropanol for 20 minutes followed by washing with DCM and NMP. After synthesis the resin was washed with DCM, and the peptide was cleaved from the resin by a 3 hour treatment

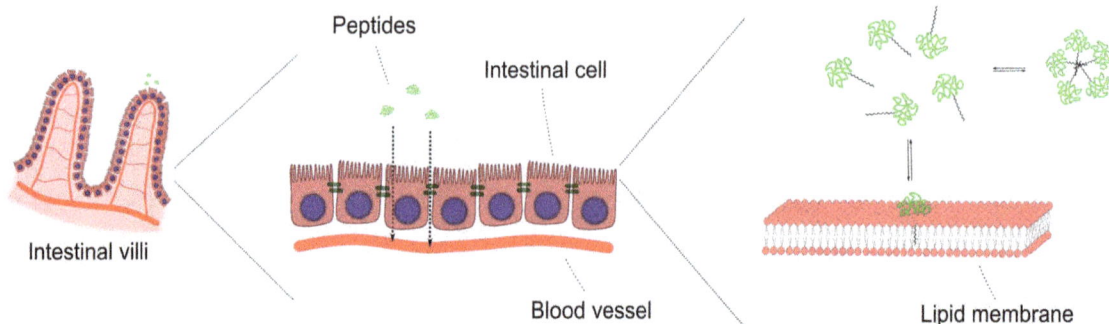

Figure 1. Schematic illustration of the study objective. The membrane interaction and *in vitro* permeability of acylated GLP-2 is investigated using an intestinal cell model and model lipid membranes.

with TFA/TIPS/H$_2$O (95/2.5/2.5) followed by precipitation with diethylether. The peptide was dissolved in a suitable solvent (e.g. 1:1 MeCN : H$_2$O) and purified by preparative HPLC on an XBridge c18 column (Waters), using a gradient from 10% MeCN/ Buffer (10 mM TRIS and 15 mM (NH$_4$)$_2$SO$_4$, pH 7.3) to 50% MeCN over 40 min, flow rate 60 mL/min. The fractions were analysed by a combination of UPLC and LCMS methods, and the appropriate fractions were pooled and transferred to TFA salt for lyophilization, using a gradient from 10% MeCN/0.1% TFA/ MQ to 60% MeCN. The peptides were finally quantified using a chemiluminescent nitrogen detector, as previously described by Fujinari et al. [29].

The isoelectric point of GLP-2 is approximately 4, with a theoretical charge of -4 at neutral pH, and the analogues are expected to be very similar.

Liposome preparation and characterization

POPC lipid films were prepared by evaporation from chloroform:methanol (9:1 v/v) using a gentle nitrogen flow. The residual organic solvent was evaporated in vacuum overnight and the lipid films were rehydrated to 25 mM in HEPES buffer (10 mM HEPES, 150 mM NaCl, pH 7.4) at room temperature with frequent vigorous agitation for 1 hour. The multilamellar lipid suspensions were subjected to 10 cycles of freeze-thawing (isopropanol/dry-ice and 40°C water bath) and extruded to 100 nm liposomes by passing it 21 times through a 100 nm polycarbonate filter in a manual extruder (Avanti Polar Lipids). [30]

The liposome hydrodynamic size after extrusion was measured in a ZetaPALS Zeta Potential Analyzer (Brookhaven Instruments Corporation), after dilution to 50 μM in sterile filtered HEPES buffer. In all experiments the liposome diameter was 130 $\pm$ 5 nm with PDI 0.1.

The lipid concentration in liposome suspensions after extrusion was determined by phosphorous analysis, as previously described by Rouser et al. [31]. Briefly, the phospholipid sample was degraded with heat and perchloric acid, reacted with ammonium molybdate and reduced by ascorbic acid. Absorbance of the resulting molybdenum oxides (molybdenum blue) was measured at 812 nm, and lipid sample concentration was determined from a phosphate standard row. All samples were measured in triplicates. A typical final concentration after extrusion at 25 mM was 22.5 $\pm$ 0.1 mM.

Tryptophan fluorescence measurements

Peptide self-association. Fluorescence measurements were carried out using an OLIS SLM8000 fluorescence spectrometer equipped with excitation and emission monochromators and polarizers. Tryptophan was excited at 280 nm and emission scans were aquired from 300 to 400 nm with 2 nm step size, 2 s integration time and slit widths 16 nm. Quartz cuvettes with an excitation pathway of 10 mm and an emission pathway of 4 mm were used with 1 mL sample volume. Peptide samples in the concentration range 0.5–100 μM were prepared directly in the cuvette from 100 μM stock solutions in sterile filtered HEPES buffer. Temperature was maintained at 37°C via an external water bath, and the solutions were mixed by magnet stirring in the cuvette.

The recorded spectra were fitted to obtain the peak position and intensity, as described by Burstein et al. [32], using an iterative least-squares fit (built-in in Gnuplot). Self-association of the peptides causes a blue-shift and increase in tryptophan maximum fluorescence, due to increased hydrophobicity near the tryptophan

residue, and the blue-shift is used as an indicator of self-association.

Liposome partitioning. Fluorescence measurements were conducted similarly to self-association experiments, with integration time 4 s, slit widths 8 nm (excitation) and 16 nm (emission) and polarizers set to 90° (excitation) and 0° (emission).

The peptide solution was prepared directly in the cuvette and was subsequently titrated with a 20 mM liposome suspension. The peptide was used at a concentration (2 μM) below its self-association concentration, to avoid any signal from self-association.

The fluorescence was initially measured as full wavelength scans (300–400 nm), where the wavelength yielding the largest change with liposome addition was identified (344 nm). Subsequently, the fluorescence was measured as time-scans at 344 nm, and the fluorescence after each liposome addition measured for 5 minutes (increased and stabilized rapidly) and averaged over the last 3 minutes. It was verified that the fluorescence did not change substantially after the 5 minute measurement (up to 1.5 h).

Liposome partitioning model

The membrane partitioning model is described in detail by Etzerodt et al. [33]. Briefly, the tryptophan fluorescence after the $i'th$ addition of liposome (F_i) depends on the lipid concentration after the $i'th$ addition ($C_{lip,i}$) and the partition coefficient (K) through:

$$F_i = F_0 + (F_{inf} - F_0) \cdot \frac{K \cdot C_{lip,i}}{C_w + K \cdot C_{lip,i}} \qquad (1)$$

where F_0 is the initial fluorescence of peptide alone, C_w is the concentration of water (55 M) and F_{inf} is the fluorescence at 'full' partitioning, i.e. after infinite liposome addition, which is a fitted value.

eq. 1 is reorganized to yield

$$\frac{F_i - F_0}{F_{inf} - F_0} = \frac{K \cdot C_{lip,i}}{C_w + K \cdot C_{lip,i}} \qquad (2)$$

which is equal to the concentration of peptide bound in the membrane C_M divided by the total concentration of peptide C_{tot}, i.e. the fraction of peptide that is membrane-bound. eq. 2 was used for plotting (the left hand side) and fitting (the right hand side), where the fit was an iterative least-squares fit with respect to K and F_{inf} (Gnuplot).

The partition coefficient K can be converted to the standard Gibbs free energy of partitioning ($\Delta G°$) through:

$$\Delta G° = -R \cdot T \cdot ln(K) \qquad (3)$$

where R is the gas constant (8.314 J/(K mol)) and T is the temperature (310 K). For simplicity, the absolute value of $\Delta G°$ is displayed in graphs.

The partition coefficient K is unitless, but can be converted to the more commonly used molar partition coefficient (in units M^{-1}) through dividing by the molarity of water (55 M).

Addition of liposomes to the peptide solution causes scattering of excitation and emission light, which yields artifacts in the experimental data that are accounted for by Ladokhin et al. [34]. To minimize these effects the fluorescence was measured using cross-polarized light settings, where the excitation light was horizontally polarized and the emission was measured vertically polarized. This ensures that only the emitted light is measured, as scattering does not alter the polarization, whereas fluorescence randomizes the polarization. However, scattering still causes a decrease in measured fluorescence, which is corrected for by a correction factor Θ, measured by titrating the non-partitioning probe L-Trp with liposomes:

$$\Theta = \frac{F_{L-Trp,i}}{F_{L-Trp,0}} \qquad (4)$$

where $F_{L-Trp,i}$ is the fluorescence of L-Trp after the $i'th$ addition of liposome and $F_{L-Trp,0}$ is the initial fluorescence before liposome addition.

Θ was fitted to a polynomial, which was used to correct the measured peptide fluorescence during titration with liposomes.

All fluorescence measurements were normalized for peptide concentration, including the slight concentration decrease during titration.

Interaction with cells

Luciferase assay. A BHK cell line (GLP-2R BHK) was previously modified to stably express the human GLP-2 receptor, which controls the expression of firefly luciferase [26,27].

GLP-2R cells were cultured in DMEM with 10% FCS, 100 U penicillin, 100 μg/mL streptomycin, 1 mM Na-Pyruvate, 250 nM methotrexate, 1 mg/ml geneticin and 0.4 mg/ml hygromycin. The assay was performed in DMEM without phenol red, containing 10 mM HEPES, 1% glutamax and 1 mg/ml OVA.

For experiments, cells were seeded in 96-well plates at 20.000 cells/well (100 μL/well) and incubated overnight. The medium was removed and the cells were washed once and replenished with 50 μL/well assay medium. The test solution (containing peptide) was diluted in HBSS buffer (containing 10 mM HEPES and 1 mg/ml OVA, pH 7.4) and 50 μL/well was added to the cells. The cells were incubated for 3 h (at 37°C and 5% CO_2) and the test compound was removed. 100 μL/well HBSS buffer was added along with 100 μL/well luciferase substrate, the plate was sealed and incubated at room temperature for 30 minutes. The luminescence (Relative Luminescence Units, RLU) was measured in a topcounter (Packard Topcount) and depends on the peptide concentration as:

$$RLU = A + \frac{B-A}{1+10^{((logEC_{50}-x)\cdot C)}} \qquad (5)$$

where x is log(concentration) of the peptide in M, and A, B, C and EC_{50} are fitting parameters [35].

The peptide test solutions were diluted to fall within the dynamic range of the assay (approximately 1–100 pM), and on each plate with test solutions a peptide standard row was included, which was fitted according to eq. 5 (using GraphPad Prism).

Caco-2 cells. Caco-2 cells (passage 40–65) were cultured routinely [11] in DMEM with 10% (v/v) FBS, 1% (v/v) nonessential amino acids, 100 U penicillin, 0.1 mg/mL streptomycin and 2 mM L-glutamine. The Caco-2 cells were seeded at a density of $1\cdot10^5$ cells/well on 12 well Transwells plates and grown for 14–16 days in DMEM with media change every second day. The cell layer was confirmed to consist of a single monolayer of cells by fixing and staining the cell layer, and visualizing by fluorescence microscopy.

Before transport experiments, DMEM medium was changed to HBSS buffer (containing 10 mM HEPES and 0.1% (w/v) OVA; 0.4 mL apical side and 1 mL basolateral) and left to equilibrate for 60 minutes. Buffer was replaced apically by 0.4 mL test solution at time zero, and the plates were incubated at 37°C and 5% CO_2

with gentle shaking. Test solutions contained 100 μM peptide and 0.8 μCi/mL [³H]mannitol (a permeability marker). Basolateral samples of 200 μL were taken every 15 minutes for 1 hour, and replaced by buffer. The apical test solution and basolateral samples were diluted and analyzed for peptide content with the luciferase assay and for [³H]mannitol content in a scintillation counter (Packard TopCount), after mixing 1:1 with scintillation fluid.

The integrity of cell monolayers before, during, and after experiments was verified by measuring the the translocation of radioactively labelled mannitol and the transepithelial electrical resistance (TEER), using a Milicell ERS-2 epithelial volt-ohm meter (Millipore, USA) [11]. Mannitol translocation and TEER values were not affected by addition of peptide or analogues compared to buffer.

After experiments, the cells were washed twice with buffer and replenished with medium for 24 hour recovery, or washed thrice with ice cold buffer and frozen at -80°C for cell binding and uptake analysis.

For cell binding and uptake studies, all preparations were carried out on ice, and all the buffers were ice cold. The filters were cut out of the inserts and placed in 12 well plates with the cell layer facing up. 250 μL of buffer was added and the cells were scraped off carefully with a cell scraper. The scraping was repeated, and the cell suspensions were pooled and centrifuged (13.000 rpm, 20 min, 4°C). The supernatant was analyzed for peptide content (cell uptake), and the pellet was resuspended in buffer and vortexed thoroughly. The membrane-bound peptide was recovered from the cell debris by addition of ethanol, followed by thorough vortexing and centrifugation (13.000 rpm, 20 min, 4°C). The supernatant solvent was evaporated under nitrogen-flow, and the dry peptide was dissolved in buffer and analyzed for peptide content (cell membrane binding). The cell uptake and membrane binding of peptide and analogues are displayed as the total amount of peptide in the cell layer. It was tested whether the recovery of peptide from cell debris was efficient and/or dependent on acyl chain length. Peptide or analogues were added to the cell debris pellet after the first centrifugation, taken from control cell layers with no added peptide, and after following the subsequent steps for membrane binding, the peptide recovery was measured. Essentially all added peptide was recovered, and there was no measurable difference for the different analogues.

The Caco-2 translocation of peptide or mannitol over Caco-2 layers is expressed as the apparent permeability (P_{app}), given by:

$$P_{app} = \frac{dQ}{dt} \cdot \frac{1}{A\cdot C_0} \qquad (6)$$

where $\frac{dQ}{dt}$ is the steady-state flux of peptide (pmol/s), A is the surface area of the cell monolayer (1.12 cm²), and C_0 is the initial sample concentration added to the cell layer [11].

Data analysis

Statistical analysis was carried out using GraphPad Prism, where unpaired Students t-tests were used for comparison and significant differences required p<0.05.

Results and Discussion

Characterization of peptide analogues

We synthesized and investigated native GLP-2 and the acylated analogues shown in fig. 2, where the acyl chain is conjugated to the ϵ−amino group of a lysine via a β−alanine spacer [8,36].

GLP-2   HADGSFSDEM^{10}NTILDNLAAR^{20}DFINWLIQTK^{30}ITD

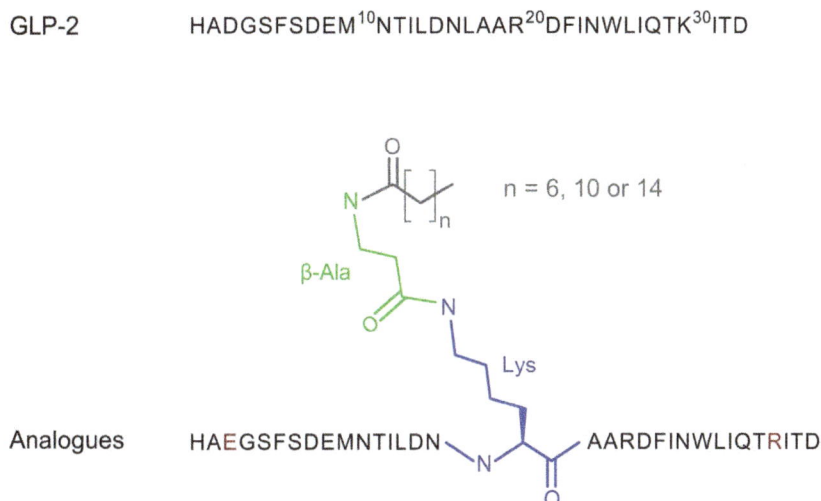

Analogues   HAEGSFSDEMNTILDN   AARDFINWLIQTRITD

Figure 2. Schematic representation of native GLP-2 and its acylated analogues. GLP-2 is acylated with c8, c12 or c16-chains (grey) at the ϵ−amino group of Lys17 (blue) via a β−alanine spacer (green). The lysine residue replaces a leucine at position 17 and the natural lysine at position 30 in GLP-2 is mutated to an arginine (red). The aspartic acid at position 3 is mutated to a glutamic acid (red) to avoid racemization during synthesis. None of these mutations caused measurable loss-of-function.

The concentration-dependent self-association of GLP-2 and its analogues was investigated by tryptophan fluorescence (fig. 3). The acyl chains confer increased hydrophobicity, causing self-assembly at lower concentrations for the acylated analogues compared to the native peptide. Furthermore, the self-association concentration decreased with increasing acyl chain length, consistent with increasing hydrophobicity.

At physiological ionic strength, the secondary structure of the c16 acylated analogue was similar to native GLP-2 (investigated by circular dichroism (CD), see fig. S1). It is worth noting that the buffer ionic strength is crucial for self-association, as no self-association was observed with 0 or 10 mM NaCl. This is consistent with electrostatic screening of the peptides's multiple negative charges as a requirement for self-association. The peptide oligomers of native GLP-2 and its c16-analogue were characterized by static and dynamic light scattering (DLS/SLS) and transmission electron microscopy (TEM), which showed that the oligomers were spherical and well-defined in size (smaller for the native peptide than the analogue), and composed of less than 10 monomers (see fig. S2).

Interaction with model lipid membranes

The partitioning of peptides into liposomes was quantified by tryptophan fluorescence, in order to investigate the effect of acylation on interactions with lipid membranes. The data presented in fig. 4 shows that acylation of GLP-2 causes increased partitioning into POPC membranes, where partitioning of native GLP-2 is below the measurable limit and is not included in the graphs. The standard Gibbs free energy of partitioning increases linearly with acyl chain length for the investigated c8, c12 and c16-chains, which is consistent with previous reports of acylated glycine [37]. It is worth noting that the slope of this linear relationship is lower for the GLP-2 analogues than for the acylated glycine analogues, as reported by Peitzsch et al. [37]. This indicates that the dependence of membrane affinity on acyl chain length is influenced by the peptide or amino acid backbone, and that the small glycine residue is more sensitive to chain length than the larger GLP-2 peptide.

It should be noted that the peptides are used at a very low concentration (2 μM) in order to avoid self-association, and that the peptide/lipid ratio is quite low. For subsequent cell experiments (and possible therapeutic applications) the peptides are used at higher concentrations, and the lipid concentration is not well-defined. Therefore, caution should be exerted when comparing liposome and cell results, and speculating on *in vivo* applications.

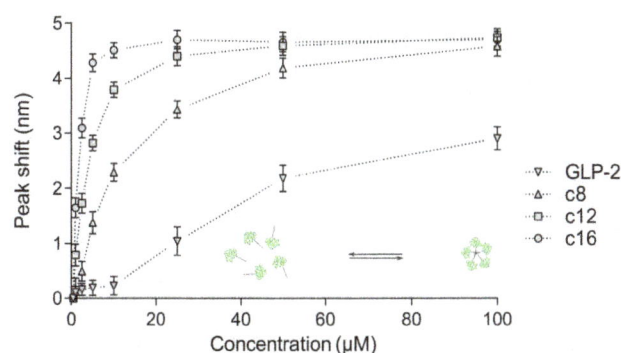

Figure 3. Concentration-dependent self-association of GLP-2 and its analogues. Increasing concentration leads to a blue-shift and increase in tryptophan maximum fluorescence, indicating self-association of the peptides. Control experiments with L-Trp, which does not self-associate, displayed no blue-shift. For the acylated analogues the self-association concentration decreases with increasing chain length, consistent with increased hydrophobicity which renders self-association more favorable. Native GLP-2 is less prone to self-association, and higher concentrations than the displayed 100 μM are required to reach full self-association. Data points are mean $\pm$ SD of 2 separate experiments, and lines are provided to guide the eye. The sketch serves as an illustration of self-association.

Receptor activation

The biological activity of the analogues was verified in a Luciferase assay (sketched in fig. 5), using a cell line stably expressing the GLP-2 receptor and a luciferase gene, which is transcribed upon receptor activation.

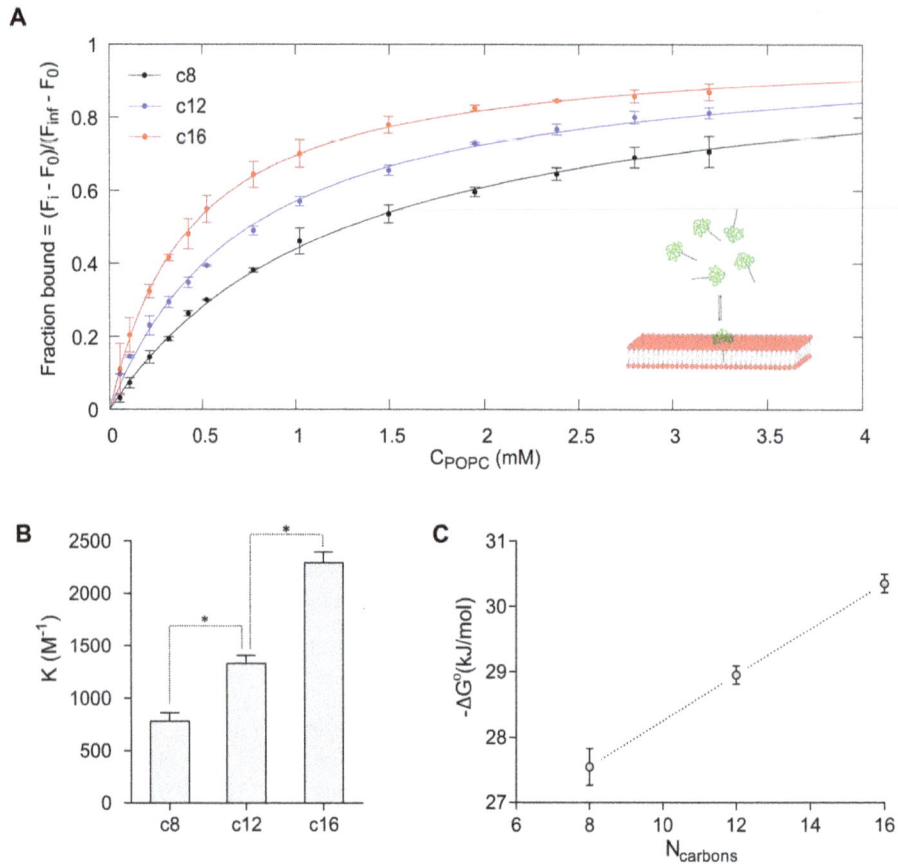

Figure 4. Binding to neutral liposomes. Addition of POPC liposomes to 2 μM peptide solutions causes an increase in tryptophan fluorescence, indicating partitioning of the peptides into the hydrophobic part of the lipid membrane. A) The peptide fluorescence during titration with POPC liposomes, fitted according to eq. 2 (solid lines), which yields the partitioning coefficients (K) shown in B). For the acylated analogues the partitioning coefficient increases with increasing chain length, and converting to the standard Gibbs free energy ($\Delta G°$) according to eq. 3 yields a linear relationship with the chain length, C). Native GLP-2 partitions very weekly into POPC liposomes, with values below the measurable limit (K<500 M^{-1}), and GLP-2 is therefore not included in the figure. Data points represent mean $\pm$ SD of 2 separate experiments and stars in B) indicate significantly different values (p<0.05). The sketch serves as an illustration of membrane-partitioning.

Alteration of therapeutic peptides is accompanied by a risk of reducing potency, and acylation is no exception. However, through a rational choice of acylation site and type and/or screening different acylations, it is often possible to limit the deleterious effects on biological activity [36]. All of the investigated acylated GLP-2 analogues display similar function compared to native GLP-2, suggesting that these acylations did not negatively impact receptor binding and activation.

The assay is employed to measure picomolar concentrations of GLP-2 and its analogues after *in vitro* experiments with Caco-2 cells, using peptide standard curves fitted according to eq. 5. Thus, the advantage of this cell based reporter assay is its very high sensitivity.

Interaction with Caco-2 cells

The Caco-2 setup is sketched in fig. 6. The translocation of peptide over time from the apical (upper) chamber to the basolateral (lower) chamber was measured, along with the amount of peptide associated with the cells after an experiment, both in the aqueous parts of the cells (uptake) and in the lipid membranes.

Peptide translocation. The translocation of GLP-2 and its analogues is presented in fig. 7, where part A shows the accumulated amount of peptide in the basolateral compartment

during the 1 hour experiment and part B shows the apparent permeability (P_{app}), calculated according to eq. 6.

The translocation depends non-trivially on the acylation chain length, as the short and medium chains (c8 and c12) increase the translocation relative to native GLP-2, but the long chain (c16) decreases it slightly. This indicates an optimum chain length where the translocation is increased by increased hydrophobicity and intermediate membrane binding, whereas the long chain causes too effective membrane insertion and strong binding, which limits translocation. For the c12 analogue, the increase in translocation through acylation is roughly a factor 1.5 compared to native GLP-2.

An alternative explanation for the decreased translocation of c16 could be increased self-association, which may limit para-cellular translocation through the concomitant increase in size. The extent to which the cellular environment affects self-association is currently unknown, as is the exact degree of peptide translocation through either the paracellular or transcellular route.

The acyl chain length may have an effect on albumin binding, and thereby circulation time, which could limit the useful acyl chain lengths to medium and long chain [8,36], but this is not fully established [4] and has not been investigated in this study.

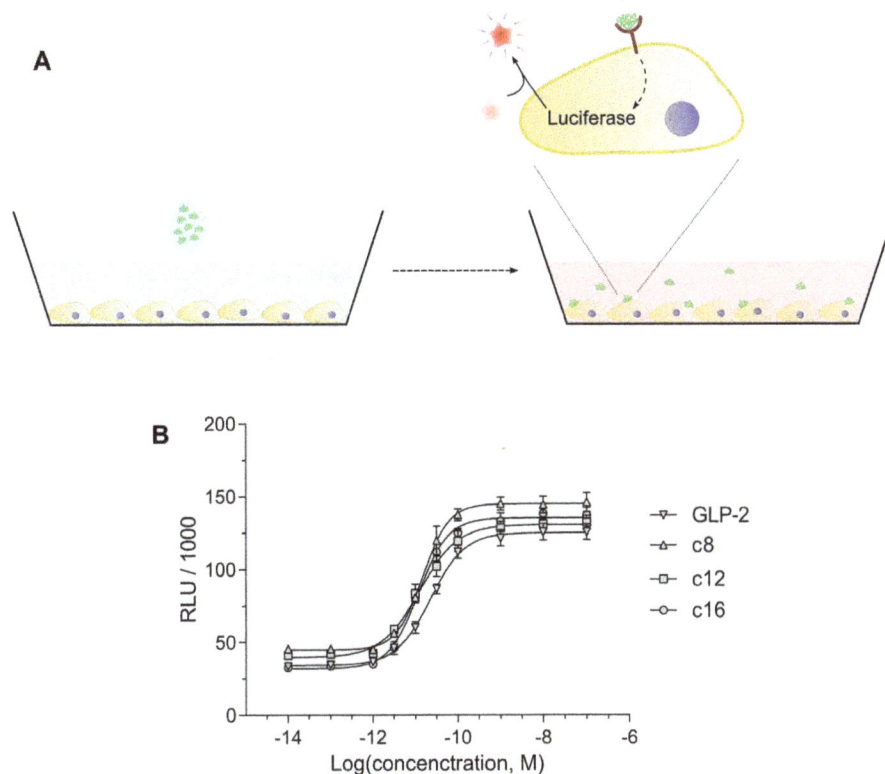

Figure 5. Schematic illustration of the Luciferase receptor activation assay and typical standard curves. A) In order to confirm biological activity of the analogues and measure picomolar concentrations of the peptides, a cell-line with the GLP-2-receptor and a luciferase reporter gene was employed. Upon receptor activation by GLP-2 or its analogues, the cells produce luciferase, which can cleave the substrate luciferin to a luminescent product. B) Representative standard curves of receptor activation (RLU) with increasing peptide concentration, for native GLP-2 and its analogues. Data points represent mean $\pm$ SD from 3 determinations, and solid lines are fitted according to eq. 5.

Cell membrane binding and uptake. The cell membrane binding and uptake of GLP-2 and its analogues is presented in fig. 8, where it is evident that both increase with acylation and increasing chain length. The cell membrane binding qualitatively resembles the liposome membrane binding presented in fig. 4C, showing that the chain-length dependence for binding to neutral model membranes is a valid predictor for cell membrane affinity, despite the added complexity of the biological environments compared to a simplified model membrane. It should be noted that cell studies employ a higher concentration of peptide than liposome binding studies, which may cause differences in peptide self-association behavior.

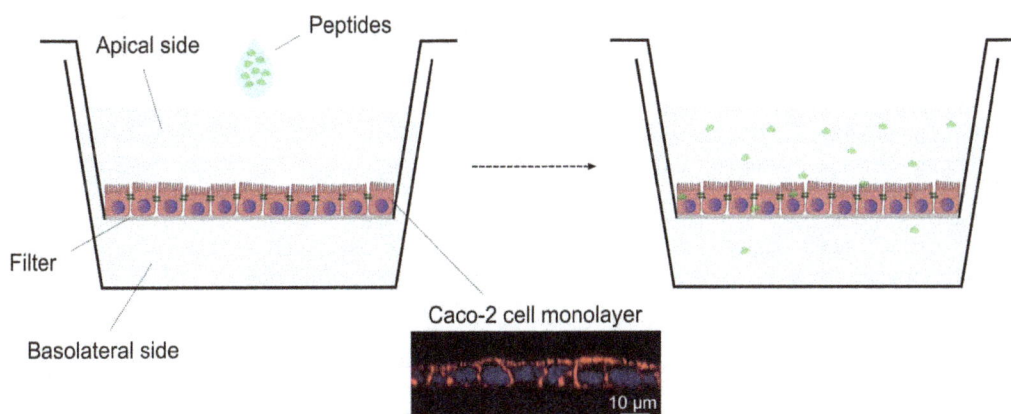

Figure 6. Schematic illustration of the Caco-2 setup. The Caco-2 intestinal model is composed of Caco-2 cells grown in a monolayer on a semipermeable filter support that separates two solution chambers. The peptide of interest is added to the top chamber (modeling the apical side of the intestine, i.e. the intestinal lumen), and the translocated peptide is sampled from the lower chamber over time (modeling the basolateral side of the intestine). Subsequently, the cells are analyzed for peptide content, both in the aqueous parts of the cells (uptake) and in the lipid membranes. The microscopy image in the insert shows a Caco-2 monolayer on the filter support, after fixing and staining for cell nucleus (blue) and actin (red).

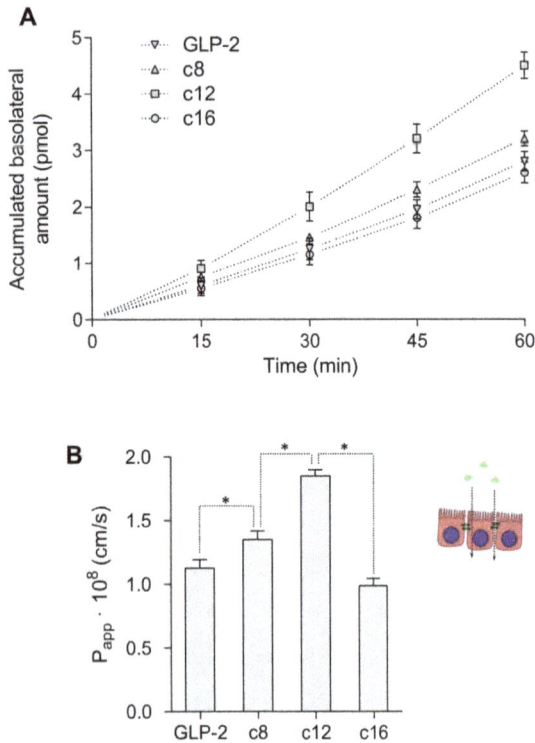

Figure 7. Translocation of GLP-2 and its analogues across Caco-2 cell monolayers. A) Accumulated amount of peptide in the basolateral chamber during the 1 hour experiment. B) Apparent permeability (P_{app}) of the peptides, calculated according to eq. 6. Data points represent mean $\pm$ SD of 3 separate experiments, each with 4 repeats, and stars indicate significant differences (p<0.05).

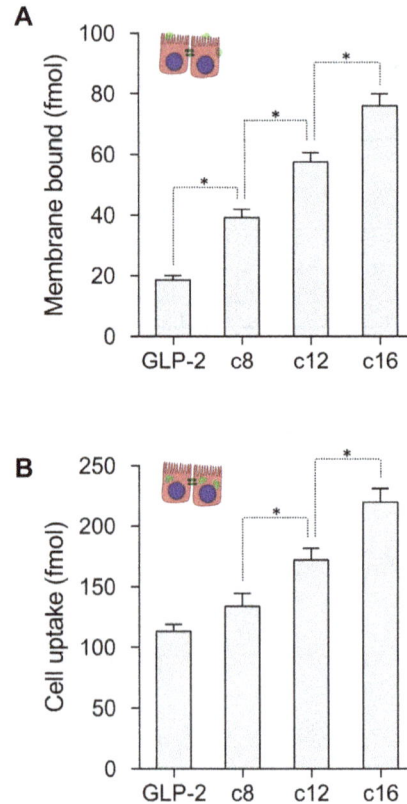

Figure 8. Cell membrane binding and uptake. The amount of peptide bound in the cell membrane A) and the amount of peptide in the aqueous parts of the cell B) increase with acylation length, indicating a larger interaction with cells. For the acylated analogues the membrane binding is roughly linear with chain length, and resembles the observed binding to model membranes shown in fig. 4C. Data points represent mean $\pm$ SD of 3 separate experiments, each with 4 repeats, and stars indicate significantly different values (p<0.05).

We speculate that the increase in cell uptake with increasing acyl chain length is caused by increased cell membrane binding, where peptide bound to the cell membrane is subsequently taken up more readily.

The observed cell binding qualitatively supports the hypothesis that the decrease in cell translocation for long chain acylation is caused by their increased membrane interactions. The decrease in translocation for c16-acylation compared to c12-acylation is larger than expected from the increased membrane binding, however, this may be explained by the added effect of increased uptake that could lead to intracellular sequestration.

If the membrane binding is a powerful determinant for cell translocation, the liposome partitioning coefficient may be used as a predictor hereof, but this should be investigated in further detail, using other peptides or acylations.

Absorption enhancers. The two absorption enhancers EGTA (paracellular) and SDS (transcellular) were employed to assess the effect of different enhancement mechanisms on the acyl chain length dependence.

We hypothesize that paracellular enhancers will have little effect on the acyl chain length dependence, i.e. enhance transport of the peptide and all its analogues to a similar extent, whereas transcellular enhancers that fluidize the membrane may have an increased effect on the long-chain acylated analogue by altering the membrane affinity and/or dynamics of membrane insertion and detachment.

A previously performed dose-response experiment was used to determine the appropriate concentration of absorption enhancers, that yielded increased mannitol transport and a reversible decrease

in TEER with full recovery after 24 hours, and the results using the optimal concentrations are presented in fig. 9A and B. The translocation of GLP-2 and its analogues in the presence of EGTA or SDS is presented in fig. 9C. For each peptide, the increased transport in the presence of enhancer was compared to the transport of the peptide alone as shown in fig. 9D, which emphasizes the differences between the two types of enhancer. For EGTA the increase is similar for peptide and analogues, and the dependence on acyl chain length is retained, whereas for SDS the increase is greater for the c16-acylation. These results support the hypothesis that the fluidization of cell membranes caused by SDS are beneficial for the long chain acylation, possibly due to altered membrane insertion. This could be verified by investigating liposome partitioning or cell membrane binding in the presence of SDS. However, despite the added benefit of SDS for the c16-acylation, the translocation of the c16 peptide remains lower than for the c12 peptide, suggesting that the acylation length is the primary determining factor for optimizing translocation. EGTA and SDS are employed as representatives of paracellular and transcellular enhancers, and other enhancers of these types are expected to elicit similar effects.

In conclusion, EGTA and SDS at the concentrations employed increase the translocation while retaining the acyl chain length dependence, with a medium chain length (c12) yielding the highest translocation.

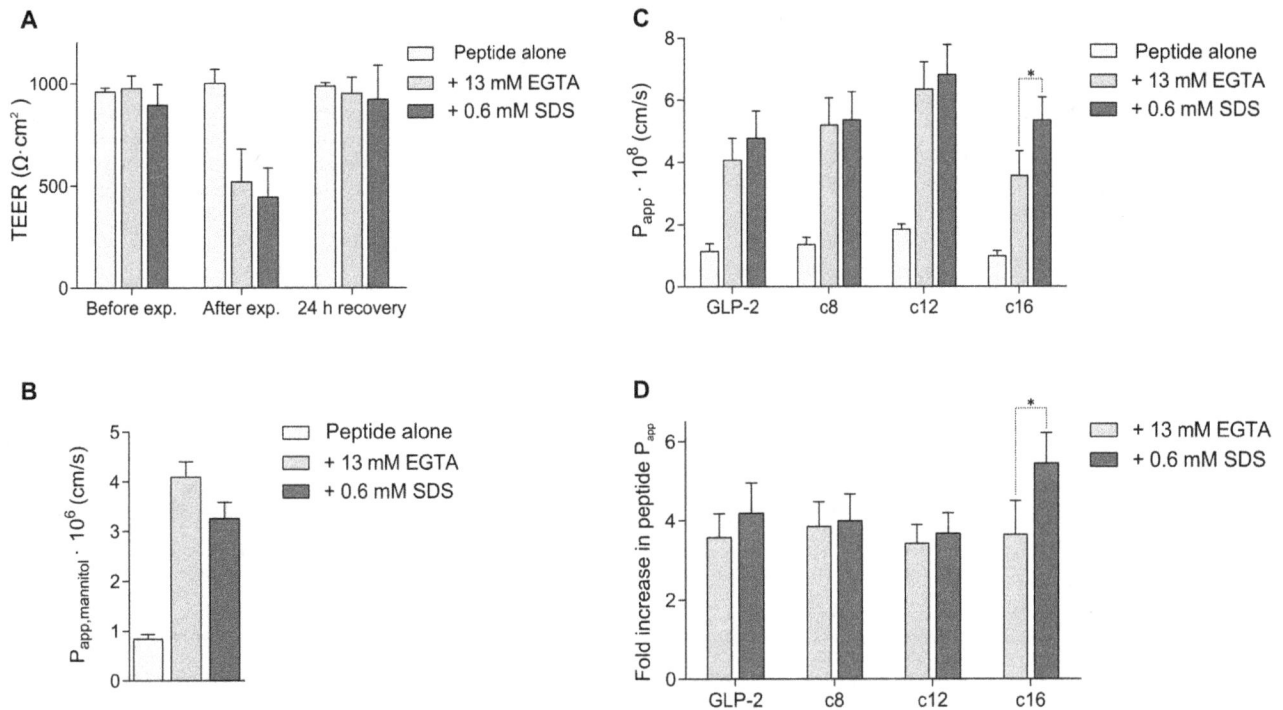

Figure 9. Effect of absorption enhancers on translocation of GLP-2 and its analogues. A) TEER values for Caco-2 cell monolayers after experiments with absorption enhancers EGTA (paracellular) or SDS (transcellular), and after 24 hour recovery in cell medium. B) Mannitol permeability in the presence of EGTA or SDS. C) Peptide permeability in the presence of EGTA or SDS. D) Fold increase in peptide permeability with enhancer, compared to each peptide alone. All data points represent mean $\pm$ SD of 2 separate experiments, each with 4 repeats, and stars indicate significant differences ($p < 0.05$).

Conclusions and Perspectivation

We have synthesized and investigated a systematic series of acylated GLP-2 analogues, in order to establish how the membrane binding correlates to *in vitro* intestinal permeability. We find that increasing acyl chain length causes increased self-association and binding to lipid and cell membranes, whereas translocation across intestinal cells displays a non-linear dependence on chain length. Short and medium chains improve translocation compared to the native peptide, but long chain acylation does not. We explain this correlation by an initial benefit for translocation for shorter chains through increased interaction with the cell membrane, which reverts to a hindrance for long chains, i.e. the analogues get stuck in the cell membrane. Measurements of liposome binding may be used to predict the optimal acylation chain length (e.g. the partitioning coefficient of approximately 1300 M^{-1} for c12-GLP-2), which can be further investigated by using other peptides or acylation types.

The translocation of peptide and analogues increases in the presence of both paracellular and transcellular absorption enhancers. The acyl chain length dependence persists for the paracellular enhancer and partially for the transcellular enhancer, with an increased benefit for the long chain acylation. This can be explained on the basis of a membrane fluidizing effect of the transcellular enhancer, which alters the dynamics and strength of membrane insertion, however, this requires further investigation. For both enhancers the medium chain acylation (c12) yields the highest translocation, approximately 1.5 times the native peptide.

The presented results suggest that rational acylation of GLP-2 increases intestinal absorption, and may benefit the oral delivery route. However, this should be verified *in vivo*, e.g. by investigating the intestinal absorption in an animal model following in situ administration to the intestine. Dosing directly to the rat intestine circumvents the esophagus and stomach, where the peptide would most likely require further stabilization to remain intact, *e.g.* through encapsulation and/or enteric coating.

In order to assert whether the effect of acylation on GLP-2 is general and can be used to benefit oral delivery of other therapeutic peptides, we are currently investigating other acylated peptides. GLP-2 exerts is therapeutic function locally in the intestine, which eases oral delivery and limits the challenging requirement for long circulation times in the blood stream. However, other therapeutic peptides that function systemically or at sites apart from the intestine may benefit thrice from acylation, *i.e.* on intestinal absorption, enzymatic stability, and circulation life time.

Supporting Information

Figure S1   Secondary structure of native and acylated GLP-2. Circular dichroism spectra of GLP-2 and its c16 analogue in buffers with different ionic strength (0–150 mM NaCl). The secondary structure of native GLP-2 and its acylated c16 analogue is different at low ionic strength (as previously reported in [38]), but similar at physiological ionic strength. It should be noted that self-association alters the secondary structure, and as described in the main text, the self-association behavior is affected by acylation. The employed concentration 150 μM was chosen in order to compare to [38], and at this concentration the peptides are expected to be self-associated.

Figure S2 Size of selected peptide oligomers. TEM (middle) and DLS (bottom) show that the native peptide and its c16-analogue forms oligomers with radius 2.4 ± 0.1 nm and 2.8 ± 0.1 nm, respectively. SLS-measurements show that the native peptide oligomers are composed of approximately 4 peptide monomers, whereas the c16 analogue oligomers are larger and composed of around 7–10 monomers. The top sketch serves as an illustration of peptide oligomers.

Materials S1 Materials and Methods for Cicular Dichroism (CD), Dynamic and Static Light Scattering (DLS/SLS) and Transmission Electron Microscopy (TEM).

Acknowledgments

We thank Holger Martin Strauss for helpful discussions regarding study design and manuscript preparation; Jonas Henriksen for assistance with the liposome partitioning assay; Lars Thim for providing native GLP-2 and a modified c16-analogue; Anja Knudsen for assistance with peptide synthesis; Anette Sams for providing hGLP-2R cells; and Lisette G. Nielsen for assistance with Caco-2 and hGLP-2R cells.

Author Contributions

Conceived and designed the experiments: ST LL SB TLA ULR. Performed the experiments: ST. Analyzed the data: ST LL SB TLA ULR. Contributed reagents/materials/analysis tools: ST LL ULR. Wrote the paper: ST LL SB TLA ULR.

References

1. Dasgupta P, Singh A, Mukherjee R (2002) N-terminal acylation of somatostatin analog with long chain fatty acids enhances its stability and anti-proliferative activity in human breast adenocarcinoma cells. Biol Pharm Bull 25: 29–36.
2. Yuan L, Wang J, Shen WC (2005) Reversible Lipidization Prolongs the Pharmacological Effect, Plasma Duration, and Liver Retention of Octreotide. Pharm Res 22: 220–227.
3. Gozes I, Bardea A, Reshef A, Zamostiano R, Zhukovsky S, et al. (1996) Neuroprotective strategy for Alzheimer disease: intranasal administration of a fatty neuropeptide. Proc Natl Acad Sci U S A 93: 427–32.
4. Kurtzhals P, Havelund S, Kiehr B, Larsen UD, Ribel U, et al. (1995) Albumin binding of insulins acylated. Biochem J 312: 725–731.
5. Havelund S, Plum A, Ribel U, Jonassen I, Vølund A, et al. (2004) The mechanism of protraction of insulin detemir, a long-acting, acylated analog of human insulin. Pharm Res 21: 1498–504.
6. Zhang L, Bulaj G (2012) Converting peptides into drug leads by lipidation. Curr Med Chem 19: 1602–18.
7. Wang J, Chow D, Heiati H, Shen WC (2003) Reversible lipidization for the oral delivery of salmon calcitonin. J Control Release 88: 369–80.
8. Knudsen LB, Nielsen PF, Huusfeldt PO, Johansen NL, Madsen K, et al. (2000) Potent derivatives of glucagon-like peptide-1 with pharmacokinetic properties suitable for once daily administration. J Med Chem 43: 1664–9.
9. Wang J, Shen WC (2000) Gastric retention and stability of lipidized BowmanBirk protease inhibitor in mice. Int J Pharm 204: 111–116.
10. Morishita M, Peppas NA (2006) Is the oral route possible for peptide and protein drug delivery? Drug discov today 11: 905–10.
11. Hubatsch I, Ragnarsson EGE, Artursson P (2007) Determination of drug permeability and prediction of drug absorption in Caco-2 monolayers. Nature protocols 2: 2111–9.
12. Artursson P, Palm K, Luthman K (2001) Caco-2 monolayers in experimental and theoretical predictions of drug transport. Adv Drug Delivery Rev 46: 27-43.
13. Uchiyama T, Kotani A, Tatsumi H (2000) Development of novel lipophilic derivatives of DADLE (leucine enkephalin analogue): Intestinal permeability characteristics of DADLE derivatives in rats. Pharm Res 17.
14. Rogers J, Wong A (1980) The temperature dependence and thermodynamics of partitioning of phenols in the n-octanol-water system. Int J Pharm 6: 339–348.
15. Balon K, Riebesehl B, Müller B (1999) Drug liposome partitioning as a tool for the prediction of human passive intestinal absorption. Pharm Res 16: 882–888.
16. Krämer S (1999) Absorption prediction from physicochemical parameters. Pharm Sci Technol To 2: 373–380.
17. Drucker DJ (1999) Glucagon-like Peptide 2. Trends in Endocrinology & Metabolism 10: 153–156.
18. Orskov C, Holst JJ, Knuhtsen S, Baldissara FGA, Poulsen SS, et al. (1986) Glucagon-Like Peptides GLP-1 and GLP-2, Predicted Products of the Glucagon Gene, Are Secreted Separately from Pig Small Intestine but Not Pancreas. Endocrinology 119: 1467–1475.
19. Wallis K, Walters JRF, Forbes A (2007) Review article: glucagon-like peptide 2-current applications and future directions. Alimentary pharmacology & therapeutics 25: 365–72.
20. Jeppesen PB (2012) Teduglutide, a novel glucagon-like peptide 2 analog, in the treatment of patients with short bowel syndrome. Therap Adv Gastroenterol 5: 159–71.
21. Hartmann B, Johnsen AH, Orskov C, Adelhorst K, Thim L, et al. (2000) Structure, measurement, and secretion of human glucagon-like peptide-2. Peptides 21: 73–80.
22. Tavares W, Drucker DJ, Brubaker PL (2000) Enzymatic- and renal-dependent catabolism of the intestinotropic hormone glucagon-like peptide-2 in rats. Am J Physiol Endocrinol Metab 278: E134–139.
23. te Welscher YM, Chinnapen DJF, Kaoutzani L, Mrsny RJ, Lencer WI (2014) Unsaturated glycoceramides as molecular carriers for mucosal drug delivery of GLP-1. J Control Release 175: 72–8.
24. Deli MA (2009) Potential use of tight junction modulators to reversibly open membranous barriers and improve drug delivery. Biochim Biophys Acta 1788: 892–910.
25. Sakai M, Imai T, Ohtake H, Otagiri M (1998) Cytotoxicity of Absorption Enhancers in Caco-2 Cell Monolayers. J Pharm Pharmacol 50: 1101–1108.
26. Thulesen J, Knudsen LB, Hartmann B, Hastrup S, Kissow H, et al. (2002) The truncated metabolite GLP-2 (3-33) interacts with the GLP-2 receptor as a partial agonist. Regul Pept 103: 9–15.
27. Sams A, Hastrup S, Andersen M, Thim L (2006) Naturally occurring glucagon-like peptide-2 (GLP-2) receptors in human intestinal cell lines. Eur J Pharmacol 532: 18–23.
28. Palasek SA, Cox ZJ, Collins JM (2007) Limiting racemization and aspartimide formation in microwave-enhanced Fmoc solid phase peptide synthesis. J Pept Sci 13: 143–8.
29. Fujinari E, Courthaudon L (1992) Nitrogen-specific liquid chromatography detector based on chemiluminescence: Application to the analysis of ammonium nitrogen in waste water. J Chromatogr 592: 209–214.
30. Jesorka A, Orwar O (2008) Liposomes: technologies and analytical applications. Annu Rev Anal Chem 1: 801–32.
31. Rouser G, Siakotos AN, Fleischer S (1966) Quantitative analysis of phospholipids by thin-layer chromatography and phosphorus analysis of spots. Lipids 1: 85–6.
32. Burstein EA, Abornev SM, Reshetnyak YK (2001) Decomposition of protein tryptophan fluorescence spectra into log-normal components. I. Decomposition algorithms. Biophys J 81: 1699–709.
33. Etzerodt T, Henriksen JR, Rasmussen P, Clausen MH, Andresen TL (2011) Selective acylation enhances membrane charge sensitivity of the antimicrobial peptide mastoparan-x. Biophys J 100: 399–409.
34. Ladokhin AS, Jayasinghe S, White SH (2000) How to measure and analyze tryptophan fluorescence in membranes properly, and why bother? Anal Biochem 285: 235–45.
35. Naylor LH (1999) Reporter gene technology: the future looks bright. Biochem Pharmacol 58: 749–757.
36. Madsen K, Knudsen LB, Agersoe H, Nielsen PF, Thøgersen H, et al. (2007) Structure-activity and protraction relationship of long-acting glucagon-like peptide-1 derivatives: importance of fatty acid length, polarity, and bulkiness. J Med Chem 50: 6126–32.
37. Peitzsch RM, McLaughlin S (1993) Binding of acylated peptides and fatty acids to phospholipid vesicles: pertinence to myristoylated proteins. Biochemistry 32: 10436–43.
38. Pinholt C, Kapp SJ, Bukrinsky JT, Hostrup S, Frokjaer S, et al. (2012) Influence of acylation on the adsorption of GLP-2 to hydrophobic surfaces. Int J Pharm: 1–9.

5

Glucolipotoxicity Impairs Ceramide Flow from the Endoplasmic Reticulum to the Golgi Apparatus in INS-1 β-Cells

Enida Gjoni[1], Loredana Brioschi[1], Alessandra Cinque[1], Nicolas Coant[2], M. Nurul Islam[3], Carl K. -Y. Ng[3], Claudia Verderio[4], Christophe Magnan[2], Laura Riboni[1], Paola Viani[1], Hervé Le Stunff[2], Paola Giussani[1]*

1 Department of Medical Biotechnology and Translational Medicine, Università di Milano, LITA Segrate, Milano, Italy, 2 Unité Biologie Fonctionnelle et Adaptative –UMR CNRS 8251, Université PARIS- DIDEROT (7), Paris, France, 3 School of Biology and Environmental Science and UCD Earth Institute, University College Dublin, Belfield, Ireland, 4 Department of Medical Biotechnology and Translational Medicine, CNR Institute of Neuroscience, Universita' di Milano, Milano, Italy

Abstract

Accumulating evidence suggests that glucolipotoxicity, arising from the combined actions of elevated glucose and free fatty acid levels, acts as a key pathogenic component in type II diabetes, contributing to β-cell dysfunction and death. Endoplasmic reticulum (ER) stress is among the molecular pathways and regulators involved in these negative effects, and ceramide accumulation due to glucolipotoxicity can be associated with the induction of ER stress. Increased levels of ceramide in ER may be due to enhanced ceramide biosynthesis and/or decreased ceramide utilization. Here, we studied the effect of glucolipotoxic conditions on ceramide traffic in INS-1 cells in order to gain insights into the molecular mechanism(s) of glucolipotoxicity. We showed that glucolipotoxicity inhibited ceramide utilization for complex sphingolipid biosynthesis, thereby reducing the flow of ceramide from the ER to Golgi. Glucolipotoxicity impaired both vesicular- and CERT-mediated ceramide transport through (1) the decreasing of phospho-Akt levels which in turn possibly inhibits vesicular traffic, and (2) the reducing of the amount of active CERT mainly due to a lower protein levels and increased protein phosphorylation to prevent its localization to the Golgi. In conclusion, our findings provide evidence that glucolipotoxicity-induced ceramide overload in the ER, arising from a defect in ceramide trafficking may be a mechanism that contributes to dysfunction and/or death of β-cells exposed to glucolipotoxicity.

Editor: Stephan Neil Witt, Louisiana State University Health Sciences Center, United States of America

Funding: This work was supported by grants from the University of Milan PUR to PG, grants from the Italian Ministry of University and Scientific and Technological Research PRIN to PV, and grants from Science Foundation Ireland (SFI/06/RFP/GEN034 and SFI/08/RFP/EOB1087) to CK-YN. This project was partly supported by grants from Centre National de la Recherche Scientifique (CNRS) and Agence Nationale de la Recherche (ANR-06-JCJC-0040) to HLS. NC received a postdoctoral fellowship from the Université Paris Diderot and the French Society of Nutrition (SFN). The funders had no role in study design, data collection and analysis, decision to publish, or preparation of the manuscript.

* Email: paola.giussani@unimi.it

Introduction

Glucolipotoxicity is defined as the condition in which the combined action of elevated glucose and free fatty acid (FFA) levels synergizes in exerting deleterious effects on pancreatic β-cell function and survival [1–3]. Accumulating evidence suggests that this condition acts as a key pathogenic component in type II diabetes, contributing to β-cell dysfunction and death during the development of this disease (reviewed in [4]). In agreement, chronic exposure of β-cells to supraphysiological levels of glucose and free fatty acids (FFAs) has been shown to be cytotoxic and cause β-cell dysfunction and failure [5]. Palmitate, a major FFA species in which β-cells might be exposed to *in vivo* [6], is particularly potent in reducing β-cell viability of clonal and primary rodent β-cells, as well as in human islets [7–9]. Hyperglycaemia has been shown to potentiate the negative effects of high levels of saturated FFAs on pancreatic β-cells [1,2]. While palmitate and other saturated FFAs exhibit low toxicity at low glucose concentrations, they have been shown to synergize with elevated glucose concentrations to promote β-cell apoptosis, both in the β-cell line INS-1 and in human islets [10–12].

Several mechanisms have been proposed for glucolipotoxicity-induced β-cell dysfunction and failure, and, among them, endoplasmic reticulum (ER) stress and elevations of the proapoptotic sphingolipid ceramide (Cer) appear to play key roles. Additionally, these two processes appear to be strictly connected [13–15].

Several enzymes of Cer metabolism have been shown to be involved in regulating its levels in β-cells in response to lipotoxicity and/or glucolipotoxicity. In particular, serine palmitoyltransferase (SPT) and ceramide synthase (CerS), both residing in the endoplasmic reticulum (ER), and involved in Cer biosynthesis [12,16,17], as well as neutral sphingomyelinase (N-SMase), involved in Cer degradation [18,19], have emerged as important regulators of elevated Cer levels. In addition, the over-expression of glucosyl-ceramide synthase, which converts Cer into glucosyl-ceramide

(GlcCer), has been shown to prevent β-cell apoptosis [20]. Altogether these data suggest that the accumulation of Cer in the ER compartment of β-cells is crucial in determining β-cell fate, *i.e.*, survival or death.

While the accumulation of Cer at the ER consequent to glucolipotoxicity appears to be critically involved in the induction of ER stress (reviewed in [15]), the dysregulation of Cer metabolism alone may not necessarily lead to Cer accumulation in the ER unless Cer traffic from the ER to the Golgi is inhibited. Cer synthesized in the ER is transferred to the Golgi where it is subsequently converted to sphingomyelin (SM), GlcCer and more complex glycosphingolipids (GSLs) [21]. Evidence to date indicates that there are two pathways by which Cer is transported from the ER to the Golgi: a protein-mediated transport, by the soluble ceramide transfer protein CERT (for SM formation) [22–25], and a CERT-independent vesicular traffic (for the biosynthesis of SM or GlcCer) [23,25,26]. The two modes of Cer transport coexist separately contributing to the regulation of Cer metabolism and levels in cells. For example, hyperphosphorylation of a serine repeat motif of CERT impairs its binding to the ER and Golgi membranes, thereby inhibiting Cer transfer from the ER to the Golgi [22,27,28]. Additionally, nitric oxide or the overexpression of sphingosine-1-phosphate phosphohydrolase 1 (SPP1) inhibits Cer vesicular traffic [26,29], resulting in Cer accumulation in the ER. Notwithstanding, our knowledge on the effect of glucolipotoxicity on Cer transport is scarce.

The aim of this investigation was to determine if the transport mechanisms of Cer from the ER to the Golgi are involved in the deleterious effects of glucolipotoxicity in β-cells, and to gain a further understanding of the relationship between Cer accumulation and ER stress. We demonstrate, using INS-1 cells as a model, which can be expanded to quantities sufficient for diverse experimentation, that palmitate and elevated glucose administration induced a rapid and potent inhibitory effect on the mechanisms of Cer transport, resulting in the accumulation of Cer at the ER.

Material and Methods

Materials

All reagents were of analytical grade unless otherwise stated. The tissue culture medium RPMI 1640 was purchased from Lonza (Basel, Switzerland). L-glutamine, sodium pyruvate solution, penicillin/streptomycin, dimethyl sulfoxide (DMSO), palmitate, glucose, Hepes, bovine serum albumin fraction V (BSA), fatty acid free-BSA, 3-[4,5-dimethylthiazol-2-yl]2,5-diphenyl tetrazolium bromide (MTT), leupeptin, aprotinin, wortmannin (Wm), Thapsigargin (Tg), Kodak Biomax film, HPLC grade water, tetrahydrofuran (THF), methanol, LC-MS grade water, formic acid and ammonium formate were purchased from Sigma-Aldrich (St. Louis, MO, USA). Fetal calf serum (FCS) was from Euroclone (Pero, Milano, Italy). LY294002 was from Cayman Chemical (Ann Arbor, MI, USA). Lipofectamine 2000 and the Stealth RNAi were from Invitrogen (Carlsbad, CA, USA). D-erythro-[3-^{3}H]sphingosine (Sph) (19.7 Ci/mmol), was from PerkinElmer Life Science (Boston, MA, USA). Pepstatin was from Roche Applied Sciences (Mannheim, Germany). High performance thin layer chromatography (HPTLC) silica gel plates were from Merck (Darmstadt, Germany). The Golgi marker Texas red wheat germ agglutinin (WGA), 6-((N-(7-nitrobenz-2-oxa-1,3-diazol-4-yl) amino) hexanoyl) sphingosine (NBD-C$_6$Cer) and N-(4,4,-difluoro-5-,7-dimethyl-bora-3a,4a-diaza-sindacene- 3-pentanoyl) sphingosine (BODIPY-C$_5$Cer) were from Life Technologies (Italy). The antibodies recognizing Phospho-Akt (Ser473) were from Cell Signaling Technology, Inc. (Danvers, MA, USA); polyclonal antibodies against Cer transfer protein (CERT) from Bethyl Laboratories (Montgomery, TX, USA). Primary mouse monoclonal anti-phospho-serine, goat anti-GRP78 and rabbit anti-GAPDH antibodies, and secondary HRP-conjugated anti-rabbit or anti-goat antibodies were from Santa Cruz Biotechnology (Santa Cruz, CA, USA). Secondary anti-mouse HRP-conjugated antibody, SuperSignal WestPico Chemioluminescent Substrate and SuperSignal WestFemto Maximum Sensitivity Substrate were from Thermo Scientific (Rockford, IL, USA). Ceramide/Sphingoid Internal Standard Mixture I from Avanti Polar Lipids (Alabaster, Alabama, USA) was used for quantitative analysis. The plasmid of CERT tagged with green fluorescent protein (CERT-GFP) was kindly provided by Dr. Maria Antonietta De Matteis, Telethon Institute of Genetics and Medicine, Napoli (Italy).

Cell culture conditions

Rat insulinoma INS-1 cells, kindly provided by Merck–Serono, were grown in RPMI 1640 medium buffered with 10 mM Hepes containing 10% (v/v) FCS, 2 mM L-glutamine, 1 mM sodium pyruvate, 50 μM 2-mercaptoethanol and 100 units/ml penicillin/streptomycin at 37°C in an atmosphere of 5% CO$_2$ and 95% humidified air. Before each experiment, INS-1cells plated at 2×10^5 cell/cm^2 were cultured for 24 h in RPMI 1640 plus 10% FCS. Cells were then cultured in the presence of 5 mM or 30 mM glucose with or without 0.4 mM palmitate for 12 h; incubation in the presence of 30 mM glucose and 0.4 mM palmitate mimics glucolipotoxicity conditions. When indicated, the cells were preincubated with 20 μM LY294002, or 10 nM Wm or 0.1 μM Tg for 30 min. Palmitate was administered to the cells as a conjugate with fatty-acid-free BSA. Briefly, dried aliquots of palmitate in ethanol were dissolved in PBS containing 5% (w/v) BSA to obtain a 4 mM stock solution. The molar ratio of FFAs to BSA was 5:1. The FFA stock solutions were diluted in RPMI 1640 medium supplemented with 1% FCS to obtain a 0.4 mM final concentration at a fixed concentration of 0.5% BSA.

Analysis of cell viability

Cell viability was determined by MTT assay. INS-1 cells were plated and grown on a glass coverslip and cultured in the presence of 5 mM or 30 mM glucose with or without 0.4 mM palmitate for 12 h, as previously described, or for 24 hours. At the end of the treatments, the medium was replaced by MTT dissolved in fresh medium (0.8 mg/ml) for 4 hours. The formazan crystals were then solubilized in isopropanol/formic acid (95:5 v/v) for 10 minutes and the absorbance (570 nm) was measured using a microplate reader (Wallack Multilabel Counter, Perkin Elmer, Boston, MA, USA).

RNA interference

Small interfering RNA (siRNA) duplexes for rat CERT (Gene accession number XM 345143.1) S87, S522 and control non-targeting siRNAs (scrambled sequences of S87 and S522 oligonucleotides) described in [23] were used. We used Stealth RNAi, the chemically modified synthetic RNAi duplexes that virtually eliminate the induction of non-specific cellular stress response, and that also improve the specific, effective knock-down of gene expression. INS-1 cells plated at 2×10^5 cell/cm^2 were maintained for 24 h in RPMI 1640 plus 10% FCS and then transfected in the same medium with a 1:1 (by mol) mix of S87 + S522 (si-CERT) or the non-targeting corresponding sequences (si-CT) using LipofectAMINE 2000 according to the manufacturer's protocol. The final concentration of siRNA–lipofectamine duplex mixture was 100 nM. All the experiments were performed 72 h after transfection.

Plasmid Transfection

INS-1 cells were plated at 2×10^5 cell/cm^2 on a glass coverslip and grown in RPMI 1640 supplemented with 10% FCS until they were 50–70% confluent. Then cells were transfected with expression plasmid encoding the protein CERT tagged with GFP (CERT-GFP) or pcDNA3.1 empty vector using the Lipofectamine 2000 reagent according to the manufacturer's directions.

[³H]Sphingosine metabolism

INS-1 cells, si-control (siCT) and si-CERT (siCERT) INS-1 cells cultured in the presence of 5 mM or 30 mM glucose with or without 0.4 mM palmitate for 12 h were pulsed with [³H]Sph (0.3 μCi/ml), for 1 h maintaining the treatment conditions. All experiments were performed at 37°C. Stock solutions of [³H]Sph in absolute ethanol were prepared and added to fresh medium. In all cases, the final concentration of ethanol never exceeded 0.1% (v/v). At the end of pulse, cells were washed twice with phosphate-buffered saline (PBS) at 4°C, harvested and submitted to lipid extraction and partitioning as previously described [30]. The methanolized organic phase and the aqueous phase were analyzed by HPTLC using chloroform/methanol/water (55:20:3 by vol) and chloroform/methanol/0.2% CaCl$_2$ (55:45:10 by vol) as solvent system respectively. Digital autoradiography of HPTLC plates was performed with Beta-Imager 2000 (Biospace, France) and the radioactivity associated with individual lipids was measured using the software provided with the instrument. The [³H]-labeled sphingolipids were recognized and identified as previously described [30].

Liquid chromatography-tandem mass spectrometry (LC-MS/MS) protocol for lipid extraction and quantitation

Cellular lipids were extracted from INS-1 cells according to Shaner et al. [31] with modifications. Briefly, freeze-dried INS-1 cells (2 million) were transferred to a 5 ml glass tube, spiked with 10 μl of internal standard (12.5 μM Ceramide/Sphingoid Internal Mix I), and extracted with 2 ml of chloroform:methanol (1:2, v/v) following brief sonication and constant agitation in a 50°C water bath for 2 h. After cooling to room temperature, 200 μl of 1 M KOH in methanol was added and incubated for 2 h at 37°C. After cooling, 15 μl of glacial acetic acid were added to neutralize the extract. Following centrifugation, the supernatant was transferred to a new glass tube and the extraction was repeated with a further 2 ml of chloroform:methanol (1:2, v/v). The resulting supernatants were pooled and dried under a stream of N$_2$. The lipids were resuspended in 200 μl of mobile phase (mobile phase A: mobile phase B, 1:1, v/v). After centrifugation, aliquots were used for LC-MS analysis. Inorganic phosphate (Pi) content of extracts were determined according to van Veldhoven and Mannaerts [32]. Peak areas were used for quantitation by comparison with the peak areas of internal standards.

Chromatographic separation of lipids was performed with an HPLC system consisting of a binary pump, auto sampler, column oven (1200 RRLC, Agilent Technologies, http://www.chem.agilent.com). The lipid molecules were separated using a GeminiNX C$_{18}$ analytical column (2.0 mm I.D. x 100 mm, particle size 3 μm, Phenomenex, http://www.phenomenex.com). Column oven and auto sampler temperatures were maintained at 45°C and 4°C, respectively. The mobile phase consists of solvent A

(15 mM ammonium formate (pH 4.0):MeOH:THF, 5:2:3) and solvent B (15 mM ammonium formate (pH 4.0):MeOH:THF, 1:2:7). Elution was performed at a flow rate of 0.30 ml min^{-1} in a binary gradient mode. The initial composition of mobile phase was 65:35 (A:B), linearly changed to 70:30 (A:B) over 12 min, and maintained this composition over 22 min, changed to initial composition 65:35 (A:B) over 1 min, followed by 7 min of column re-equilibration. Column eluant was directed to waste for the initial 1 min.

The HPLC system was coupled online to an Agilent 6460 triple quadrupole mass spectrometer (Agilent Technologies) equipped with a Jet Stream ion source. Data were recorded in positive ionization mode using electrospray ionization with nitrogen as the nebulizing gas. The gas temperature and flow rate was 350°C and 10 l min^{-1}, and the sheath gas temperature and flow rate was 360°C and 12 l min^{-1}, respectively. The ESI needle voltage was adjusted to 4000 V in positive mode and optimum fragmentor voltages and collision energies were assigned by analysis of reference compounds (SM (d18:0/12:0); Cer (d18:1/12:0); GlcCer (d18:1/12:0); S1P (d18:1-P)) in selected ion and product ion scanning mode to determine multiple-reaction monitoring (MRM) conditions and mass spectrometric structural studies. MRM detection was applied using nitrogen as the collision gas.

Analysis of the Intracellular Distribution of Fluorescent Ceramides

INS-1 cells were plated and grown on a glass coverslip and cultured as previously described. At the end of the treatments, the cells were loaded with 2.5 μM BODIPY-C$_5$-Cer or NBD-C$_6$-Cer (as 1:1 complex with fatty acid free BSA) in RPMI 1640 at 4°C for 30 min [23]. After loading, the cells were incubated 30 min at 37°C in RPMI 1640 containing 5 mM or 30 mM glucose ± 0.4 mM palmitate and fixed with 0.5% glutaraldehyde solution in PBS for 10 min at 4°C. The specimens were immediately observed and analyzed with a fluorescence microscope (Olympus BX-50) equipped with a fast high resolution charge-coupled device camera (Colorview 12) and an image analytical software (Analysis from Soft Imaging System GmbH).

Analysis of Intracellular Localization of CERT-GFP by Confocal microscopy

INS-1 cells plated at 2×10^5 cell/cm^2 were grown on a glass coverslip and maintained 24 h in RPMI 1640 plus 10% FCS. The cells were then transfected with the plasmid, CERT-GFP using lipofectamine 2000 according to manufacturer's instructions. 24 h after transfection, cells were treated with 5 mM or 30 mM glucose ±0.4 mM palmitate for 12 h and fixed with 0.5% glutaraldehyde solution in PBS for 10 min at 4°C. The cells were then permeabilized with 0.2% Triton X-100 for 30 min at room temperature and stained with WGA-texas red. The specimens were analyzed with a confocal microscope (Leica SP5).

Immunoblotting

Phosho-Akt and GRP78 immunoblotting were performed on INS-1 cells lysed with lysis buffer (20 mM Tris-HCl pH 7.4, 150 mM NaCl, 1% NP-40, 10 mM sodium fluoride, 1 mM EDTA, 10 mM Na$_4$P$_2$O$_7$, 1 mM Na$_3$VO$_4$, and the protease inhibitor cocktail). Solubilized proteins were centrifuged at 14,000×g at 4°C for 10 min. Supernatants were subjected to 10% SDS polyacrylamide gel electrophoresis and transferred to nitrocellulose membranes. Membranes were blocked for 1 h at room temperature in Tris-buffered saline (10 mM Tris-HCl, pH 7.4, 140 mM NaCl) containing 0.1% Tween-20 (TBS-T) and

5% skim milk, and then incubated with primary antibodies against phosho-Akt overnight at 4°C or against GRP78 1 h at room temperature. Membranes were washed in TBS-T, and bound antibodies visualized with horseradish peroxidase-coupled secondary antibodies (Santa Cruz Biotechnology) and chemiluminescent substrate. The relative intensities of bands were quantified by densitometry.

CERT immunoblotting were performed using wild type or si-control and si-CERT transfected cells lysed with CERT buffer (10 mM Tris-HCl pH 7.4, 0.25 mM sucrose, 0.5 mM phenyl-methylsulfonyl fluoride, 10 μg/ml aprotinin, 5 μg/mL leupeptin, 5 μg/mL pepstatin), processed and analyzed as previously described [23]. The membranes were stripped 30 minutes at 50°C in 2% SDS, 100 mM DTT, 0.5 M Tris-HCl pH 6.8, washed in TBS-T and incubated 1 hour in Tris-buffered saline (10 mM Tris-HCl, pH 7.4, 140 mM NaCl) containing 0.1% Tween-20 (TBS-T) and 5% BSA and then incubated with the primary antibody against phospho-serine 2 h at room temperature. Membranes were washed in TBS-T and bound antibodies visualized with horseradish peroxidase-coupled secondary antibodies (Santa Cruz Biotechnology) and chemiluminescent substrate.

RNA isolation, reverse transcription and Real-Time PCR

INS-1 cells were plated and grown on a glass coverslip and cultured as previously described. At the end of the treatments, total RNA was isolated from INS-1 cells with the RNeasy mini kit and treated with the RNase-free DNAse I. One microgram of RNA was reverse transcribed using the iScript cDNA synthesis kit according to manufacturer's instructions. Real-Time PCR was performed using the iQ5 Real-Time PCR detection system (Biorad Laboratories, Hercules, CA, USA). Specific SYBR green expression assays (SYBR green super mix) for CERT and TBP (TATA-box-binding protein) were carried out. Simultaneous amplification of the target sequences was carried out as follows: 3 minutes at 95°C, 50 cycles 95°C 10 sec, 59°C 40 sec and 60°C 30 sec and 1 cycle of 60°C 3 minutes. Results were analyzed using the iQ5 optical system software (Biorad Laboratories, Hercules, CA, USA). Relative gene expression was determined using the $2^{-\Delta\Delta Ct}$ method [33]. Data were normalized to TBP expression (used as endogenous control) and INS-1 G5 cells were used as calibrator.

Sphingomyelin synthase activity

INS-1 cells were plated and treated as described above. At the end of the treatments, the cells were loaded with 2.5 μM NBD-C_6-Cer (as 1:1 complex with fatty acid free BSA) in RPMI 1640 at 4°C for 30 min. After loading, the cells were incubated 15 or 30 min at 37°C in RPMI 1640 with 5 mM glucose or 30 mM glucose ±0.4 mM palmitate. At the end of the incubation, cells were immediately put at 4°C to stop the enzymatic reaction; lipids were extracted with chloroform -methanol [34] and separated by thin-layer chromatography (TLC) using chloroform/methanol/0.1 M KCl (1:2:0.8 [vol/vol/vol]) as the developing solvent. Fluorescence-labeled sphingomyelin was quantified with a luminescence spectrometer (LS50B PerkinElmer).

Other methods

Total protein amount was assayed with the Comassie Blue based Pierce reagent, using BSA fraction V as standard. Radioactivity was measured by liquid scintillation counting.

Statistical analysis

Statistical significance of differences was determined by one-way ANOVA.

Results

Effect of palmitate and glucose on [³H]Sph metabolism in INS-1 cells

Treatment for 12 h with 0.4 mM palmitate (G5P4), 30 mM glucose (G30), or 0.4 mM palmitate plus 30 mM glucose (G30P4) did not exert a toxic effect on INS-1 control cells treated with 5 mM glucose (G5) (Fig. 1a, left panel). A similar effect of palmitate on cell viability was observed when assessing the protein content of each dish, demonstrating that at 12 hours there is no toxic effect on INS-1 cells (results not shown). After 24 hours, palmitate or 30 mM glucose alone exerted no toxicity. We also observed that 30 mM glucose increased INS-1 cell numbers, in agreement with previously published report [12] (Fig. 1a). In contrast, co-administration of 0.4 mM palmitate and 30 mM glucose for 24 hours reduced INS-1 cell viability by 53% (Fig. 1a right panel). Western blot analysis of GRP78, a ER stress marker, showed that a 12 h co-treatment with 0.4 mM palmitate and 30 mM glucose induced about 60% increase in the amount of GRP78 compared to 5 mM glucose with or without 0.4 mM palmitate and 30 mM glucose in INS-1 cells (Fig. 1b), suggesting that glucolipotoxic conditions induce ER stress. In the positive control, we observed that 0.1 μM thapsigargin in the presence of 5 mM glucose doubled GRP78 levels in INS-1 cells (Fig. 1b). To evaluate the effects of palmitate and high glucose concentrations on Cer utilization for the biosynthesis of SM and GSLs in INS-1 cells, we studied Cer metabolism using [³H]-sphingosine as a metabolic precursor as it is rapidly internalized in the cells and efficiently N-acylated to Cer, which in turn is converted to SM, glucosylceramide and complex GSLs. We performed short pulse experiments to monitor the utilization of newly synthesized Cer for the biosynthesis of SM and GlcCer [26]. In all cases, [³H]Sph was mainly metabolized to N-acylated compounds, mostly represented by Cer, SM and, in lower amounts, GSLs (Fig. 1c); the extent of N-acylation being similar in all conditions. Treatment with palmitate in the presence of 5 mM glucose significantly modified the distribution of radioactivity between the different Sph metabolites. Our results showed that palmitate induced [³H]Cer accumulation was associated with a decrease of [³H]SM levels (Fig. 1c). Treatment with 30 mM glucose by itself also reduced [³H]Cer conversion to [³H]SM. Moreover, high glucose levels strongly potentiated the reduced utilization of Cer for SM biosynthesis induced by 0.4 mM palmitate, and also significantly reduced GSL biosynthesis (Fig. 1c). The percent increase of [³H]Cer was 23% and 14%, respectively in 5 mM glucose plus palmitate, and in 30 mM glucose treated cells; in cells treated with 30 mM glucose plus palmitate [³H]Cer was 54% higher (Fig. 1c) than in cells treated with 5 mM glucose. Conversely, the reduction in [³H]SM was 32% and 23% in cells treated separately with 0.4 mM palmitate and 30 mM glucose, respectively, and 64% in glucolipotoxic conditions, demonstrating that co-treatment with 30 mM glucose and palmitate has a greater effect than the sum of the individual treatments (palmitate or 30 mM glucose) (Fig. 1c). Taken together, these results showed that in pancreatic β-cells, glucolipotoxicity can regulate the use of Cer for the biosynthesis of complex sphingolipids in the Golgi apparatus.

Effect of palmitate and glucose on ceramide, sphingomyelin and glucosylceramide molecular species

Next, we evaluated the effect of palmitate treatment on Cer, SM and GlcCer mass levels, and the levels of their metabolic molecular species by LC/MS/MS. Our data showed that palmitate, in the presence of 5 mM glucose as well as 30 mM glucose alone, did not significantly alter Cer, SM and GlcCer mass levels in INS-1 cells

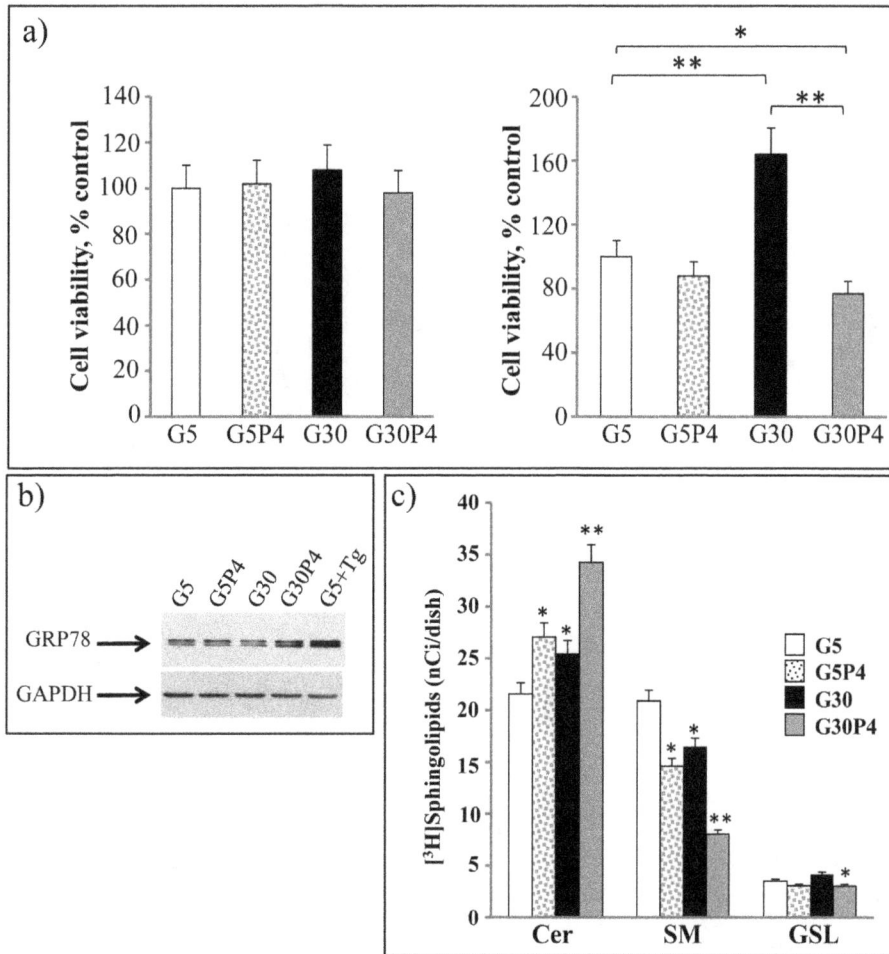

Figure 1. Palmitate and glucose regulate the use of Cer for the biosynthesis of complex sphingolipids in INS-1 cells. a) Cells were treated for 12 h (left panel) or 24 h (right panel) with 0.4 mM palmitate (P4) or without palmitate in the presence of 5 mM or 30 mM glucose. Cell viability was assessed by the MTT assay. Results are expressed as percentage of cell viability with respect to 5 mM glucose-treated cells (100%). Data are the mean ± S.D. of three independent experiments. *, $p<0.05$; ** p, <0.01. b) INS-1 cells were treated with 5 mM or with 30 mM glucose ±0.4 mM palmitate and harvested in lysis buffer for immunoblot analysis of GRP78 and GAPDH levels as described in experimental procedures. INS-1 cells were pretreated 30 min ±0.1 μM thapsigargin (Tg). Equal amounts of protein from homogenates were analyzed by immunoblotting with an anti-GRP78 antibody and an anti-GAPDH antibody. c) Cells were treated for 12 h ±0.4 mM palmitate in the presence of 5 mM or 30 mM glucose and then pulsed with 0.3 μCi/ml [C3-^{3}H]sphingosine for 1 h. At the end of pulse, cells were harvested and submitted to lipid extraction and partitioning. The methanolized organic phase and the aqueous phase were analyzed by HPTLC and digital autoradiography of HPTLC (see experimental procedures). G5, 5 mM glucose; G5P4, 5 mM glucose+0.4 mM palmitate; G30, 30 mM glucose; G30P4, 30 mM glucose+0.4 mM palmitate. Data are the mean ± S.D. of at least three independent experiments. *p<0.05 for Ceramide and Sphingomyelin G5P4 or G30 compared with G5 and for GSLs G30P4 compared with G30; **p <0.01 for Cer and SM G30P4 compared with G30.

(Fig. 2); In contrast, palmitate with high glucose levels (30 mM) promoted an increase in Cer mass levels with a concomitant decrease in the mass levels of SM but not that of GlcCer. Our data show that glucolipotoxic conditions led to an increase in saturated ceramides, the greatest increase being observed in C18:0-Cer and C22:0-Cer (Fig. 2). In addition, this condition mostly reduced the levels of saturated SM and, more specifically, of C18:0-SM and C24:0-SM (Fig. 2). In contrast, glucolipotoxicity did not appear to alter significantly the levels of the different GlcCer molecular species (Fig. 2). We also evaluated the effect of palmitate treatment on S1P levels in INS-1 cells. Our data showed that at low levels of glucose (5 mM), palmitate was unable to increase significantly S1P levels in INS-1 cells. However, after a 12 h of treatment, palmitate in the presence of 30 mM glucose promoted an increase in S1P levels (Fig. 2) in agreement with the findings of Verét et al. [35].

Altogether, these data suggest that glucolipotoxicity induced an increase in Cer levels and a reduction of SM levels in pancreatic β-cells, but did not appear to significantly affect the amounts of GlcCer.

Effect of palmitate and glucose on Sphingomyelin Synthase activity

We then evaluated if the decrease in SM levels induced by glucolipotoxicity is due to the inhibition of SM synthase (SMS) activity using NBD-C$_6$-Cer, a fluorescently labeled ceramide that is an efficient substrate for SM synthases. Treatment with palmitate in the presence of different glucose concentrations was unable to alter the activity of SMS in INS-1 cells (data not shown), suggesting that the activities of enzymes responsible for the

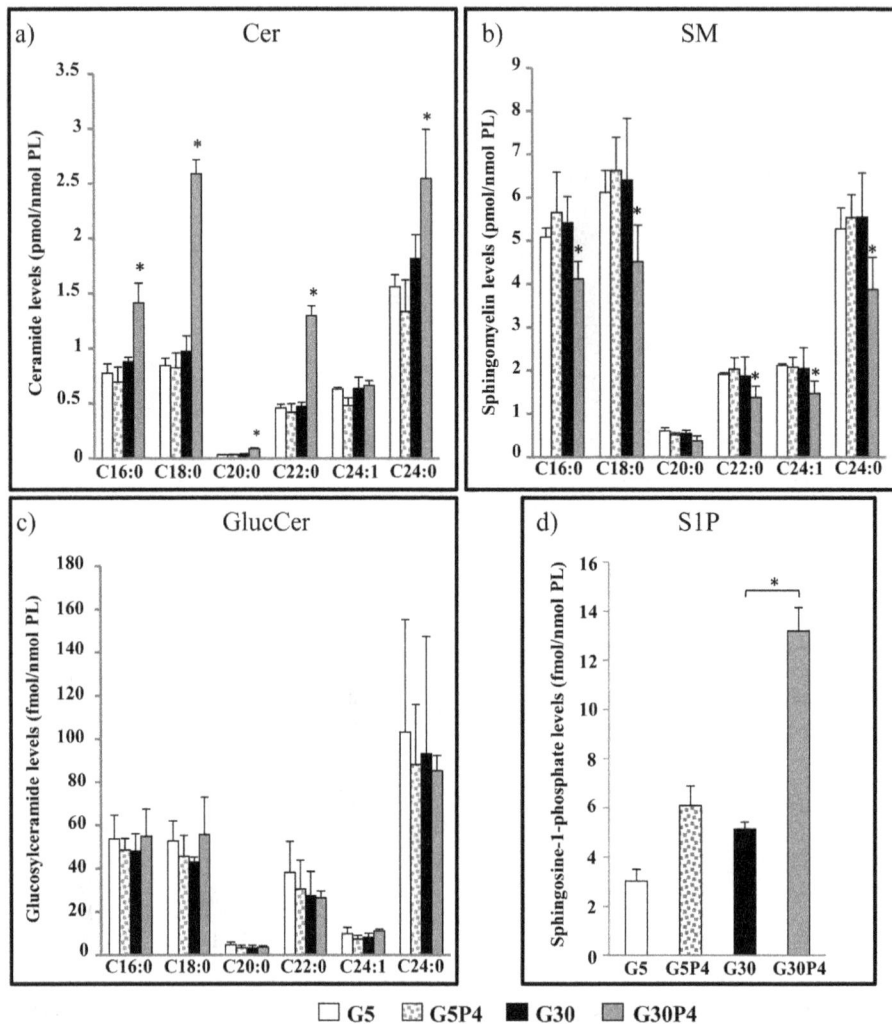

Figure 2. Chain-length specificity of ceramide, sphingomyelin and glucosylceramide in response to palmitate and high concentrations of glucose in INS-1 cells. Cells were incubated with 0.4 mM palmitate in the presence of 5 mM (G5) or 30 mM (G30) glucose for 12 h. Levels of N-acyl chain lengths of Cer, SM and GlcCer were determined by LC–MS/MS. Levels of S1P in INS-1 cells were also determined by LC-MS/MS measurement. Results are expressed as pmol/nmol of phospholipids (PL) for Cer and SM and as fmol/nmol PL for GlcCer and S1P and are means ± S.D. for three independent experiments. *$p < 0.05$ vs G5 except for S1P *$p < 0.05$ vs G30.

biosynthesis of SM were not affected by glucolipotoxic conditions in INS-1 cells.

Effect of palmitate and glucose on intracellular distribution of BODIPY-C_5Cer and NBD-C_6Cer

We then tested the possibility that glucolipotoxicity inhibited the synthesis of complex sphingolipids, mainly represented by SM, by affecting the transport of Cer synthesized in the ER to the Golgi apparatus (where SM and GSL biosynthesis occurs). The transport of natural Cer from the ER to the Golgi apparatus can be qualitatively evaluated from the analysis of BODIPY-C_5-Cer redistribution in cells [34]. In 5 mM and 30 mM glucose-treated INS-1 cells, most of the fluorescence accumulated in the perinuclear region (Fig. 3a), which is representative of the Golgi apparatus. In INS-1 cells treated with 0.4 mM palmitate together with 5 mM glucose, fluorescence was also observed in the perinuclear region but to a lesser extent compared to control cells suggesting a partial defect in Cer traffic. (Fig. 3a). In contrast, co-

administration of 0.4 mM palmitate and 30 mM glucose strongly reduced fluorescence accumulation in the Golgi apparatus region (Fig. 3a), suggesting an impairment of ceramide flow from the ER to the Golgi apparatus as a result of glucolipotoxicity in pancreatic β-cells. In INS-1 cells, the presence of thapsigargin (Tg) which induced ER stress in β-cells (Fig. 1b) [14] mimics the effect of high glucose together with 0.4 mM palmitate by strongly reducing the fluorescence accumulation in the Golgi apparatus region (Fig. 3a). The effect of thapsigargin was not affected by the presence of glucose or palmitate. In contrast, when cells were labeled with NBD-C_6Cer, which selectively localizes at the Golgi apparatus [34], 5 mM and 30 mM glucose in the presence or absence of 0.4 mM palmitate with or without Tg did not modify the accumulation of NBD fluorescence in the perinuclear Golgi region (Fig. 3b). Altogether, these results suggest that glucolipotoxicity induce an impairment of ceramide flow from the ER to the Golgi apparatus in pancreatic β-cells.

Effect of palmitate and glucose on CERT levels and activation

A recent report suggests that inhibition of CERT-mediated Cer transport can exacerbate repression of pro-insulin gene expression induced by long term treatment (48 h) with palmitate in INS-1 cells [36]. On this basis, we evaluated if a shorter treatment (12 h) of INS-1 cells with palmitate and glucose can affect the levels and the activation status of the protein CERT. Western blot analysis showed that both 0.4 mM palmitate or 30 mM glucose alone did not significantly alter the total amount of the CERT protein (Fig. 4a and b) whereas co-treatment with 0.4 mM palmitate and 30 mM glucose induced a 65% reduction in the total amount of CERT in INS-1 cells (Fig. 4a and b). Moreover, using an antibody against phospho-serine according to Guo and co-workers [36], we found that palmitate or 30 mM glucose alone did not modify significantly the amount of phospho-CERT. Interestingly, co-administration of 0.4 mM palmitate and 30 mM glucose doubled the levels of phospho-CERT in INS-1 cells (Fig. 4a and c). In

order to assess how palmitate plus high glucose regulates CERT levels, CERT expression was evaluated by quantitative PCR in INS-1 cells. The results obtained (Fig. 4d) show that palmitate in the presence of 30 mM glucose induced a 80% reduction in steady-state levels of CERT transcripts, thus demonstrating that glucolipotoxicity regulates CERT expression by inhibiting its synthesis.

Effect of palmitate and glucose on CERT subcellular localization

The phosphorylated form of CERT should have the PH domain, the docking site for the Golgi apparatus, covered by the START domain and should therefore not be able to colocalize with the Golgi apparatus [37]. With this premise, we evaluated the ability of CERT to localize at the Golgi apparatus in INS-1 cells as a measure of CERT activity. To this purpose, we analysed the co-localization of over-expressed CERT-GFP with the Golgi marker WGA in INS-1 cells. The images obtained (Fig. 5a) showed that

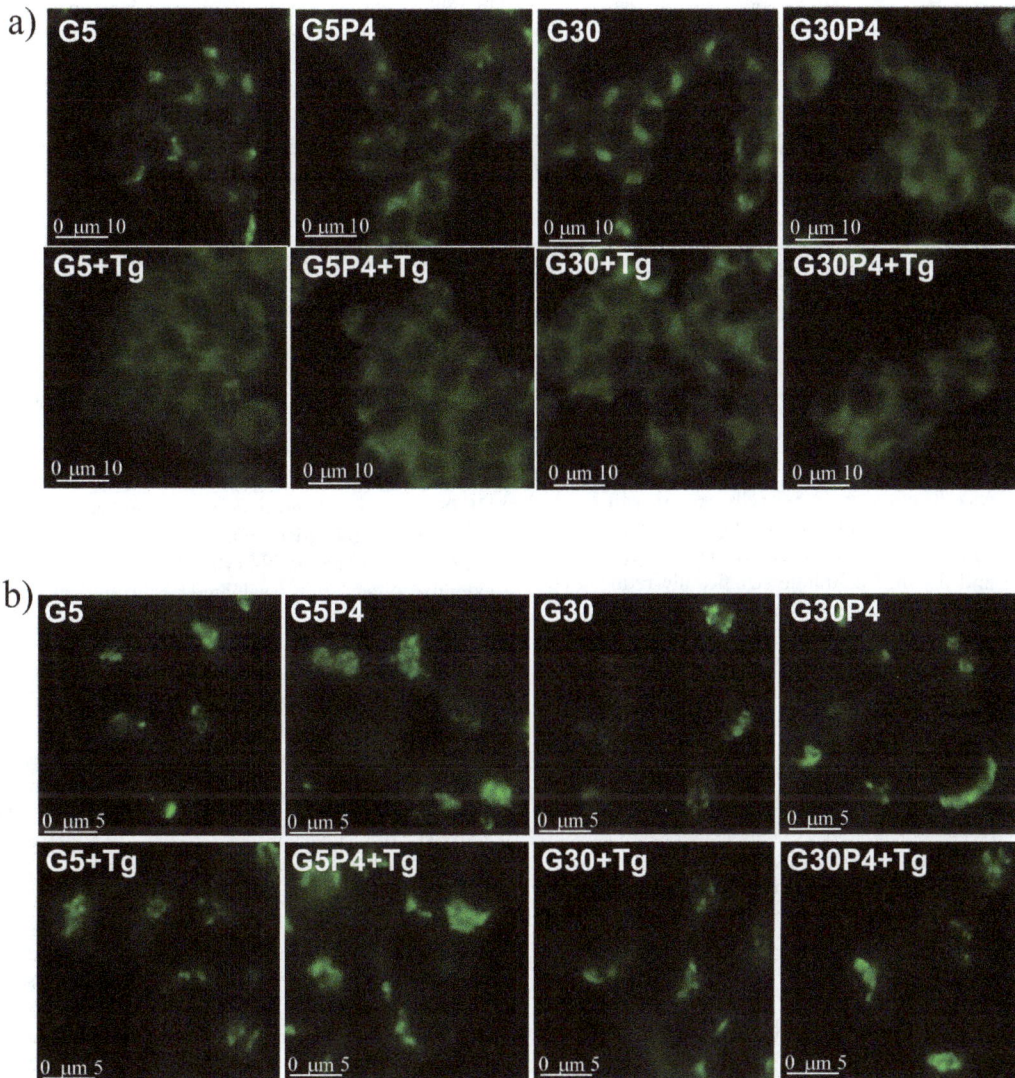

Figure 3. Palmitate and glucose impairs ceramide flow from the ER to the Golgi apparatus in INS-1 cells. INS-1 cells grown on a glass coverslip were pretreated 30 min ± 0.1 μM thapsigargin. At the end of the pretreatment, the cells were treated with or without palmitate in the presence of 5 mM or 30 mM glucose for 12 h and then incubated with a) 2.5 μM BODIPY-C$_5$Cer or b) 2.5 μM NBD-C$_6$Cer as BSA complex 1:1 (m/m) in DMEM for 30 min at 4°C; labeled cells were incubated at 37°C for 30 min and analyzed. All images were processed and printed identically.

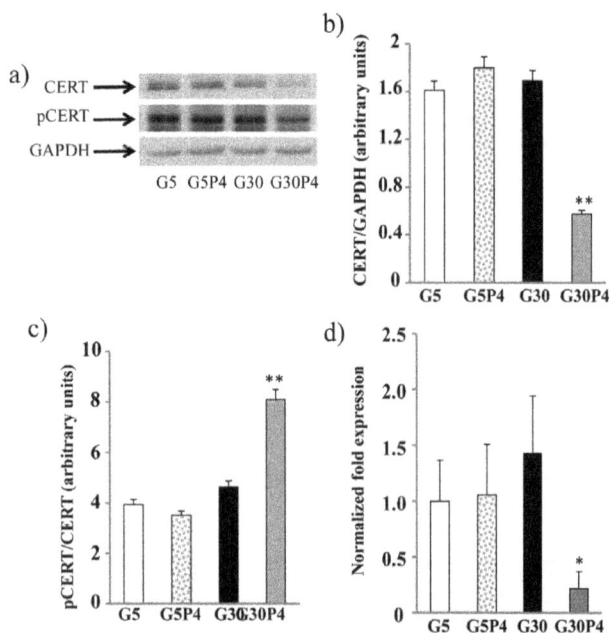

Figure 4. Palmitate and glucose regulate CERT expression and activation in INS-1 cells. a) INS-1 cells were harvested in lysis buffer as described in material and methods. Equal amounts of protein from homogenates were analyzed by immunoblotting with an anti-CERT antibody, an anti-phosphoserine and an anti-GAPDH antibody; b) the amount of CERT expressed was determined by densitometric quantitation and normalized for GAPDH **p<0.01 for G30+palmitate compared with G30; c) the amount of pCERT expressed was determined by densitometric quantitation and normalized for CERT **p<0.01 for G30+palmitate compared with G30; d) Relative expression of CERT assessed by Real-Time PCR. Results are expressed as fold-change relative to G5 *p<0.05 G30+palmitate cells vs G30. Values are mean ± SD of three independent experiments.

CERT and WGA co-localized in INS-1 cells treated with 5 mM glucose in the presence or the absence of 0.4 mM palmitate, or treated with 30 mM glucose. In contrast, co-treatment with 30 mM glucose and 0.4 mM palmitate significantly reduced co-localization between CERT and WGA in INS-1 cells (Fig. 5a) according to the Pearson's colocalization coefficients (Fig. 5b). Moreover when INS-1 cells were treated with 5 mM glucose in the presence or absence of 0.4 mM palmitate, or treated with 30 mM glucose alone, CERT and WGA co-localize in more than 90% of the cells analysed (Fig. 5c). In contrast, in cells treated with 30 mM glucose in the presence of 0.4 mM palmitate, co-localization of CERT and WGA was detectable in only 8% of the cells (Fig. 5c). Altogether, our results (Fig. 4 and 5) demonstrated that glucolipotoxicity can impair CERT-mediated Cer transport by reducing both CERT levels and residual CERT activity through its phosphorylation, a condition that reduces the capacity of CERT to bind ceramide and prevent its localization to the Golgi apparatus.

Effect of palmitate and glucose on [³H]Sph metabolism in INS-1 cells silenced for CERT

On the basis of the results obtained, we next evaluated if CERT silencing could mimic the defect in Cer utilization induced by glucolipotoxicity. We examined the effects of palmitate with 30 mM glucose on [³H]Sph metabolism in control and CERT-down-regulated INS-1 cells. We set up optimal conditions to silence CERT and showed in the western blots that a ≥90%

reduction in CERT expression was achieved when cells were transfected with a mixture of S87 and S522 (1:1 ratio) (Fig. 6a). As expected, down regulation of CERT promoted a significant but not complete reduction of Cer conversion to SM without modifying the amount of synthesized GSLs (Fig. 6b). Palmitate together with 30 mM glucose decreased Cer utilization for the synthesis of SM and GSLs in control cells (siCT) but also in siCERT ones (Fig. 6b). In particular, our results demonstrated that glucolipotoxicity induced a 52% decrease in [³H]SM in siCT cells and 65% in siCERT cells compared to cells treated with 30 mM glucose in the absence of palmitate. These results suggest that glucolipotoxicity, in addition to its effect on CERT, can also affect vesicular-mediated Cer transport in pancreatic β-cells.

Effect of PI3K/Akt on [³H]Sph metabolism in INS-1 cells treated with palmitate and glucose

We previously demonstrated [38] that the PI3K/Akt pathway regulates the vesicular traffic of Cer in glioma cells, and we were interested in the present study to examine if this was also the case in INS-1 cells. We evaluated if a 12 h treatment with palmitate and high glucose levels was able to regulate the PI3K/Akt pathway in INS-1 cells. Western blot analysis with an antibody specific for Akt phosphorylated at Ser473, demonstrated that 30 mM glucose did not modify pAkt levels but 0.4 mM palmitate alone decreased pAkt levels and 0.4 mM palmitate together with 30 mM glucose induced a marked decrease of pAkt levels in INS-1 (Fig. 7a).

To investigate the effect of PI3K/Akt on Cer metabolism in INS-1 cells, we utilized LY294002 and Wm as pharmacological inhibitors of PI3K. We initially set up working concentrations to specifically inhibit PI3K. To do this, we evaluated the capacity of LY294002 and Wm to inhibit the PI3K/Akt pathway. Immunoblot analysis demonstrated that both 20 μM LY294002 and 10 nM Wm strongly reduced Akt activation (data not shown). INS-1 cells, treated with 5 mM or 30 mM glucose with or without palmitate, were pulsed with [³H]Sph in the presence or absence of LY294002 or Wm, and these last two molecules did not alter the [³H]Sph uptake, determined as the radioactivity measured in the total lipid extract (data not shown). In all cases, [³H]Sph was mainly metabolized to N-acylated compounds (Fig. 7b–e) and the extent of N-acylation (evaluated as the sum of tritiated Cer, SM and GSLs) was always very similar in the control and treated cells. However treatment with LY294002 or Wm strongly modified the radioactivity distribution among the different Sph metabolites both in the presence of 5 mM or 30 mM glucose (Fig. 7b and 7c). In 5 mM glucose+LY294002 or 5 mM glucose+Wm treated INS-1 cells, the radioactivity associated with [³H]Cer was 21% and 17% higher, respectively than that in 5 mM treated cells, with a concomitant reduction of both [³H]SM and [³H]GSL levels (39% and 29%, respectively in the presence of LY294002 and 32% and 27% respectively in the presence of Wm compared to 5 mM treated cells) (Fig. 7b). Similarly, in cells treated with 30 mM glucose, LY294002 and Wm promoted about 30% reductions in synthesized SM and 20% decrease in GSL compared to 30 mM glucose treated cells (Fig. 7c). Taken together these results suggest that the PI3K/Akt pathway can regulate the metabolic utilization of Cer through the modulation of the vesicular traffic of Cer, favouring the maintenance of low Cer levels in INS-1 cells even under conditions (high glucose) of reduced utilization of Cer for the biosynthesis of complex sphingolipids. In cells treated with 5 mM glucose+palmitate, LY294002 or Wm slightly reduced the amount of synthesized SM and GSL (Fig. 7d). In cells treated with high glucose in the presence of palmitate (characterized by a reduced utilization of Cer for the biosynthesis of complex

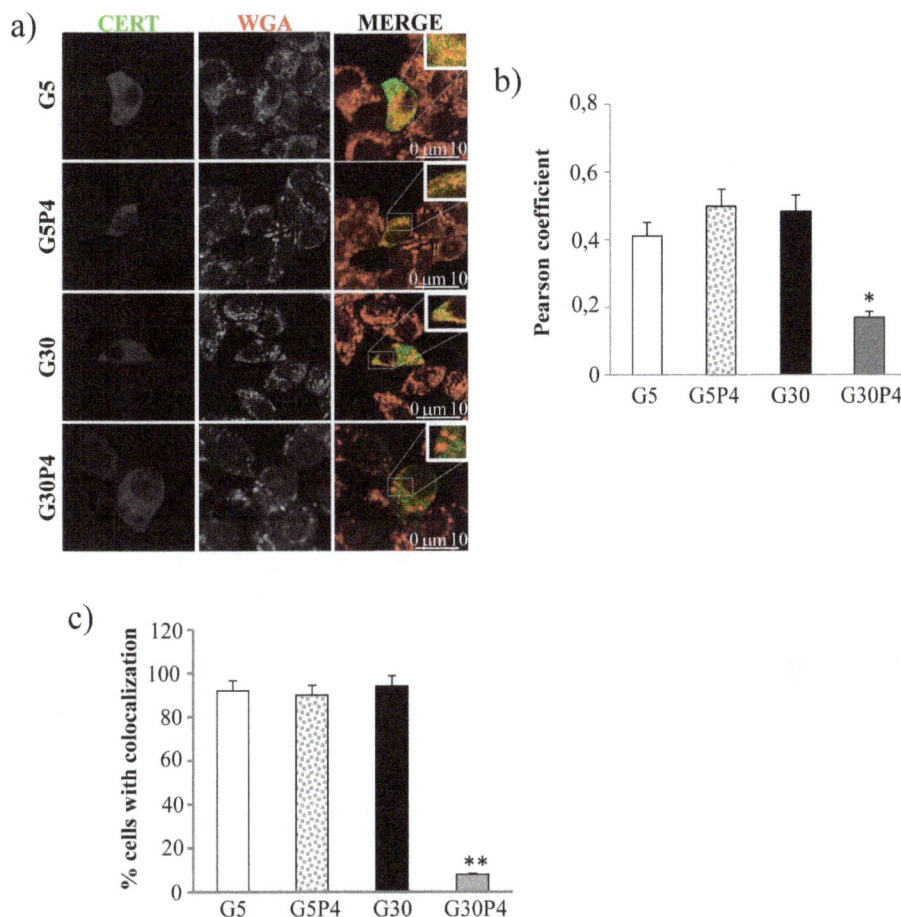

Figure 5. Palmitate and glucose prevent colocalization of CERT and Golgi apparatus in INS-1 cells. INS-1 cells grown on a glass coverslip were transfected with the plasmid CERT-GFP as described in experimental procedures. 24 h later, the cells were treated with or without palmitate in the presence of 5 mM or 30 mM glucose for 12 h. Cells were then fixed and immunostained with WGA texas red-conjugated, a specific marker for the Golgi apparatus. a) Representative confocal microscopy images are shown; all images were processed and printed identically. b) The co-localization between CERT and WGA has been quantified through the Image J software and reported as Pearson colocalization coefficient. *p<0.05 G30+palmitate cells vs G30. c) The percentage of cells with co-localization of CERT and WGA was determined. The data are means ± the SD. **p<0.01 for G30+palmitate compared with G30.

sphingolipids as shown in Fig. 1), LY294002 or Wm did not further reduce the amount of synthesized SM and GSL (Fig. 7e). Altogether, these data suggest that palmitate inhibits vesicular-mediated Cer traffic through down-regulation of the PI3K/Akt pathway, and this effect is potentiated in glucolipotoxic conditions, thereby contributing to the accumulation of Cer in pancreatic β-cells.

Discussion

The most relevant result obtained by this investigation is that in pancreatic β-cells, glucolipotoxicity impairs Cer traffic from the ER to the Golgi apparatus, thus promoting the accumulation of Cer in the ER. As a model for mimicking glucolipotoxicity, we incubated INS-1 cells with 0.4 mM palmitate in the presence of high glucose concentrations (30 mM). This condition resulted in a 60% reduction in cell viability after a 24 h treatment. Studies using [3H]Sph as a metabolic precursor showed that glucolipotoxicity strongly reduced the utilization of newly synthesized Cer mainly for the biosynthesis of SM and, to a lesser extent, for synthesis of GlcCer and complex GSLs. This occurred after a 12 h treatment, when all INS-1 cells are still viable but show a

significant ER stress response (Fig. 1b). It is noteworthy that palmitate-induced reduction of Cer utilization for complex sphingolipid biosynthesis was strongly potentiated by high glucose levels. As a consequence, the observation of increased Cer levels is in agreement with previously published data [11,35]. We show for the first time that this increase was associated with a significant reduction of total SM but not of GlcCer levels in INS-1 cells. Glucolipotoxic conditions does not significantly modify GlcCer mass level but this does not exclude that the amount of complex GSLs could be reduced. Moreover we cannot exclude that glucosylceramide synthase is increased by palmitate as Boslem and co-workers [20] have shown that palmitate increase GlcCer in MIN6 β-cells. Interestingly, glucolipotoxicity lowered only saturated SM species in β-cells, favouring the accumulation of saturated Cer species such as C18:0-Cer, which are pro-apoptotic in β-cells [12]. The decrease in SM biosynthesis suggests that glucolipotoxicity could affect SM synthase activity and/or the availability of its substrate by inhibiting Cer traffic from the ER to the Golgi. However, the capacity of INS-1 β-cells to synthesize SM from a diffusible substrate such as NBD-Cer is maintained, indicating that SM synthase activity is preserved during glucolipotoxic conditions. The analysis of intracellular distribution of

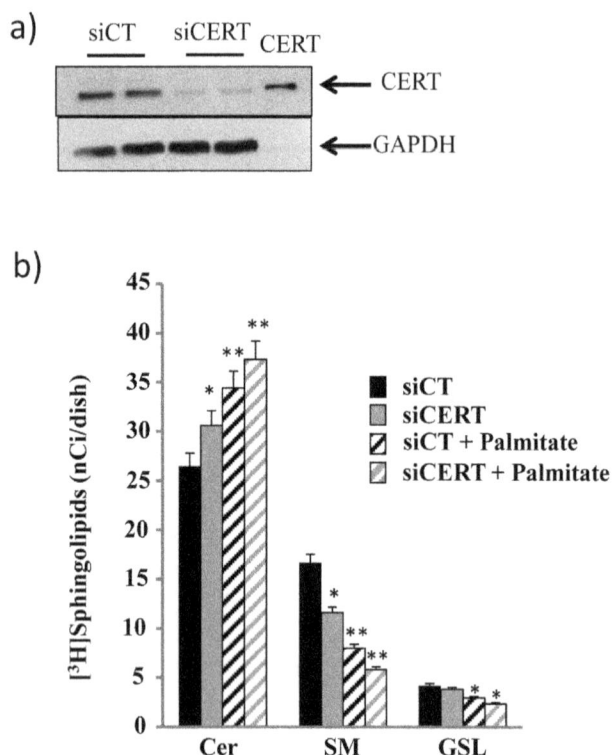

Figure 6. Palmitate and glucose affect vesicular-mediated Cer transport. a) Cells were transfected with a mix of S87 and S522 siRNA for CERT (siCERT) and the corresponding non-targeting corresponding sequences as control (siCT) and harvested in lysis buffer 72 h after transfection. 40 μg of protein from homogenate fractions and 2.4 ng of recombinant CERT (CERT) were analyzed by immunoblotting with polyclonal antibody anti-CERT and monoclonal anti-GAPDH. b) INS-1 cells down-regulated for CERT were treated for 12 h with or without palmitate in the presence of 30 mM glucose. Then the cells were pulsed with 0.3 μCi/ml [C3-^{3}H]sphingosine for 1 h and processed and analyzed as described in the legend of Fig. 1. Data are mean ± S.D. of at least three independent experiments. *p<0.05 for siCERT compared with siCT and in GSL for siCT+palmitate vs siCT and siCERT+palmitate vs siCERT; **p<0.01 for siCT+palmitate compared with siCT and for siCERT+palmitate compared with siCERT.

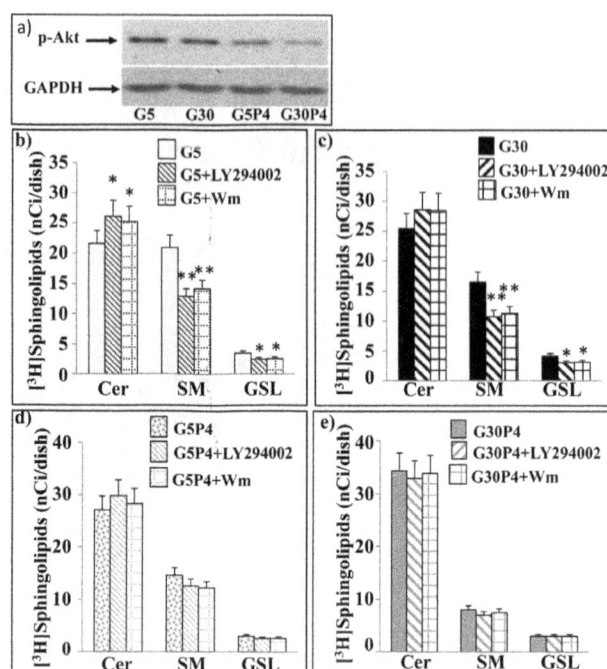

Figure 7. Palmitate and glucose inhibit vesicular-mediated Cer traffic through downregulation of PI3K/Akt pathway. a) INS-1 cells were treated with 5 mM or with 30 mM glucose ±0.4 mM palmitate and harvested for immunoblot analysis of phospho-Akt and GAPDH levels as described in experimental procedures. INS-1 cells were pretreated 30 min with or without 20 μM LY294002. The cells were then **b)** treated for 12 h with 5 mM glucose in the presence or absence of LY294002 or of Wm; **c)** treated for 12 h with 30 mM glucose in the presence or absence of LY294002 or Wm; **d)** treated for 12 h with 5 mM glucose plus 0.4 mM palmitate in the presence or absence of LY294002 or Wm; **e)** treated for 12 h with 30 mM glucose plus 0.4 mM palmitate in the presence or absence of LY294002 or Wm. Then cells were pulsed 1 h with [^{3}H]Sph in the absence (*opened and dotted bars*) or presence of 20 μM LY294002 (*striped bars*) or 10 nM Wm (*square bars*). At the end of pulse, cells were harvested and submitted to lipid extraction and analyzed as described in the legend of Fig. 1. All values are the mean ± S.D. of at least three individual experiments. *p<0.05 for Cer and GSLs G5+LY294002 and G5+Wm compared with G5 and for GSLs G30+LY294002 and G30+Wm compared with G30; **p<0.01 for SM G5+LY294002 and G5+Wm compared with G5 and G30+LY294002 and G30+Wm compared with G30.

BODIPY-C$_5$-Cer, which mimics the intracellular movements of naturally occuring Cer [26,34,39], provided evidence that glucolipotoxicity impairs the intracellular traffic of Cer from ER to the Golgi apparatus in INS-1 β-cells, supporting the idea that, under glucolipotoxic conditions, this mechanism can contribute to the accumulation of Cer at the ER. Recent studies have shown that specific accumulation of Cer at the ER due to palmitate can induce ER stress and apoptosis of pancreatic β-cells [14,19,20,34]. This phenomenon could be alleviated by over-expression of glucosylceramide synthase in β-cells, an enzyme which transforms ceramide into glucosylceramide, thereby preventing excessive accumulation of ceramide in the ER [20]. Importantly, our results enforce the crucial role of Cer accumulation at the ER in induced apoptosis and the loss of β-cell mass that contributes to the development of type II diabetes. Moreover, our results confirmed that glucolipotoxicity increased total S1P levels in INS-1 cells in agreement with Véret and co-workers [35] who showed that palmitate treatment not only induce anti-apoptotic signals but also pro-apoptotic signals such as the production of S1P. Evidence from published literature [29] demonstrated that the decrease in S1P levels is associated with Cer traffic impairment, leading us to

hypothesize that increased S1P levels are unlikely to be involved in the initiation of defective ceramide transport. Additionally, the increased in S1P levels are also unlikely to be sufficient to restore normal ceramide trafficking in order to avoid ceramide accumulation to counteract the effect of glucolipotoxicity.

Two main mechanisms are involved in ceramide transport from the ER to the Golgi apparatus: a protein-mediated transport that is mediated by CERT [23,24], and a CERT-independent vesicular transport pathway [23,26]; our data demonstrate that both of these pathways are strongly inhibited by glucolipotoxicity. In relation to CERT, a 12 h treatment of β-cells with glucolipotoxic conditions but not the treatment with high glucose and palmitate separately, induced a significant decrease in the total amount of the CERT protein and steady-state transcript levels, suggesting a reduced rate of CERT protein synthesis in glucolipotoxic conditions. In this respect, Granero et al. identified a human-specific TNFα-responsive promoter for CERT [40] as a possible mechanism of transcriptional CERT regulation. However, we cannot exclude the possibility that, during glucolipotoxic

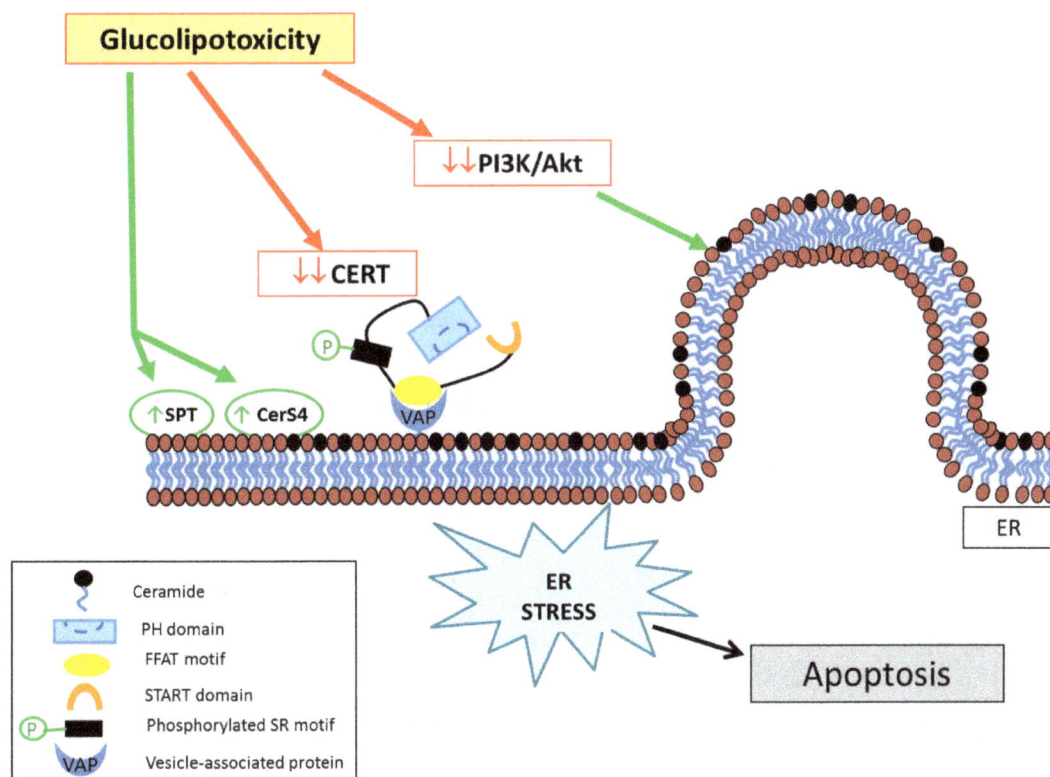

Figure 8. Schematic representation of the model showing the involvement of ceramide traffic in ER stress induced by glucolipotoxicity. Glucolipotoxicity impairs CERT- and vesicular-mediated Cer traffic. Glucolipotoxicity decrease the amount of active CERT significantly decreasing a) the total amount of the protein and b) the phosphorylation of CERT SR motif that is no longer able to localize at the Golgi apparatus. Moreover glucolipotoxicity inhibits PI3K/Akt pathway that could in turn impairs vesicular trafficking of Cer from the ER to the Golgi apparatus. Both transport systems contribute to the accumulation of Cer at the ER, thereby inducing ER stress. Furthermore ceramide synthase 4 (CerS4) [12] and serine palmitoyltransferase (SPT) [16,17], both residing in the endoplasmic reticulum (ER), have been shown to be involved in regulating Cer levels in β-cells in response to lipotoxicity and/or glucolipotoxicity.

stress, CERT could also be cleaved by caspase-3, an enzyme highly activated by glucolipotoxicity in β-cells [12] as has been proposed in Hela cells under pro-apoptotic stress [41]. Moreover we also found that, under glucolipotoxic conditions, residual CERT protein is highly phosphorylated and is no longer able to localize at the Golgi apparatus in INS-1 cells. CERT specifically targets the Golgi apparatus through its PH domain, which selectively recognizes phosphatidyl-inositol-4-phosphate (PI-4P) at the Golgi, and its hyperphosphorylation results in the repression of PI4P binding activity of the PH domain [37]. Palmitate or glucose alone did not modify either phosphorylation or CERT localization. However, a recent paper [36] demonstrated that short term exposure (3h) to palmitate increased phosphorylation of CERT in INS-1 cells and this was associated with the inhibition of insulin gene expression. This appears to suggest that acute and chronic exposure to palmitate might differentially affect CERT activity. Overall, glucolipotoxicity strongly decreases the amount of active CERT and the amount of remaining CERT is mostly inactive. Altogether, these data demonstrate that glucolipotoxicity impairs the CERT-mediated Cer traffic from the ER to the Golgi apparatus, thus supporting the idea that, under glucolipotoxic conditions, this mechanism contributes to the accumulation of Cer at the ER. Therefore, CERT is important not only in the regulation of the insulin gene expression [36], but also in the predisposition to β-cell death.

The vesicle-mediated pathway of Cer with the transport of a thousand lipid molecules per transfer step could provide a feasible mechanism to compensate for CERT defect in the bulk of SM biosynthesis. We demonstrated that in INS-1 cells where expression of CERT is silenced (a condition able to mimic the effect of glucolipotoxicity on CERT), glucolipotoxic conditions are still able to further decrease the metabolic utilization of Cer for the synthesis of SM and GSLs, lending credence that vesicular traffic of Cer is impaired. Additionally, we observed in INS-1 cells that the PI3K/Akt pathway regulates Cer metabolism by controlling the vesicular transport of Cer between the ER and the Golgi apparatus, similar to glioma cells [38]. Glucolipotoxic conditions promote a strong inhibition of the PI3K/Akt pathway, consistent with previous studies showing that long-term exposure to glucolipotoxic conditions and/or high palmitate decreases the PI3K/Akt pathway in pancreatic β-cells [42–44]. Our experiments also showed that palmitate alone partially decreased pAkt levels and this was associated with the inhibition of Cer vesicular traffic. The role of the PI3K/Akt pathway in mediating the effects of palmitate and glucolipotoxicity on Cer vesicular flow is confirmed by the evidence that the PI3K inhibitors, LY294002 and Wm, which are not effective in glucolipotoxic conditions, have a slight effect on Cer flow in cells treated with palmitate in the presence of 5 mM glucose, probably because palmitate alone only partially inhibits pAkt. It is worth noting that the reduced Cer utilization is higher than the simple sum of the effects induced by the two single nutrients separately. The results of our study support the idea that glucolipotoxicity inhibits the PI3K/Akt pathway in pancreatic β-cells, which in turn inhibits the vesicular trafficking of

Cer. On the other hand, 30 mM glucose decreased the biosynthesis of SM but does not modify pAkt levels and Cer traffic suggesting that this effect on SM could be associated with increased N-SMase activity, consistent with previous papers [18,19] demonstrating that high glucose levels increased N-SMase activity.

Altogether, our data demonstrate that the CERT- and vesicular-mediated Cer trafficking pathways can separately contribute to the control of sphingolipid metabolism and Cer levels in INS-1 cells, thus participating in regulating the accumulation of ER-associated Cer involved in the regulation of pancreatic β-cell function and death during type II diabetes [14,20]. In support to this new idea, Guo and co-workers recently showed that down-regulation of CERT by specific siRNA potentiated palmitate-induced inhibition of insulin gene expression in pancreatic β-cells [36]. Moreover, these findings suggest that an impairment in ER to Golgi Cer traffic can act synergistically, leading to enhanced *de novo* Cer biosynthesis [12,16], resulting in accumulation of Cer in the ER in response to glucolipotoxicity (Fig. 8).

Further understanding of the mechanisms that regulate the accumulation of Cer at the ER will be important for developing new strategies to prevent type II diabetes. Moreover, the capacity of the PI3K/Akt pathway to regulate sphingolipid metabolism may also be pathologically relevant in β-cells if we consider that the PI3K/Akt pathway plays a crucial role in the control of β-cell mass and function by modulating a dynamic balance of proliferation, cell size and apoptosis [45].

Acknowledgments

We thank Dr. Maria Antonietta De Matteis, for the CERT-GFP plasmid, and Dr. Suhas Shinde for PL analysis.

Author Contributions

Conceived and designed the experiments: PG HLS CK-YN. Performed the experiments: PG HLS CK-YN EG AC NC MNI CV LB. Analyzed the data: PG HLS CK-YN MNI CM PV LR LB. Wrote the paper: PG HLS CK-YN PV.

References

1. Weir GC, Laybutt DR, Kaneto H, Bonner-Weir S, Sharma A (2001) Beta-cell adaptation and decompensation during the progression of diabetes. Diabetes 50 Suppl 1: S154–159.
2. Prentki M, Joly E, El-Assaad W, Roduit R (2002) Malonyl-CoA signaling, lipid partitioning, and glucolipotoxicity: role in beta-cell adaptation and failure in the etiology of diabetes. Diabetes 51 Suppl 3: S405–413.
3. Véret J, Bellini L, Giussani P, Ng C, Magnan C, et al. (2014) Roles of Sphingolipid Metabolism in Pancreatic β Cell Dysfunction Induced by Lipotoxicity. *J Clin Med* 3: 646–662.
4. Poitout V, Amyot J, Semache M, Zarrouki B, Hagman D, et al. (2010) Glucolipotoxicity of the pancreatic beta cell. Biochim Biophys Acta 1801: 289–298.
5. Briaud I, Harmon JS, Kelpe CL, Segu VB, Poitout V (2001) Lipotoxicity of the pancreatic beta-cell is associated with glucose-dependent esterification of fatty acids into neutral lipids. Diabetes 50: 315–321.
6. Richieri GV, Kleinfeld AM (1995) Unbound free fatty acid levels in human serum. J Lipid Res 36: 229–240.
7. Karaskov E, Scott C, Zhang L, Teodoro T, Ravazzola M, et al. (2006) Chronic palmitate but not oleate exposure induces endoplasmic reticulum stress, which may contribute to INS-1 pancreatic beta-cell apoptosis. Endocrinology 147: 3398–3407.
8. Laybutt DR, Preston AM, Akerfeldt MC, Kench JG, Busch AK, et al. (2007) Endoplasmic reticulum stress contributes to beta cell apoptosis in type 2 diabetes. Diabetologia 50: 752–763.
9. Cunha DA, Hekerman P, Ladriere L, Bazarra-Castro A, Ortis F, et al. (2008) Initiation and execution of lipotoxic ER stress in pancreatic beta-cells. J Cell Sci 121: 2308–2318.
10. Pinget M, Boullu-Sanchis S (2002) [Physiological basis of insulin secretion abnormalities]. Diabetes Metab 28: 4S21–32.
11. El-Assaad W, Buteau J, Peyot ML, Nolan C, Roduit R, et al. (2003) Saturated fatty acids synergize with elevated glucose to cause pancreatic beta-cell death. Endocrinology 144: 4154–4163.
12. Veret J, Coant N, Berdyshev EV, Skobeleva A, Therville N, et al. (2011) Ceramide synthase 4 and de novo production of ceramides with specific N-acyl chain lengths are involved in glucolipotoxicity-induced apoptosis of INS-1 beta-cells. Biochem J 438: 177–189.
13. Eizirik DL, Cardozo AK, Cnop M (2008) The role for endoplasmic reticulum stress in diabetes mellitus. Endocr Rev 29: 42–61.
14. Lang F, Ullrich S, Gulbins E (2011) Ceramide formation as a target in beta-cell survival and function. Expert Opin Ther Targets 15: 1061–1071.
15. Back SH, Kaufman RJ (2012) Endoplasmic reticulum stress and type 2 diabetes. Annu Rev Biochem 81: 767–793.
16. Shimabukuro M, Higa M, Zhou YT, Wang MY, Newgard CB, et al. (1998) Lipoapoptosis in beta-cells of obese prediabetic fa/fa rats. Role of serine palmitoyltransferase overexpression. J Biol Chem 273: 32487–32490.
17. Kelpe CL, Moore PC, Parazzoli SD, Wicksteed B, Rhodes CJ, et al. (2003) Palmitate inhibition of insulin gene expression is mediated at the transcriptional level via ceramide synthesis. J Biol Chem 278: 30015–30021.
18. Lei X, Zhang S, Bohrer A, Barbour SE, Ramanadham S (2012) Role of calcium-independent phospholipase A(2)beta in human pancreatic islet beta-cell apoptosis. Am J Physiol Endocrinol Metab. United States. pp. E1386–1395.
19. Lei X, Zhang S, Emani B, Barbour SE, Ramanadham S (2010) A link between endoplasmic reticulum stress-induced beta-cell apoptosis and the group VIA Ca2+-independent phospholipase A2 (iPLA2beta). Diabetes Obes Metab 12 Suppl 2: 93–98.

20. Boslem E, MacIntosh G, Preston AM, Bartley C, Busch AK, et al. (2011) A lipidomic screen of palmitate-treated MIN6 beta-cells links sphingolipid metabolites with endoplasmic reticulum (ER) stress and impaired protein trafficking. Biochem J 435: 267–276.
21. Kumagai K, Yasuda S, Okemoto K, Nishijima M, Kobayashi S, et al. (2005) CERT mediates intermembrane transfer of various molecular species of ceramides. J Biol Chem 280: 6488–6495.
22. Hanada K, Kumagai K, Tomishige N, Yamaji T (2009) CERT-mediated trafficking of ceramide. Biochim Biophys Acta 1791: 684–691.
23. Giussani P, Colleoni T, Brioschi L, Bassi R, Hanada K, et al. (2008) Ceramide traffic in C6 glioma cells: evidence for CERT-dependent and independent transport from ER to the Golgi apparatus. Biochim Biophys Acta 1781: 40–51.
24. Hanada K, Kumagai K, Yasuda S, Miura Y, Kawano M, et al. (2003) Molecular machinery for non-vesicular trafficking of ceramide. Nature 426: 803–809.
25. Riboni L, Giussani P, Viani P (2010) Sphingolipid transport. Adv Exp Med Biol 688: 24–45.
26. Viani P, Giussani P, Brioschi L, Bassi R, Anelli V, et al. (2003) Ceramide in nitric oxide inhibition of glioma cell growth. Evidence for the involvement of ceramide traffic. J Biol Chem. United States. pp.9592–9601.
27. Yamaji T, Kumagai K, Tomishige N, Hanada K (2008) Two sphingolipid transfer proteins, CERT and FAPP2: their roles in sphingolipid metabolism. IUBMB Life 60: 511–518.
28. Voelker DR (2009) Genetic and biochemical analysis of non-vesicular lipid traffic. Annu Rev Biochem 78: 827–856.
29. Giussani P, Maceyka M, Le Stunff H, Mikami A, Lepine S, et al. (2006) Sphingosine-1-phosphate phosphohydrolase regulates endoplasmic reticulum-to-golgi trafficking of ceramide. Mol Cell Biol 26: 5055–5069.
30. Riboni L, Viani P, Bassi R, Giussani P, Tettamanti G (2000) Cultured granule cells and astrocytes from cerebellum differ in metabolizing sphingosine. J Neurochem 75: 503–510.
31. Shaner RL, Allegood JC, Park H, Wang E, Kelly S, et al. (2009) Quantitative analysis of sphingolipids for lipidomics using triple quadrupole and quadrupole linear ion trap mass spectrometers. J Lipid Res 50: 1692–1707.
32. Van Veldhoven PP, Mannaerts GP (1987) Inorganic and organic phosphate measurements in the nanomolar range. Anal Biochem 161: 45–48.
33. Livak KJ, Schmittgen TD (2001) Analysis of relative gene expression data using real-time quantitative PCR and the 2(-Delta Delta C(T)) Method. Methods 25: 402–408.
34. Pagano RE, Martin OC, Kang HC, Haugland RP (1991) A novel fluorescent ceramide analogue for studying membrane traffic in animal cells: accumulation at the Golgi apparatus results in altered spectral properties of the sphingolipid precursor. J Cell Biol 113: 1267–1279.
35. Veret J, Coant N, Gorshkova IA, Giussani P, Fradet M, et al. (2013) Role of palmitate-induced sphingoid base-1-phosphate biosynthesis in INS-1 beta-cell survival. Biochim Biophys Acta 1831: 251–262.
36. Guo J, Zhu JX, Deng XH, Hu XH, Zhao J, et al. (2010) Palmitate-induced inhibition of insulin gene expression in rat islet beta-cells involves the ceramide transport protein. Cell Physiol Biochem 26: 717–728.
37. Kumagai K, Kawano M, Shinkai-Ouchi F, Nishijima M, Hanada K (2007) Interorganelle trafficking of ceramide is regulated by phosphorylation-dependent cooperativity between the PH and START domains of CERT. J Biol Chem 282: 17758–17766.
38. Giussani P, Brioschi L, Bassi R, Riboni L, Viani P (2009) Phosphatidylinositol 3-kinase/AKT pathway regulates the endoplasmic reticulum to golgi traffic of ceramide in glioma cells: a link between lipid signaling pathways involved in the control of cell survival. J Biol Chem 284: 5088–5096.

39. Fukasawa M, Nishijima M, Hanada K (1999) Genetic evidence for ATP-dependent endoplasmic reticulum-to-Golgi apparatus trafficking of ceramide for sphingomyelin synthesis in Chinese hamster ovary cells. J Cell Biol 144: 673–685.

40. Granero F, Revert F, Revert-Ros F, Lainez S, Martinez-Martinez P, et al. (2005) A human-specific TNF-responsive promoter for Goodpasture antigen-binding protein. FEBS J 272: 5291–5305.

41. Chandran S, Machamer CE (2012) Inactivation of ceramide transfer protein during pro-apoptotic stress by Golgi disassembly and caspase cleavage. Biochem J 442: 391–401.

42. Kim SJ, Winter K, Nian C, Tsuneoka M, Koda Y, et al. (2005) Glucose-dependent insulinotropic polypeptide (GIP) stimulation of pancreatic beta-cell survival is dependent upon phosphatidylinositol 3-kinase (PI3K)/protein kinase B (PKB) signaling, inactivation of the forkhead transcription factor Foxo1, and down-regulation of bax expression. J Biol Chem 280: 22297–22307.

43. Martinez SC, Tanabe K, Cras-Meneur C, Abumrad NA, Bernal-Mizrachi E, et al. (2008) Inhibition of Foxo1 protects pancreatic islet beta-cells against fatty acid and endoplasmic reticulum stress-induced apoptosis. Diabetes 57: 846–859.

44. Wrede CE, Dickson LM, Lingohr MK, Briaud I, Rhodes CJ (2002) Protein kinase B/Akt prevents fatty acid-induced apoptosis in pancreatic beta-cells (INS-1). J Biol Chem 277: 49676–49684.

45. Elghazi L, Bernal-Mizrachi E (2009) Akt and PTEN: beta-cell mass and pancreas plasticity. Trends Endocrinol Metab 20: 243–251.

6

Identification of a Novel Drug Lead That Inhibits HCV Infection and Cell-to-Cell Transmission by Targeting the HCV E2 Glycoprotein

Reem R. Al Olaby[1], Laurence Cocquerel[2], Adam Zemla[3], Laure Saas[2], Jean Dubuisson[2], Jost Vielmetter[4], Joseph Marcotrigiano[5], Abdul Ghafoor Khan[5], Felipe Vences Catalan[6], Alexander L. Perryman[7], Joel S. Freundlich[7,8], Stefano Forli[9], Shoshana Levy[6], Rod Balhorn[10]*¤, Hassan M. Azzazy[1]

1 Department of Chemistry, The American University in Cairo, New Cairo, Egypt, 2 Center for Infection and Immunity of Lille, CNRS-UMR8204/Inserm-U1019, Pasteur Institute of Lille, University of Lille North of France, Lille, France, 3 Pathogen Bioinformatics, Lawrence Livermore National Laboratory, Livermore, CA, United States of America, 4 Protein Expression Center, Beckman Institute, California Institute of Technology, Pasadena, CA, United States of America, 5 Department of Chemistry and Chemical Biology, Rutgers University, Piscataway, NJ, United States of America, 6 Department of Medicine, Stanford University Medical Center, Stanford, CA, United States of America, 7 Department of Medicine, Division of Infectious Diseases, Center for Emerging & Re-emerging Pathogens, Rutgers University-New Jersey Medical School, Newark, NJ, United States of America, 8 Department of Pharmacology and Physiology, Rutgers University-New Jersey Medical School, Newark, NJ, United States of America, 9 Department of Integrative Structural and Computational Biology, The Scripps Research Institute, La Jolla, CA, United States of America, 10 Department of Applied Science, University of California Davis, Davis, CA, United States of America

Abstract

Hepatitis C Virus (HCV) infects 200 million individuals worldwide. Although several FDA approved drugs targeting the HCV serine protease and polymerase have shown promising results, there is a need for better drugs that are effective in treating a broader range of HCV genotypes and subtypes without being used in combination with interferon and/or ribavirin. Recently, two crystal structures of the core of the HCV E2 protein (E2c) have been determined, providing structural information that can now be used to target the E2 protein and develop drugs that disrupt the early stages of HCV infection by blocking E2's interaction with different host factors. Using the E2c structure as a template, we have created a structural model of the E2 protein core (residues 421–645) that contains the three amino acid segments that are not present in either structure. Computational docking of a diverse library of 1,715 small molecules to this model led to the identification of a set of 34 ligands predicted to bind near conserved amino acid residues involved in the HCV E2: CD81 interaction. Surface plasmon resonance detection was used to screen the ligand set for binding to recombinant E2 protein, and the best binders were subsequently tested to identify compounds that inhibit the infection of Huh-7 cells by HCV. One compound, 281816, blocked E2 binding to CD81 and inhibited HCV infection in a genotype-independent manner with IC50's ranging from 2.2 µM to 4.6 µM. 281816 blocked the early and late steps of cell-free HCV entry and also abrogated the cell-to-cell transmission of HCV. Collectively the results obtained with this new structural model of E2c suggest the development of small molecule inhibitors such as 281816 that target E2 and disrupt its interaction with CD81 may provide a new paradigm for HCV treatment.

Editor: Vladimir N. Uversky, University of South Florida College of Medicine, United States of America

Funding: This work was conducted as part of the first authors PhD thesis work. This work was supported by a Yousif Jameel PhD Fellowship from The American University in Cairo awarded to Reem Al Olaby. The funders had no role in study design, data collection and analysis, decision to publish, or preparation of the manuscript.

Competing Interests: Reem Al Olaby, Dr. Hassan Azzazy and Dr. Rod Balhorn are co-inventors on a patent application related to the described work that has been submitted by The American University in Cairo, Cairo, Egypt. The title for the application is Ligands That Target Hepatitis C Virus E2 Protein. None of the other coauthors have competing interests.

* Email: rod@shaltech.com

¤ Current address: SHAL Technologies Inc., Livermore, CA, United States of America

Introduction

Hepatitis C virus (HCV) is a global public health problem [1] in which nearly 85% of affected individuals have acute HCV infections and exhibit no symptoms. In addition, more than three-quarters of these cases will advance to chronic disease, which include liver cirrhosis and liver cancer [2]. The current standard of care treatment for HCV (Peg-interferon/Ribavirin, PR) can cause deleterious side effects, and a sustained virologic response (SVR) is achieved in less than 50% of genotype-1 patients [3]. The FDA approved protease inhibitors Telaprevir (TVR) and Boceprevir (BOC) have been shown to provide higher SVR rates in genotype 1 patients [3,4] when each is combined with PR. However the poor safety profile of TVR and BOC reported in the Week 16 analysis of the French Early Access Program suggest there is still a need for better HCV drugs [5]. The two most recent FDA approvals have been for the oral drugs Simeprevir and Sofosbuvir, inhibitors that target the HCV NS3/4A protease and polymerase, respectively [6]. Semiprevir, which needs to be administered with Ribavirin and Peg-interferon, has a number of undesirable side effects [7]. The efficacy of Semiprevir has also been shown to be diminished significantly, due to viral breakthrough (HCV RNA rebounds and becomes detectable in the patient before treatment is completed), in patients infected by HCV genotypes 4–6 containing the Q80K, R155K and D168E/V polymorphisms in the NS3 protease [7]. Recommendations for the use of Sofosbuvir indicate it should be administered with Ribavirin in HCV genotype 2 and 3 infections and that Peg-Interferon should be included in the treatment when infections involve genotypes 1 and 4. While Sofosbuvir is considered the Holy Grail in HCV treatment by some, it is recommended that treatments be limited to 12 weeks [6]. Its high cost ($1,000 USD/pill) also puts it out of reach of many HCV infected patients. This has led many of the larger pharmaceutical companies to continue developing new drugs that target one or more steps in the HCV life cycle and block virus invasion, processing of the pro-protein or replication of the viral genome.

Since its identification as the first putative receptor for HCV [8], the tetraspanin CD81 has been demonstrated to be a key player in HCV entry [9]. In particular, its large extracellular loop (CD81-LEL) is involved in the binding to the HCV envelope glycoprotein E2 [10,11]. Zhang et al. [12] elucidated a separate, additional function for CD81 in the HCV life cycle. These studies revealed that CD81-LEL is important for efficient HCV genome replication. In addition, the E2-CD81-LEL interaction has been determined to induce several immuno-modulatory effects such as the production and release of pro-inflammatory cytokine gamma interferon from T-cells. In addition, this interaction has also been shown to down regulate T-cell receptors and suppress the activity of natural killer (NK) cells [13]. Therefore, it is tempting to speculate that blocking the CD81-LEL:HCV E2 interaction might also contribute to arresting disease progression to liver cirrhosis.

Following the discovery of the E2 glycoprotein's role in HCV infection and disease progression, several approaches have been used to attempt to develop anti-HCV drugs and vaccines that target the HCV E2 glycoprotein [14–17] located on the surface of viral particles. These efforts have had to deal with challenges that relate to the genomic diversity and heterogeneity of HCV, limitations in animal models used to test vaccines and drugs, and the lack of a resolved crystal structure for the HCV E2 glycoprotein. Recently, two crystal structures have been reported for the core ectodomain of the HCV E2 protein [18,19]. Kong et al. [18] obtained the structure of amino acid residues 384–746 (E2c) by designing and expressing 41 soluble HCV E2 constructs and selecting 15 to screen against E2-specific Fab fragments in crystallization trials. Using a combination of x-ray crystallography and negative stain-electron microscopy, Kong et al. [18] discovered the structures they obtained for E2 were globular and very different from the predicted models of E2 that were created using class II fusion protein templates containing three β-sheet domains. Additionally, they were able to identify key CD81-binding residues through mutational studies. Important CD81 binding sites were determined to be in the epitope recognized by the neutralizing antibody AR3C, along one side of the β-sandwich (an isolated region of the CD81-binding loop) and a front layer consisting of loops, short helices and β-sheets [18–20]. AR3C was also found to cross-neutralize HCV genotypes by blocking CD81 binding to HCV E2 [21]. A second structure was reported for E2c (amino acid residues 492–649) by Khan et al. [19]. This new structure, which was obtained by crystallizing E2c in complex with a Fab

fragment of the mouse monoclonal antibody 2A12, is very similar to the previously reported structure. In addition to providing a second structure for the E2 core from a different HCV genotype (2a), new information was also reported on the accessibility of the E2 core amino acids within the structure using a combination of limited proteolytic degradation and deuterium exchange.

Despite the advances that have been made in the field of HCV drug development, our current drugs offer little protection against the emergence of genetic variants (escape variants) of HCV – a feature of HCV biology that complicates both drug and vaccine development. Drugs that target only one step in the HCV life cycle will be the least effective in treating patients that become infected with these emerging variants. The FDA approved drugs for HCV are good examples, as they are only effective against a subset of genotypes. In an effort to identify a suitable drug candidate that targets the majority of the existing HCV genotypes, we created an HCV E2 homology model based on the new HCV E2 core crystal structure reported by Kong et al. [18] that contains three peptide segments that were not present in the reported structure, and we have used this model to identify small molecule drug leads that target highly conserved sites on the HCV E2 glycoprotein located within the region bound by CD81. AutoDock was used to perform virtual screening runs against 1,715 small molecules and 34 of the best compounds were tested experimentally using surface plasmon resonance (Biacore T100) to identify a set of small molecules that bind to the recombinant E2 protein. The compounds showing binding activities were then tested for their ability to block HCV infection of Huh-7 cells. One compound, 281816, was found to block infection of the cells by each of the HCV genotypes and subtypes tested (1a, 1b, 2a, 2b, 4a and 6a) in a dose-dependent manner. Experiments with Huh-7 cells have shown that both mechanisms that lead to HCV infection, cell-free and cell-to-cell transmission, are abrogated by 281816. Inhibition of cell-free infection is limited to the viral attachment step, as well as interactions occurring during viral internalization and fusion; 281816 appears to have no effect on post-entry processes.

Materials and Methods

Creation of the homology model of E2 used for docking

A crystal structure of E2c deposited in the PDB under a code 4MWF was solved by Kong et al. [18] at a resolution of 2.65 Angstroms. However, upon examination of the structure file prior to docking, the set of reported atom coordinates of the protein was found to be incomplete. In addition to the coordinate file containing structural information for only 171 residues out of the 363 amino acids present in the full-length protein, structural information was missing for several peptide segments or loops within the structural core of the protein. In order to prepare a more complete version of the structure for docking, we have performed several homology modeling and structure analysis tasks using the coordinates of E2c as a template. The final structural model was created using the AS2TS system [21] based on atom coordinates from the PDB chains 4mwf_C and 4mwf_D and extensively manually edited. A structural search for similar fragments in proteins in the PDB that could be used to model missing loop regions was performed using the StralSV algorithm [22], which identifies protein structures that exhibit structural similarities despite low primary amino acid sequence similarity. The side-chain prediction was accomplished using SCWRL [23] when residue-residue correspondences did not match. Residues that were identical in the template and E2 protein were copied from the template onto the model. Potential steric clashes were identified in the unrefined model using a contact-dot algorithm in the MolProbity software package [24], and the constructed model was finished with relaxation using UCSF Chimera [25].

Virtual screen of the NCI Diversity Set III to the HCV E2 protein model

AutoDock VINA 1.1.2 (VINA) [26] was used to perform a virtual screen of the NCI Diversity Set III against the homology model that was created using the new crystal structure solved by Kong et al. [18] (PDB ID: 4MWF) as a template. The model of the protein was prepared for docking using the MolProbity Server (to add all of the hydrogen atoms and to flip the HIS/ASN/GLN residues if doing so significantly lowered the energy) and AutoDockTools 4.2 (which added the Gasteiger-Marsili charges and merged the non-polar hydrogens onto their respective heavy atoms) [27,28]. The NCI Diversity Set III library containing 1,715 models of compounds was obtained from the ZINC server (http://zinc.docking.org) [29]. The multi-molecule "mol2" files from ZINC were prepared for docking calculations using Raccoon [30], which added the Gasteiger-Marsili charges, merged the non-polar hydrogen atoms onto their respective heavy atoms, and determined which bonds should be allowed to freely rotate during the calculations, to generate the "pdbqt" docking input format.

Four different, overlapping grid boxes were used in this virtual screen to enable the docking calculations to explore almost the entire surface of this E2 model. Those amino acids missing from the E2c crystal structure whose modeled coordinates were known with the lowest degree of certainty, such as residues E454 and L456–E482 located in the large missing loop and residues G575–L580 and F586–K588 in the two other two missing segments, were not included in the boxes. By defining the boxes to exclude these residues, we were able to minimize the impact of these less accurate parts of the model on ligand docking. Since large grid boxes were used in these calculations, the "exhaustiveness" setting in VINA was increased to 20. Each calculation used 8 CPUs on the Linux cluster at Rutgers University-NJMS. The first box, which included P490, was centered at 38.829, 12.968, −40.958 (x, y, z) and had the following dimensions: 24.0×35.0×30.0 (x, y, z in Angstroms). The second grid box, which included G436, was centered at 48.401, 11.791, −14.449 and had a size of: 32.0×36.0×24.0. The third grid box, which included S528, was centered at 51.644, 25.877, −27.795 and encompassed 30.0×30.0×30.0 Angstroms^3. The fourth grid box, which was selected to include the side of E2 not covered by the previous three grid boxes, was centered at 57.777, 12.968, −34.067 and enclosed 24.0×35.0×32.0 Angstroms^3.

The docking outputs generated by VINA were processed and filtered using python scripts from Raccoon2 and Fox [30]. The top-ranked VINA mode from each docking calculation was harvested, and 17 different sets of energetic and interaction-based filters (Table 1) were investigated to harvest the most promising docking results for visual inspection. Different sites have different numbers and arrangements of hydrogen bond donors, hydrogen bond acceptors, and aromatic rings. They also have very different geometries (i.e., van der Waal surfaces and solvent accessibility patterns and percentages). Consequently, several different filters were tested for the docking results against each site in order to harvest a"reasonable" number of docked modes for visual inspection against each site. This is a subjective process, guided by extensive experience with virtual screening. If the same filters are used against each site, then for some sites (or for filters that are not restrictive enough), too many compounds are obtained for visual inspection (i.e., the process is less efficient and a larger number of false positive results are likely to occur). For other sites (or for filters that are too restrictive), an insufficient number of

compounds will be obtained for visual inspection. This would increase the chance of missing promising candidates (having too many false negatives) [31]. The following parameters were explored in the filtering process: -e indicates the minimum estimated Free Energy of Binding from the VINA score in kcal/mol, -l is the minimum ligand efficiency value in kcal/mol/heavy atom, -S is the minimum number of hydrogen bonds between the ligand and target, and -H indicates that the ligand had to form a hydrogen bond with either a backbone amino group (::N) or a backbone carbonyl oxygen (::O) of any residue in that grid box.

For the results with grid box 1, filters 12 and 13 each harvested 70 and 51 compounds, respectively. Those filtered sets were pooled together to form a set of 96 unique compounds for visual inspection. Filters 14 (which harvested 11 compounds), 15 (which harvested 21 compounds), and 1 (which harvested 34 compounds) were pooled together from the results with grid box 2, in order to identify 52 compounds for visual inspection. Similarly, for the results with grid box 3, filters 1 (which harvested 25 compounds), 14 (which identified 20 candidates), and 15 (which harvested 13 compounds) were pooled to obtain 34 compounds for visual inspection. To identify candidates in the results with grid box 4, filters 1 (which harvested 26 compounds), 14 (which harvested 19 compounds), and 15 (which harvested 14 compounds) were pooled to obtain 42 compounds.

These four different pools of potentially promising compounds were then visually inspected to select the ligands to be tested experimentally for binding to recombinant E2 protein. Both the structure of the compound and the nature of its predicted interaction with the protein were examined. Compounds were considered good hits and suitable for testing if they 1) were small (molecular weight ~200–600 Da), 2) contained a single free amine or carboxyl to facilitate their potential conjugation to other ligands, 3) were not highly charged or highly hydrophobic, 4) did not contain iodine, disulfide bonds or highly reactive functional groups, 5) did not contain multiple conjugated aromatic ring systems, 6) exhibited multiple contacts to the protein surface, and 7) had conformers that bound to the protein surface near one or more E2 amino acid residues that have been shown to participate in or be required for binding to CD81. Detergent-like molecules were avoided and only commercially available compounds were considered for screening.

Expression and purification of the HCV E2 protein Con1eE2

A construct containing a sequence encoding amino acids 384–656 of the Con1 envelope protein 2 ectodomain (eE2) [19], a genotype 1 E2 sequence, was cloned into a lentiviral expression vector containing a carboxy-terminal Protein A tag separated by a PreScission Protease cleavage consensus sequence. eE2-ProtA was stably expressed in HEK293T cells using lentiviral infection. The protein was secreted into the media and supernatants were purified using IgG Sepharose (GE Healthcare, Piscataway, NJ). eE2-ProtA was eluted with 100 mM sodium citrate and 20 mM KCl at pH 3 directly into tubes containing 1M Tris pH 9 for immediate neutralization. PreScission Protease (GE Healthcare, Piscataway, NJ) was added to the eluted sample at a ratio of 1:50 (enzyme:eE2), and the digest was then dialyzed into 20 mM HEPES pH 7.5, 250 mM NaCl, 5% glycerol. eE2 was separated from the cleaved tag and the PreScission Protease by ion exchange chromatography [19].

Table 1. Energetic and interaction-based filters used to harvest the most promising results from the ligand docking runs.

Filter	Parameter Set
1	-e −6.5 -l −0.29 -S 3
2	-e −7.0 -l −0.29 -S 3
3	-e −7.5 -l −0.29 -S 3
4	-e −8.0 -l −0.29 -S 3
5	-e −7.0 -l −0.29 -S 4
6	-e −7.5 -l −0.29 -S 4
7	-e −8.0 -l −0.29 -S 4
8	-e −6.5 -l −0.29 -S 3 -H ::N
9	-e −6.5 -l −0.29 -S 3 -H ::O
10	-e −7.0 -l −0.29 -S 3 -H ::N
11	-e −7.0 -l −0.29 -S 3 -H ::O
12	-e −7.0 -l −0.29 -S 4 -H ::N
13	-e −7.0 -l −0.29 -S 4 -H ::O
14	-e −7.0 -S 3 -H ::N
15	-e −7.0 -S 3 -H ::O
16	-e −7.0 -S 4 -H ::N
17	-e −7.0 -S 4 -H ::O

Experimental analysis of ligand binding to recombinant E2 and CD81-LEL by surface plasmon resonance (SPR) detection

A set of 34 of the ligands predicted by AutoDock to bind to E2 were tested experimentally to determine if they bound to recombinant E2 protein immobilized on a chip using surface plasmon resonance detection. The SPR analyses were performed using a Biacore T100 workstation (GE Healthcare, NJ, USA) and recombinant HCV E2 protein. 1 µM HCV E2 was diluted into 10 mM sodium acetate buffer pH 5 and immobilized for 15 min at a flow speed of 5 µl/min onto a CM5 sensor chip using amine coupling (EDC-NHS). Approximately 10,000 response units (RU) of protein were immobilized on the chip. His-CD81-LEL (Bioclone Inc., San Diego, CA) binding to HCV E2 was tested as a positive control prior to injecting the ligands to confirm the E2 protein was functional and would bind CD81-LEL. In a typical experiment with CD81, 1 µl of his-CD81 (50 nM) in 114 µl PBS was injected into channel 2 and 106.4 RUs of CD81 bound to the E2 on the chip. This was followed by testing the binding of the 34 virtual screening hits where the ligands were prepared as 200 µM solutions in PBS and they were introduced to the protein using a pre-programmed 3 min association and 1 min dissociation interval. The response was measured at two time points during dissociation, 10 and 50 seconds, to obtain information on the rate of ligand dissociation from E2.

Two single cycle kinetic studies were also performed to compare the binding of 281816 to the recombinant E2 and his-tagged CD81-LEL proteins. In both studies, the proteins were diluted to a concentration of 1 µM in 10 mM sodium acetate buffer pH 4.5 and immobilized for 15 min on a CM5 sensor chip using amine coupling (EDC-NHS). Data on the kinetics and affinity of 281816 binding was obtained by flowing five concentrations of the 281816 ligand (2.5 µM, 7.4 µM, 22.2 µM, 66.7 µM and 200 µM) over the chip sequentially at a flow rate of 30 µl/min. Equilibrium binding curves were generated for each protein and the data were fitted using a monovalent binding model to determine the Kd for 281816 binding to E2 and CD81-LEL.

HCV infection assays

Pseudotyped retroviral particles harboring HCV envelope proteins (HCVpp) from different genotypes were produced as described previously [32,33] with plasmids kindly provided by F.L. Cosset, J. Ball, and R. Bartenschlager. A plasmid encoding the feline endogenous virus RD114 glycoprotein [34] was used for the production of RD114pp. Both HCVpp and RD114pp expressed *Firefly* luciferase.

The cell culture-produced HCV particles (HCVcc) used in this study were based on the JFH1 strain [35] and were prepared as described previously [36,37]. They were engineered to express the A4 epitope, titer-enhancing mutations and *Gaussia* luciferase [36,37].

To identify ligands that inhibit HCV infection, Huh-7 cells were seeded in 96-well plates and treated the day after with six different concentrations of each ligand diluted in DMSO in duplicate using a Zephyr automated liquid handling workstation (Caliper BioSciences, Hopkinton, MA). The final concentration of DMSO (1%) was adjusted to be the same for all ligand concentrations. Cells treated with DMSO were used as negative controls. Cells treated with different concentrations of anti-CD81 antibody (JS-81 from BD Pharmingen, San Jose, CA) 1 hr before infection, were also used as positive controls. The third day, RD114pp, HCVpp or HCVcc were inoculated and incubated for 30 hr at 37°C. *Firefly* and *Gaussia* luciferase assays were performed as indicated by the manufacturer (Promega, San Luis Obispo, CA).

The analysis of the effect of the 281816 ligand on Huh-7 infection by HCVpp bearing envelope proteins from different genotypes was performed in 24-well plates using the method described above. This ligand was also screened for toxicity to the cells using the MTS (3-(4,5-dimethylthiazol-2-yl)-5-(3-carboxy-methoxyphenyl)-2-(4-sulfophenyl)-2Htetrazolium) assay [38] and was found to not be toxic under the conditions used in the infection assays.

Inhibition of recombinant E2 binding to native CD81

Two different assays were performed to test for the inhibition of E2 binding to CD81 by 281816. In a cell binding assay, the human B cell line Raji (ATCC, Manassas, VA), which expresses high levels of CD81 on its surface [39,40], was used to determine if ligand 281816 inhibits the binding of HCV-E2 protein to native human CD81. Purified HCV-E2 protein (4 µg) was pre-incubated with 1,5,15, 50, 100 or 400 µM of the ligand 281816 for 25 min at RT. After pre-incubation the E2-ligand complex was added to the cells and incubated for 25 min. The complexes were washed from the cells and 0.5 µg of mouse anti E2 antibody (clone H53) was added followed by a FITC-conjugated anti-mouse antibody (Southern Biotechnology, Birmingham, AL). The cells were washed, fixed with 3% paraformaldehyde, and analyzed by flow cytometry (BD FACSCalibur, software: Cell Quest Pro). The mean fluorescence intensity (MFI) was calculated using Flowjo software (TreesStar, www.flowjo.com).

The second test used an ELISA assay to determine if E2 binding to a recombinant CD81 protein is inhibited by the presence of ligand 281816. In this assay, a 96 well plate was coated with GST-tagged human CD81-LEL (5 µg/ml) overnight as previously described [10], then washed with PBS, 0.5% Triton X-100 and blocked with 2% milk in PBS for 1 hr. HCV E2 protein (5 µg/ml) was pre-incubated with different concentrations of 281816 for 30 min before adding to the plate, then HCV-E2 protein (with or without the ligand) was added to the GST-tagged human CD81-LEL coated plate and incubated for 1 hr at room temperature to allow the protein to bind. To detect HCV-E2 binding, a primary mouse anti-E2 antibody (H53 clone, 5 µg/ml) was added and incubated for 1 hr followed by a secondary goat anti-mouse-horse radish peroxidase (HRP) antibody (Southern Biotechnology Associates, Birmingham, AL) diluted 1:5000. Substrate was added (citrate buffer pH 4.0, 3.5 µl hydrogen peroxide and 100 µl 2,2'-azino-bis(3-ethylbenzothiazoline-6-sulphonic acid)) and the absorbance was measured at 405 nm.

Inhibition of anti-CD81 5A6 antibody binding to CD81-LEL

To determine if ligand 281816 binds to the E2 binding site on CD81, a competition binding experiment was run using the ligand and an anti-CD81 antibody (5A6 clone) that has been shown previously to block E2 binding [57,58]. In this assay, a 96 well plate was coated with GST-tagged human CD81-LEL overnight and then blocked with 2% milk in PBS for 1 hr at RT. The plate was then incubated for 40 min with 281816 (1 µM) or PBS as a control. The indicated concentrations of mouse anti-human CD81 antibody (5A6) were added to the plate and incubated for an additional 1 hr, followed by anti-mouse IgG-HRP. The absorbance was then measured at 405 nm.

HCVcc cell-to-cell transmission assay

Cell-to-cell transmission was measured as described previously [41,42]. Briefly, Huh-7 cells were seeded on coverslips and infected at low multiplicity of infection with HCVcc for 2 hr at 37°C. After washing, cells were cultured in medium containing neutralizing anti-E2 antibody (3/11; 50 µg/ml) to block cell-free transmission and 281816 at the indicated concentrations. Cells cultured in the presence of DMSO or Epigallocatechin-3-gallate (EGCG, 50 µM) [42] were used as negative and positive controls of inhibition, respectively. Three days post-infection, cells were fixed with formalin solution (formaldehyde 4%, Sigma, St Louis, MO) and stained by indirect immunofluorescence using the anti-E1 monoclonal antibody A4 and Alexa555-conjugated anti-mouse immunoglobulins. Cell-to-cell transmission was quantified by counting the number of infected cells per focus. As a control, cells were infected with HCVcc pre-incubated with 50 µg/ml 3/11 antibody to confirm cell-free transmission is blocked by the antibody as reported previously [41].

Kinetics of entry

Cells treated with 281816 at 10 µM or with DMSO were infected with HCVcc for 1 hr at 4°C (attachment/binding period). Virus was removed, cells were washed with medium and incubated again for 1 hr at 4°C (post-attachment/binding period). Cells were then washed and incubated for 1 hr at 37°C (endocytosis/fusion period). Lastly, cells were washed and incubated in complete culture medium for 21 hr. Infection levels were monitored by measuring luciferase activities. To confirm 281816 was not toxic to the cells, an MTS assay [38] was also performed after incubating the cells with 10 µM 281816 for the same lengths of time (1 hr, 2 hr or 3 hr) the ligand was exposed to the cells in the entry experiments.

Antibodies

Mouse anti-E1 A4 [43], anti-E2 H53 [44] and rat anti-E2 3/11 [45] were produced in vitro using a MiniPerm apparatus (Heraeus). FITC-conjugated and Alexa555-conjugated anti-mouse immunoglobulins were obtained from Southern Biotechnology (Birmingham, AL) and Jackson Immunoresearch (West Grove, PA), respectively.

Results

Structural model of E2

In order to maximize the likelihood that these experiments would lead to the discovery of small molecules that bind to E2 and block E2's binding to CD81, we created a homology model of the core of the E2 protein to use as our docking target. This model was created using the HCV genotype 1a protein sequence NP_751921.1, which corresponds to isolate H77, and the crystal structure of E2c as the primary template (PDB entry: 4MWF). Using a model, rather than the E2c crystal structure, was important because the reported crystal structure of E2c has three large gaps in which atom coordinates for 57 amino acids, or one quarter of the E2c structure, is missing. The coordinates listed in PDB chains 4mwf_C and 4mwf_D provide structural information for only 169 and 171 residues respectively out of the 363 amino acids present in the full-length E2 protein. Within each of the deposited PDB chains, three stretches of amino acid sequence (large loop P453-P491 containing 39 amino acids, T542-G547 or V574-N577, and F586-R596) are missing from the structure (Figure 1A). Docking to structures lacking such a large proportion of their amino acids can be problematic because the missing peptide segments are usually located on the protein's surface, and the underlying amino acid residues packed in the interior of the protein are exposed and incorrectly presented as the surface during the docking. Unfortunately, similar regions are also not present in the crystal structure of the genotype 2a HCV E2c protein (PDB chain: 4nx3_D) reported by Kahn et al. [19] which provides atom coordinates for only 119 amino acids. Structural superposition of 4mwf_C and 4nx3_D (Figure 1B) shows strong conformational similarities between the experimentally solved structures of the E2 proteins with a root mean square deviation of 1.07 Angstroms measured on 98 residues for which distances

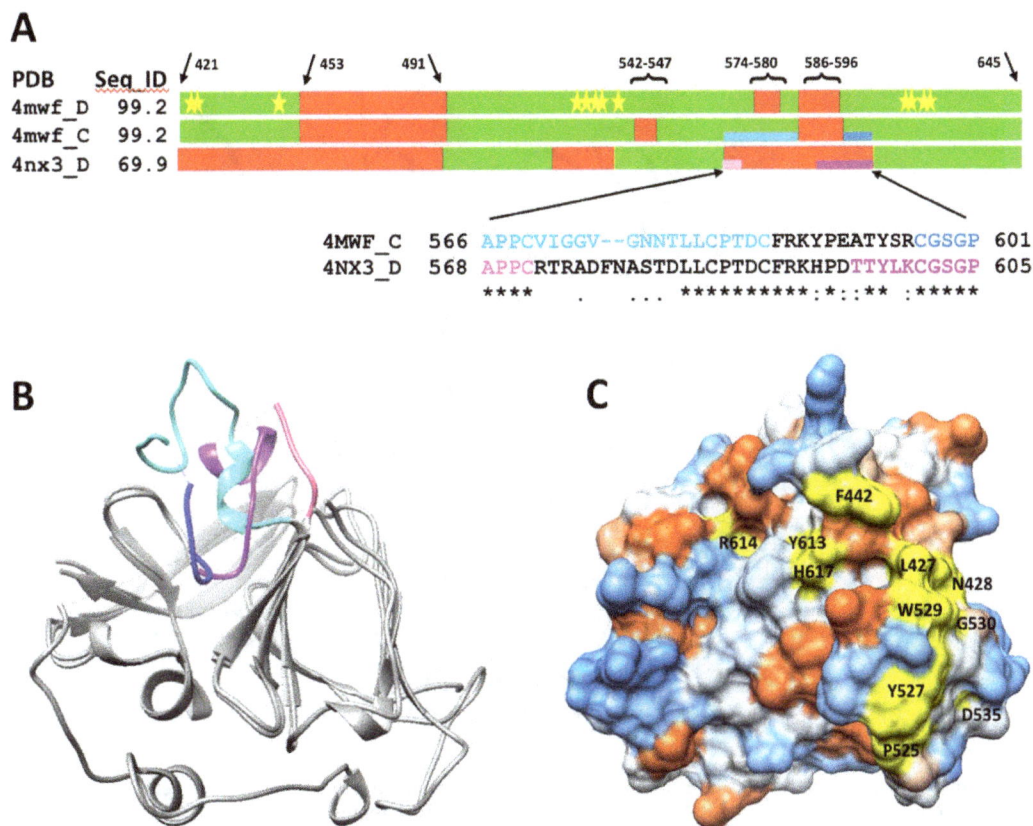

Figure 1. Comparison of structural templates used for modeling the HCV E2c protein. (A) Bar representation of E2 sequence showing the structural similarities between crystal structures 4MWF chains C and D (E2c structure, genotype 1a), and 4NX3 chain D (genotype 2a). Regions reported in the coordinates span amino acid residues from H421 to N645. The percent sequence identities between amino acid sequences taken from coordinates and corresponding sequence fragments from HCV E2 protein of genotype 1a are shown in the column Seq_ID. In green are colored regions where structural deviations are below 3 Ångstroms measured as Cα-Cα distances between corresponding residues from the superimposed structures. In red are regions where structural data is missing or deviations are greater than 3 Angstroms. The locations of amino acid residues that have been reported to be important for E2 binding to CD81 are marked with yellow stars. (B) Structural superposition of 4mwf_C and 4nx3_D shows strong conformational similarities between experimentally solved structures of E2 proteins for which the level of sequence identity is 69%. In blue and purple are colored structural fragments where two structures 4mwf_C (566–601; Blue: light-dark) and 4nx3_D (568–605; Purple: light-dark) significantly differ. (C) Surface presentation of the 4mwf_D structure showing the amino acid residues identified to be important for E2 binding to CD81 (yellow). The other amino acid residues are color coded with the most hydrophilic residues being colored blue, the most hydrophobic residues colored red orange, and intermediate residues colored white.

between corresponding Cα atoms are under 3 Angstroms. The most significant structural deviations are observed in the region 566–601 (numbering from 4mwf_C) which corresponds to the region that also exhibits the greatest variation in sequence (see sequence alignment in Figure 1A).

Exhaustive structure similarity searches of 90 residue structural fragments of E2 conducted using the entire PDB database (255,302 PDB chains) revealed that no additional structural homologs could be found at the level of calculated structure similarities by LGA score [46] higher than LGA_S = 45%, suggesting that the HCV E2 protein represents a novel fold in the current PDB. Thus, the modeling of the structure of the insertions needed to fill in missing regions in the experimentally solved structures and to complete the model was a difficult task, and it was completed with a very low degree of confidence. By applying a combination of structural modeling and analysis methods to the E2 crystal structure (see Materials and Methods section), we were able to construct a model (Figure 2) that contains the 57 amino acids that are missing in the E2c structure, including an amino acid known to be critical for E1 binding (W487), key

amino acids known to participate in CD81 binding (Figure 1C), as well as the exact sequence for the HCV genotype 1a E2 protein. Three regions of the protein that have been identified by others to be critical for E2 binding to CD81 [47–49] are contained in the model in their entirety. Currently, however, only three of the twenty-one Region 1 amino acids (H421–N423) are present in the model. A comparison of our model to the two E2c structures (see bar plots in Figure 1A and superposition of the E2c structure and the model in Figure 2) shows the main core regions are, as one would expect, very similar. The differences that are observed in the core region are small and appear to reflect only minor local deviations between experimentally solved structural templates. The large region that does differ corresponds to the missing peptide segments.

Ligands predicted to bind to CD81 binding sites on E2

Five ligand-binding sites on the HCV E2 homology model (Figure 3) were identified by docking the National Cancer Institute's Diversity Set III library of ligands to the E2 model. Each of these sites is associated with or positioned next to one or

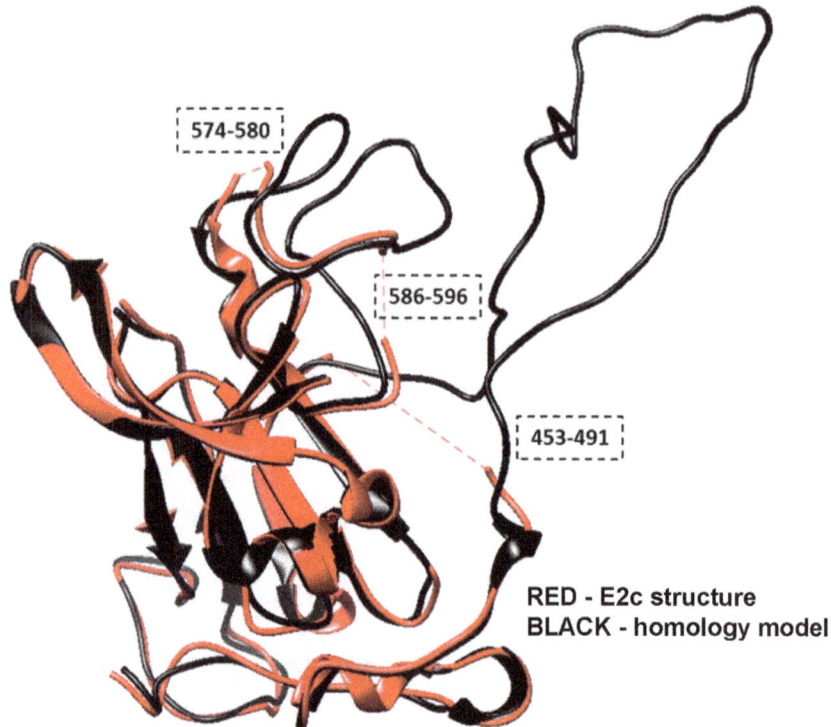

Figure 2. Comparison of the crystal structure of E2c with the homology model. Structural superposition between E2c crystal structure from the PDB chain 4mwf_D (red) and the homology model (black) is illustrated using a ribbon representation. The crystal structure and homology model overlap in most of the regions, except the fragments where coordinates in the experimental structure are missing (red dashed lines).

more of the amino acid or peptide sequences that have been identified by others to either participate in E2 binding to CD81, E2 binding to E1, or to be important for HCV infectivity. While the accuracy of the structure of the modeled segments missing from E2c may be low, the docking and visual inspection processes focused on the regions of the target that were based on the crystal structure. The majority of the amino acids that make up or surround each of the cavities used for ligand docking and the neighboring amino acids that play a role in E2 binding to CD81 are all located in the core region of E2. The structure of this region of the model is known with high confidence, as it is essentially identical to the two recent crystal structures of the E2 protein core determined by two different groups [18,19]. The locations of the grid boxes were also defined in such a manner that the amino acids in the modeled segments missing from the crystal structure of E2c would only be marginally considered during the docking. Only those residues in close contact with the core E2c structure were included in the boxes. In this way, the regions of the homology model with the least well-defined structures had a minimal impact on the docking results.

The first sequence of importance is the peptide segment Q412–N423 that was identified to bind to the broadly neutralizing antibody AP33 [20,50]. Alanine mutagenesis studies have shown all of the amino acids in this region appear to be important for HCV infectivity [48]. The model used in this study currently contains only three of the amino acids that correspond to this segment, H421, I422 and N423. Sequence 2 spans the second hyper-variable domain of E2, extending from amino acid Y474 to R492 [13,47–49]. The majority of amino acids in this sequence have been shown to have no effect on E2 binding to CD81 when mutated [51], but antibodies binding to this region of the protein do inhibit HCV infectivity [49] and CD81 binding [50]. One

amino acid located within sequence 2, W487, does however appear to be critical for E2 binding to E1. This amino acid is the first residue in one of the WHY motifs that have been reported to play a role in E1:E2 dimerization [47]. The third sequence spans amino acids S522–G551 [20,47–49] and the fourth sequence of importance is comprised of amino acids P612–P619 [49]. Mutations of residues Y527, W529, D535, Y613, R614, W616, H617 and Y618 in these two regions have all been shown to eliminate E2 binding to CD81 [47,49]. Mutating all but three of these amino acids (D535, R614 and W616) appears to eliminate specific interactions with CD81. W616 is the first amino acid in another WHY motif that is located in a region (G600–C620) that has been shown to be involved in fusion [52]. Alanine mutagenesis of D535, R614 and W616 was found to disrupt the structure of the AR3A epitope and indirectly impact CD81 binding [49].

These five binding sites were used to guide to our selection of the top virtual screening hits to be tested experimentally for binding to recombinant E2 protein. All five sites are cavities in the protein surface that would be expected to be accessible to ligands because they contain or are surrounded by amino acid residues known to participate in E2 binding to CD81 or they are located within the epitopes of antibodies that inhibit HCV infectivity or block CD81 binding. While there is still some debate regarding the importance of the entire regions bound by neutralizing antibodies, amino acid mutagenesis studies have provided a great deal of insight into those amino acids located within the epitopes that participate in E2 binding to CD81. Based on this information, we have used the set of amino acids W420–I422, S424, G523, Y527, W529, G530, D535, P612–R614 and W616–P619, whose mutation has been shown to eliminate E2 binding to CD81, to identify locations within these sites (Figure 3) where ligand binding would be expected to disrupt E2's ability to bind to CD81.

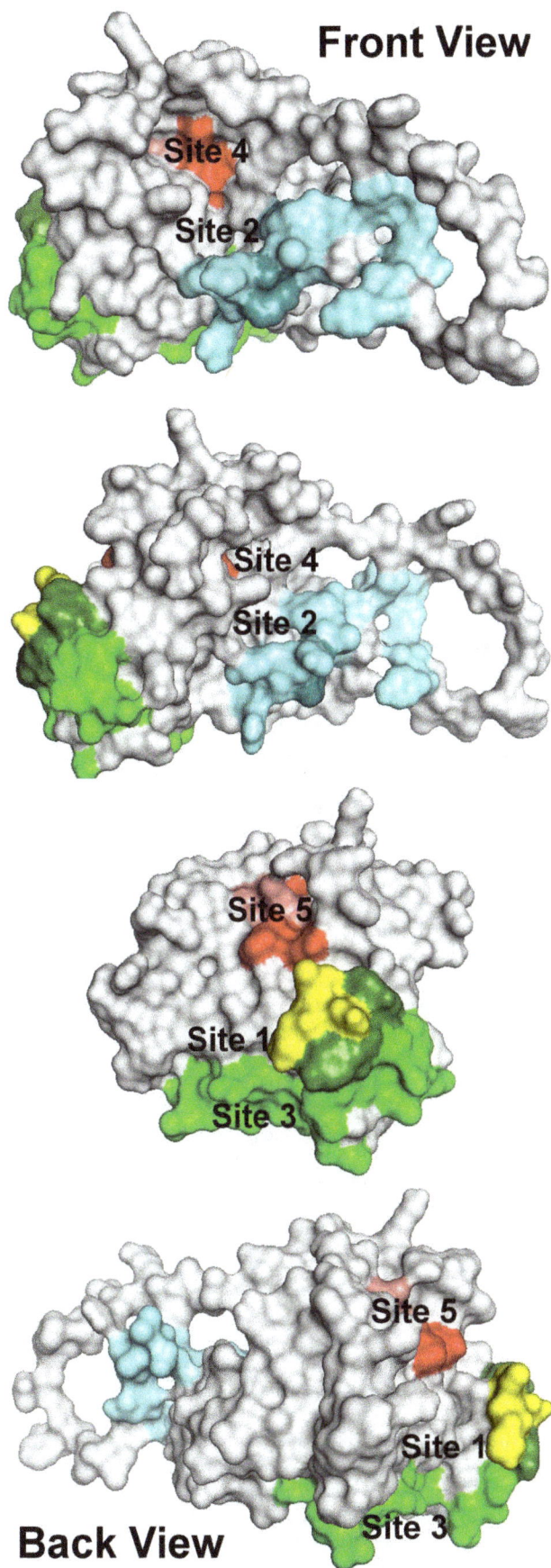

Front View

Back View

Figure 3. Location of ligand-binding sites on the E2 homology model used to select ligands for testing. Each of these sites either covers or is located immediately adjacent to amino acids or peptide segments of the E2 protein known to be important for HCV infectivity. H421–N423 (yellow): each amino acid in this region is important for infectivity. Antibodies binding to amino acids Y474–R492 (light cyan) have been shown to prevent infectivity, but this region of the protein has no effect on E2 binding CD81. W487 (dark cyan) is a key amino acid that is involved in E2 binding to E1. S522–G551 (light green) and Y527 and W529 (dark green) are critical for E2 binding to CD81. Site 4: P612, Y613, and H617–P619 (red) are critical for E2 binding to CD81; mutations to R614–W616 (pink) disrupt the structure of the region. The four views show the structure as it is rotated counterclockwise from left to right. Movie S1 shows the rotating structure.

Thirty-four of the highest scoring ligands were selected from the docking run for experimental analysis. Docked conformations of each of the ligands were predicted to bind to one or more of these five binding sites. The best ligands were considered to be those that exhibited the lowest free energy of binding and were predicted to interact with or bind nearby one or more of the E2 amino acids within the sites that were reported to be critical for E2 binding to CD81. The free energy of binding predicted for the best bound ligand conformations, shown in Table 2, ranged from −6.2 to −8.7 kcal/mol. Additional criteria used to select among the group of ligands predicted to bind include the number of contact points/interactions (such as hydrogen bonds, salt bridges, van der Waals interactions) with amino acids in the model (the larger number of contacts or interactions the better) and the chemical structure of the ligands (preference is given to those that contain a free amino or carboxyl group that is exposed to solvent). Ligands with free amino or carboxyl groups can easily be linked to other ligands to create higher affinity or more selective second-generation inhibitors. Compounds that have been reported previously to be highly toxic were excluded.

Experimental confirmation of ligand binding to HCV E2

Each of the 34 ligands was tested experimentally using surface plasmon resonance (SPR) detection (on a Biacore T100 instrument) to determine if it would bind to recombinant HCV E2 protein and to obtain an assessment (relative to the other ligands) of how well it binds. Twenty-three of the molecules provided a positive change in response units (RUs) indicating they bound to the E2 protein immobilized on the chip (Table 3). The measured responses for the ligands that bound varied from 54 to 276 RUs. Data was also obtained on the rate of ligand dissociation by measuring the amount of ligand remaining bound at two time points, dissociation 1 (10 seconds) and dissociation 2 (50 seconds), during the rinsing of the chip with buffer (Figure 4). The majority of the ligands dissociated quickly, as one might expect for small molecules that bind to the surface of a protein. A few, such as ligands 121861, 4429, 158413, 81462, and 57103, exhibited slower off rates when compared to others.

Inhibition of HCV entry

The 23 compounds that were observed to bind to recombinant E2 protein were then tested to determine if they would block HCV infection of Huh-7 cells. Pseudotyped retroviral particles harboring the envelope protein of an endogenous feline retrovirus (RD114pp) were first used to determine the specificity and the safety of molecules. We excluded from a further characterization the molecules for which the half maximal inhibitory concentration (IC50) against RD114pp was lower than 10 μM or the molecules that significantly increased RD114pp infection (Table 4). The remaining ligands were next tested against pseudotyped retroviral

Table 2. Ligands predicted to bind to the HCV E2 protein by blind docking of the NCI Diversity Set III small molecule library to the HCV E2 structural model and their predicted free energies of binding.

Ligand ID (NSC#)	Free Energy of Binding (kcal/mol)	Ligand ID	Free Energy of Binding (kcal/mol)
670283	−7.69	211490	−8.7
86467	−7.47	113486	−6.26
639174	−7.81	144694	−7.27
81462	−6.81	4429	−7.3
403379	−7.58	133071	−7.5
213700	−7.89	163910	−7.4
359472	−7.91	54709	−7.3
146554	−7.67	135618	−8.7
204232	−8.54	281254	−6.5
281816	−8.64	319990	−7.4
308835	−8.4	369070	−6.3
60785	−7.48	59620	−7.3
84100	−6.99	38968	−3.9
158413	−7.9	171303	−5.8
57103	−6.36	228155	−8.7
121861	−8.16	13316	−6.8
3076	−7.71	117268	−7.6

particles harboring genotype-2a HCV envelope proteins (HCVpp 2a), cell culture produced HCV particles (HCVcc) or RD114pp. As a positive control, an anti-CD81 antibody (JS-81) was included in the assays. One compound, 281816, showed an inhibitory effect on both HCVpp and HCVcc infection with IC50's of 1.02 μM and 3.95 μM, respectively (Table 4 and Figure 5A), indicating that this molecule inhibits the entry step of the HCV lifecycle, probably through a specific effect on the virus's interaction with CD81. Huh-7 cell toxicity was not observed over the range of ligand 281816 concentrations tested in these assays.

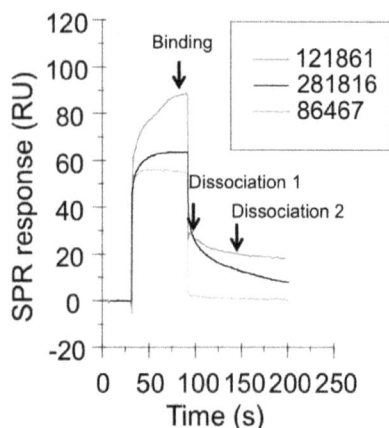

Figure 4. Surface plasmon resonance sensorgrams of ligands binding to recombinant E2 protein (Biacore T100). This figure shows sensorgrams (binding and dissociation plots) for three of the ligands that bound to the recombinant E2 protein immobilized on a CM5 chip, 281816 (black), 86467 (green) and 121861 (red), and the three reference points that are used to measure the binding and dissociation (dissociation 1 and dissociation 2) of the compound expressed in response units (RU).

To determine if 281816 would inhibit HCV genotypes other than 2a, a series of infection assays was performed with HCVpp bearing envelope proteins from a number of different HCV genotypes. Interestingly, 281816 was found to be equally effective in inhibiting Huh-7 infection by all the HCV genotypes tested (1a, 1b, 2a, 2b, 4a and 6a, Figure 5B). The IC50 values ranged from 2.2 μM to 4.6 μM (Table 5).

To confirm that 281816 inhibits HCV entry with no further effect on post-entry steps, 281816 (10 μM) was added at different time points (Figure 6A) before (−2 to 0 hr, b), during (0 to 2 hr, c), or after (2 to 24 hr, d) inoculation of Huh-7 cells with HCVcc, as previously described [53]. Cells treated with dimethylsulfoxide (DMSO) and cells treated continuously (−2 to 24 hr, a) with 281816 were used as controls. The results clearly showed that 281816 significantly inhibits HCVcc infection when present during virus infection (Figure 6A, c). The decrease in HCVcc infection that was observed in condition b is likely to be due either to some 281816 remaining bound to the cell after the washing step or its entering into the cells and acting on the entry step (Figure 6A, b). Similarly, a slight decrease was also observed in condition d (Figure 6A), which is likely related to 281816 acting on the entry of the remaining particles (those entering after 2 hr). Together, these results confirm that 281816 inhibits the entry step of HCV lifecycle.

After attachment to the cell surface and binding to entry factors, HCV virions are internalized by clathrin-mediated endocytosis [54,55]. Following internalization, HCV is transported to early endosomes along actin stress fibers, where fusion seems to take place [55,56]. To determine which step in HCV entry is impaired by 281816, we administered the ligand at different intervals during the early phase of infection. Virus attachment and binding were performed at 4°C (Figure 6B, Steps 1 and 2), Then, cells were shifted to 37°C to allow endocytosis and fusion (Figure 6B, Step 3). Cells treated with JS-81 were used as controls. The addition of 281816 during step 2 and step 3 led to the strongest inhibition of

Table 3. Magnitude of surface plasmon resonance binding response obtained for the 23 ligands that were identified to bind to recombinant E2 protein immobilized on a CM5 sensor chip.

Ligand ID (NSC#)	Binding (RU)	Dissociation 1 (RU)	Dissociation 2 (RU)
670283	54.3	4	1.4
86467	54.9	1.9	0.8
639174	55.4	2.3	0.6
81462	57.2	9.2	6.5
403379	58	2.8	1.1
213700	62	3.1	0.8
359472	62	2.5	0.8
146554	63.4	3.1	0.8
204232	63.4	2.5	0.4
281816	64.5	3.7	0.9
308835	64.8	7.1	5.2
60785	70.4	2.8	0.6
84100	71.2	4.2	2.2
158413	71.2	10.3	8.5
57103	81.6	11.4	2.5
121861	88.4	26.1	20.4
117268	88.5	4.1	1.2
3076	92.2	3.2	1.6
211490	102.9	6.1	2.1
113486	104.7	7	2.6
144694	118.8	6	2.3
4429	155.3	28.9	14.2
133071	276.3	1.8	−2

The rate of ligand dissociation is assessed by measuring the response units at two time points (10 sec and 50 sec) after the chip with bound ligand is rinsed with buffer.

HCV infection, as strong as the one observed when 281816 was present during all three steps. We also observed a significant inhibition of HCV infection when 281816 was added during the early attachment/binding steps (Figure 6B, Step 1). An MTS assay performed with 10 μM 281816 for each length of time the cells were treated with 281816 (1 hr, 2 hr, and 3 hr) showed the compound was not cytotoxic to the cells under the conditions used in the assay (Figure 7). Together, these results indicate that 281816 inhibits HCV infection by acting on more than the first (attachment/binding) step of viral entry. These data suggest the ligand also affects interactions during HCV internalization and fusion.

Blocking of E2 binding to CD81

Ligand 281816 was originally selected for testing based on the prediction by docking that it would bind to a site on the HCV E2 protein where CD81 binds. The infection assay conducted with Huh-7 cells demonstrated 281816 is effective in inhibiting the entry step in the HCV life cycle. To confirm that the binding of 281816 to E2 inhibits the HCV E2-CD81 interaction, flow cytometry was used to monitor the binding of a recombinant form of the E2 protein to native CD81 overexpressed on Raji cells as a function of 281816 concentration. The results in Figure 8 show binding of the E2 protein to Raji cells is inhibited by 281816 in a dose dependent manner. Using a second technique (an ELISA-based assay), we observed a similar dose-dependent effect of 281816 on the inhibition of the E2 protein binding to recombinant

CD81-LEL immobilized on micro titer plates (Figure 9). While an IC50 for 281816 blocking the binding of E2 to CD81 could not be determined from the flow cytometry data, the ELISA results indicate the IC50 is in the range of 0.2–0.5 μM.

A blind docking experiment with 281816 has also suggested the ligand may bind to several sites on CD81, including one that is located within the region bound by E2. 281816 binding to CD81 or other cellular proteins could explain the 281816 retention observed in washed cells in the HCV entry experiments. Such binding would likely be of little consequence, unless the ligand were to be bound within the E2 binding site on CD81 and were to block the E2:CD81 interaction by targeting both proteins.

To determine if ligand 281816 also binds to CD81, surface plasmon resonance was used to test for 281816 binding to recombinant CD81-LEL protein immobilized on a chip. As shown in Figure 10, 281816 does bind to CD81-LEL. A kinetic analysis of this binding has shown that the ligand binds to CD81-LEL (Kd = 57 μM) almost as well as it binds to E2 (Kd = 41 μM). However, a competition ELISA experiment that used the anti-CD81 antibody, 5A6, that blocks E2 binding to CD81 (Figure 11) revealed that 281816 does not bind to the E2 binding site on CD81. In this experiment, CD81-LEL was immobilized on a micro titer plate and the binding of the 5A6 antibody was monitored in the presence of 1 μM 281816 as a function of antibody dilution. The antibody 5A6 has been shown previously to bind to the same site on CD81 recognized by E2 [57,58] with an affinity (75 nM) [59] about 1/10th that of E2 (4–10 nM) [60,61].

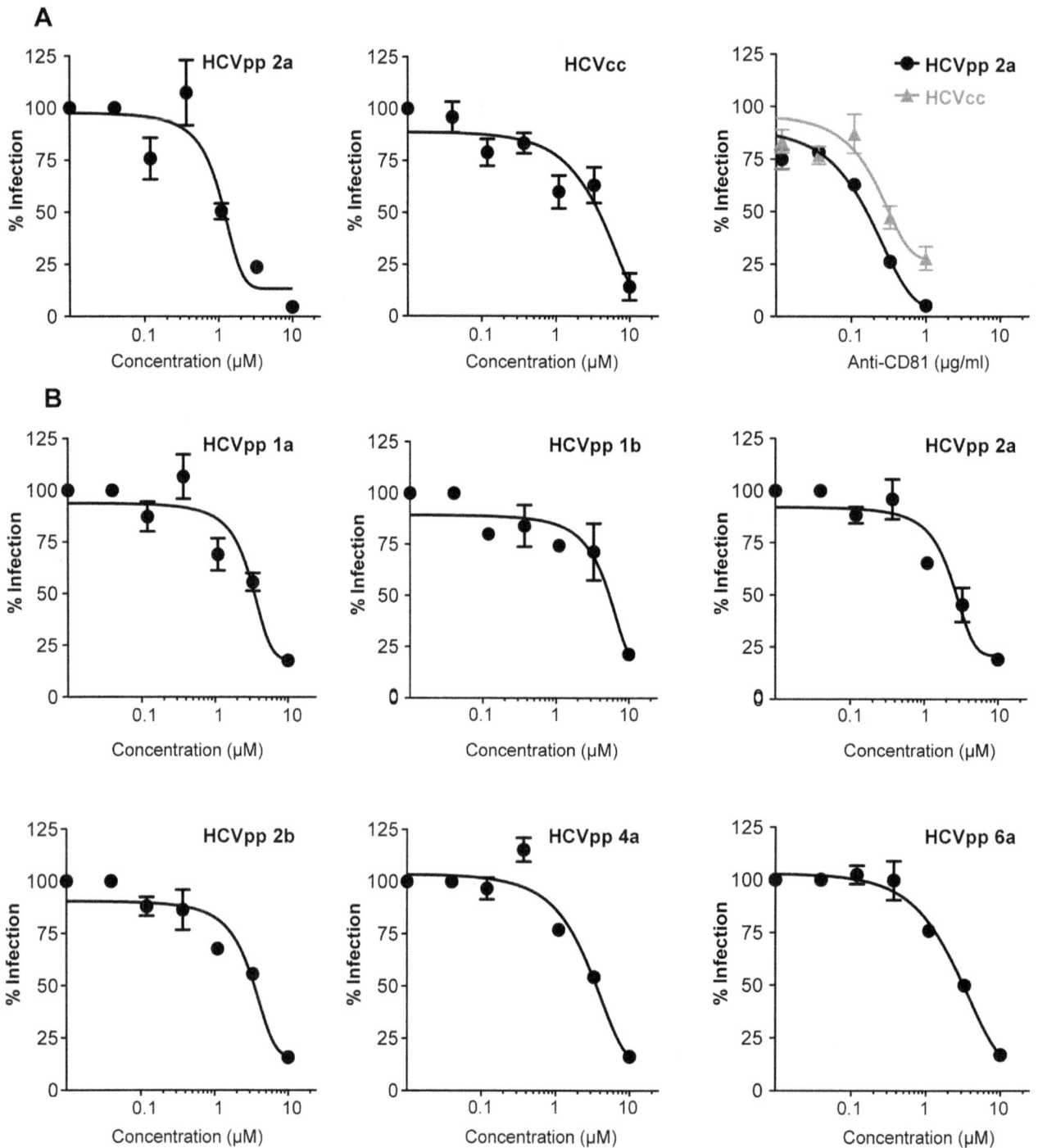

Figure 5. 281816 inhibits HCV entry in a genotype-independent manner. (A) Huh-7 cells in 96-well plates were pre-treated with 281816 (left and middle panels) or anti-CD81 antibody (right panel) at the indicated concentrations and then infected with HCVpp 2a or HCVcc. (B) Huh-7 cells in 24-well plates were pre-treated with 281816 at the indicated concentrations and infected with HCVpp expressing envelope proteins from the indicated genotype. After 30 hr of infection, cells were lysed and luciferase activities quantified. HCVpp infections were normalized to RD114pp infections.

Working at the same concentration of 281816 (1 μM) that provided the best inhibition of E2 binding to CD81-LEL, 281816 did not inhibit the 5A6 antibody binding to CD81 even when the antibody concentration was reduced to 1/2000th the concentration of the ligand (Figure 11). The amount of 5A6 antibody bound to CD81 remained the same in the presence and absence of the ligand, demonstrating that 281816 does not bind sufficiently well to the E2 binding site on CD81 to block 5A6 binding.

Table 4. The IC50 values obtained for the 23 ligands screened for their ability to inhibit HCVcc, HCVpp and RD114pp infection of Huh-7 cells.

Ligand ID (NSC#)	IC50 (µM)		
	RD114pp	HCVpp	HCVcc
670283	3	ND	ND
86467	>10	>10	>10
639174	0.03	ND	ND
81462	>10	>10	>10
403379	>10	>10	>10
213700	>10	>10	>10
359472	>10	>10	>10
146554	>10	ND	ND
204232	>10	>10	>10
281816	**>10**	**1.02**	**3.95**
308835	>10	>10	>10
60785	3.5	ND	ND
84100	>10	>10	>10
158413	>10	>10	>10
57103	0.3	ND	ND
121861	>10	>10	>10
117268	0.1	>10	>10
3076	0.25	ND	ND
211490	0.5	ND	ND
113486	>10	>10	>10
144694	>10	>10	>10
4429	>10	>10	>10
133071	0.10	ND	ND
Anti-CD81	>10*	0.17*	0.36*

ND refers to molecules that were not assayed because the molecule was not specific for HCV (it inhibited RD114pp infection). *IC50 values for anti-CD81 antibody are in µg/ml.

281816 abrogates HCV cell-to-cell transmission

In addition to cell-free infection, HCV can also be transmitted to neighboring cells via cell-to-cell contact by a mechanism that is not completely understood [41,42,62]. Indeed, HCV is transmitted in the presence of monoclonal antibodies, such as the anti-E2 antibody 3/11, or patient-derived antibodies that are able to neutralize virus-free infectivity [42,62]. Since cell-to-cell transmission has been suggested to be a major route of transmission for HCV [41], we next analyzed the effect of 281816 on this process.

For this purpose, Huh-7 cells were infected at low multiplicity of infection with HCVcc for 2 hr and then cultured with neutralizing anti-E2 antibody (3/11), which blocks infection by free particles as shown in Figure 12 [41], and in the presence of 281816 (1 µM and 10 µM). Cells cultured in the presence of 3/11 and solvent (DMSO) or Epigallocatechin-3-gallate (EGCG, 50 µM) [42] were used as negative and positive controls of inhibition, respectively. Three days post-infection, cells were fixed and foci of infected cells were visualized by immunofluorescence. Cell-to-cell transmission

Table 5. Genotype independent inhibition of HCVpp infection of Huh-7 cells by ligand 281816.

Subtypes	IC50 (µM)
HCVpp 1a	2.95
HCVpp 1b	4.66
HCVpp 2a	2.22
HCVpp 2b	2.93
HCVpp 4a	3.44
HCVpp 6a	3.30

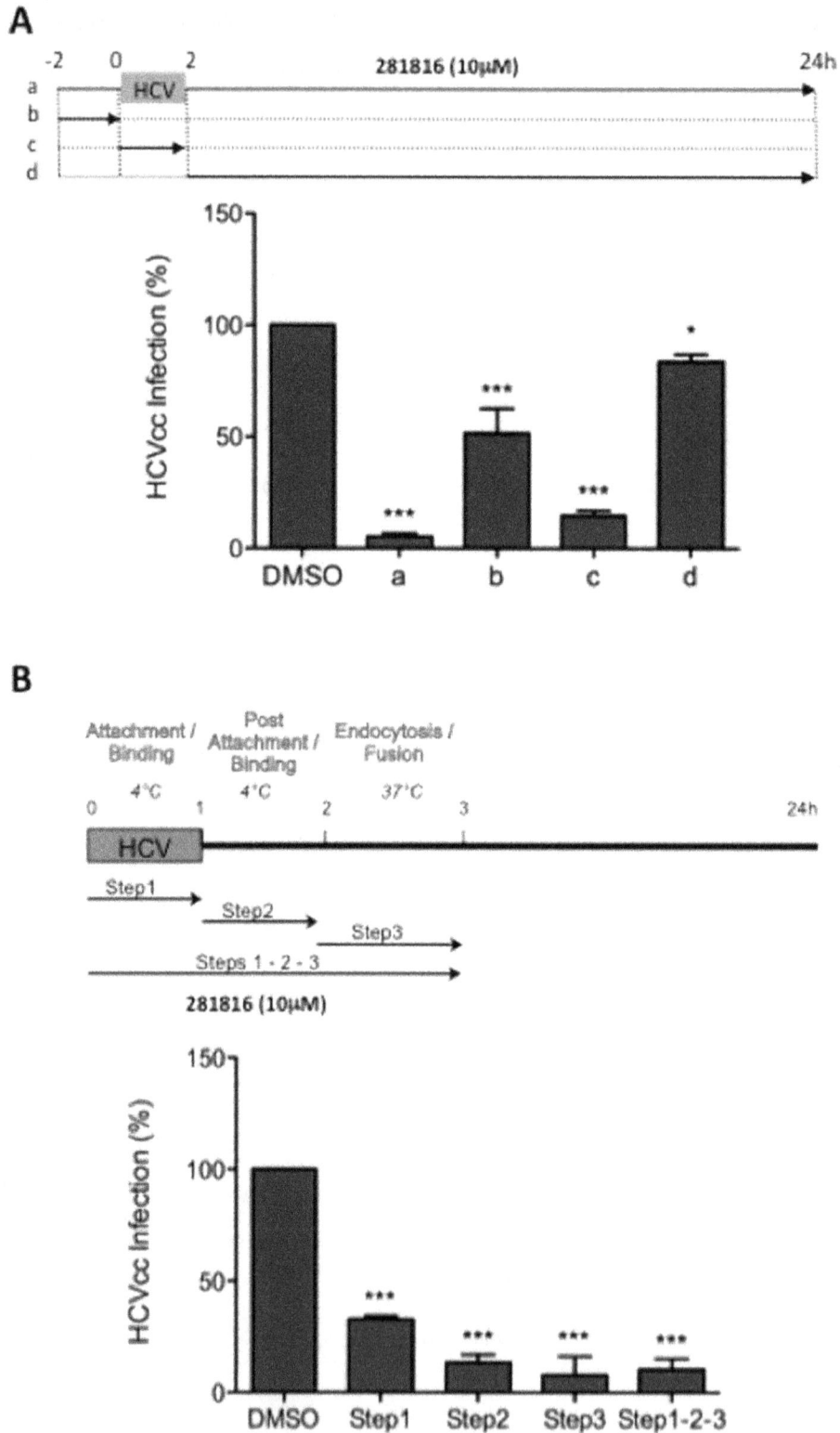

Figure 6. 281816 inhibits HCV entry. (A) Huh-7 cells in 24-well plates were treated at different time points with 281816 at 10 μM and infected with HCVcc for 2 hr at 37°C. 281816 was added full-time during the experiment (a), 2 hr before virus inoculation (b), 2 hr during virus inoculation (c), or full-time after virus inoculation (d). (B) Huh-7 cells were infected with HCVcc for 1 hr at 4°C (Step 1: attachment/binding), then virus was removed and cells incubated again at 4°C for 1 hr (Step 2: post-attachment/binding). Finally, cells were shifted at 37°C for 1 hr (Step 3: endocytosis/fusion) and left at 37°C for 21 hr. 281816 was added at 10 μM either during the Step 1, Step 2, Step 3 or Steps 1-2-3. * and *** indicate p values below 0.05 and 0.0001, respectively.

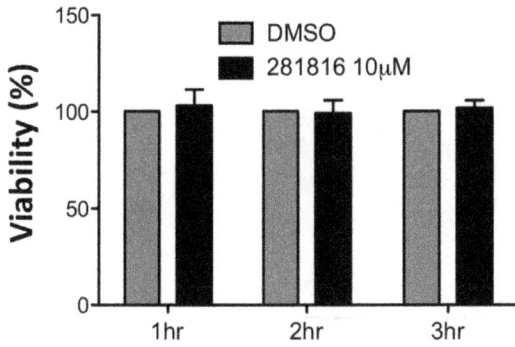

Figure 7. Viability of Huh-7 cells treated with 281816 in the HCV entry experiments. An MTS assay [38] was used to determine the viability of cells treated with 10 μM 281816 in DMSO (and DMSO alone, as a control) for 1 hr, 2 hr, or 3 hr under the same conditions used in the HCV entry experiments. 281816 is not toxic under any of the conditions used in this assay. There were no significant differences between the 281816 treated and control samples (p values <0.05).

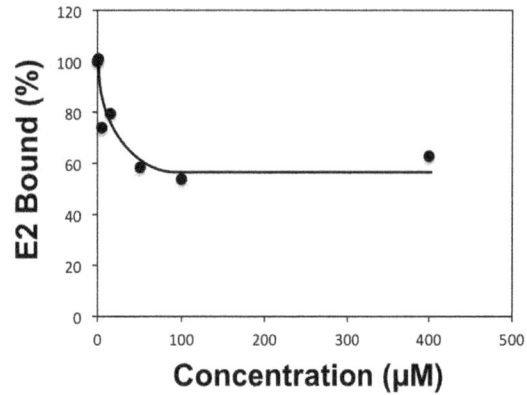

Figure 8. 281816 inhibition of HCV E2 protein binding to native CD81 on Raji cells. Flow cytometry was used to quantify recombinant HCV E2 protein binding to native CD81 over-expressed on Raji cells. Binding of the recombinant E2 protein to native CD81 on the surface of Raji cells was detected using the mouse monoclonal E2 antibody clone H53 followed by staining with a secondary FITC anti-mouse antibody. E2 binding is inhibited by 281816 in a dose-dependent manner up to 100 μM.

was measured by counting the number of infected cells per focus. The results showed that 281816 led to a significant reduction of the number of infected cells per focus in a dose-dependent manner (Figure 13). Together, these results indicate that 281816 also inhibits cell-to-cell transmission of HCV.

Discussion

While it has been known for some time that the E2 envelope glycoprotein plays an important role in the life cycle of HCV, we are only now beginning to learn details about the structure of the E2 and how it functions. This has been attributed to the challenging intrinsic properties of the protein, such as the presence of multiple flexible loops, its tendency to form disulfide aggregates in solution and the high level of N-linked glycosylation, all of which make it difficult to determine it's structure. Neutralizing antibody epitope analyses and mutation studies, in contrast, have provided a great deal of information about the regions of the E2 protein and specific amino acids that participate in CD81 binding and are important for HCV infectivity [9].

The recent determination of two HCV E2 protein core crystal structures [18,19] and our use of the deposited coordinates to create a new homology model of the protein's structure containing the majority of conserved amino acids and peptide segments known to be important for viral invasion of hepatocytes has made it possible to use computational docking and structure-based drug design methods to begin developing anti-HCV drugs that target the conserved regions of E2 and block its interaction with host receptors. Our docking of a library of diverse small molecules to this homology model led to the identification of a set of ligands that were predicted to bind to sites near key amino acids known to participate in CD81 or E1 binding or to block HCV infection, and 23 of the 34 compounds were confirmed by experiment to bind to recombinant E2 protein. When these 23 ligands were tested for activity in blocking HCV infection of Huh-7 cells, only ligand 281816 was found to inhibit HCV infection using both HCVcc and HCVpp based assays. Upon analyzing the activity spectrum of HCV using HCVpp bearing envelope proteins from different HCV genotypes (1a, 1b, 2a, 2b, 4a and 6a), 281816 was found to inhibit the infection of all tested genotypes with IC50's ranging from 2.2 μM to 4.6 μM (Table 5), indicating that this small molecule inhibits HCV infection in a genotype-independent manner. Ligand 281816 was also observed to block the binding

of HCV E2 protein to CD81-LEL protein and to Raji cells expressing CD81.

The docking experiments conducted with 281816 identified the two binding sites on E2 shown in Figure 14. One cluster of 281816 conformers bound deep inside a cavity positioned directly above Y618 and P619, two amino acids in site 4 (Figure 3) that are

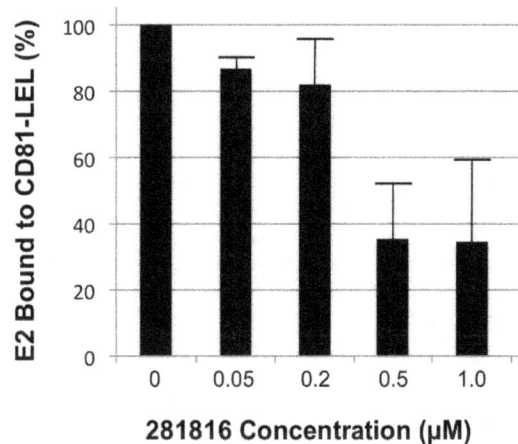

Figure 9. Ligand 281816 inhibits HCV-E2 binding to recombinant CD81-LEL. Binding of recombinant E2 protein to GST-tagged human CD81-LEL immobilized on a 96 well plate was determined using an ELISA assay. The plate was coated with GST-tagged human CD81-LEL (5 μg/ml) overnight as previously described [10], HCV E2 protein (5 μg/ml) was pre-incubated with different concentrations of 281816 for 30 min before adding to the plate, and the HCV-E2 protein (with or without the ligand) was then added to the GST-tagged human CD81-LEL coated plate and incubated for 1 hr. HCV-E2 binding was detected using a primary mouse anti-E2 antibody (H53 clone) and a secondary goat anti-mouse-HRP antibody by measuring the absorbance at 405 nm. The results, which are plotted as percent of E2 protein bound to CD81-LEL relative to E2 binding observed in the absence of the ligand (buffer control), show a dose-dependent effect of 281816 on the inhibition of the E2 protein binding to immobilized recombinant CD81-LEL. P values for the 0.05, 0.2, 0.5 and 1.0 μM 281816 samples are 0.0069, 0.0195, 0.0006 and 0.0009 respectively.

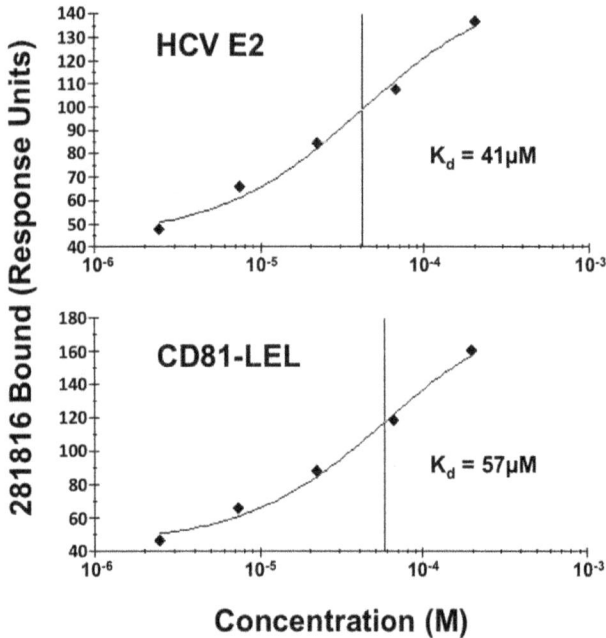

Figure 10. Single cycle kinetics of 281816 binding to recombinant HCV E2 and his-tagged CD81-LEL. Using surface plasmon resonance detection, ligand 281816 is observed to bind to both HCV E2 and CD81-LEL proteins immobilized on a CM5 chip. Analyses of the binding kinetics were used to obtain dissociation constants for 281816 binding to recombinant E2 ($K_d = 41$ μM) and recombinant his-tagged CD81-LEL ($K_d = 57$ μM) protein.

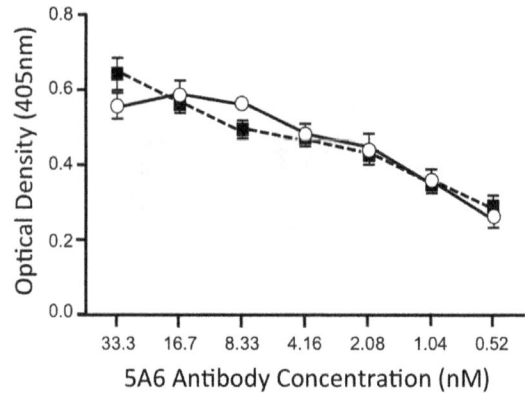

Figure 11. Ligand 281816 does not compete with binding of an anti-CD81 antibody to human CD81-LEL. Binding of anti-CD81 monoclonal antibody 5A6 to recombinant GST-tagged human CD81-LEL protein was determined by ELISA. Serial dilutions of 5A6 antibody were incubated with the CD81-LEL protein in the presence of 1 μM ligand 281816 (black squares) or PBS as a control (open circles). The amount of 5A6 antibody bound to CD81 remained the same in the presence and absence of the ligand, demonstrating that 281816 does not bind sufficiently well to the E2 binding site on CD81 to block 5A6 binding, even when the antibody concentration was reduced to 1/2000th the concentration of the ligand.

known to contribute to E2's binding to CD81 [47]. The two strongest 281816 ligand binding modes are shown bound to this site. 281816 was also predicted to bind to a shallow cavity on the opposite side of the protein. These conformers were predicted to bind to site 1 near residues V515, G517, P515 and H421–N423. H421–N423 is part of a larger segment of E2 that has been shown to bind to CD81 [19]. As expected, the ligand positioned above Y618 and P619 in the deeper cavity was predicted to bind more strongly to this region of the protein (free energy of binding of the best bound ligand = −8.64 kcal/mol) than when it was bound to the shallow cavity on the other side of the protein (free energy of binding = −6.39 kcal/mol).

A subset of the 281816 conformers in the cluster observed to bind near site 4 overlapped into site 2 and bound immediately adjacent to D481–P490, part of the epitope targeted by antibodies that block HCV infectivity and E2 binding to CD81 [50]. W487, a residue within this peptide segment whose mutation has been shown to disrupt E2:E1 dimerization [63], is also located near site 2. Other conformers in the cluster binding to site 1 also overlapped into site 5 and bound near amino acid residues P612 and Y613. These docking results illustrate one interesting and unique feature of the 281816 ligand; a number of its conformers are predicted to bind immediately above or next to both exposed faces of the P612–P619 amino acid residues that are known to participate in E2 binding to CD81 [47].

One factor that can have a significant impact on the accessibility of these sites to the binding of 281816 and other ligands is the oligomeric state of the native E2 protein. Analyses of the E2 protein and its complexes have shown that the protein exists in several different oligomeric states [43,64–73]. These include non-covalent heterodimers of E1 and E2, large disulfide cross-linked E2 complexes and aggregates, as well as monomers and disulfide

linked E2 homodimers. E1E2 non-covalent heterodimers are formed in infected cells [70–72], and it has been proposed that the two proteins remain as a complex on the virus surface, perhaps covalently linked through an intermolecular disulfide bond located in their transmembrane regions [69]. Large covalent complexes containing E2 stabilized by multiple disulfide bonds have also been observed to be associated with the surface of infective virions [71]. It has been hypothesized that the disulfide crosslinks in these large complexes may contribute to the structural stability of the virion. It has also been proposed that these large complexes may play a role in budding. Other large disulfide cross-linked aggregates of E2 have been found in recombinant E2 expression systems [74–78] and in the endoplasmic reticulum, but these aggregates do not

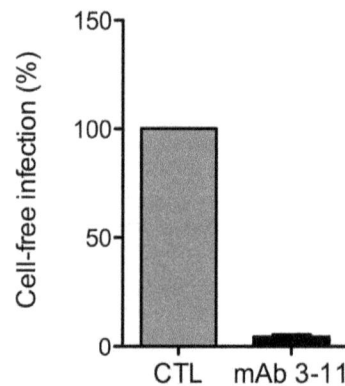

Figure 12. Cell-free infection of Huh-7 cells by HCVcc is blocked by the anti-E2 antibody 3/11. HCVcc were pre-incubated for 1 hr with neutralizing anti-E2 monoclonal antibody 3/11 at a concentration of 50 μg/ml and next inoculated to Huh-7 cells for 2 hr. Three days post-infection, cells were fixed, stained with the mouse anti-E1 antibody A4 and Alexa555 conjugated anti-mouse IgG, and the number of infected cells was counted. The results show the anti-E2 antibody 3/11 effectively blocks cell-free infection of Huh-7 cells by HCVcc.

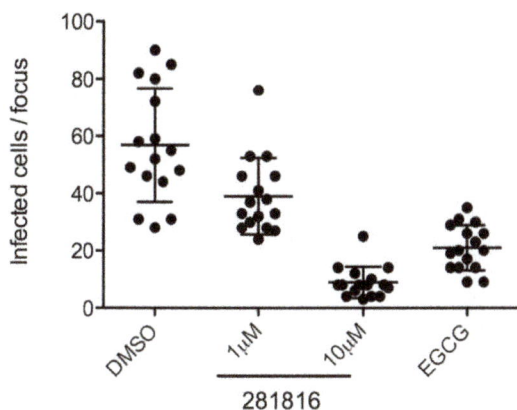

Figure 13. 281816 blocks HCV cell-to-cell transmission. Huh-7 cells were seeded on coverslips and infected with HCVcc for 2 hr at 37°C. Cells were then washed and cultured for 72 hr at 37°C in culture medium containing the 3/11 neutralizing antibody (50 µg/ml) in presence or in absence of 281816 at indicated concentrations. Cells cultured in presence of DMSO or EGCG at 50 µM were used as controls. The number of infected cells per focus was determined by A4 indirect immunofluorescence. The results show treatment with 281816 significantly reduces the number of infected cells per focus in a dose-dependent manner. Mean p values were below 0.001 for 1 µM 281816 and below 0.0001 for the 10 µM 281816 and EGCG treatment groups.

appear to represent a functional or biologically relevant oligomeric state.

E2 is known to be heavily glycosylated [69,79,80]. In the HCVcc system E2 has been reported to have 11 glycosylated sites that collectively account for nearly half the mass of the protein [69,79,80]. Glycosylation of a number of these sites has been shown to prevent the binding of neutralizing antibodies to E2 and to block E2 binding to CD81 [81], Although the ligands identified in this study are much smaller than antibodies or CD81, the presence of the glycans could also prevent small molecules from binding to the protein's surface if the glycosylated amino acid residues are located close to the ligand binding site.

While the oligomeric structure of E2 on the surface of the HCV virus is not known, the E2 present in the virus and pseudoparticles does bind to both recombinant CD81-LEL and the native CD81 receptor present on hepatocytes. Since each of the five ligand binding sites on E2 we used in our docking experiments contain or are located immediately adjacent to amino acid residues that are known to participate in E2 binding to CD81, ligands targeting these sites should have access to bind in the cavities. In addition, a number of these binding sites are located within epitopes recognized by antibodies that inhibit HCV infectivity, providing an additional confirmation of the accessibility of the sites. Only one of the 11 known glycosylated amino acid residues, E2N7, is located near a ligand docking site (Site 3). Our docking studies have identified 281816 ligand conformers that are predicted to bind to sites 1 and 4 and a few that overlap into sites 2 and 5, but none are expected to bind to Site 3. This suggests that the binding

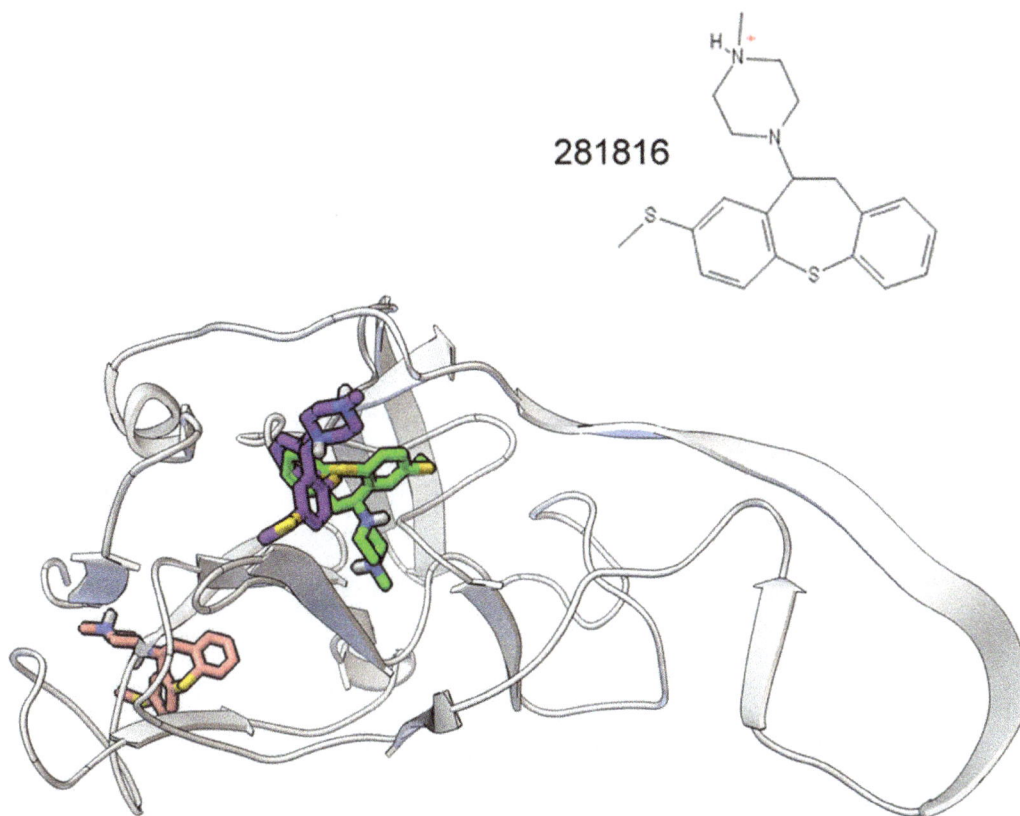

Figure 14. Relative location of 281816 binding sites 1 and 4 on HCV E2. 281816 (structure, top) is predicted to bind to two sites on the E2 protein. Two conformers of 281816 with the lowest free energy of binding are shown bound to site 4. The conformer with the lowest free energy of binding to site 1 is also shown. A video showing the surface structure of the E2 homology model with the three 281816 conformers bound that rotates 360° can be found in Movie S1.

of 281816 and its analogs should not be affected by glycosylation. The limited proteolysis and deuterium exchange experiments conducted with the E2 protein core and reported by Khan et al. [19] also indicate that each of the five ligand docking/binding sites is accessible and exposed to solvent – an important prerequisite for ligand binding.

To probe more deeply into the inhibition of the infection process by 281816, experiments were conducted to determine if the inhibition of cell-free infection by 281816 might be limited to viral entry, which step in the entry process might be affected by the compound, and what impact, if any, 281816 might have on cell-to-cell transmission of HCV. Analyses of Huh-7 cells inoculated with HCVcc before, during or after treatment with 281816 revealed the compound only blocks HCV entry and does not inhibit post-entry processes in the HCV life cycle. A kinetic analysis of the effect of 281816, coupled with a temperature block to endocytosis and fusion, was used to examine the cell-free entry steps in more detail and showed 281816 inhibits not only the initial attachment/binding step, but it also has an effect on interactions that occur later during viral internalization and fusion. Ligand 281816 was also observed to abrogate the cell-to-cell transmission of HCV. 281816 treatment of Huh-7 cells cultured in the presence of the anti-E2 neutralizing antibody 3/11 not only led to a dose dependent reduction in the number of cells forming foci, but it was found to be more effective in blocking cell-to-cell transmission that the Epigallocatechin-3-gallate [42] used as a positive control.

The observation that some 281816 remained bound to cells after washing in the HCV entry experiments suggests that 281816 may bind to other cellular proteins. This should not be surprising, since this compound has been reported to have other activities [82–87]. A blind docking experiment performed with 281816 also predicted the ligand could bind to CD81. Three potential binding sites were identified, one located within the E2 binding site on CD81 and two others in regions that would not be expected to impact E2 binding. Collectively these observations suggested the exciting possibility that 281816 might play a dual role in blocking E2 binding to CD81 by binding not only to the CD81 binding site on E2, but also by binding to the E2 binding site on CD81. In support of this idea, results obtained in an SPR binding study showed 281816 bound to CD81 almost as well as it bound to E2. However, a subsequent competition experiment conducted with 281816 and a monoclonal antibody (5A6) known to bind to the E2 binding site on CD81 revealed that the ligand did not compete with the antibody. One micromolar 281816, which effectively blocks E2 binding to CD81-LEL and inhibits viral invasion of Huh-7 cells, had no effect on antibody 5A6 binding to CD81-LEL.

In addition to identifying a promising new small molecule drug lead for treating HCV that targets the E2 glycoprotein, this study also demonstrates the utility of our new E2 homology model in the discovery of small molecules that bind to important sites on E2. By targeting sites containing amino acid residues identified by others to participate in CD81 binding and CD81-dependent processes that impact HCV infectivity, a small molecule was identified that not only blocks E2 binding to CD81 and the cell-free entry process, but it is also effective in blocking the cell-to-cell transmission of HCV – the predominant mechanism of transmission that contributes to the persistence of infections [41] but for which the precise mechanism needs to be defined. Recently, it has been shown that exosomes produced by HCV infected hepatic cells can transfer viral RNA to plasmacytoid dendritic cells [88] and might transmit infection to naïve hepatic cells [89]. Although several entry factors have been implicated in this process, the viral determinants, entry factor requirements and molecular mechanisms involved in cell-to-cell transmission route still need to be

further characterized. In particular, the role played by CD81 has remained controversial with studies reporting HCV cell-to-cell transmission as a CD81-dependent pathway [41,90,91], whereas others demonstrated a CD81-independent transmission [56,62,92]. However, a recent study has highlighted the coexistence of CD81-dependent and CD81-independent cell-to-cell transmission [93]. The inhibition of cell-to-cell transmission of HCV by 281816, which blocks E2 binding to CD81, is consistent with other reports of a CD81-dependent cell-to-cell transmission process [93–94] that can be blocked by anti-CD81 antibodies [41,94] and soluble CD81 [56], both of which also block E2 binding to CD81. While it is possible that E2 binding to CD81 may play a role in the cell-to-cell transmission of HCV, it is also possible the 281816 that binds to CD81, which does not inhibit E2 binding, may have a totally unrelated effect that impacts the interaction of CD81 with other proteins or molecular structures in the tetraspanin web [95,96] and blocks fusion related events involving CD81 that occur during the cell-to-cell transmission process.

281816, known as methiothepin or 1-methyl-4-(3-methylsulfanyl-5,6-dihydrobenzo[b] [1]benzothiepin-5-yl)piperazine, is also interesting because it has been determined previously to block dopamine [82] and serotonin [83] receptors and has been reported to inhibit a number of other biological activities, which include the binding or entry of two other unrelated viruses into cells (Lassa [84], Marburg [85]), *Plasmodium falciparum* proliferation [86], and *Mycoplasmodium tuberculosis* infections [87]. Numerous structural analogs of 281816 have been tested and shown to be effective in treating a wide variety of neurological diseases (schizophrenia [97,98], Parkinson and dementia-related psychoses [99,100]), bipolar disorders [101,102], and depression [103,104]. While we have not found experimental studies that report the membrane permeability of 281816, the logP (log of octanol/water partition coefficient) has been calculated to be 4.14, which indicates the compound is likely to exhibit good membrane permeability. Since the logP is less than 5, according to Chris Lipinski/Pfizer's Rule of 5 the compound could also be orally active, as are a number of 281816 structural analogs (octoclothepine, loxapine, amoxapine, clozapine, quetiapine, olanzapine, and amitriptyline) that have been used to treat a variety of neurological disorders.

Acknowledgments

We thank Lucie Feneant for providing the HCVpp used in this work. Biacore support and instrument use was provided by the Protein Expression Center at the California Institute of Technology, Pasadena, CA. We thank Leslie P. Michelson and Ryan Novosielski of the Rutgers University Office of Information Technology, High Performance and Research Computing, for developing and maintaining the Linux cluster at Rutgers University-NJMS and for assisting us with its use. The ligands tested in this study were provided by the National Cancer Institute through its NCI/Developmental Therapeutics Program (DTP) Open Chemical Repository (http://dtp.cancer.gov).

Author Contributions

Conceived and designed the experiments: RRA LC AZ JD JV JM JSF SF SL RB HMA. Performed the experiments: RRA AZ LS AGK FVC ALP SF. Analyzed the data: RRA LC AZ JV AGK FVC ALP SF RB. Contributed reagents/materials/analysis tools: RRA JD JV JM JSF SL HMA. Contributed to the writing of the manuscript: RRA LC AZ RB.

References

1. Anwar MI, Rahman M, Hassan MU, Iqbal M (2013) Prevalence of active hepatitis C virus infections among general public of lahore, Pakistan. Virol J 10(1): 351-422X-10-351. 10.1186/1743-422X-10-351; 10.1186/1743-422X-10-351.

2. Blackard JT, Shata MT, Shire NJ, Sherman KE (2008) Acute hepatitis C virus infection: A chronic problem. Hepatology 47(1): 321–331. 10.1002/hep.21902.

3. Zeuzem S, Berg T, Moeller B, Hinrichsen H, Mauss S, et al. (2009) Expert opinion on the treatment of patients with chronic hepatitis C. J Viral Hepat 16(2): 75–90. 10.1111/j.1365-2893.2008.01012.x.

4. Marks KM, Jacobson IM (2012) The first wave: HCV NS3 protease inhibitors telaprevir and boceprevir. Antivir Ther 17(6 Pt B): 1119–1131. 10.3851/IMP2424; 10.3851/IMP2424.

5. Colombo M, Fernandez I, Abdurakhmanov D, Ferreira PA, Strasser SI, et al. (2013) Safety and on-treatment efficacy of telaprevir: The early access programme for patients with advanced hepatitis C. Gut. 10.1136/gutjnl-2013-305667; 10.1136/gutjnl-2013-305667.

6. Asselah T (2014) Sofosbuvir for the treatment of hepatitis C virus. Expert Opin Pharmacother 15(1): 121–130. 10.1517/14656566.2014.857656; 10.1517/14656566.2014.857656.

7. Lenz O, Vijgen L, Berke JM, Cummings MD, Fevery B, et al. (2013) Virologic response and characterization of HCV genotype 2-6 in patients receiving TMC435 monotherapy (study TMC435-C202). J Hepatol 58(3): 445–451. 10.1016/j.jhep.2012.10.028; 10.1016/j.jhep.2012.10.028.

8. Pileri P, Uematsu Y, Campagnoli S, Galli G, Falugi F, et al. (1998) Binding of hepatitis C virus to CD81. Science 282: 938.

9. Feneant L, Levy S, Cocquerel L (2014) CD81 and hepatitis C virus (HCV) infection. Viruses 6: 535–572. doi:10.3390/v6020535.

10. Higginbottom A, Quinn ER, Kuo CC, Flint M, Wilson LH, et al. (2000) Identification of amino acid residues in CD81 critical for interaction with hepatitis C virus envelope glycoprotein E2. J Virol 74(8): 3642–3649.

11. Drummer HE, Wilson KA, Poumbourios P (2002) Identification of the hepatitis C virus E2 glycoprotein binding site on the large extracellular loop of CD81. J Virol 76(21): 11143–11147.

12. Zhang YY, Zhang BH, Ishii K, Liang TJ (2010) Novel function of CD81 in controlling hepatitis C virus replication. J Virol 84(7): 3396–3407. 10.1128/JVI.02391-09.

13. Ahlenstiel G (2013) The natural killer cell response to HCV infection. Immune Netw 13(5): 168–176. 10.4110/in.2013.13.5.168.

14. El-Awady MK, Tabll AA, El-Abd YS, Yousif H, Hegab M, et al. (2009) Conserved peptides within the E2 region of hepatitis C virus induce humoral and cellular responses in goats. Virol J 6: 66-422X-6-66. 10.1186/1743-422X-6-66; 10.1186/1743-422X-6-66.

15. Carlsen TH, Scheel TK, Ramirez S, Foung SK, Bukh J (2013) Characterization of hepatitis C virus recombinants with chimeric E1/E2 envelope proteins and identification of single amino acids in the E2 stem region important for entry. J Virol 87(3): 1385–1399. 10.1128/JVI.00684-12; 10.1128/JVI.00684-12.

16. Li YP, Kang HN, Babiuk LA, Liu Q (2006) Elicitation of strong immune responses by a DNA vaccine expressing a secreted form of hepatitis C virus envelope protein E2 in murine and porcine animal models. World J Gastroenterol 12(44): 7126–7135.

17. Ray R, Meyer K, Banerjee A, Basu A, Coates S, et al. (2010) Characterization of antibodies induced by vaccination with hepatitis C virus envelope glycoproteins. J Infect Dis 202(6): 862–866. 10.1086/655902; 10.1086/655902.

18. Kong L, Giang E, Nieusma T, Kadam RU, Cogburn KE, et al. (2013) Hepatitis C virus E2 envelope glycoprotein core structure. Science 342(6162): 1090–1094. 10.1126/science.1243876.

19. Khan AG, Whidby J, Miller MT, Scarborough H, Zatorski AV, et al. (2014) Structure of the core ectodomain of the hepatitis C virus envelope glycoprotein 2. Nature. 10.1038/nature13117.

20. Law M, Maruyama T, Lewis J, Giang E, Tarr AW, et al. (2008) Broadly neutralizing antibodies protect against hepatitis C virus quasispecies challenge. Nat Med 14(1): 25–27. 10.1038/nm1698.

21. Zemla A, Zhou CE, Slezak T, Kuczmarski T, Rama D, et al. (2005) AS2TS system for protein structure modeling and analysis. Nucleic Acids Res 33(Web Server issue): W111-5. 10.1093/nar/gki457.

22. Zemla AT, Lang DM, Kostova T, Andino R, Ecale Zhou CL (2011) StralSV: assessment of sequence variability within similar 3D structures and application to polio RNA-dependent RNA polymerase, BMC Bioinformatics 12: 226.

23. Krivov GG, Shapovalov MV, Dunbrack Jr RL (2009) Improved prediction of protein side-chain conformations with scwrl4. Proteins 77: 778.

24. Chen VB, Arendall 3rd WB, Headd JJ, Keedy DA, Immormino RM, et al. (2010) MolProbity: all-atom structure validation for macromolecular crystallography, Acta Crystallogr D Biol Crystallogr 66: 12.

25. Pettersen EF, Goddard TD, Huang CC, Couch GS, Greenblatt DM, et al. (2004) UCSF chimera-a visualization system for exploratory research and analysis, J Comput Chem 25: 1605.

26. Trott O, Olson AJ (2010) AutoDock VINA: Improving the speed and accuracy of docking with a new scoring function, efficient optimization, and multi-threading. J Comput Chem 31(2): 455–461. 10.1002/jcc.21334.

27. Chen VB, Arendall WB, Headd JJ, Keedy DA, Immormino RM et al. (2010) MolProbity: All-atom structure validation for macromolecular crystallography, Acta Cryst D66: 12.

28. Morris GM, Huey R, Lindstrom W, Sanner MF, Belew RK, et al. (2009) AutoDock4 and AutoDockTools4: Automated docking with selective receptor flexibility. J Comput Chem 30(16): 2785–2791. 10.1002/jcc.21256.

29. Irwin JJ, Sterling T, Mysinger MM, Bolstad ES, Coleman RG (2012) ZINC: A free tool to discover chemistry for biology. J Chem Inf Model 52: 1757.

30. Forli S Raccoon. Available: http://autodock.scripps.edu/resources/raccoon. Accessed 2013. Molecular Graphics Laboratory, The Scripps Research Institute, La Jolla, CA, 2010.

31. Perryman AL, Santiago DN, Forli S, Santos-Martins D, Olson AJ (2014) Virtual screening with AutoDock VINA and the common pharmacophore engine of a low diversity library of fragments and hits against the three allosteric sites of HIV integrase: participation in the SAMPL4 protein-ligand binding challenge. J Comput Aided Mol Des 28: 429-41. doi: 10.1007/s10822-014-9709-3.

32. Bartosch B, Bukh J, Meunier JC, Granier C, Engle RE, et al. (2003) In vitro assay for neutralizing antibody to hepatitis C virus: Evidence for broadly conserved neutralization epitopes. Proc Natl Acad Sci U S A 100(24): 14199–14204. 10.1073/pnas.2335981100.

33. Op De Beeck A, Voisset C, Bartosch B, Ciczora Y, Cocquerel L, et al. (2004) Characterization of functional hepatitis C virus envelope glycoproteins. J Virol 78(6): 2994–3002.

34. Sandrin V, Boson B, Salmon P, Gay W, Negre D, et al. (2002) Lentiviral vectors pseudotyped with a modified RD114 envelope glycoprotein show increased stability in sera and augmented transduction of primary lymphocytes and CD34+ cells derived from human and nonhuman primates. Blood 100(3): 823–832. 10.1182/blood-2001-11-0042.

35. Wakita T, Pietschmann T, Kato T, Date T, Miyamoto M, et al. (2005) Production of infectious hepatitis C virus in tissue culture from a cloned viral genome. Nat Med 11(7): 791–796. 10.1038/nm1268.

36. Rocha-Perugini V, Montpellier C, Delgrange D, Wychowski C, Helle F, et al. (2008) The CD81 partner EWI-2wint inhibits hepatitis C virus entry. PLOS One 3(4): e1866. 10.1371/journal.pone.0001866; 10.1371/journal.-pone.0001866.

37. Delgrange D, Pillez A, Castelain S, Cocquerel L, Rouille Y, et al. (2007) Robust production of infectious viral particles in huh-7 cells by introducing mutations in hepatitis C virus structural proteins. J Gen Virol 88(Pt 9): 2495–2503. 10.1099/vir.0.82872-0.

38. Malich G, Markovic B, Winder C (1997) The sensitivity and specificity of the MTS tetrazolium assay for detecting the in vitro cytotoxicity of 20 chemicals using human cell lines, Toxicology 124: 179.

39. Ferrer M, Yunta M, Lazo PA (1998) Pattern of expression of tetraspanin antigen genes in Burkitt lymphoma cell lines. Clin Exp Immunol 113: 346–352.

40. Luo RF, Zhao S, Tibshirani R, Myklebust JH, Sanyal M, et al. (2010) CD81 protein is expressed at high levels in normal germinal center B cells and in subtypes of human lymphomas. Human Pathol 41(2): 271–280.

41. Vausselin T, Calland N, Belouzard S, Descamps V, Douam F, et al. (2013) The antimalarial ferroquine is an inhibitor of hepatitis C virus. Hepatology 58: 86–97. 10.1002/hep.26273.

42. Calland N, Albecka A, Belouzard S, Wychowski C, Duverlie G, et al. (2012) (−)-Epigallocatechin-3-gallate is a new inhibitor of hepatitis C virus entry. Hepatology 55(3): 720–729. 10.1002/hep.24803; 10.1002/hep.24803.

43. Dubuisson J, Hsu HH, Cheung RC, Greenberg HB, Russell DG, et al. (1994) Formation and intracellular localization of hepatitis C virus envelope glycoprotein complexes expressed by recombinant vaccinia and sindbis viruses. J Virol 68(10): 6147–6160.

44. Cocquerel L, Meunier JC, Pillez A, Wychowski C, Dubuisson J (1998) A retention signal necessary and sufficient for endoplasmic reticulum localization maps to the transmembrane domain of hepatitis C virus glycoprotein E2. J Virol 72: 2183–2191.

45. Flint M, von Hahn T, Zhang J, Farquhar M, Jones CT, et al. (2006) Diverse CD81 proteins support hepatitis C virus infection. J Virol 80(22): 11331–11342. 10.1128/JVI.00104-06.

46. Zemla A (2003) LGA: A method for finding 3D similarities in protein structures. Nucleic Acids Res 31(13): 3370–3374.

47. Owsianka AM, Timms JM, Tarr AW, Brown RJ, Hickling TP, et al. (2006) Identification of conserved residues in the E2 envelope glycoprotein of the

hepatitis C virus that are critical for CD81 binding. J Virol 80(17): 8695–8704. 10.1128/JVI.00271-06.

48. Roccasecca R, Ansuini H, Vitelli A, Meola A, Scarselli E, et al. (2003) Binding of the hepatitis C virus E2 glycoprotein to CD81 is strain specific and is modulated by a complex interplay between hypervariable regions 1 and 2, J Virology 77: 1856.

49. Rothwangl KB, Manicassamy B, Uprichard SL, Rong L (2008) Dissecting the role of putative CD81 binding regions of E2 in mediating HCV entry: Putative CD81 binding region 1 is not involved in CD81 binding. Virol J 5: 46-422X-5-46. 10.1186/1743-422X-5-46; 10.1186/1743-422X-5-46.

50. Lavillette D, Pecheur EI, Donot P, Fresquet J, Molle J, et al (2007) Characterization of fusion determinants points to the involvement of three discrete regions of both E1 and E2 glycoproteins in the membrane fusion process of hepatitis C virus, J Virology 81: 8752.

51. Tarr AW, Owsianka AM, Timms JM, McClure CP, Brown RJ, et al. (2006) Characterization of the hepatitis C virus E2 epitope defined by the broadly neutralizing monoclonal antibody AP33, Hepatology 43: 592.

52. Goueslain L, Alsaleh K, Horellou P, Roingeard P, Descamps V, et al (2010) Identification of GBF1 as a cellular factor required for hepatitis C virus RNA replication, J Virology 84: 773.

53. Blanchard E, Belouzard S, Goueslain L, Wakita T, Dubuisson J, et al. (2006) Hepatitis C virus entry depends on clathrin-mediated endocytosis. J Virol 80(14): 6964–6972. 10.1128/JVI.00024-06.

54. Meertens L, Bertaux C, Dragic T (2006) Hepatitis C virus entry requires a critical postinternalization step and delivery to early endosomes via clathrin-coated vesicles. J Virol 80(23): 11571–11578. 10.1128/JVI.01717-06.

55. Coller KE, Berger KL, Heaton NS, Cooper JD, Yoon R, et al. (2009) RNA interference and single particle tracking analysis of hepatitis C virus endocytosis. PLOS Pathog 5(12): e1000702. 10.1371/journal.ppat.1000702; 10.1371/journal.ppat.1000702.

56. Timpe JM, Stamataki Z, Jennings A, Hu K, Farquhar MJ, et al. (2008) Hepatitis C virus cell-cell transmission in hepatoma cells in the presence of neutralizing antibodies. Hepatology 47(1): 17–24. 10.1002/hep.21959.

57. Flint M, Maidens C, Loomis-Price LD, Shotton C, Dubuisson J, et al. (1999) Characterization of hepatitis C virus E2 glycoprotein interaction with a putative cellular receptor, CD81. J Virol 73 (8): 6235–6244.

58. VanCompernolle SE, Wiznycia AV, Rush JR, Dhanasekaran M, Baures PW, et al. (2003) Small molecule inhibition of hepatitis C virus E2 binding to CD81. Virol 314: 371–380.

59. Delandre C, Penabaz TR, Passarelli AL, Chapes SK, Clem RJ (2009) Mutation of juxtamembrane cysteines in the tetraspanin CD81 affects palmitoylation and alters interaction with other proteins at the cell surface. Exp Cell Res 315(11): 1953–1963. Doi:10.1016/j.yexcr.2009.03.013.

60. Takayama H, Chelikani P, Reeves PJ, Zhang S, Khorana HG (2008) High-level expression, single-step immunoaffinity purification and characterization of human tetraspanin membrane protein CD81. Plos One 3(6) e2314.

61. Rosa D, Campagnoli S, Moretto C, Guenzi E, Cousens L, et al. (1996) A quantitative test to estimate neutralizing antibodies to the hepatitis C virus: Cytofluorimetric assessment of envelope glycoprotein 2 binding to target cells. Proc Natl Acad Sci USA 93: 1759–1763.

62. Witteveldt J, Evans MJ, Bitzegeio J, Koutsoudakis G, Owsianka AM, et al. (2009) CD81 is dispensable for hepatitis C virus cell-to-cell transmission in hepatoma cells. J Gen Virol 90(Pt 1): 48–58. 10.1099/vir.0.006700-0; 10.1099/vir.0.006700-0.

63. Yi M, Nakamoto Y, Kaneko S, Yamashita T, Murakami S (1997) Delineation of regions important for heteromeric association of hepatitis C virus E1 and E2. Virology 231: 119.

64. Op de Beeck A, Cocquerel L, Dubuisson J (2001) Biogenesis of hepatitis C virus envelope glycoproteins. J Gen Virol 82: 2589–2595.

65. Dubuisson J, Rice CM (1996) Hepatitis C virus glycoprotein folding: disulfide bond formation and association with calnexin. J Virol 20(2): 778–86.

66. Grakoui A, Wychowski C, Lin C, Feinstone SM, Rice CM (1993) Expression and identification of hepatitis C virus polyprotein cleavage products. J Virol 67(3): 1385–1395.

67. Lanford RE, Notvall L, Chavez D, White R, Frenzel G, et al. (1993) Analysis of hepatitis C virus capsid, E1, and E2/NS1 proteins expressed in insect cells. Virol 197(1): 225–35.

68. Ralston R, Thudium K, Berger K, Kuo C, Gervase B, et al. (1993) Characterization of hepatitis C virus envelope glycoprotein complexes expressed by recombinant vaccinia viruses. J Virol 67(11): 6753–6761.

69. Whidby J, Mateu G, Scarborough H, Demeler B, Grakoui A, et al. (2009) Blocking hepatitis C virus infection with recombinant form of envelope protein 2 ectodomain. J Virol 83(21): 11078–11089.

70. Drummer HE, Poumbourios P (2004) Hepatitis C virus glycoprotein E2 contains a membrane-proximal heptad repeat sequence that is essential for E1E2 glycoprotein heterodimerization and viral entry. J Biol Chem 279: 30066–30072.

71. Vieyres G, Thomas X, Descamps V, Duverlie G, Patel AH, et al. (2010) Characterization of the envelope glycoproteins associated with infectious hepatitis C virus. J Virol 84(19): 10159–68. doi: 10.1128/JVI.01180-10.

72. Op De Beeck A, Montserret R, Duvet S, Cocquerel L, Cacan R, et al. (2000) The transmembrane domains of hepatitis C virus envelope glycoproteins E1 and E2 play a major role in heterodimerization. J Biol Chem 275: 31428–31437. doi: 10.1074/jbc.M003003200.

73. Deleersnyder V, Pillez A, Wychowski C, Blight K, Xu J, et al. (1997) Formation of native hepatitis C virus glycoprotein complexes. J Virol 71(1): 697–704.

74. Brazzoli M, Helenius A, Foung SK, Houghton M, Abrignani S, et al. (2005) Folding and dimerization of hepatitis C virus E1 and E2 glycoproteins in stably transfected CHO cells. Virol 332: 438–453.

75. Liu J, Zhu L, Zhang X, Lu M, Kong Y, et al. (2001) Expression, purification, immunological characterization and application of Escherichia coli-derived hepatitis C virus E2 proteins. Biotechnol Appl Biochem 34: 109–119.

76. Martinez-Donato G, Capdesuner Y, Acosta-Rivero N, Rodriguez A, Morales-Grillo J, et al. (2007) Multimeric HCV E2 protein obtained from Pichia pastoris cells induces a strong immune response in mice. Mol Biotechnol 35: 225–235.

77. Rodriguez-Rodriguez M, Tello D, Yelamos B, Gomez-Gutierrez J, Pacheco B, et al. (2009) Structural properties of the ectodomain of hepatitis C virus E2 envelope protein. Virus Res 139: 91–99.

78. Yurkova MS, Patel AH, Fedorov AN (2004) Characterization of bacterially expressed structural protein E2 of hepatitis C virus. Protein Expr Purif 37: 119–125.

79. Goffard A, Callens N, Bartosch B, Wychowski C, Cosset FL, et al. (2005) Role of N-linked glycans in the functions of hepatitis C virus envelope glycoproteins. J Virol 79: 8400–8409.

80. Helle F, Goffard A, Morel V, Duverlie G, McKeating J, et al. (2007) The neutralizing activity of anti-hepatitis C virus antibodies is modulated by specific glycans on the E2 envelope protein. J Virol 81(15): 8101–8111.

81. Helle F, Vieyres G, Elkrief L, Popescu, Wychowski C, et al. (2010) Role of N-linked glycans in the functions of hepatitis C virus envelope proteins incorporated into infectious virions. J Virol 84(22): 11905–11915.

82. HTS assay for allosteric agonists of the human D1 dopamine receptor: Primary screen for antagonists. NIH Molecular Libraries Probe Production Network. BioAssay AID 488983. Accessed on March 21, 2014. Available: https://pubchem.ncbi.nlm.nih.gov/assay/assay.cgi?aid=488983.

83. Antagonists at human 5-hydroxytryptamine receptor 5-htle. Extracted from literature and IUPHAR database. BioAssay 624232. Accessed on March 21, 2014. Available: https://pubchem.ncbi.nlm.nih.gov/assay/assay.cgi?aid=624232.

84. qHTS for inhibitors of binding or entry into cells for Lassa virus. NIH Molecular Libraries Probe Production Network. BioAssay 540256. Accessed on March 21, 2014. Available: https://pubchem.ncbi.nlm.nih.gov/assay/assay.cgi?aid=540256.

85. qHTS for inhibitors of binding or entry into cells for Marburg virus. NIH Molecular Libraries Probe Production Network. BioAssay 540256. Accessed on March 21, 2014. Available: https://pubchem.ncbi.nlm.nih.gov/assay/assay.cgi?aid=720532.

86. qHTS for inhibitors of Plasmodium falciparum proliferation. NIH National Institute of Allergy and Infectious Diseases, Xinzhuan Su. BioAssay 504749. Accessed on March 21, 2014. Available: https://pubchem.ncbi.nlm.nih.gov/assay/assay.cgi?mid=504749_53.

87. High throughput screen to identify inhibitors of Mycobacterium tuberculosis H37Rv. Southern Research Institute, Birmingham, AL. E. Lucile White. BioAssay AID 1332. Accessed on March 21, 2014. Available: https://pubchem.ncbi.nlm.nih.gov/assay/assay.cgi?aid=1332.

88. Dreux M, Garaigorta U, Boyd B, Decembre E, Chung J, et al. (2012) Short range exosomal transfer of viral RNA from infected cells to plasmacytoid dendritic cells triggers innate immunity. Cell Host Microbe 12(4): 558–570. doi:10.1016/j.chom.2012.08.010.

89. Ramakrishnaiah V, Thumann C, Fofana I, Habersetzer F, Pan Q, et al. (2013) Exosome-mediated transmission of hepatitis C virus between human hepatoma Huh7.5 cells. Proc Natl Acad Sci USA 110(32): 13109–13113.

90. Russell R, Meunier J, Takikawa S, Faulk K, Engle RE, et al. (2008) Advantages of a single-cycle production assay to study cell culture-adaptive mutations of hepatitis C virus. Proc Natl Acad Sci USA 105(11): 4370–4375.

91. Potel J, Rassam P, Montpellier C, Kaestner L, Werkmeister E, et al. (2013) EWI-2wint promotes CD81 clustering that abrogates Hepatitis C virus entry. Cell Microbiol 15(7): 1234-52. doi: 10.1111/cmi.12112. Epub 2013 Feb 16.

92. Jones CT, Catanese MT, Law LMJ, Khetani SR, Syder AJ, et al. (2010) Real-time imaging of hepatitis C virus infection using a fluorescent cell-based reporter system. Nat Biotechnol 28(2): 167–171. doi:10.1038/nbt.1604.

93. Catanese MT, Loureiro J, Jones CT, Dorner M, von Hahn T, et al. (2013) Different requirements for scavenger receptor class B Type 1 in hepatitis C virus cell-free versus cell-to-cell transmission. J Virol 87 (15): 8282–8293.

94. Fofana I, Xiao F, Thumann C, Turek M, Zona L, et al. (2013) A novel monoclonal anti-CD81 antibody produced by genetic immunization efficiently inhibits hepatitis C virus cell-to-cell transmission. PLOS One 8(5): e64221.

95. Rubinstein E, Le Naour F, Lagaudriere-Gesbert C, Billard M, Conjeaud H, et al. (1996) CD9, CD63, CD81, and CD82 are components of a surface tetraspan network connected to HLA-DR and VLA integrins. Eur J Immunol 26(11): 2657–2665.

96. Shoham T, Rajapaksa R, Kuo C, Haimovich J, Levy S (2006) Binding of the tetraspanin web: Distinct structural domains of CD81 function in different cellular compartments. Mol Cell Biol 26(4): 1373–1385. doi:10.1128/MCB.26.4.1373–1385.2006.

97. Uhlir F, Stevkova K, Kanczuka V (1975) A comparison of oxypertine Winthrop and clorothepin (Clothepin Spofa) in schizophrenic psychoses. Activitas Nervosa Superior 17(4): 215.

98. Allen MH, Feifel D, Lesem MD, Zimbroff DL, Ross R, et al. (2011) Efficacy and safety of loxapine for inhalation in the treatment of agitation in patients with schizophrenia: a randomized, double-blind, placebo-controlled trial. J Clin Psychiatry 72(10): 1313–1321.

99. Maher AR, Theodore G (2012) Summary of the comparative effectiveness review on off-label use of atypical antipsychotics. J Manag Care Pharm 18(5 Suppl B): S1–20.

100. Weintraub D, Chen P, Ignacio RV, Mamikonyan E, Kales HC (2011) Patterns and trends in antipsychotic prescribing for Parkinson disease psychosis. Arch Neurol 68(7): 899–904. doi: 10.1001/archneurol.2011.139.

101. Kwentus J, Riesenberg RA, Marandi M, Manning RA, Allen MH, et al. (2012) Rapid acute treatment of agitation in patients with bipolar I disorder: a multi-center, randomized, placebo-controlled clinical trial with inhaled loxapine. Bipolar Disord 14(1): 31–40.

102. Thase ME (2008) Quetiapine monotherapy for bipolar depression. Neuropsychiatric Disease and Treatment 4(1): 21–31.

103. Ban TA, Fujimori M, Petrie WM, Ragheb M, Wilson WH (1982) Systematic studies with amoxapine, a new antidepressant. Int Pharmacopsychiatry 17(1): 18–27.

104. Hormazabal L, Omer LM, Ismail S (1985) Cianopramine and amitriptyline in the treatment of depressed patients – a placebo-controlled study. Psychopharmacology (Berl) 86(1–2): 205–8.

7

High Throughput Identification of Monoclonal Antibodies to Membrane Bound and Secreted Proteins Using Yeast and Phage Display

Lequn Zhao[1], Liang Qu[1], Jing Zhou[1], Zhengda Sun[1], Hao Zou[1], Yunn-Yi Chen[2], James D. Marks[1]*, Yu Zhou[1]*

1 Department of Anesthesia and Perioperative Care, University of California San Francisco, San Francisco General Hospital, San Francisco, California, United States of America, 2 Departments of Pathology & Laboratory Medicine, University of California San Francisco, San Francisco, California, United States of America

Abstract

Antibodies are ubiquitous and essential reagents for biomedical research. Uses of antibodies include quantifying proteins, identifying the temporal and spatial pattern of expression in cells and tissue, and determining how proteins function under normal or pathological conditions. Specific antibodies are only available for a small portion of the proteome, limiting study of those proteins for which antibodies do not exist. The technologies to generate target-specific antibodies need to be improved to obtain high quality antibodies to the proteome at reasonable cost. Here we show that renewable, validated, and standardized monoclonal antibodies can be generated at high throughput, without the need for antigen production or animal immunizations. In this study, 60 protein domains from 24 selected secreted proteins were expressed on the surface of yeast and used for selection of phage antibodies, over 400 monoclonal antibodies were identified within 3 weeks. A subset of these antibodies was validated for binding to cancer cells that overexpress the target protein by flow cytometry or immunohistochemistry. This approach will be applicable to many of the membrane-bound and the secreted proteins, 20–40% of the proteome, accelerating the timeline for Ab generation while reducing the cost.

Editor: Dimiter S. Dimitrov, CIP, NCI-Frederick, NIH, United States of America

Funding: Research reported in this publication was supported by NCI of the National Institutes of Health under award number P50 CA58207. The content is solely the responsibility of the authors and does not necessarily represent the official views of the National Institutes of Health. The funder had no role in study design, data collection and analysis, decision to publish, or preparation of the manuscript.

Competing Interests: The authors have declared that no competing interests exist.

* Email: zhoue@anesthesia.ucsf.edu (YZ); marksj@anesthesia.ucsf.edu (JDM)

Introduction

Availability of antibodies (Abs) strongly determines which proteins of the proteome are studied [1]. Over half the human proteome is not annotated, and functional Abs are not reliably available for these proteins. Even when monoclonal or polyclonal Abs are commercially available, a high proportion of these Abs show either poor specificity or fail to recognize their targets [2–6]. For example, a recent editorial by Michel et al. highlighted the lack of target specificity for 49 Abs against 19 subtypes of GPCRs [7]. An additional problem is lot-to-lot variability in Ab specificity, including monoclonal Abs (mAbs) made via hybridoma technology, resulting in inconsistent assay results [5].

Among the proteome, the secretome includes membrane-bound and extracellular proteins that are processed through the secretory pathway [8]. Secreted proteins are involved in a myriad of normal functions [8–11], as well as in disease processes [12,13]. This class of proteins is extensively studied for their roles in the pathogenesis of disease, as diagnostic and prognostic biomarkers and as targets of therapeutics [12,14]. As of May 2014, 39 of the 40 FDA approved Abs target proteins in a subset of the human secretome [15–17]. This is also true of the majority of the more than 338 therapeutic Abs under clinical development. Secreted proteins are ideal candidates for a high throughput recombinant Ab (rAb) generation platform because they are frequently implicated in disease pathogenesis, and because expression and purification of these types of proteins for use in Ab generation is challenging. Secreted proteins generally do not fold properly in the bacterial cytosol, necessitating use of the bacterial secretion system for expression. The presence of multiple disulfide bonds in the extracellular proteins and in the extracellular domains of type 1 and type 2 membrane proteins is typical, and their large size makes expression yields in bacteria frequently too low to be useful [18]. This can be partially overcome by expressing isolated protein domains. Although expression in either insect or mammalian cells is often required but these are difficult systems to automate and expression yields are variable [19]. Multi-pass transmembrane proteins are even more difficult to express and purify. Because of the large hydrophobic transmembrane domains, they must be harvested from membrane fractions and purified in the presence of detergents [20,21]. It is not uncommon for them to denature during purification making recognition of the native conformation unlikely. Furthermore, many of these proteins are evolutionarily conserved, limiting the robustness of the immune response when the protein is used as an immunogen [22].

Yeast display is an attractive platform for generating antigens for phage Ab selections as no antigen purification is required [23]. Yeast display is a robust system for displaying a variety of different proteins in their properly folded states on the yeast surface. The antigen of interest is fused to either the N- or C-terminus of the yeast Aga2 protein which disulfide bonds to the Aga1 membrane protein. A flexible linker between the antigen of interest and Aga2 ensures accessibility of the antigen to the Abs. Domains of human EGFR, T-cell receptor, NY-Eso-1, breast cancer antigens, and botulinum neurotoxin have been functionally displayed on the yeast surface and used to map Ab epitopes [24–27] [28].

Here we report a high throughput scalable approach to generate widely available, renewable, validated and standardized sets of recombinant Ab reagents (rAbs) to plasma membrane and extracellular proteins. We demonstrate the generality of our approach by generating Abs to a variety of protein classes. A key benefit of the technology is that the expensive, time-consuming and tedious task of antigen generation and purification is bypassed by displaying the antigen at high levels on the surface of yeast. Antigens expressed on yeast were used for selection of phage Abs as well as for validation and characterization. The use of phage display bypasses the low throughput, time-consuming, and expensive immunization of animals to generate polyclonal Abs or the use of hybridoma technology to generate mAbs. Moreover, the Ab genes are cloned, and the rAbs are forever renewable and can easily be formatted for expression as Ab fragments or traditional mAbs with any Fc.

Results

Display of membrane and secreted proteins on the surface of yeast cells

Sixty different cDNA from 17 single pass, 1 multi-pass transmembrane, 2 GPI-anchored, and 4 secreted proteins were displayed on the surface of yeast (Table 1). These included 14 different single domain classes, multiple domains, and full-length extracellular domains (ECD). Domains were identified based on experimentally determined or homology-modeled structures, and cloned into the yeast display vector pYD2. Display was induced and quantitated using the SV5 tag at the C-terminus of the displayed protein with 86% of the proteins showing detectable display above the background (Fig. 1a). The display level was weakly inversely correlated with the protein size (Fig. 1a). Some

protein domains were poorly expressed, such as the semaphorin domain of c-Met.

Selection of phage single chain (scFv) antibody library on yeast displayed proteins

To determine whether phage antibodies could be successfully selected on yeast-displayed antigens, a naïve human scFv multivalent phage display library [29,30] was selected on each of the 60 yeast-displayed protein domains. After three rounds of selection, the polyclonal phage outputs were analyzed for binding to the target protein domains displayed on yeast cells by flow cytometry and a percentage of antigen binding phage above 10% used as an indicator of successful antibody selection. Of 60 selections, 49 specifically bound the yeast displayed target antigen compared to the un-induced yeast cells (Fig. 1b, Fig. 2a). On average, 47% of the polyclonal phage from the selections bound the yeast-displayed antigens, with a range of 11% to 77%. Of the failed selections, 9 were from proteins that were poorly displayed. Since most proteins were represented by more than one domain, there was at least one domain from each of the 24 target antigens that displayed and resulted in antigen binding phage antibodies. The exception was the antigen NRP1, which was only displayed as full-length ECD. The correlation between protein density on the yeast surface and the percentage of target binding phage suggested that display levels greater than 50,000 copies per yeast cell are required for successful phage selections. However, good display was insufficient by itself, since some of the better-displayed protein domains failed to generate Abs, such as FGFR1 Ig domain 1, NRP1, and VEGFR2 Ig domain 2&3 (Fig. 1b).

A naïve scFv phage display library in the monovalent format (scFv) was also selected on 9 of the yeast-displayed protein domains, including ErbB2 domain1, EphA2 ECD, PECAM domain 1-2, MMP9 FN domain 1-3, and 5 CD44 variants (link, H, v6, v67, and v47) (Fig. 2b). All nine selections yielded an antigen binding polyclonal phage population.

Characterization of phage antibodies

From the multivalent phage library selections, the initial screening of 96 random clones from each of the 26 polyclonal phage outputs yielded 1152 mAbs that bound the yeast-displayed antigens. DNA sequencing revealed 162 unique sequences for Abs binding the 26 yeast-displayed domains (Table 2). Flow cytometry analysis and RCA amplification for DNA sequencing using 96-well

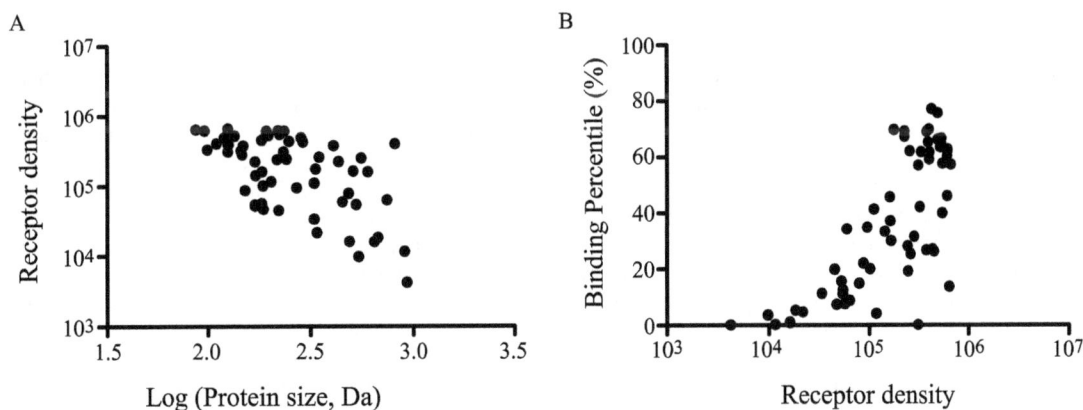

Figure 1. Yeast surface display of proteins and results of phage antibody library selections. A) The density of each protein domain on the induced yeast surface was quantitated and plotted against the domain size; B) The percentage of the polyclonal phage output binding the target antigen is shown as a function of antigen receptor density on the yeast surface.

Table 1. Secretome proteins used for yeast display and phage antibody selection.

Gene Name	Display	Domain Classes	Yeast Display Strategy (# residues displayed)
EGFR	Type 1 TM	Receptor L domain, Furin-like cysteine rich region, Furin-like repeats	4 Individual domains (185, 125, 171, 135)
HER2	Type 1 TM	Receptor L domain, Furin-like cysteine rich region, Furin-like repeats	4 Individual domains (193, 126, 170, 145)
HER3	Type 1 TM	Receptor L domain, Furin-like cysteine rich region, Furin-like repeats	4 Individual domains (187, 124, 170, 147)
HER4	Type 1 TM	Receptor L domain, Furin-like cysteine rich region, Furin-like repeats	4 Individual domains (183, 125, 170, 152)
EPHA2	Type 1 TM	Ligand Binding Domain, SAM (Sterile alpha motif), FN3 domain	Full-length ECD (510)
EPHB3	Type 1 TM	Ligand Binding Domain, SAM, FN3	Full-length ECD (527)
VEGFR2	Type 3 TM	Ig	ECD (744) & Ig2-3 (204)
FGFR1	Type 1 TM	Ig	ECD (348), Ig1 (87), Ig2-3 (218)
c-Met	Type 1 TM	Sema (semaphorin domain), PSI (Plexin repeat), IPT (Ig-like, plexins, transcription factors)	Full-length ECD & individual domain combinations (489, 908, 271, 340)
MST1R	Type 1 TM	Sema (semaphorin domain), PSI, IPT	ECD & 3 individual domain combinations (933, 647, 222, 330)
ICAM1	Type 1 TM	Ig	Ig1-5 (452) & Ig1 (99)
PECAM	Type 1 TM	Ig	Ig1-2 (236)
VCAM	Type 1 TM	Ig	Ig2-7 (673) & Ig2-3 (197)
EpCAM	Type 1 TM	Thyroglobulin type-1	Full-length ECD (242)
E-Cad	Type 1 TM	Cadherin like domain	Cad domains 1-5 (542), 1-2 (220)
CD44	Type 1 TM	Link domain	7 variant domains (149, 409, 558, 249, 291, 335, 433)
CD47	5 TM	Ig-like V-type	N-terminal ECD 1 (123)
CD73	GPI anchor	N-terminal metallophosphatase domain	Full-length ECD (523)
CD168	GPI anchor	Hyaluronan-binding fragment	63 kDa isoform (561)
MSLN	Secreted	No domain superfamily	Cleaved form (285)
MMP9	Secreted	Fibronectin type-II, Hemopexin	82 kDa (601) & FNII 1-3 (184)
TIMP1	Secreted	NTR	Full-length protein (183)
TIMP2	Secreted	NTR	Full-length protein (193) & NTR (125)
Robo1	Type 1 TM	Ig-like C2-type, Fibronectin type-III	Full-length ECD & individual domain combinations (96, 110, 330, 484, 814)

microtiter plate allowed the screening to be completed in 3 weeks. From the monovalent phage library selections, the nine successful selections yielded 345 mAbs of which 108 were unique sequences (Table 2).

Phage antibody libraries in the monovalent and multivalent forms selected on six yeast-displayed domains were also compared with regard to the selection enrichment and the number of unique antibodies. Although the multivalent fd phage library presents multiple copies of scFv potentially favoring antigen binding compared to the monovalent phagemid library, no obvious differences in enrichment of binding phage antibodies was observed (Figure 2). However, more unique sequences were identified from the phagemid library selections than from the fd library selections (Table 2), perhaps reflecting its over 10-fold greater size compared to the multivalent fd phage library.

Of the 270 unique mAbs from 35 selections (26 from the multivalent fd phage and 9 from the monovalent phagemid library), over 100 antibodies were evaluated for binding to seven of the target antigens using intact mammalian cell lines known to overexpress the target antigen. This screening identified 49 mAbs recognizing the cell surface target antigen as expressed on the mammalian cell surface. The fraction of mAbs that stained the cells specifically varied (Table 2), suggesting that in some cases the protein domains were not displayed on the yeast cell surface in their native conformations. For 19 single chain Fv (scFv) mAbs, the binding affinities were evaluated by measuring the K_D on both the yeast-displayed antigens and the antigen overexpressing cells in order to show that 1) the affinities measured on yeast are comparable to affinities on antigen in its "native" state; and 2) that the antibody affinities are in the expected range. The affinity of binding to mammalian cell surface proteins was measured using flow cytometry for 19 purified scFv antibodies binding 4 different antigens (Her2, EphA2, EphB3 and CD44). Affinities generally correlated with those measured on the yeast displayed antigen and

A. Phage selections

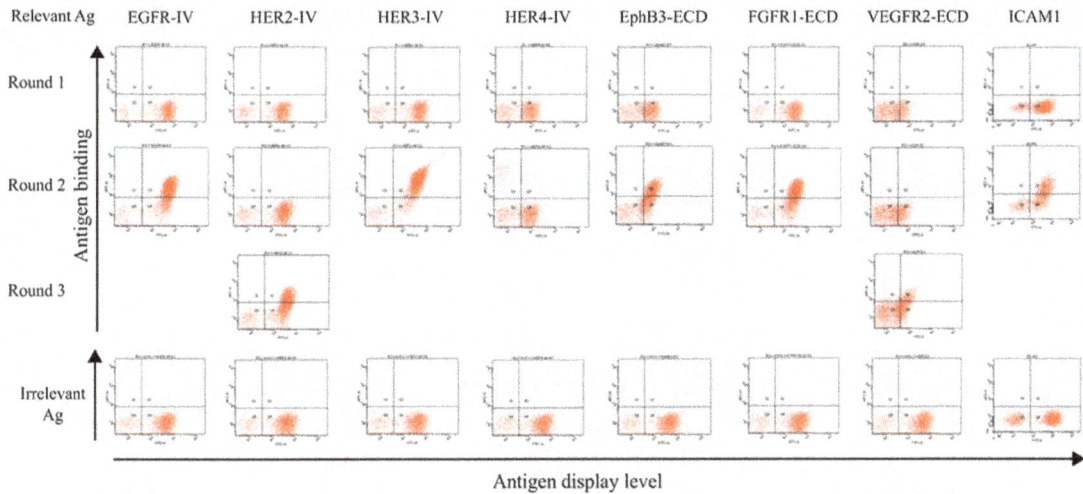

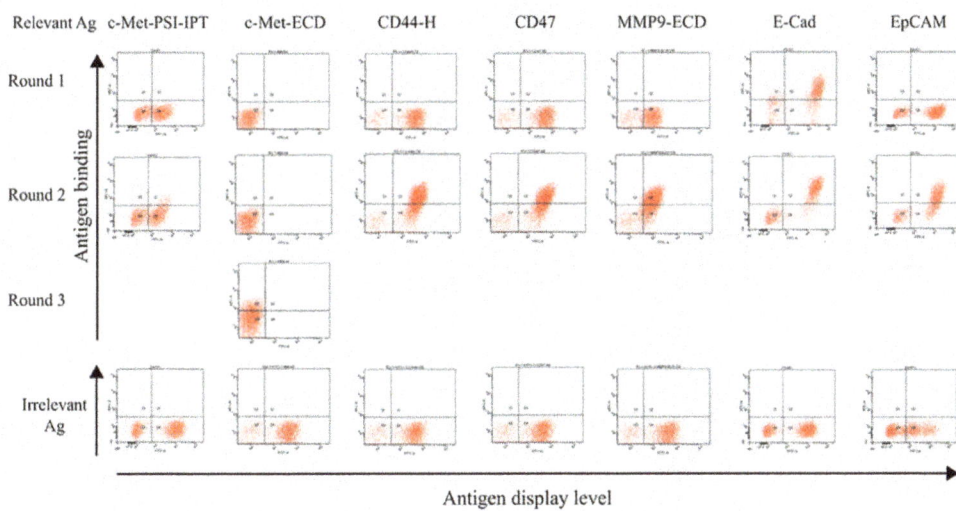

B. Phagemid selections

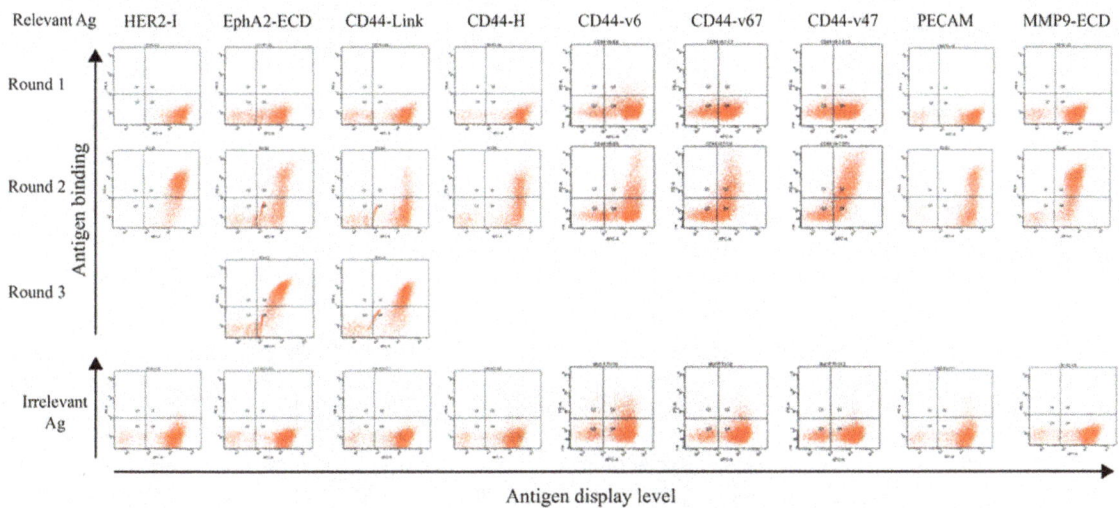

Figure 2. Selection enrichment demonstrated by flow cytometry analysis of polyclonal phage binding to target antigen expressing yeast cells. Binding of fifteen polyclonal phage Ab outputs (A) and nine phagemid Ab outputs (B) from the first, second and third round of selection compared to binding to an irrelevant yeast displayed antigen. The antigen used for selection is shown above each dot-plot.

Table 2. Results of phage antibody selections.

Multivalent phage library	Domain	Binding Abs (per 95 screened)	Unique mAbs identified	Cell binding (cell line and number of mAbs)
EGFR	III	62	1	A431 (1)
	IV	54	2	A431 (2)
ErbB2	I	90	2	AU565 (2)
	II	15	1	AU565 (1)
	III	11	1	AU565 (1)
	IV	5	6	AU565 (1)
ErbbB3	IV	59	12	N.D.
ErbbB4	IV	42	10	N.D.
EphA2	ECD	50	3	MDAMB231 (3)
EphB3	ECD	53	10	MDAMB453 (5)
VEGFR2	ECD	10	4	N.D.
FGFR1	Ig2-3	78	4	N.D.
	ECD	57	9	N.D.
c-Met	PSI-IPTs	16	4	MDAMB231 (2)
NRP1	ECD	20	2	N.D.
ICAM1	Ig1	76	14	SUM159PT (5)
	Ig1-5	30	6	SUM159PT (3)
PECAM 1	Ig1-2	21	3	N.D.
EpCAM	ECD	67	13	N.D.
E-Cad	Cad1-2	2	2	N.D.
CD44	Link	60	6	MDAMB231 (1)
	H	76	2	N.D.
CD47	ECD1	60	16	N.D.
MMP9	FN1-3	51	16	N.D.
	ECD	22		N.D.
hFc	CH2-CH3	65	13	N.D.
Monovalent phage library				
ErbB2	I	63	8	SKBR3 (5)
EphA2	ECD	57	18	MDAMB231 (10)
CD44	Link	16	2	MDAMB231 (2)
	H	40	6	MDAMB231 (1)
	V6	11	5	JIMT (1)
	V67	34	13	JIMT (1)
	V47	48	29	JIMT (4)
MMP9	FN1-3	50	17	N.D.
PECAM 1	Ig1-2	26	10	N.D.

N.D. This test was not done.
Antigen and antigen domain used for selection are indicated along with the number of binding mAbs, the number of unique mAbs, and the number of mAbs binding mammalian cells expressing the target antigen. The specific cell line used is also indicated.

12 of them were less than 30 nM (Fig. 3). The correlation coefficient for these two measurements was 0.47. Variations in some K_D values may reflect the conformational differences between the epitopes displayed on yeast surface and that on the cancer cells (Fig. 3).

Validation of antibodies

Of the unique mAbs generated, we examined those binding CD44 in detail due to the importance of this antigen in tumor biology. Expression of multiple isoforms of CD44 has been

identified in tumors [31–38]. To distinguish mAbs that are selective for different CD44 isoforms, we displayed 5 splice variants (CD44 V6, CD44 V7, CD44 V67, CD44 V4-7, CD44 V3-10), and 2 standard forms (CD44 link domain and CD44 H) on the yeast surface (Figure 4a). These yeast displayed CD44 forms were used for Ab library selection, mAb screening, as well as evaluation of Ab binding specificities and affinities. MAbs binding specifically to the V6, V7, or the constant domains with affinities ranging from 7 to 482 nM were identified from the phagemid library (Figure 4b). The CD44 domain binding profile suggested that antibody V6-2C5, V47-2B8, V47-2B12 bound CD44 v6,

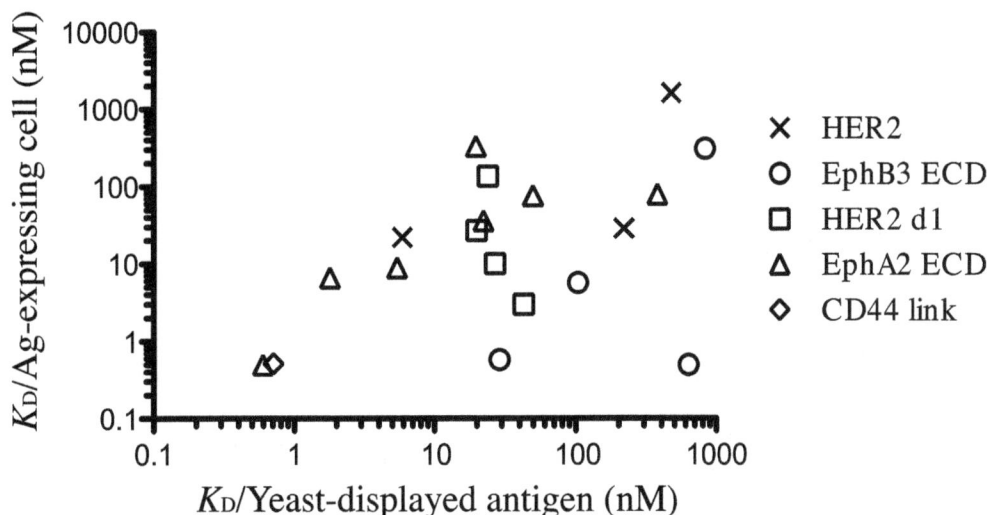

Figure 3. Binding affinities of scFv antibodies. The K_D of purified scFv antibody was measured on yeast displayed antigen domains and compared with the K_D for binding to antigen (Ag) expressing mammalian cells by flow cytometry. scFv mAbs were incubated with the yeast cells transformed and induced for antigen expression, or with the antigen overexpressing cancer cells; SKOV3 cells for HER2 mAbs, MDA-MB-231 for EphA2 and CD44 mAbs, Colo205 for EphB3 mAbs.

V47-2A6 bound CD44 v7, and V47-2G10 bound the CD44 standard form at the membrane proximal region.

The anti-CD44 phage antibodies with the highest affinities were converted to full length IgG which were used to stain breast cancer cell lines (Figure 5), and to detect specific variant in the breast tumor sections (Figure 6). Both MDA-MB-231 and JIMT-1 [39] are known to express CD44 at high level, and MDA-MB-453 cells express lower levels of CD44 [40]. The V6 specific mAb 2B12 showed membrane binding on JIMT-1 cells but not MDA-MB-231 or MDA-MB-453 cells. The link domain specific mAb F2-1A6 showed positive binding on both JIMT-1 and MDA-MB-231, but not MDA-MB-453 cells, consistent with prior flow cytometry analysis [23]. The CD44-H specific mAb showed the strongest signal, while the V7 specific mAb showed a weaker signal, and both antibodies showed some intracellular signal in MDA-MB-453 cells.

Breast tumor sections stained with three antibodies, 2A6, 2B12, and 2G10, also demonstrated unique CD44 staining patterns. The V7 and H specific Abs showed diffusing staining with epithelial cells including the normal mammary duct, the V6 specific 2B12 antibody stained a subset of the tumor epithelial cells with a distinctive membrane pattern (Figure 6). The biological relevance of V6 within the tumor tissue is unknown.

Other cell binding antibodies were also validated for cell staining and internalization. Ab internalization was evaluated using the chelated ligand-mediated internalization assay (CLIA). Specifically, the histidine-tagged scFv coupled with the Ni-NTA coated fluorescent liposome bound the target receptor on the intact mammalian cell surface at 4°C compared to that internalized into the cells at 37°C. The surface bound liposomes were removed by imidazole wash showing no fluorescence signal left for the 4°C incubation group, while the internalized liposomes remained for the 37°C incubation group (Figure 7). The ratio of fluorescence between the imidazole and the PBS wash reflected the fraction of internalized Abs. Both monoclonal scFv antibodies to ErbB2 domain 1 and EphA2 ECD were internalized into SKOV3 and MDA-MB-231 cells, respectively (Figure 7). Some mAbs showed comparable internalization to mAbs F5 and D2-

1A7, which bound ErbB2 and EphA2, respectively, but were isolated from selection for internalization into intact cells [23,41].

Breast cancer cell lines known to express the target antigens ErbB2, EphA2, EphB3, and CD44 with high and low levels were stained with the identified phage mAbs and evaluated by fluorescence microscopy. Consistent with the gene expression data [40], EphB3 antibodies stained MDA-MB-453 cells, CD44 and EphA2 antibodies were positive on MDA-MB-231, and the ErbB2 domain 1 binders showed intensive intracellular staining of ErbB2 positive AU565 cells (Figure 8).

Discussion

We have demonstrated that antigen yeast display combined with Ab phage display is a rapid method to generate highly specific mAbs to secretome proteins that recognize native proteins. Previously, we showed that the extracellular domains of EphA2 and CD44 can be displayed on the yeast cell surface and used to identify mAbs from a phage antibody library pre-selected on breast cancer cells overexpressing EphA2 and CD44 [23]. Here, we demonstrate the generality of this approach using un-selected naïve phage and phagemid antibody libraries. The approach presented here could be amenable to automation allowing high throughput antibody selection.

Yeast antigen display proved to be a robust system to display protein domains for antibody selection and screening based on the range of proteins we tested. The advantages over other antigen production system include 1) easy and fast cloning of antigen using gap repair, 2) simultaneous display of multiple domains, 3) high density of homogeneous antigens on cell surface, and 4) relatively clean background for antibody selection. Potential limitations of this method are the non-natural epitope space caused by mannosylation [42,43] and inability to express multi-pass trans-membrane proteins and GPI-linked membrane proteins.

The expression level of antigen domains was weakly correlated with the antigen size and more strongly correlated with the protein fold and the structure. Generally, the Ig fold and the Fn3 fold were better displayed in the yeast expression system, while the sema domain was difficult to display. Expression could potentially be

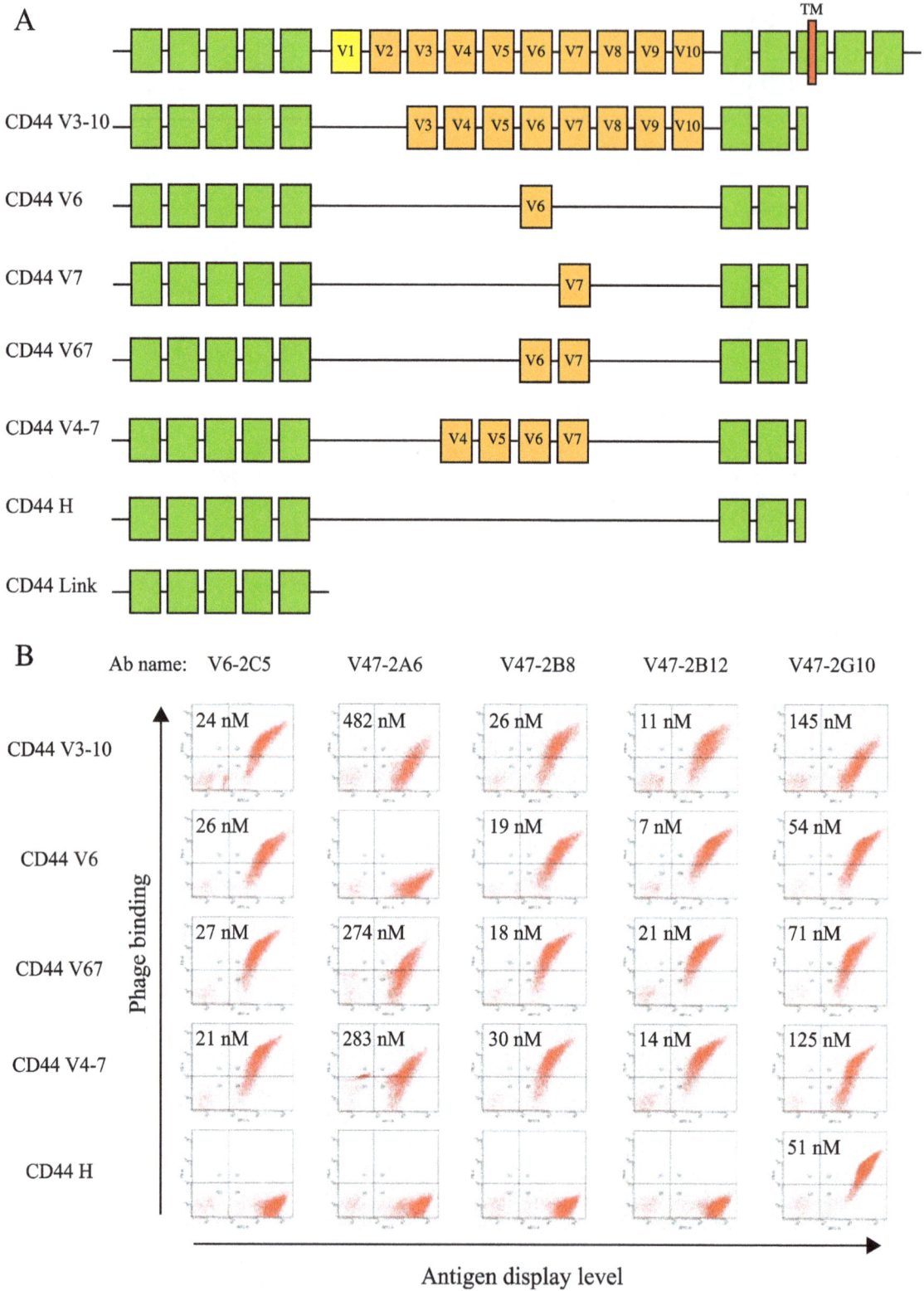

Figure 4. Characterization of five CD44 antibodies using a panel of CD44 variants displayed on the yeast cell surface. A) Diagram of the yeast displayed CD44 variant domains; B) binding specificity and affinity of anti-CD44 scFv assessed on the yeast displayed domains by flow cytometry.

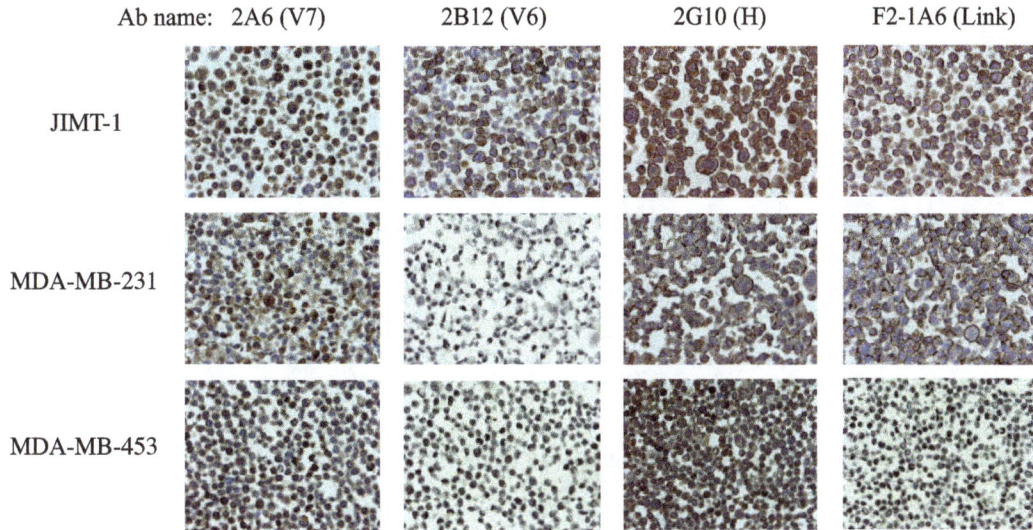

Figure 5. Immunohistochemistry staining of breast cancer cell buttons using CD44 mAbs. Breast cancer cell lines with known CD44 expression levels were processed to make paraffin embedded cell buttons, and used to evaluate anti-CD44 mAb staining patterns. Abs studied included V7 specific mAb 2A6, V6 specific mAb 2B12, CD44 standard form binding mAb 2G10, and CD44 link domain specific mAb F2-1A6.

improved with codon and domain optimization, changing the orientation of Aga2, or by including yeast chaperones [44].

The number of phage antibodies generated was directly correlated with the display level of the antigen, but not associated with the antigen size. A higher number of unique mAbs seemed to be generated from the phagemid library compared to the phage Ab library, suggesting that some Ab sequences were lost during the construction of the smaller phage library by sub-cloning from the phagemid library [29,30].

A number of the mAbs we generated were able to recognize the native antigen as presented on the cell surface. Interestingly, we identified identical mAbs that we previously identified using direct selection on intact cells, namely F5 and D2-1A7, which binds HER2 and EphA2, respectively [23,41]. This observation confirms the native structure of the antigen as it is presented on yeast and suggests similar selection pressures.

The yeast display of antigen domains allowed flexible, rapid, and easy expression of multiple antigen domains as well as direct affinity assessment by flow cytometry [28], which proved extremely useful for CD44 splice variants where the soluble

proteins are not available. We displayed seven different variant forms of CD44 on the yeast surface, and completed the phage antibody library selection and screening within a month. The resulting mAbs have affinities ranging from 7 nM to 482 nM, were differentially specific to different domains, and could stain breast tumor sections.

MAb internalization via receptor-mediated endocytosis was evaluated for HER2 and EphA2 mAbs; the majority of mAbs to the HER2 domain 1 Abs and EphA2 Abs demonstrated intracellular staining as determined by CLIA [45] and fluorescent microscopy. The results indicate that the functionality of mAbs are associated with the methods of Ab selection, and can also be approached by using the correct epitope domain of the antigens, such as the HER2 domain 1.

The described approach may be challenging in the absence of knowledge of a target's domain structure. However, the yeast system allowed us to rapidly display multiple domains with different starting and ending amino acids to search for the optimal domains. We also noticed that some antigen epitopes are difficult to recapitulate on the yeast cells as exemplified by the failure of

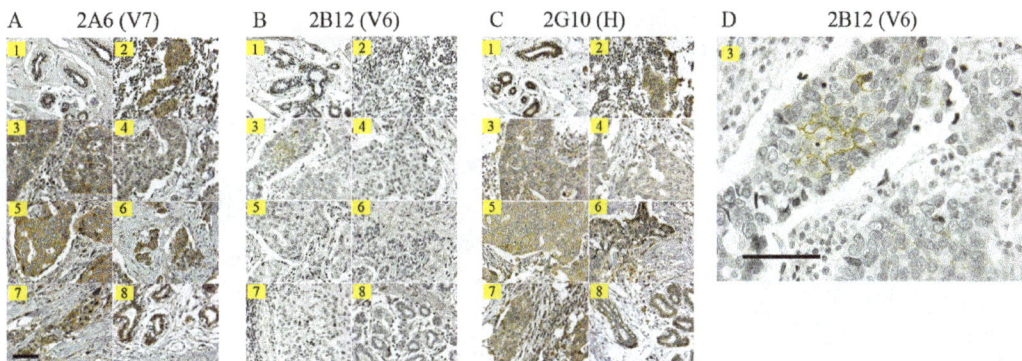

Figure 6. Breast tumor tissue array staining by immunohistochemistry with CD44 mAbs. A tissue array containing samples from 8 different patients (numbered 1–8) was evaluated after staining with different CD44 mAbs. A) V7 specific mAb 2A6, B) V6 specific mAb 2B12, C) CD44 standard from binding mAb 2G10, D) an enlarged view of tissue #3 stained with 2B12 showing distinct membrane bound CD44v6 in the center of a subset of cancerous cells. Magnification is 400x, and the scale bar is 50 μm.

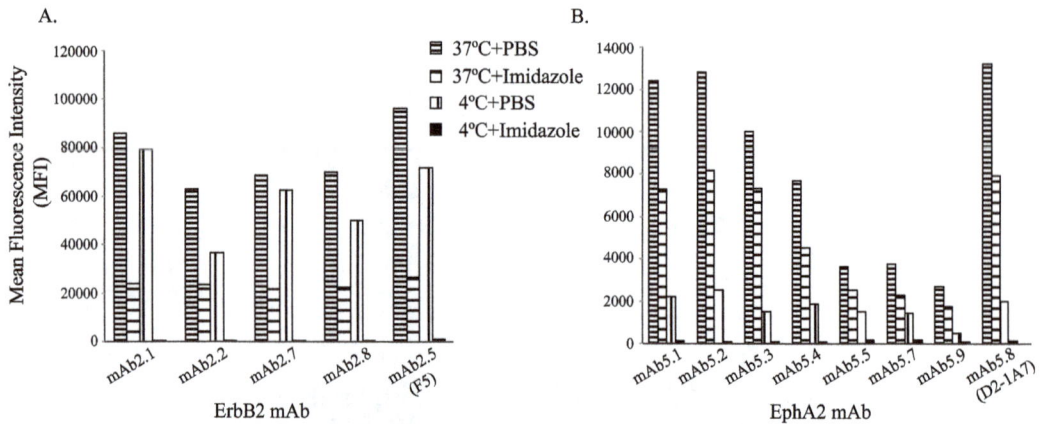

Figure 7. Internalization of mAbs into cancer cells as determined by chelated ligand-mediated internalization assay (CLIA). Ni-NTA liposomes loaded with dye were internalized into cells mediated by hexa-histidine tagged scFv mAbs prebound to the cell surface at 37°C compared to 4°C. Imidazole buffer was used to remove the surface bound dye-loading liposomes by disrupting the Ni-NTA and hexa-Histidine interaction. A) Internalization of ErbB2 mAbs into ErbB2 expressing SKBR3 cells; B) internalization of EphA2 mAbs into EphA2 expressing MDA-MB-231 cells.

mAbs isolated by using recombinant antigens to bind the yeast displayed domains. Such epitopes will definitely be missed using this approach.

We conclude that the yeast display of antigen domains and domain combinations are useful for isolating recombinant antibodies from naïve antibody libraries in a short time period starting from the antigen sequences. This platform will allow Ab discovery for a large function of the antigen epitope space, but not all the epitopes, which will need other approaches to complete Ab discovery.

Materials and Methods

Cell lines, media, antibodies and full-length cDNA clones

Breast cancer cell lines A431, AU565, Colo205, JIMT-1, MDA-MB-231, MDA-MB-453, MDA-MB-468, SKBR3, SKOV3, SUM159PT and HEK293A cells were obtained from the American Type Culture Collection (ATCC). The cell lines were cultured using conditions described previously [46]. The yeast strain EBY100 was grown in YPD medium (*Current Protocols in Molecular Biology, John Wiley and Sons, Chapter 13.1.2*). EBY100 was transfected with expression vector pYD2 [47] and was selected on SD-CAA medium [48] (*Current Protocols, Chapter 13*). The Aga2p antigen fusion was expressed on the yeast surface by induction in SG-CAA medium (identical to SD-CAA medium except the glucose is replaced by galactose) at 18°C

for 24–48 hr as described previously [49]. *E. coli* strains DH5α and TG1 were used for the preparation of plasmid DNA and the expression of soluble scFv antibodies respectively. SV5 antibody was purified from hybridoma supernatant using Protein G and directly labeled with Alexa-488 or Alexa-647 using a kit provided by the manufacturer (Invitrogen; Carlsbad, CA). Biotin conjugated rabbit anti-fd bacteriophage (Sigma), biotin conjugated anti-rabbit (Vector Labs), Streptavidin Phycoerythrin (PE) (Biosource/Invitrogen), streptavidin Texas Red (GE Healthcare), streptavidin HRP conjugate were used to detect antibodies. The full-length cDNAs were obtained from the ATCC and Open Biosystems.

Cloning of protein domain coding genes into the yeast display vector

For antigen display, the cDNA encoding the full length antigen or antigen domain was amplified by PCR using primers that anneal to the 5′ and 3′ ends of the antigen cDNA and with 25 nucleotide overlaps with the yeast display vector pYD2 digested with the restriction enzymes NcoI and NotI. The antigen DNA fragment was then cloned into NcoI-NotI-digested pYD2 vector by using gap repair [50,51]. Gap repair allows cloning of virtually all cDNA fragments including those with internal digestion sites.

Primers for cloning the target antigen gene by gap repair:

- Ag-Gap5 primer 5′-GTTGTTCTGCTAGCGGGG**C-CATGG**---3′

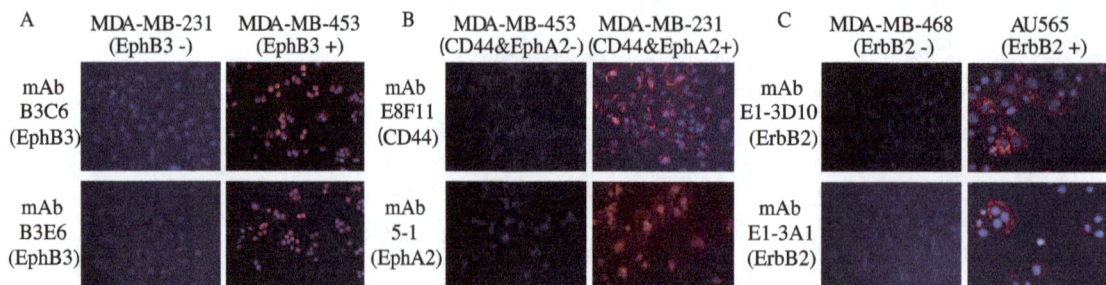

Figure 8. Binding of mAbs to cancer cell lines that overexpress the target antigens as determined by fluorescent microscopy. A) EphB3 mAbs showed positive staining on EphB3 overexpressing cell MDA-MB-453, B) CD44 and EphA2 mAbs showed positive staining on MDA-MB-231 cells, C) ErbB2 domain 1 binding mAbs showed positive staining on the ErbB2 overexpressing cell AU565.

- Ag-Gap3 primer 5′-TTCGAAGGGCCCGCCT**GCGGC-CGC**---3′

The 5′ sequence indicated anneals to Nco1-Not1 digested pYD2 DNA for cloning by gap repair, the bolded sequences are the Nco1 and Not1 sites. The dashed sequence should include approximately 24 nucleotides that anneal to the 5′ and 3′ ends of the target antigen DNA.

Display of protein domains on the surface of yeast cells

Multiple target genes were amplified by PCR in 96-well plates using high-fidelity DNA polymerase, and used to transform LiAc treated EBY100 cells together with NcoI/NotI digested vector pYD2 using the TRAFO method with gap repair in 1 ml 96-well plate [50,51]. Specifically, TRAFO mix including the vector with large enough quantity for 96 transformation was prepared and used to resuspend the EBY100 yeast cells followed by aliquoting 359 μl of TRAFO-treated cells into each well that contains 1 μl of the target gene PCR fragment. After incubation for 30 min at 30°C followed by 45 min at 42°C, the cells were pelleted by centrifugation at 3000 rpm for 2 min, the transformation mix removed, cells washed once in 1 ml of sterile water, 1/10 of the cells transferred to a new 1 ml deep 96-well plate containing 600 μl of SD-CAA media, and cultured for 24 hrs at 30°C. Dilution of untransformed yeast cells was repeated by growing 1/10 of 24 hr culture in a new 1 ml 96-well plate containing 600 μl of SD-CAA media followed by induction culture in SG-CAA media for 24 hrs at 18°C.

Quantitation of protein density on yeast surface by flow cytometry

A fraction of the induced yeast cells were transferred into a V-bottom 96-well plate, washed once by resuspending in 100 μl of FACS buffer, followed by centrifugation, aspiration of the supernatant, resuspension in 50 μl of FACS buffer, and stained by incubation with 50 μl of 1 μg/mL Alexa-647 labeled anti-SV5 IgG diluted in FACS buffer for 1 hr at 4°C. After washing the stained yeast cells once in 200 μl of FACS buffer, the cell fluorescence was measured in a FACS LSRII flow cytometer using Quantum™ Alexa Fluor 647 bead (Bangs Laboratories Inc.) as reference to calculate the protein density on each cell.

Selection of phage antibody library against yeast displayed protein domains

A phage display naïve human single chain library was used for target specific antibody selections [29,30]. The induced yeast cells displaying an human Fc domain were used to deplete the non-specific binders by incubating 1×10^{13} phage particles with 10^9 yeast cells for 2 hr at 4°C. The filtered supernatant containing the depleted phage library was then aliquoted into each well of 1 ml 96-well plate containing about 1×10^7 yeast cells displaying individual target protein domains for 1 h at 4°C. Yeast cells were washed ten times with cold PBS and pelleted by centrifugation. The bound phage antibodies were eluted by incubating yeast cells with 100 μl of 100 mM triethylamine, neutralized with 50 μl of 1 M Tris-HCl (pH 7.4), and used to infect exponentially growing E. coli TG1 as described previously [52]. Specifically, half of the phage eluent was transferred to a new 1 ml 96-well plate, incubated with 1 ml of E. coli TG1 cells for 30 min at 37°C, and 1/10 of the infected TG1 transferred to a new 1 ml 96-well plate containing 2YT/Amp/2% glucose, and cultured for 16 hrs at 30°C. A small fraction of the overnight culture was transferred to another 1 ml 96-well plate containing 150 μl of 2YT/Amp/0.1% glucose using 96-pin transfer device followed by culture for 2.5 hr at 37°C, infection with VCSM13 for 30 min at 37°C, and culture in additional of 450 μl of 2YT/Amp for 16 hrs at 30°C. In the case of fd phage library, the TG1 infected phage eluent was cultured in 2YT/Tet for 16 hrs at 30°C.

The resulting phage output of each target was harvested, precipitated using 20% PEG6000/1.5 M NaCl, resuspended in 100 μl of cold PBS, and subjected to another round of selection to the corresponding target protein domain.

For the second round of selection, the 100 μl of output phage from the round 1 selection was depleted in 1 ml 96-well plate by incubating with 10^8 yeast cells displaying an human Fc domain for 2 hr at 4°C followed by centrifugation to pellet the yeast cells. The supernatant of each well was transferred to a new 1 ml 96-well plate, and mixed with 1×10^7 of yeast cells displaying the corresponding target protein domains for 1 h at 4°C. The rest of the selection procedure was the same as the first round. If needed, the third round of selection can be performed following the same procedure as the second round of selection. When the selection was completed, each selection output can be plated for monoclonal phage analysis.

Evaluation of phage antibody binding to yeast-displayed protein domains

Polyclonal phage antibodies from each round of selections were evaluated by flow cytometry for binding to the corresponding target protein domains displayed on yeast cells. In detail, the supernatant of phage culture in 96-well plate was transferred to V-bottom 96-well plate and incubated with yeast cells displaying the corresponding target antigen for 2 hrs at 4°C. The bound phage antibodies were detected by incubating cells with biotin conjugated anti-fd antibody (1 μg/ml) for 30 min at 4°C and streptavidin-PE followed by flow cytometry analysis. The binding of phage antibodies to the yeast displayed target antigens were represented by the percentile of yeast population in the Q2 quadrant of the dot plot.

Characterization of phage antibodies

After two or three rounds of selection, 96 individual colonies from each target specific selection were isolated and the phage prepared by growing the single colonies in 96-well microtiter plates as described [53]. Binding of each phage antibody to yeast displayed antigen was determined by incubation of 10^5 yeast cells with 100 μl phage supernatant diluted in FACS buffer (PBS with 1 mM MgCl$_2$, 0.1 mM CaCl$_2$ and 0.3% BSA) for 2 h at 4°C in conical 96-well microtiter plates, followed by incubation with biotinylated anti-fd antibody and streptavidin-phycoerythrin conjugate, and analyzed using a FACS LSRII (Becton Dickinson). The number of unique phage antibodies was determined by 96-well RCA reaction followed by DNA sequencing.

Expression of soluble antibodies as scFv and IgG

The scFv coding genes from the identified phage antibodies were subcloned from the phage-display pHEN1 [52] via NcoI-NotI into expression vector pSyn1 for scFv expression [54], or AbVec with rabbit Fc for IgG expression [55]. scFvs were produced in the periplasm of E. coli strain TG1, and purified by osmotic shock and immobilized metal affinity chromatography, as reported previously [54]. IgG proteins were secreted in the media by transfected HEK293A cells as described elsewhere [55], and purified by Protein A affinity chromatography (GE Healthcare). The homogeneity and purity of the protein preparations were verified by SDS-PAGE stained with coomassie blue; protein concentrations were measured by micro-bicinchoninic acid assay (Pierce, Rockford, IL).

Binding affinity assay by flow cytometry

Each scFv was incubated with 1×10^5 yeast cells induced to express the target protein domain for 1 hr-16 hrs at the indicated concentration, or the target antigen overexpressing cancer cells [28,56]. Cell binding was performed at 4°C in PBS containing 0.3% BSA in volume sufficient to maintain constant antibody concentration for equilibrium conditions. After two washes with 200 μl of PBS, bound scFv was detected by the addition of 100 μl (1 μg/ml) of biotinylated His probe (Santa Cruz Biotech.) and streptavidin-PE (Biosource/Invitrogen), and the yeast displayed Ag co-stained with Alexa 647 labeled anti-SV5. After incubating 30 minutes at 4°C, the cells were washed twice and resuspended in PBS containing 4% paraformaldehyde. Fluorescence was measured by flow cytometry in a FACS LSRII (Becton Dickinson), and median fluorescence intensity (MFI) was calculated using Cellquest software (Becton Dickinson). Equilibrium constants were determined as described [57], except that values were fitted to the equation $MFI = MFI_{min} + MFI_{max} * [Ab]/(K_D+[Ab])$ using Kaleidagraph (Synergy Software).

Antibody internalization assay

The anti-ErbB2 and anti-EphA2 scFv antibodies were evaluated for the ability to internalize into human breast cancer cells SKBR3 and MDA-MB-231 cells, respectively, by chelated ligand-mediated internalization assay (CLIA) as described previously [45]. Briefly, the cells were grown to 80-90% confluence in the media recommended by ATCC supplemented with 10% fetal calf serum (FCS) and harvested by trypsinization, seeded at 10,000/well in 96-well plates, and incubated overnight at 37°C. The next day, Ni-NTA liposomes (0-1 mM total phospholipid) were incubated for 4 hrs with the cells along with the (His) 6-containing scFv (20 μg/mL unless otherwise indicated) in 100 μL tissue culture media supplemented with 10% FCS under four different conditions. The conditions are 37°C followed by PBS wash, 37°C with 250 mM phosphate buffered imidazole wash, 4°C with PBS wash, and 4°C with 250 mM phosphate buffered imidazole wash. Cells were then trypsinized, washed, and the fluorescence measured using a FACS LSRII (Becton Dickinson).

Immunohistochemistry

Three breast cancer cell lines, MDA-MB-231, MDA-MB-453, and JIMT-1, were used to make paraffin embedded cell button at UCSF tissue core, and stained with CD44 antibodies following standard IHC protocol.

De-identified human breast tumor tissue samples were obtained from University of California, San Francisco, Department of Pathology archives. Samples were collected with consent under UCSF IRB# 11-06720 "Retrospective Study of Adjunctive Pathologic Diagnostic Features, Biomarkers Expression, and Molecular Alterations in Early Breast Neoplasms and in Morphologic Mimics". The UCSF Mount Zion Panel Institutional Review Board approved the use of this archived patient tissue samples in research and waived the need of consent.

Breast tumor tissue samples were arrayed, and cut to five-micrometer sections. After antigen retrieval, the samples were stained with three different anti-CD44 IgG antibodies with rabbit Fc, 2A6, 2B12, and 2G10, respectively, followed by staining with biotin-conjugated anti-rabbit secondary antibody (Vector Labs Co.), ABC complex (Vector Labs Co.), and standard DAB color development (Vector Labs Co.) sequentially. A pathologist (YYC) evaluated the staining result.

Immunofluorescence

MDA-MB-231, MDA-MB-453, MDA-MB-468, and AU565 cells were grown on coverslips to 70% of confluence in 12 well-plates and incubated with 10^{11} phage antibodies for three hours at 37°C. The coverslips were washed once with PBS, twice with PEM (80 mM Potassium PIPES (pH 6.8), 5 mM EGTA (pH 7), 2 mM MgCl$_2$), and fixed with PEM containing 4% (W/V) paraformaldehyde for 30 min on ice. Cells were quenched with 0.1 M NH$_4$Cl, permeabilized with 0.5% Triton X-100, and blocked with 5% non-fat dry milk in TBS-T buffer overnight at 4°C. After blocking endogenous biotin with Avidin-Biotin Kit (Lab Vision), intracellular phages were detected with biotinylated anti-fd polyclonal antibody (Sigma) and streptavidin Texas Red (GE healthcare). Coverslips were inverted on a slide on mounting medium and microscopic images were taken with a Zeiss fluorescence microscope (Zeiss, Germany).

Acknowledgments

Research reported in this publication was supported by *NCI* of the National Institutes of Health under award number P50 CA58207.

Author Contributions

Conceived and designed the experiments: YZ JDM. Performed the experiments: ZS HZ LQ LZ JZ YZ YYC. Analyzed the data: YZ JDM YYC. Contributed reagents/materials/analysis tools: YYC. Contributed to the writing of the manuscript: YZ JDM.

References

1. Isserlin R, Emili A (2007) Nine steps to proteomic wisdom: A practical guide to using protein-protein interaction networks and molecular pathways as a framework for interpreting disease proteomic profiles. Proteomics Clin Appl 1: 1156–1168.

2. Grimsey NL, Goodfellow CE, Scotter EL, Dowie MJ, Glass M, et al. (2008) Specific detection of CB1 receptors; cannabinoid CB1 receptor antibodies are not all created equal! J Neurosci Methods 171: 78–86.

3. Jensen BC, Swigart PM, Simpson PC (2009) Ten commercial antibodies for alpha-1-adrenergic receptor subtypes are nonspecific. Naunyn Schmiedebergs Arch Pharmacol 379: 409–412.

4. Jositsch G, Papadakis T, Haberberger RV, Wolff M, Wess J, et al. (2009) Suitability of muscarinic acetylcholine receptor antibodies for immunohisto-chemistry evaluated on tissue sections of receptor gene-deficient mice. Naunyn Schmiedebergs Arch Pharmacol 379: 389–395.

5. Pozner-Moulis S, Cregger M, Camp RL, Rimm DL (2007) Antibody validation by quantitative analysis of protein expression using expression of Met in breast cancer as a model. Lab Invest 87: 251–260.

6. Saper CB (2005) An open letter to our readers on the use of antibodies. J Comp Neurol 493: 477–478.

7. Michel MC, Wieland T, Tsujimoto G (2009) How reliable are G-protein-coupled receptor antibodies? Naunyn Schmiedebergs Arch Pharmacol 379: 385–388.

8. Klee EW (2008) The zebrafish secretome. Zebrafish 5: 131–138.

9. Caneparo L, Huang YL, Staudt N, Tada M, Ahrendt R, et al. (2007) Dickkopf-1 regulates gastrulation movements by coordinated modulation of Wnt/beta catenin and Wnt/PCP activities, through interaction with the Dally-like homolog Knypek. Genes Dev 21: 465–480.

10. Merritt WM, Sood AK (2007) Markers of angiogenesis in ovarian cancer. Dis Markers 23: 419–431.

11. Pickart MA, Klee EW, Nielsen AL, Sivasubbu S, Mendenhall EM, et al. (2006) Genome-wide reverse genetics framework to identify novel functions of the vertebrate secretome. PLoS One 1: e104.

12. Karagiannis GS, Pavlou MP, Diamandis EP (2010) Cancer secretomics reveal pathophysiological pathways in cancer molecular oncology. Mol Oncol 4: 496–510.

13. Baggetta R, De Andrea M, Gariano GR, Mondini M, Ritta M, et al. (2010) The interferon-inducible gene IFI16 secretome of endothelial cells drives the early steps of the inflammatory response. Eur J Immunol 40: 2182–2189.

14. Xue H, Lu B, Lai M (2008) The cancer secretome: a reservoir of biomarkers. J Transl Med 6: 52.

15. Nelson AL, Dhimolea E, Reichert JM (2010) Development trends for human monoclonal antibody therapeutics. Nat Rev Drug Discov 9: 767–774.

16. Reichert JM (2009) Probabilities of success of antibody therapeutics. MAbs 1: 387–389.

17. Dimitrov DS, Marks JD (2009) Therapeutic antibodies: current state and future trends–is a paradigm change coming soon? Methods Mol Biol 525: 1–27, xiii.

18. Wagner S, Bader ML, Drew D, de Gier JW (2006) Rationalizing membrane protein overexpression. Trends Biotechnol 24: 364–371.

19. Tate CG, Haase J, Baker C, Boorsma M, Magnani F, et al. (2003) Comparison of seven different heterologous protein expression systems for the production of the serotonin transporter. Biochim Biophys Acta 1610: 141–153.

20. Ren H, Yu D, Ge B, Cook B, Xu Z, et al. (2009) High-level production, solubilization and purification of synthetic human GPCR chemokine receptors CCR5, CCR3, CXCR4 and CX3CR1. PLoS One 4: e4509.

21. Sarramegna V, Talmont F, Demange P, Milon A (2003) Heterologous expression of G-protein-coupled receptors: comparison of expression systems from the standpoint of large-scale production and purification. Cell Mol Life Sci 60: 1529–1546.

22. Goding J (1983) Monoclonal Antibodies: principles and practice.

23. Zhou Y, Zou H, Zhang S, Marks JD (2010) Internalizing cancer antibodies from phage libraries selected on tumor cells and yeast-displayed tumor antigens. J Mol Biol 404: 88–99.

24. Cochran JR, Kim YS, Olsen MJ, Bhandari R, Wittrup KD (2004) Domain-level antibody epitope mapping through yeast surface display of epidermal growth factor receptor fragments. J Immunol Methods 287: 147–158.

25. Johns TG, Adams TE, Cochran JR, Hall NE, Hoyne PA, et al. (2004) Identification of the epitope for the epidermal growth factor receptor-specific monoclonal antibody 806 reveals that it preferentially recognizes an untethered form of the receptor. J Biol Chem 279: 30375–30384.

26. Piatesi A, Howland SW, Rakestraw JA, Renner C, Robson N, et al. (2006) Directed evolution for improved secretion of cancer-testis antigen NY-ESO-1 from yeast. Protein Expr Purif 48: 232–242.

27. Aggen DH, Chervin AS, Insaidoo FK, Piepenbrink KH, Baker BM, et al. (2010) Identification and engineering of human variable regions that allow expression of stable single-chain T cell receptors. Protein Eng Des Sel.

28. Levy R, Forsyth CM, LaPorte SL, Geren IN, Smith LA, et al. (2007) Fine and domain-level epitope mapping of botulinum neurotoxin type A neutralizing antibodies by yeast surface display. J Mol Biol 365: 196–210.

29. Sheets MD, Amersdorfer P, Finnern R, Sargent P, Lindquist E, et al. (1998) Efficient construction of a large nonimmune phage antibody library: the production of high-affinity human single-chain antibodies to protein antigens. Proc Natl Acad Sci U S A 95: 6157–6162.

30. Huie MA, Cheung MC, Muench MO, Becerril B, Kan YW, et al. (2001) Antibodies to human fetal erythroid cells from a nonimmune phage antibody library. Proc Natl Acad Sci USA 98: 2682–2687.

31. Khan SA, Cook AC, Kappil M, Gunthert U, Chambers AF, et al. (2005) Enhanced cell surface CD44 variant (v6, v9) expression by osteopontin in breast cancer epithelial cells facilitates tumor cell migration: novel post-transcriptional, post-translational regulation. Clin Exp Metastasis 22: 663–673.

32. Al-Hajj M, Wicha MS, Benito-Hernandez A, Morrison SJ, Clarke MF (2003) Prospective identification of tumorigenic breast cancer cells. Proc Natl Acad Sci U S A 100: 3983–3988.

33. Berner HS, Nesland JM (2001) Expression of CD44 isoforms in infiltrating lobular carcinoma of the breast. Breast Cancer Res Treat 65: 23–29.

34. Yang Q, Liu Y, Huang Y, Huang D, Li Y, et al. (2013) Expression of COX-2, CD44v6 and CD147 and relationship with invasion and lymph node metastasis in hypopharyngeal squamous cell carcinoma. PLoS One 8: e71048.

35. Shiozaki M, Ishiguro H, Kuwabara Y, Kimura M, Mitsui A, et al. (2011) Expression of CD44v6 is an independent prognostic factor for poor survival in patients with esophageal squamous cell carcinoma. Oncol Lett 2: 429–434.

36. Zavrides HN, Zizi-Sermpetzoglou A, Panousopoulos D, Athanasas G, Elemenoglou I, et al. (2005) Prognostic evaluation of CD44 expression in correlation with bcl-2 and p53 in colorectal cancer. Folia Histochem Cytobiol 43: 31–36.

37. Watanabe O, Kinoshita J, Shimizu T, Imamura H, Hirano A, et al. (2005) Expression of a CD44 variant and VEGF-C and the implications for lymphatic metastasis and long-term prognosis of human breast cancer. J Exp Clin Cancer Res 24: 75–82.

38. Cho EY, Choi Y, Chae SW, Sohn JH, Ahn GH (2006) Immunohistochemical study of the expression of adhesion molecules in ovarian serous neoplasms. Pathol Int 56: 62–70.

39. Olsson E, Honeth G, Bendahl PO, Saal LH, Gruvberger-Saal S, et al. (2011) CD44 isoforms are heterogeneously expressed in breast cancer and correlate with tumor subtypes and cancer stem cell markers. BMC Cancer 11: 418.

40. Spellman PT (2006) Transcription profiling of 51 human breast cancer cell lines.

41. Poul M-A, Becerril B, Nielsen UB, Morrison P, Marks JD (2000) Selection of internalizing human antibodies from phage libraries. J Mol Biol 301: 1149–1161.

42. Gerngross TU (2004) Advances in the production of human therapeutic proteins in yeasts and filamentous fungi. Nat Biotechnol 22: 1409–1414.

43. Hamilton SR, Bobrowicz P, Bobrowicz B, Davidson RC, Li H, et al. (2003) Production of complex human glycoproteins in yeast. Science 301: 1244–1246.

44. Robinson AS, Hines V, Wittrup KD (1994) Protein disulfide isomerase overexpression increases secretion of foreign proteins in Saccharomyces cerevisiae. Biotechnology (N Y) 12: 381–384.

45. Nielsen UB, Kirpotin DB, Pickering EM, Drummond DC, Marks JD (2006) A novel assay for monitoring internalization of nanocarrier coupled antibodies. BMC Immunol 7: 24.

46. Neve RM, Chin K, Fridlyand J, Yeh J, Baehner FL, et al. (2006) A collection of breast cancer cell lines for the study of functionally distinct cancer subtypes. Cancer Cell 10: 515–527.

47. Razai A, Garcia-Rodriguez C, Lou J, Geren IN, Forsyth CM, et al. (2005) Molecular evolution of antibody affinity for sensitive detection of botulinum neurotoxin type A. J Mol Biol 351: 158–169.

48. (2008) Current Protocols in Molecular Biology: Wiley Interscience.

49. Feldhaus MJ, Siegel RW, Opresko LK, Coleman JR, Feldhaus JM, et al. (2003) Flow-cytometric isolation of human antibodies from a nonimmune Saccharomyces cerevisiae surface display library. Nat Biotechnol 21: 163–170.

50. Gietz RD, Schiestl RH (1991) Applications of high efficiency lithium acetate transformation of intact yeast cells using single-stranded nucleic acids as carrier. Yeast 7: 253–263.

51. Orr-Weaver TL, Szostak JW (1983) Yeast recombination: the association between double-strand gap repair and crossing-over. Proc Natl Acad Sci U S A 80: 4417–4421.

52. Marks JD, Hoogenboom HR, Bonnert TP, McCafferty J, Griffiths AD, et al. (1991) By-passing immunization. Human antibodies from V-gene libraries displayed on phage. J Mol Biol 222: 581–597.

53. O'Connell D, Becerril B, Roy-Burman A, Daws M, Marks JD (2002) Phage versus phagemid libraries for generation of human monoclonal antibodies. J Mol Biol 321: 49–56.

54. Schier R, Bye J, Apell G, McCall A, Adams GP, et al. (1996) Isolation of high-affinity monomeric human anti-c-erbB-2 single chain Fv using affinity-driven selection. J Mol Biol 255: 28–43.

55. Wrammert J, Smith K, Miller J, Langley WA, Kokko K, et al. (2008) Rapid cloning of high-affinity human monoclonal antibodies against influenza virus. Nature 453: 667–671.

56. Zhou Y, Goenaga AL, Harms BD, Zou H, Lou J, et al. (2012) Impact of intrinsic affinity on functional binding and biological activity of EGFR antibodies. Mol Cancer Ther 11: 1467–1476.

57. Benedict CA, MacKrell AJ, Anderson WF (1997) Determination of the binding affinity of an anti-CD34 single-chain antibody using a novel, flow cytometry based assay. J Immunol Methods 201: 223–231.

8

Characterization of the Nuclear Import Mechanism of the CCAAT-Regulatory Subunit Php4

Md. Gulam Musawwir Khan, Jean-François Jacques, Jude Beaudoin, Simon Labbé*

Département de Biochimie, Faculté de Médecine et des Sciences de la Santé, Université de Sherbrooke, Sherbrooke, Québec, Canada

Abstract

Php4 is a nucleo-cytoplasmic shuttling protein that accumulates in the nucleus during iron deficiency. When present in the nucleus, Php4 associates with the CCAAT-binding protein complex and represses genes encoding iron-using proteins. Here, we show that nuclear import of Php4 is independent of the other subunits of the CCAAT-binding complex. Php4 nuclear import relies on two functionally independent nuclear localization sequences (NLSs) that are located between amino acid residues 171 to 174 (KRIR) and 234 to 240 (KSVKRVR). Specific substitutions of basic amino acid residues to alanines within these sequences are sufficient to abrogate nuclear targeting of Php4. The two NLSs are biologically redundant and are sufficient to target a heterologous reporter protein to the nucleus. Under low-iron conditions, a functional GFP-Php4 protein is only partly targeted to the nucleus in *imp1Δ* and *sal3Δ* mutant cells. We further found that cells expressing a temperature-sensitive mutation in *cut15* exhibit increased cytosolic accumulation of Php4 at the nonpermissive temperature. Further analysis by pull-down experiments revealed that Php4 is a cargo of the karyopherins Imp1, Cut15 and Sal3. Collectively, these results indicate that Php4 can be bound by distinct karyopherins, connecting it into more than one nuclear import pathway.

Editor: Javier Marcelo Di Noia, Institut de Recherches Cliniques de Montréal (IRCM), Canada

Funding: M.G.M.K. was recipient of a Studentship from the Faculté de médecine et des sciences de la santé of the Université de Sherbrooke. This study was supported by the Natural Sciences and Engineering Research Council of Canada (Grants MOP-238238-2010-15 and MOP-396029-2010-DAS to S.L). The funders had no role in study design, data collection and analysis, decision to publish, or preparation of the manuscript.

Competing Interests: The authors have declared that no competing interests exist.

* Email: Simon.Labbe@USherbrooke.ca

Introduction

In eukaryotic cells, the nucleus is a membrane-enclosed organelle that physically separates genetic material and transcriptional machinery from cytoplasm. Although proteins are translated in the cytoplasm, several of them play important roles in the nucleus. In order to accomplish their cellular function, they must be imported into the nucleus. The way that proteins can be transported into and out of the nucleus is through large protein assemblies denoted nuclear pore complexes (NPCs) [1]. Although some proteins smaller than ~40–0 kDa can passively diffuse through NPCs, most of the proteins with functions in the nucleus are actively transported by specific soluble carrier proteins called karyopherins (Kaps) [2,3]. The orientation of transport through NPCs is determined by short signal sequences within proteins or cargoes. The nuclear localization signal (NLS) triggers proteins into the nucleus, whereas the nuclear export signal (NES) fosters the transport of proteins into the cytoplasm [4]. Kaps are responsible for the vast majority of protein flow through NPCs. Kaps are classified in two families: Kap α (also known as importin α) and Kap β (also known as importin β) [5]. Kap α is an adaptor protein that recognizes two classes of NLSs, which are also called classical NLSs [6]. One class, denoted monopartite NLS, is composed of a single cluster of basic amino acid residues, whereas the second class, termed bipartite NLS, possesses two clusters of basic amino acid residues separated by a 10–2-amino acid spacer. Furthermore, there are two types of monopartite NLSs. The first type has at least four consecutive basic amino acid residues in its primary structure, whereas the second type possesses the degenerate consensus sequence of K(K/R)X(K/R) [6]. To be transported in the nucleus, a protein containing a classical NLS is recognized by a Kap α. Subsequently, a Kap β1 binds the Kap α-cargo-complex to mediate its transport across NPCs. Kap β1 interacts with both Kap α-cargo-complex and NPC proteins (nucleoporins), thereby targeting the cargo to the NPC for its translocation into the nucleus [6,7]. Numerous proteins contain nonclassical NLSs. These proteins bind directly and specifically to different Kap β1 homologs that constitute the Kap β family [2]. Kap β1 is unique among the Kap β family in its use of Kap α as an adaptor protein. Other members of the Kap β family bind their substrates directly [2]. The dissociation of Kap β-cargo complexes is under the control of the GTPase Ran. Inside the nucleus, Ran-nucleotide guanine triphosphate (GTP) binds to Kap β-cargo complexes, resulting in the dissociation and release of cargoes into the nucleus [8].

In the fission yeast *Schizosaccharomyces pombe*, Imp1 and Cut15 are two members of the Kap α family [9]. In the case of Kap βs, twelve candidates have been annotated from the *S. pombe* Genome Project [10]. Although the majority of them have not yet been characterized, Kap95 is predicted to be the ortholog of *S. cerevisiae* Kap95, which is a Kap β1 [2,11]. Recent studies have also shown that Kap104 is a Kap β2-type receptor, which mediates nuclear import of proline-tyrosine (PY)-NLS cargoes [12]. Unlike classical NLSs, PY-NLS consensus sequence corresponds to [basic/hydrophobic]-X_n-[R/H/K]-X_{2-5}-PY [12,13].

Iron-regulatory transcription factors play fundamental roles by controlling expression of multiple genes encoding proteins involved in iron homeostasis. In the model organism *S. pombe*, regulation of iron homeostasis is mainly controlled by two iron-responsive proteins, the GATA-binding transcription factor Fep1 and the CCAAT-regulatory subunit Php4 [14]. When iron levels exceed those needed by the cells, Fep1 binds to GATA-type *cis*-acting elements and represses the expression of a number of genes involved in iron transport and intracellular iron utilization [15,16]. In contrast, Fep1 is unable to bind chromatin in response to iron deficiency [17]. This situation leads to transcriptional activation of the Fep1 regulon, which includes the *php4+* gene [18,19]. During iron starvation, Php4 is synthesized and coordinates the iron-sparing response by repressing many genes encoding iron-using proteins [19]. At the molecular level, Php4 regulates its target genes by recognition of the CCAAT-binding complex which is constituted of Php2, Php3 and Php5. The Php2/3/5 heterotrimer binds CCAAT *cis*-acting elements whereas Php4 lacks DNA-binding activity. Php4 is responsible for the ability of the Php complex to repress transcription as a consequence of its association with the heteromeric complex [18,19]. As for Fep1 orthologs, Php4-like proteins are widely distributed in other fungal species. *Saccharomyces* species is the only group that lacks Php4 and Fep1 orthologs [20].

Studies have shown that the monothiol glutaredoxin Grx4 is a binding partner of Php4 and that it plays an essential role in inhibiting Php4 function when cells undergo a shift from iron-limiting to iron-replete conditions [21,22]. Under conditions of iron abundance, Php4 is exported from the nucleus to the cytoplasm. The nuclear export of Php4 requires both exportin Crm1 and Grx4 [21]. Consistently, disruption of the *grx4+* gene (*grx4Δ*) results in Php4 being constitutively active and invariably located in the nucleus. Although the mechanism by which Grx4 communicates the high concentrations of iron to Php4 remains unclear, deletion mapping analysis revealed that the thioredoxin (TRX) domain of Grx4 interacts strongly and constitutively with Php4 [22]. Further analysis has revealed that, in response to iron repletion, the glutaredoxin (GRX) domain of Grx4 associates with Php4. A putative mechanism for Grx4-mediated inhibition of Php4 function would be that the Php4-GRX domain iron-dependent association disrupts the Php4/Php2/Php3/Php5 heteromeric complex, leading to Php4 release and its subsequent export from the nucleus to the cytoplasm by Crm1.

Exported Php4 is observed in the cytosol. However, when external growth conditions change and cells are exposed to iron-poor conditions, it follows that nuclear localization of Php4 should be re-established via its import to the nucleus. To address this issue, we have characterized the mechanism of cytosolic-to-nuclear import of Php4. In response to iron deficiency, nuclear import of Php4 occurred and deletion of *php2+*, *php3+* and *php5+* (*php2Δ php3Δ php5Δ*) did not cause any defects in its nuclear localization. Protein function analysis identified two independent and biologically redundant NLSs within Php4. Each NLSs was sufficient to target an unrelated reporter protein to the nucleus. Disruption of *imp1Δ* or *sal3Δ* gene caused GFP-Php4 to partly mislocalize to the cytoplasm under low-iron conditions. Similarly, in cells containing a temperature-sensitive mutation of *cut15*, GFP-Php4 was mistargeted to the cytoplasm at the nonpermissive temperature. Further analysis by pull-down experiments showed that Php4 interacted with Imp1, Cut15 and Sal3 in *S. pombe*. Collectively, our findings show that Php4 possesses two nuclear targeting

sequences that are used by different Kaps for its nuclear import in response to iron starvation.

Materials and Methods

Strains and growth media

S. pombe strains used in this study are listed in Table 1. Cells were grown in yeast extract medium plus supplements (YES) containing 0.5% yeast extract, 3% glucose, and 225 mg/l of adenine, histidine, leucine, uracil and lysine. Strains for which plasmid transformation was required were grown in synthetic Edinburgh minimal medium (EMM) lacking specific amino acids required for plasmid selection and maintenance [23]. Cells constitutively expressing a *GFP-php4+* allele were seeded to an A_{600} of 0.2, grown to mid-logarithmic phase (A_{600} of 0.5) and then treated with either 2,2'-dipyridyl (Dip, 250 μM) or $FeCl_3$ (100 μM), or were left untreated for 3 h, unless otherwise stated. When the wild-type or mutant *php4* alleles were expressed under the control of the *nmt1+* promoter, induction of transcription was initiated by removal of thiamine to cells grown to an A_{600} of 0.2. After 12 h of induction, cells were incubated with Dip (250 μM) or $FeCl_3$ (100 μM) for 3 h. In contrast, to prevent expression of *php4* alleles, cells were grown in the presence of thiamine (15 μM or 45 μM), unless otherwise indicated. In the case of *cut15-85ts* cells expressing a thermolabile Cut15, cells were grown at the permissive temperature (25°C) to an A_{600} of ~0.4. The cells were then shifted to 36°C for 1 h and then further incubated at 36°C in the presence of Dip (250 μM) or $FeCl_3$ (100 μM) for an additional 3 h.

Plasmids

pJK-194*prom*php4+*-GFP-php4+* plasmid has been described previously [21]. Plasmids pJKGFP-^{1}Php4^{88}, pJKGFP-^{1}Php4^{144}, pJKGFP-^{1}Php4^{179}, pJKGFP-^{1}Php4^{218}, pJKGFP-^{152}Php4^{295}, pJKGFP-^{188}Php4^{295}, pJKGFP-^{219}Php4^{295}, and pJKGFP-^{245}Php4^{295} were created by cloning different truncated versions of the *php4+* gene into pJK-194*prom*php4+*-GFP-php4+*. Different lengths of *php4+* were generated by PCR using primers that contained SalI and Asp718 restriction sites at their ends. After amplification, purified DNA fragments were digested with these two enzymes and then swapped into the corresponding sites of pJK-194*prom*php4+*-GFP-php4+*, generating a series of plasmids bearing deletions within different regions of *php4+*. To create *php4* mutant alleles K171A/R172A/I173/R174A, K214A/I215/R216A/K217A/R218A, and K234A/S235/V236/K237A/R238A/V239A/R240A, the plasmid pJK-194*prom*php4+*-GFP-php4+* was used in conjunction with the overlap extension method [24]. Primers were designed to ensure the presence of nucleotide substitutions that gave rise to the above-mentioned mutations. Using two additional oligonucleotides corresponding to the start and stop codons of the ORF of *php4+*, overlap extension PCR allowed generation of *php4-K171A/R172A/I173/R174A*, *php4-K214A/I215/R216A/K217A/R218A*, and *php4-K234A/S235/V236/K237A/R238A/V239A/R240A* alleles. These mutant alleles were used to replace the equivalent wild-type *php4+* DNA segment in pJK-194*prom*php4+*-GFP-php4+*. Similarly, overlap extension PCR was used to generate additional *php4* mutants that included different combinations of *K171A/R172A/I173A/R174A* mutations with *K214A/I215/R216A/K217A/R218A* or/and *K234A/S235/V236/K237A/R238A/V239A/R240A* mutations. Plasmid pSP-1178nmt-GST-GFP [25] was digested with SpeI and SacI restriction enzymes and used to join annealed synthetic DNA fragments encoding wild-type versions of SV40 NLS and Pap1

Table 1. S. pombe strain genotypes.

Strain	Genotype	Source or reference
FY435	h⁺ his7–366 leu1-32 ura4-Δ18 ade6-M210	[48]
AMY17	h⁺ his7–366 leu1-32 ura4-Δ18 ade6-M210 php4Δ::loxP	[21]
GKY1	h⁺ his7–366 leu1-32 ura4-Δ18 ade6-M210 php4Δ::loxP php2Δ::loxP php3Δ::loxP php5Δ::loxP	This study
GKY2	h⁺ his7::loxP leu1-32 ura4-Δ18 ade6-M210 kap104Δ::natMX6 php4Δ::KANʳ	This study
GKY3	h⁺ his7–366 leu1-32 ura4-Δ18 ade6-M210 imp1Δ::loxP php4Δ::KANʳ	This study
GKY4	h⁺ his7–366 leu1-32 ura4-Δ18 ade6-M210 sal3Δ::loxP php4Δ::KANʳ	This study
GKY5	h⁺ his7–366 leu1-32 ura4-Δ18 ade6-M210 imp1Δ::loxP sal3Δ::loxP php4Δ::KANʳ	This study
GKY6	h⁺ his7Δ::loxP leu1-32 ura4Δ::loxP ade6Δ::loxP php4Δ::KANʳ	This study
GKY7	h⁺ his7Δ::loxP leu1-32 ura4Δ::loxP ade6Δ::loxP cut15-85 php4Δ::KANʳ	This study
GKY8	h⁺ his7–366 leu1-32 ura4-Δ18 ade6-M210 php4Δ::loxP imp1⁺-TAP::KANʳ	This study
GKY9	h⁺ his7–366 leu1-32 ura4-Δ18 ade6-M210 php4Δ::loxP sal3⁺-TAP::KANʳ	This study
GKY10	h⁺ his7–366 leu1-32 ura4-Δ18 ade6-M210 php4Δ::loxP cut15⁺-TAP::KANʳ	This study

NES [26–28]. Wild-type $php4^+$ coding regions corresponding to amino acid residues 160–190, 188–224, and 219–246 were isolated by PCR and cloned downstream of and in-frame to *GST-GFP* fusion genes, generating plasmids pSP-1178nmt-GST-GFP-^{160}Php4^{190}, pSP-1178nmt-GST-GFP-^{188}Php4^{224}, and pSP-1178nmt-GST-GFP-^{219}Php4^{246}, respectively. Similarly, these $php4^+$ coding regions (amino acid residues 160–190, 188–224, and 219–246) were amplified from plasmids pJK-194*prom*php4*⁺-GFP-php4-K171A/R172A/I173/R174A*, pJK-194*prom*php4*⁺-GFP-php4-K214A/I215/R216A/K217A/R218A*, and pJK-194*prom*php4*⁺-GFP-php4-K234A/S235/V236/K237A/R238A/V239A/R240A* to create plasmids pSP-1178nmt-GST-GFP-mutant^{160}Php4^{190}, pSP-1178nmt-GST-GFP-mutant^{188}Php4^{224}, and pSP-1178nmt-GST-GFP-mutant^{219}Php4^{246}, respectively. The wild-type $php4^+$ coding region corresponding to amino acid residues 160–246 was amplified by PCR using primers designed to generate SpeI and SacI sites at each extremity of the PCR product. The DNA fragment was inserted into the corresponding sites of pSP-1178nmt-GST-GFP. The resulting plasmid, named pSP-1178nmt-GST-GFP-^{160}Php4^{246}, was subsequently used to create three additional plasmids harboring *K171A/R172A/I173/R174A, K234A/S235/V236/K237A/R238A/V239A/R240A* or *K171A/R172A/I173/R174A/K234A/S235/V236/K237A/R238A/V239A/R240A* substitutions.

RNase protection analysis

Total RNA was extracted using a hot phenol method as described previously [29]. In the case of RNase protection assays, RNA (15 μg per reaction) was hybridized and digested with RNase T1 as described previously [19]. Riboprobes derived from plasmids pSK*isa1*⁺ and pSK*act1*⁺ [18] were used to detect *isa1*⁺ and *act1*⁺ transcripts, respectively. Plasmids were linearized with BamHI for subsequent antisense RNA labeling with [α-^{32}P]UTP and T7 RNA polymerase. *act1*⁺ mRNA was probed as an internal control for normalization during quantification of RNase protection products.

Fluorescence microscopy analysis

Fluorescence microscopy was performed as described previously [30]. Both fluorescence and differential interference contrast images (Nomarski) of cells were obtained using a Nikon Eclipse E800 epifluorescent microscope (Nikon, Melville, NY) equipped with a Hamamatsu ORCA-ER digital cooled camera (Hamamatsu, Bridgewater, NJ). Samples were analyzed using a 1,000X magnification with the following filters: 520 to 550 nm (YFP), 465 to 495 nm (GFP), and 340 to 380 nm (Hoechst 33342). Cell fields shown in this study represent a minimum of five independent experiments.

TAP pull-down experiments

For pull-down experiments, we created $php4\Delta$ null strains in which the TAP coding sequence was integrated at the chromosomal locus of $imp1^+$, $cut15^+$, or $sal3^+$. These integrations were performed using a PCR-based gene fusion approach as described previously [31], except that pFA6a-kanMX6-CTAP2 [32] was used to amplify the TAP coding sequence. The method allowed homologous integration of TAP at the chromosomal locus of $imp1^+$, $cut15^+$, or $sal3^+$, thereby replacing wild-type allele by $imp1^+$-TAP, $cut15^+$-TAP or $sal3^+$-TAP allele. To determine whether Php4 interacted with Imp1, Cut15 or Sal3 in S. pombe, $php4\Delta$ $imp1^+$-TAP, $php4\Delta$ $cut15^+$-TAP, or $php4\Delta$ $sal3^+$-TAP cells were transformed with pBPade6⁺-nmt41x-GFP-php4⁺. The cells were grown to mid-logarithmic phase in a thiamine-free medium and then treated with Dip (250 μM) for 3 h. Total cell lysates were prepared as described previously [33], except that PMSF (1 mM) was directly added to cell cultures 10 min before cell lysis. Preparation of IgG-Sepharose 6 Fast-Flow beads (GE Healthcare) and coupling of proteins to beads were carried out as described previously [33]. After end-over-end mixing for 30 to 60 min at 4°C, the beads were washed four times with lysis buffer (1 ml each time) and then transferred to a fresh microtube prior to a final wash. The immunoprecipitates were resuspended in sodium dodecyl sulfate loading buffer (60 μl), heated for 5 min at 95°C and proteins resolved by electrophoresis on 9% sodium dodecyl sulfate-polyacrylamide gels. The following antibodies were used for Western blotting analysis of Imp1-TAP, Cut15-TAP, Sal3-TAP, GFP-Php4 and α-tubulin: polyclonal anti-mouse IgG antibody (1:500) (ICN Biomedicals); monoclonal anti-GFP antibody B-2 (1:500) (Santa Cruz Biotechnology) and monoclonal anti-α-tubulin antibody

(1:5000) (clone B-5-1-2; Sigma-Aldrich). Following incubation with primary antibodies, membranes were washed and incubated with the appropriate horseradish peroxidase-conjugated secondary antibodies (1:5000) (Amersham Biosciences), developed with enhanced chemiluminescence (ECL) reagents (Amersham Biosciences) and visualized by chemiluminescence.

Results

Localization of Php4 to the nucleus in a Php2/Php3/Php5-independent manner under low-iron conditions

As we have previously shown, functional GFP-Php4 localized in the cytoplasm of cells under iron-sufficient conditions (Fig. 1) [21]. Conversely, GFP-Php4 accumulated in the nucleus when cells underwent a transition from iron-sufficient to iron-limiting conditions (Fig. 1) [21]. To further investigate the mechanism by which GFP-Php4 was imported in the nucleus, we tested whether Php2, Php3, and Php5 were required for its nuclear accumulation in response to iron starvation. To perform these experiments, *php4Δ* and *php2Δ php3Δ php4Δ php5Δ* mutant strains were transformed with an integrative plasmid harboring a *GFP-php4+* allele constitutively expressed from a GATA-less *php4+* promoter. Cells expressing GFP-Php4 were grown under basal conditions to mid-logarithmic phase and then treated with the iron chelator Dip or with FeCl₃ for 3 h. Results showed that in the presence of Dip, GFP-Php4 accumulated in the nucleus of both *php4Δ* and *php2Δ php3Δ php4Δ php5Δ* mutant strains (Fig. 1). In contrast, when these strains were treated with FeCl₃, GFP-Php4 was observed primarily in the cytoplasm (Fig. 1). As we have previously observed, GFP alone displayed a pancellular-fluorescence pattern, regardless of cellular iron status (Fig. 1) [21]. Taken together, these results indicated that GFP-Php4 localizes to the nucleus in iron-starved cells in a Php2/Php3/Php5-independent manner. Conversely, in iron-replete cells, GFP-Php4 exhibits a distinct distribution pattern that is cytoplasmic.

Mapping NLSs of Php4

To begin to characterize regions within Php4 responsible for nuclear localization, we created a series of N- and C-terminal deletions and fused GFP to the N terminus of each truncated protein (Fig. 2A). *php4Δ* cells expressing these truncated versions of GFP-Php4 were analyzed by fluorescence microscopy to identify which mutants localized to the nucleus. Truncated GFP-¹Php4⁸⁸, in which the last 207 amino acid residues of Php4 were deleted exhibited a pancellular-fluorescence pattern under both low and high iron concentrations (Fig. 2B), suggesting that GFP-¹Php4⁸⁸ was able to passively enter and exit the nucleus. In the case of GFP-¹Php4¹⁴⁴, results showed that it primarily accumulated in the cytoplasmic region of *php4Δ* cells (Fig. 2B). This finding was consistent with the presence of a NES encompassing amino acid residues 93–100 [21]. GFP-¹Php4¹⁷⁹ and GFP-¹Php4²¹⁸ were located in the nucleus under both iron-limiting and iron-replete conditions (Fig. 2B). Although their nuclear location was independent of the cellular iron status, these observations were consistent with the interpretation of the presence of at least one NLS encompassing a common minimal region composed of amino acid residues 144–179. One reason that may explain the absence of iron-mediated nuclear export of GFP-¹Php4¹⁷⁹ and GFP-¹Php4²¹⁸ is the fact that these chimeric proteins miss part of the C-terminal region (positions 152 to 254) of Php4. Previous structure-function studies have revealed that the association of the GRX domain of Grx4 and Php4 depends on the presence of this region (Php4 152–254) [22]. Furthermore, it is known that the GRX domain-Php4 association is required for the iron-mediated inhibition of Php4 that leads to its recruitment by Crm1 (via Php4 NES 93–100), and its subsequent export out of the nucleus to the cytoplasm [21]. Deletion of amino acid residues 1 to 151, 1 to 187, and 1 to 218 from the N-terminus to generate GFP-¹⁵²Php4²⁹⁵, GFP-¹⁸⁸Php4²⁹⁵, and GFP²¹⁹Php4²⁹⁵ did not affect nuclear localization. Due to the absence of NES, GFP-¹⁵²Php4²⁹⁵, GFP-¹⁸⁸Php4²⁹⁵ and GFP²¹⁹Php4²⁹⁵ were located exclusively in

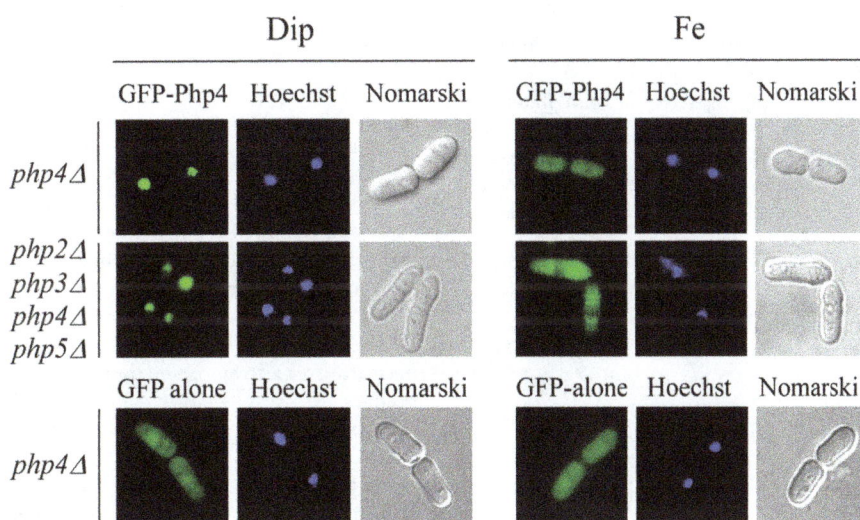

Figure 1. Iron-regulated nucleo-cytoplasmic trafficking of *Php4* is independent of *Php2, Php3* and *Php5* proteins. *php4Δ* or *php2Δ php3Δ php4Δ php5Δ* mutant cells were transformed with an integrative vector expressing GFP alone or a *GFP-php4+* allele under the control of a GATA-less *php4+* promoter. Transformed cells were treated with either Dip (250 μM) or FeCl₃ (Fe) (100 μM) for 3 h. Nuclear DNA was visualized by Hoechst staining whereas Nomarski optics (Nomarski) was used to reveal cell morphology. For simplicity, only *php4Δ* cells transformed with GFP alone are shown because fluorescent images of *php2Δ php3Δ php4Δ php5Δ* cells were identical. The results shown are representative of five independent experiments.

the nucleus, regardless of cellular iron status. However, further deletion of 26 amino acid residues in GFP^{219}Php4^{295} to generate GFP^{245}Php4^{295}, nullified its ability to localize exclusively in the nucleus. Instead, GFP^{245}Php4^{295} exhibited pancellular localization in iron-starved and iron-replete cells (Fig. 2B), revealing loss of signal to promote active entry of Php4 into the nucleus. Taken together, these results were consistent with the interpretation that regions of Php4 from amino acid residues 144 to 179, 188 to 245, and 219 to 245 are sufficient to mediate nuclear import and accumulation of Php4.

Mutation of three predicted NLSs of Php4

In light of these observations, we sought to identify amino acid residues in regions 144 to 179, 188 to 245, and 219 to 245 of Php4 that could serve as NLSs. One of the characteristic features of a classical NLS is a degenerate consensus sequence of K(K/R)X(K/R) (where X indicates any amino acid residue) [6]. Analysis of Php4 using the NLS Mapper [34] prediction program highlighted three short regions containing positively charged residues that matched or partially matched the consensus K(K/R)X(K/R) motif. The first potential NLS, ^{171}KRIR^{174} (amino acid residues 171–174) was found in region 144 to 179, whereas the second ^{214}KIRKR^{218} (amino acid residues 214–218) and the third ^{234}KSVKRVR^{240} (amino acid residues 234–240) putative NLSs were located in region 188 to 245. In the case of region 219 to 245, it contained only the ^{234}KSVKRVR^{240} motif. To determine a functional NLS within Php4 that directs nuclear localization, we first mutated three positively charged amino acids, K^{171}, R^{172}, and R^{174} to Ala in full-length Php4 to generate Php4-N1. We also examined the effect of mutating K^{214}, R^{216}, K^{217}, and R^{218} (Php4-N2) or K^{234}, K^{237}, R^{238}, V^{239}, and R^{240} (Php4-N3) to Ala on the ability of Php4 to localize to the nucleus (Fig. 3A). Results showed that Php4-N1, Php4-N2, and Php4-N3 mutants were efficiently targeted to the nucleus under iron starvation conditions, whereas their localization was predominantly cytoplasmic under high levels of iron (Fig. 3B). Iron-dependent nuclear-cytoplasmic trafficking of these mutants was similar to that of wild-type GFP-Php4 fusion protein (Fig. 3B). Subsequently, we combined the mutated residues within Php4-N3 with mutations in Php4-N2 (generating Php4-N4) or with mutations in Php4-N1 (generating Php4-N6) or with mutations in Php4-N1 and Php4-N2 (creating Php4-N7) (Fig. 3A). Similarly, mutated residues within Php4-N1 were combined with mutations in Php4-N2 to generate Php4-N5 mutant. php4Δ cells expressing Php4-N4 displayed nuclear accumulation following treatment with Dip. In contrast, Php4-N4 was exported out of the nucleus to the cytoplasm when cells had been treated with iron (Fig. 3B). Microscopy analysis showed that iron-starved cells expressing the php4-N5 allele appeared to have less nuclear accumulation than wild-type protein or Php4-N1, -N2, -N3, and N4 mutants. On the other hand, cells harboring Php4-N6 and Php4-N7 did not show obvious nuclear accumulation under iron deprivation conditions (Fig. 3B). Under elevated iron levels, Php4-N5, Php4-N6, and Php4-N7 remained in the cytoplasm as observed in the case of wild-type GFP-Php4 protein. Taken together, these results revealed that Php4 harbors two functionally redundant NLSs, ^{171}KRIR^{174} and ^{234}KSVKRVR^{240}, which could mediate nuclear import of Php4 independently.

Because nuclear import is prerequisite to Php4 function, we hypothesized that mutations in Php4-N6 mutant (^{171}AAIA^{174} and ^{234}ASVAAAA^{240}) would cause loss of Php4 function as well as produce cells defective in repression of the Php4 regulon in

response to iron starvation. Indeed, cells expressing mutant php4-N6 allele exhibited elevated isa1^{+} mRNA levels that were virtually not repressed by iron starvation (Fig. 4A). In fact, steady-state levels of isa1^{+} mRNA under low iron conditions were increased at least ~7-fold above the levels of wild-type or a strain expressing a functional GFP-Php4 protein that was treated with Dip (Fig. 4B). In contrast, isa1^{+} transcript levels were down-regulated under conditions of iron starvation in cells expressing the wild-type Php4 protein or Php4-N1, −N2, and -N3 mutants. In the case of the Php4-N5 mutant (^{171}AAIA^{174} and ^{214}AIAAA^{218}), its mutations resulted in a ~2-fold increase in the expression of the isa1^{+} gene in the presence of low iron concentrations when compared to the levels observed in iron-starved cells expressing the wild-type GFP-Php4. Nonetheless, the levels of isa1^{+} expression in the Php4-N5 mutant were still much lower under low iron than those under basal or iron-replete conditions (Fig. 4, A and B). Because the absence of Php4 led to a constitutive expression of iron-using genes, php4Δ mutant cells are known to be hypersensitive to low iron conditions (lack of optimization of iron utilization when iron is limited) (Fig. 4C). Results consistently showed that php4Δ cells expressing the php4-N6 allele exhibited poor growth on low iron medium in comparison to wild-type cells (Fig. 4C). In contrast, cells expressing the wild-type GFP-Php4 protein or Php4-N1, -N2, -N3, and -N5 mutants were able to grow on medium containing Dip (Fig. 4C). Taken together, these results indicated that Php4 nuclear localization is necessary for Php4-mediated repressive transcriptional regulation of gene expression.

Two NLSs trigger nuclear import by themselves

To assess whether NLS regions of Php4 had the ability to trigger nuclear import, Php4 160–190, Php4 188–224, and Php4 219–246 fragments were fused to GST-GFP, which was used as a reporter protein in sufficiency experiments [35]. In addition, we examined the effect of mutating K^{171}, R^{172}, and R^{174} to Ala in Php4 160–190 (mutant 160–190), K^{214}, R^{216}, K^{217}, and R^{218} to Ala in Php4 188–224 (mutant 188–224), and K^{234}, K^{237}, R^{238}, V^{239}, and R^{240} to Ala in Php4 219–246 (mutant 219–246) (Fig. 5A). GST-GFP-Php4 160–190 (wild-type and mutant), GST-GFP-Php4 188–224 (wild-type and mutant), and GST-GFP-Php4 219–246 (wild-type and mutant) fusion alleles were expressed under the control of the thiamine-regulatable promoter [36]. This system allowed us to induce cellular pools of the above-mentioned fusion proteins and assess the effect of the presence of a given NLS (^{171}KRIR^{174}, ^{214}KIRKR^{218}, or ^{234}KSVKRVR^{240}) on their localization. Cells expressing GST-GFP-Php4 160-190 and GST-GFP-Php4 219–246 exhibited nuclear accumulation, whereas their mutant derivatives displayed a pancellular-fluorescence pattern in a manner similar to GST-GFP alone (Fig. 5B). In the case of GST-GFP-Php4 188–224, its location was cytoplasmic as well as nuclear, irrespective of the presence or absence of the basic residues K^{214}, R^{216}, K^{217}, and R^{218} (Fig. 5B). Controls for nuclear import and export were GST-GFP-SV40NLS and GST-GFP-Pap1NES, respectively. Results showed that reporter proteins tested in sufficiency experiments were unaffected by cellular iron status (Fig. 5B). Furthermore, immunoblot analyses revealed that reporter proteins were stable and intact under the conditions analyzed (Figure S1 in File S1). Taken together, the data revealed that Php4 contains intrinsic determinants involved in nuclear import of the protein. Indeed, the Php4 160–190-(^{171}KRIR^{174}) and Php4 219–246-(^{234}KSVKRVR^{240}) regions function as transferable NLS sequence when fused with a reporter protein.

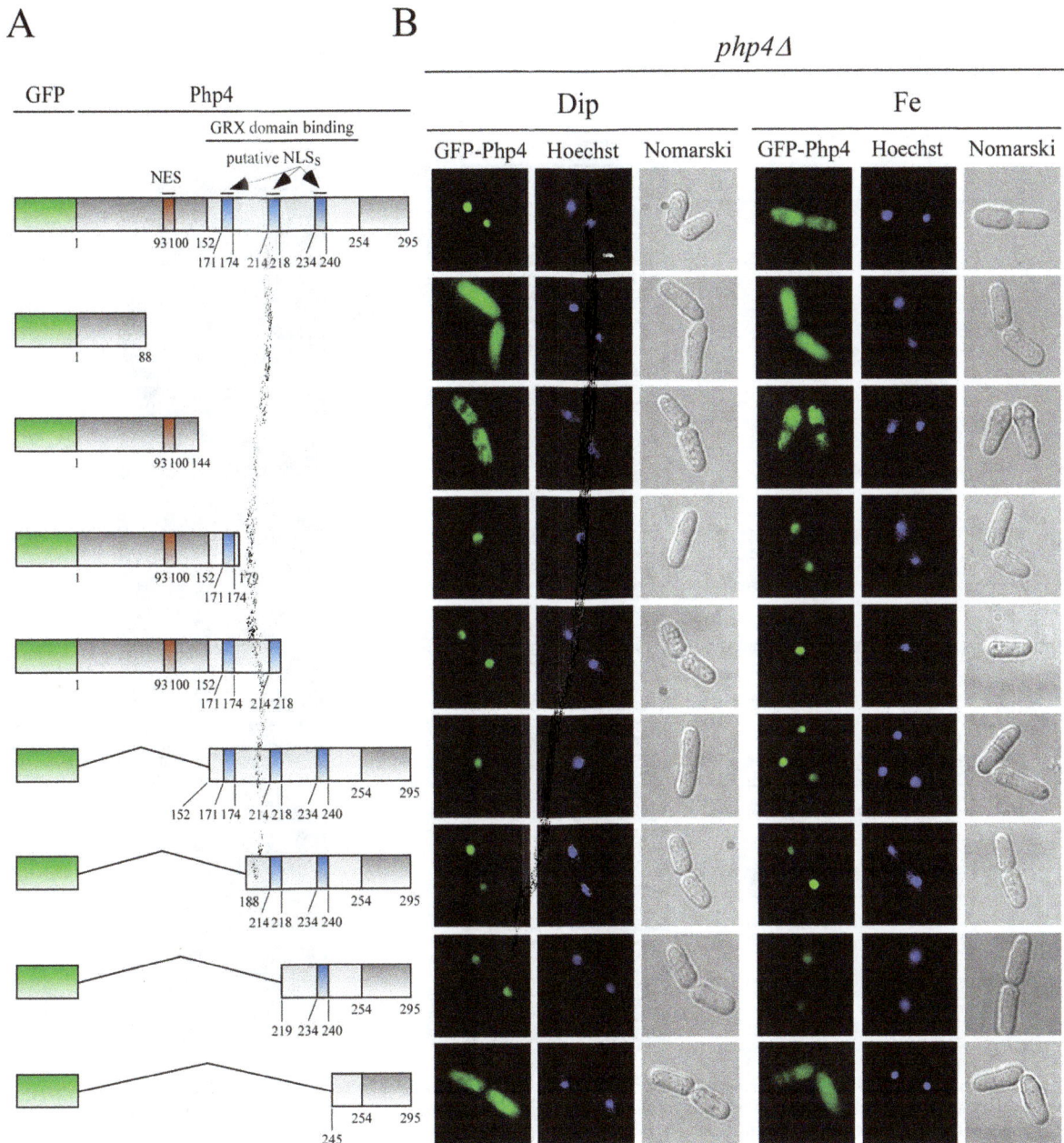

Figure 2. Distinct regions of *Php4* are required for its nuclear localization. *A*, Schematic representation of the GFP-Php4 fusion protein and different GFP-Php4 fusion derivatives. The red box indicates the nuclear export signal (NES) found in Php4 (residues 93–100). Blue boxes represent putative nuclear localization signals (NLSs) that were identified in Php4 (residues 171–174, residues 214–218 and residues 234–240). The segment encompassing residues 152–254 (light-grey box) is a C-terminal region of Php4 required for interaction with the GRX domain of Grx4, which is required for iron-mediated exportation of Php4. The green box represents the GFP coding sequence. The amino acid sequence numbers refer to the positions relative to the first amino acid of Php4. *B*, *php4Δ* cells expressing the indicated fusion alleles under the control of a GATA-less *php4*[+] promoter were incubated in the presence of Dip (250 µM) or FeCl₃ (Fe) (100 µM). After 3 h, cells were examined by fluorescence microscopy to visualize GFP-Php4 and its different fusion derivatives. Hoechst staining revealed nuclear DNA whereas Nomarski optics was used to monitor cell morphology. The results shown are representative of five independent experiments.

To further validate the observation that Php4 contained two functionally redundant NLSs, we expressed and analyzed a segment of Php4 comprising amino acid residues 160 to 246 using the GST-GFP reporter system (Fig. 6A). Amino acids K^{171}, R^{172}, and R^{174} were substituted by Ala in Php4 160–246 to generate Php4-N8. We converted the K^{234}, K^{237}, R^{238}, V^{239}, and R^{240} residues to Ala to generate Php4-N9. We also

combined the mutated residues within Php4-N8 with those in Php4-N9 to generate the Php4-N10 mutant (Fig. 6A). Fluorescence microscopy analysis showed that *php4Δ* cells expressing mutant *GST-GFP-php4-N8* and *GST-GFP-php4-N9* alleles accumulated Php4 in the nucleus in a manner comparable to that of the wild-type *GST-GFP-*[160]*Php4*[246] fusion protein (Fig. 6B). When both clusters of mutated residues were

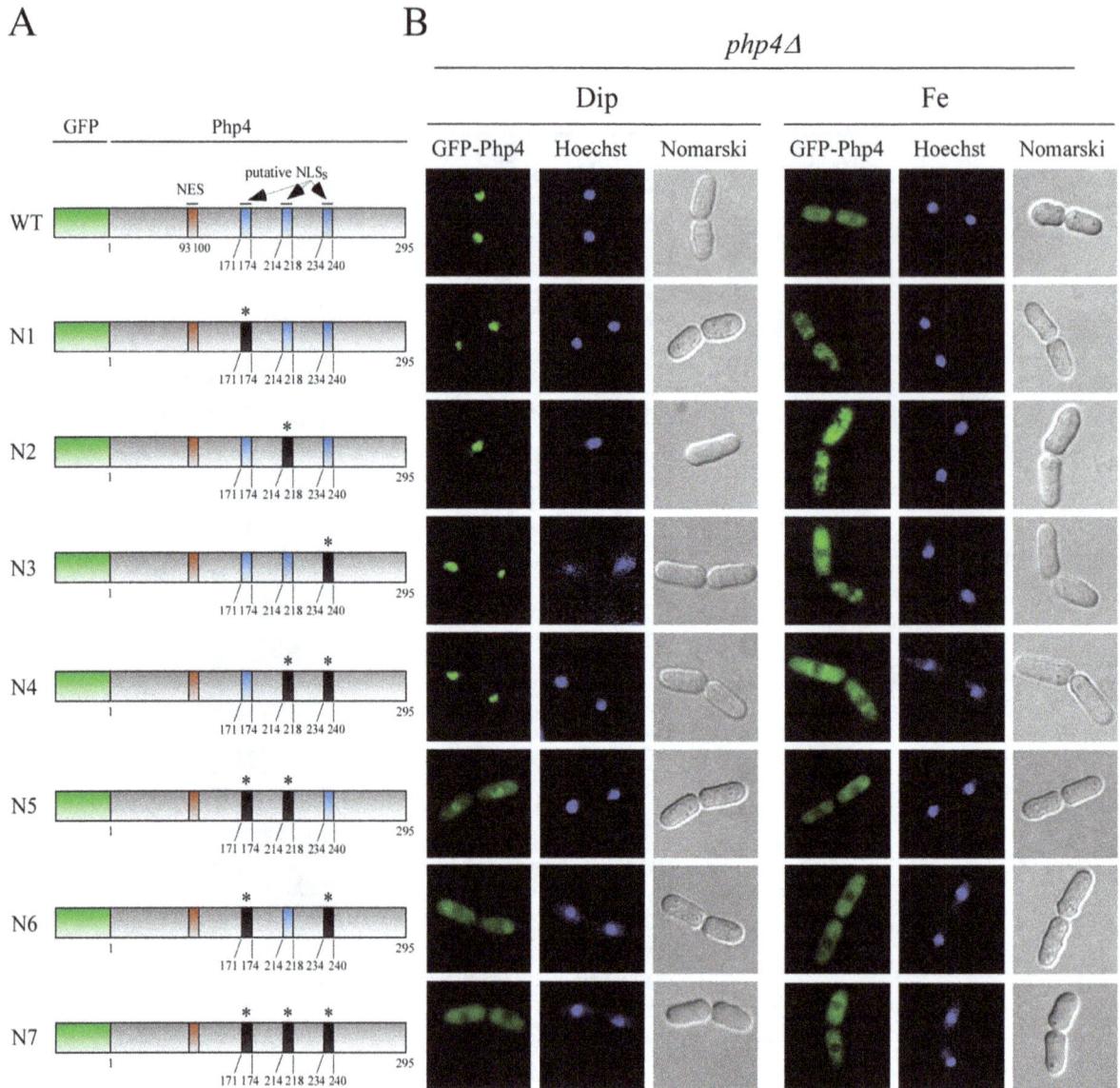

Figure 3. Two regions of *Php4* encompassing amino acid residues 171 to 174 and 234 to 240 are involved in targeting *Php4* to the nucleus. *A*, Schematic illustration of wild-type (WT) and mutant versions (N1 to N7) of GFP-Php4 fusion protein. Green, red and blue boxes represent GFP coding sequence, NES, and putative NLS, respectively. Black boxes (marked with an asterisk) indicate mutated NLS. The amino acid residues of Php4 are numbered relative to its initiator codon. *B*, Fluorescence microscopy was used to visualize cellular location of GFP-Php4 and its mutant derivatives that were expressed in *php4Δ* cells. When indicated, cultures were treated with Dip (250 μM) or FeCl₃ (Fe) (100 μM) for 3 h. Cells were stained using Hoechst to visualize nuclear DNA, whereas Nomarski optics was used to monitor cell morphology. The results shown are representative of five independent experiments.

combined, GST-GFP-php4-N10 was not efficiently targeted to the nucleus, showing primarily cytosolic fluorescence and to less extent some pancellular distribution (Fig. 6B). Western blot analysis of cell extracts showed that the chimeric proteins were present at their expected size (Fig. 6C). Collectively, the results showed that the two NLS regions of Php4 (^{171}KRIR^{174} and ^{234}KSVKRVR^{240}) are functionally redundant in the context of the truncated protein comprising amino acid residues 160 to 246. However, while the ^{234}KSVKRVR^{240} element is fully functional in the truncated protein, this element is not fully competent to mediate nuclear import in the context of the full protein.

Involvement of α- and β-karyopherins in import of Php4

Due to the fact that the two NLSs found in Php4 contained the degenerate consensus sequence of K(K/R)X(K/R), we concluded that both represented short basic classical NLSs [6]. To be transported in the nucleus, a protein containing a classical NLS is recognized by an importin α (karyopherin α or Kapα) protein, which serves as an adaptor. Subsequently, a karyopherin β1 (Kapβ1 or importin β1) binds the importin-α-cargo-complex to mediate its transport across the nuclear pore. Imp1 and Cut15 are the two importin α proteins in *S. pombe*. These two import adaptors have both unique and common binding cargoes [9]. To test whether the nuclear import of Php4 required Imp1, we disrupted the *imp1⁺* gene (*imp1Δ*) and

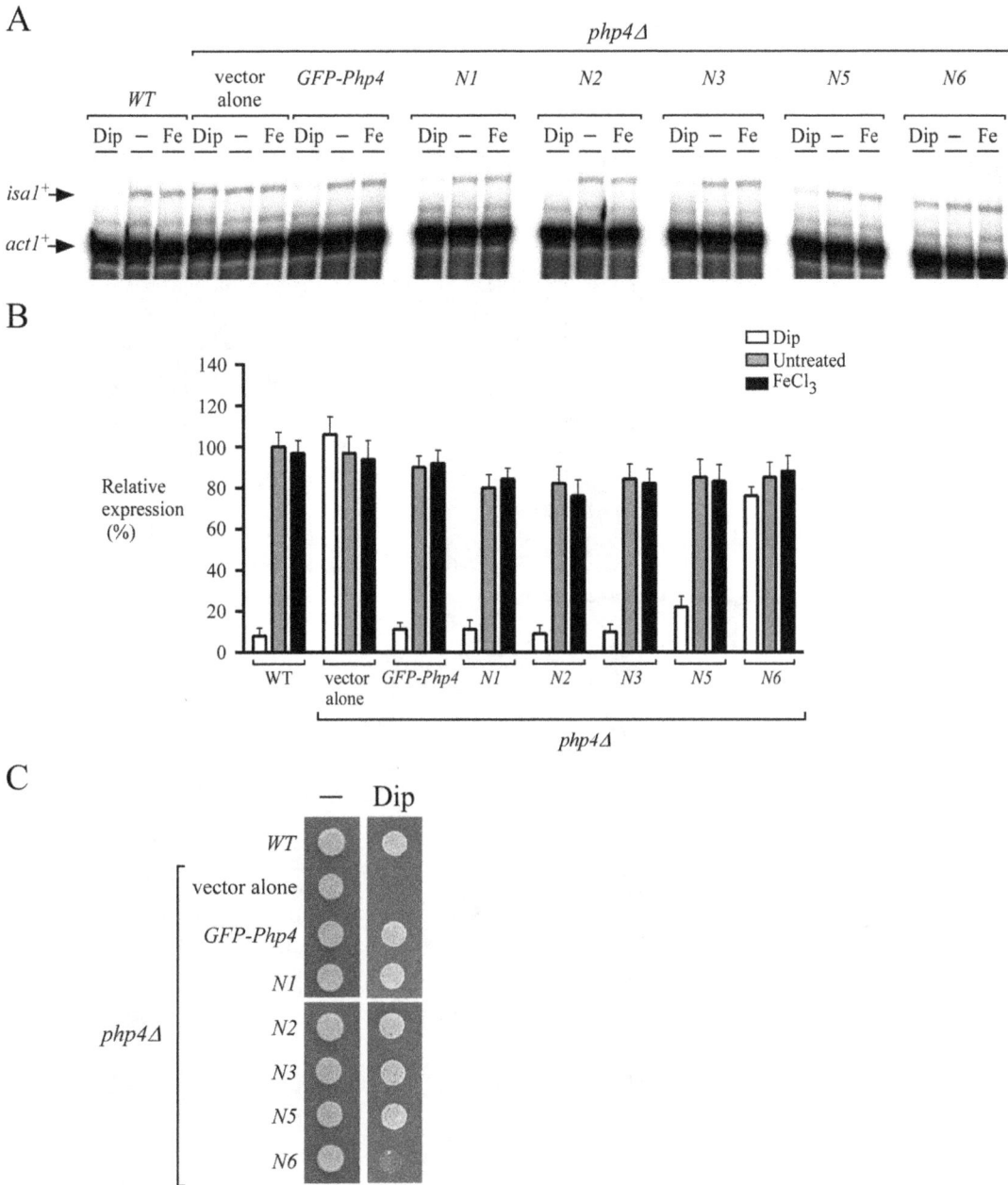

Figure 4. Php4 NLSs are required for Php4-mediated repressive function. A, Cells carrying a disrupted php4Δ allele were transformed with an empty plasmid (vector alone) or plasmids expressing GFP-php4+, GFP-php4+-N1, GFP-php4+-N2, GFP-php4+-N3, GFP-php4+-N5, and GFP-php4+-N6. Transformed cells were grown under basal (−), iron-deficient conditions (250 μM Dip) or excess iron (100 μM FeCl₃) (Fe). After total RNA extraction, isa1+ and act1+ steady-state mRNA levels were analyzed by RNase protection assays. Results shown are representative of three independent experiments. B, Quantification of isa1+ levels after treatments shown in panel A. Data are shown as the mean of triplicate ± standard deviations. C, Wild-type (WT) and php4Δ cells expressing the indicated wild-type or mutant GFP-php4 allele were spotted onto YES medium containing none (−) or 140 μM Dip and incubated at 30°C for 5 days. A php4Δ mutant containing an empty vector (vector alone) was used as a control strain known to be hypersensitive to Dip.

determined the effect on the localization of GFP-Php4. Results showed that under conditions of iron starvation, the absence of Imp1 caused a partial mislocalization of GFP-Php4 to the cytoplasm, although a nuclear accumulation of GFP-Php4 was still observed to some extent (Fig. 7A). cut15+ is essential for cell growth and our approach was to used cut15–85 cells expressing a thermolabile Cut15 in which a GFP-php4+ allele was previously integrated. At the permissive temperature (25°C) in

which case Cut15 is functional, GFP-Php4 accumulated in the nuclei of iron-starved cells (Fig. 7B). However, incubation of iron-starved cells at the nonpermissive temperature (36°C) resulted in an alteration of GFP-Php4 nuclear localization and the GFP-Php4 signal was detected to both the cytoplasm and nucleus (Fig. 7B). Control experiments showed that GFP-Php4 was localized exclusively in the cytoplasm of wild-type and cut15–85 strains when these transformed cells were incubated

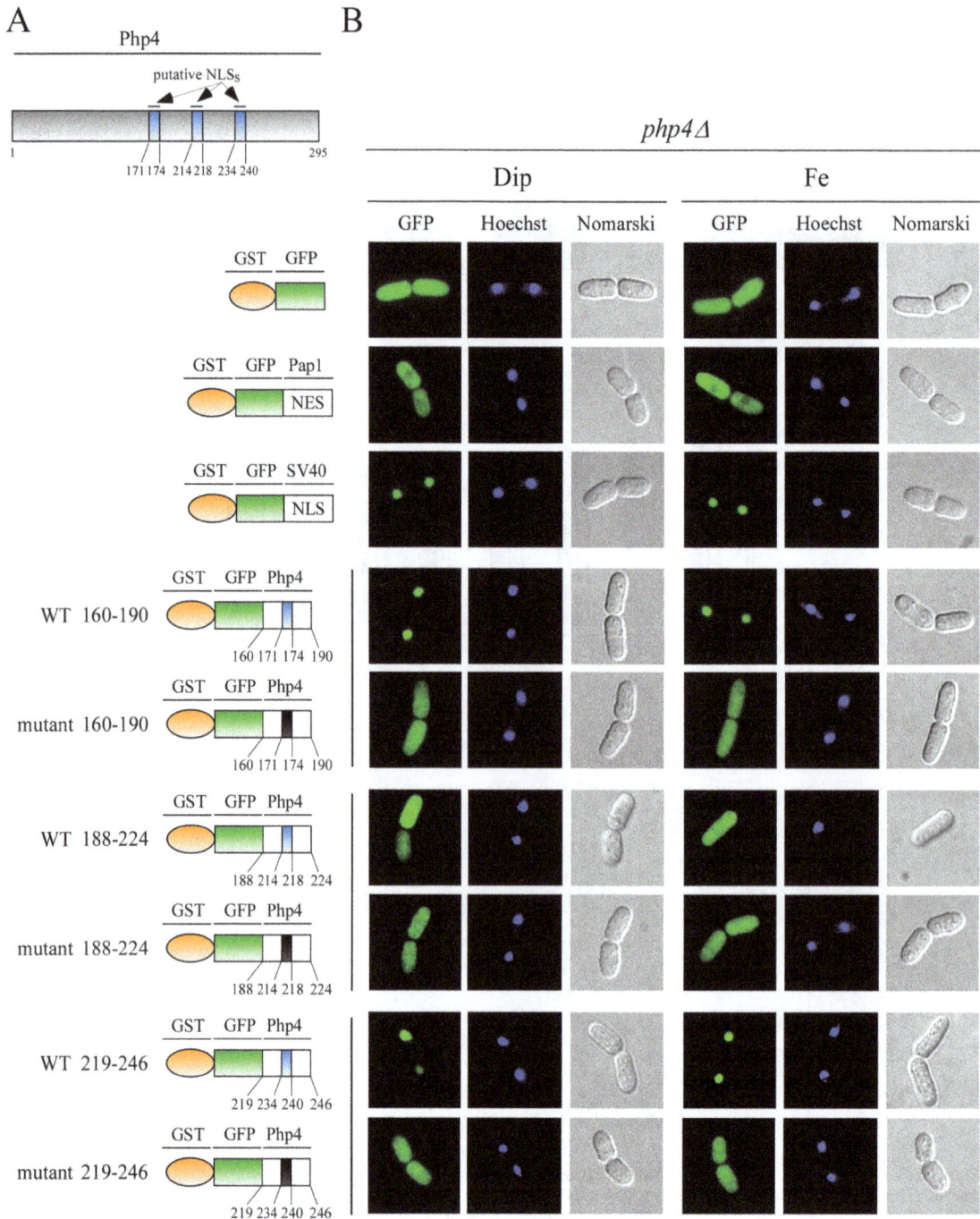

Figure 5. Amino acid fragments 160–190 and 219–246 of *Php4* contain nuclear import activity. *A*, Schematic representation of Php4 and several GST-GFP fusion reporter proteins containing NES or NLS regions of different proteins such as Pap1, SV40, and Php4. Color codes are, orange (GST), green (GFP), blue (putative Php4 NLS) and black (mutated NLS). *B*, Shown are representative *php4Δ* cells expressing GST-GFP, GST-GFP-Pap1NES, GST-GFP-SV40NLS, GST-GFP-Php4^{160}NLS^{190}, GST-GFP-Php4^{160}mutantNLS^{190}, GST-GFP-Php4^{188}NLS^{224}, GST-GFP-Php4^{188}mutantNLS^{224}, GST-GFP-Php4^{219}NLS^{246}, and GST-GFP-Php4^{219}mutantNLS^{246}, respectively. Cultures were grown in thiamine-free media for 12 h. After 3 h treatment in the presence of Dip (250 μM) or FeCl$_3$ (Fe) (100 μM), cells were analyzed by fluorescence microscopy for GFP. As controls, nuclear DNA was visualized by Hoechst staining and cell morphology by Nomarski optics. The results shown are representative of five independent experiments.

in the presence of iron under both temperature conditions (Fig. 7).

S. cerevisiae iron-responsive regulator Aft1 undergoes nucleo-cytoplasmic shuttling in response to changes in intracellular iron concentration in a manner analogous to Php4 [37,38]. Aft1 accumulates in the nucleus upon iron starvation, whereas high iron concentrations result in nuclear export. Nuclear import of Aft1 is mediated by the Kapβ1 Pse1, which is a putative ortholog of *S. pombe* Sal3 [11,38]. Based on this fact, we deleted the *sal3+* gene (*sal3Δ*). Results showed that disruption of Sal3 caused a partial mislocalization of GFP-Php4 to the cytoplasm under low levels of iron, suggesting that Sal3 also participated in nuclear

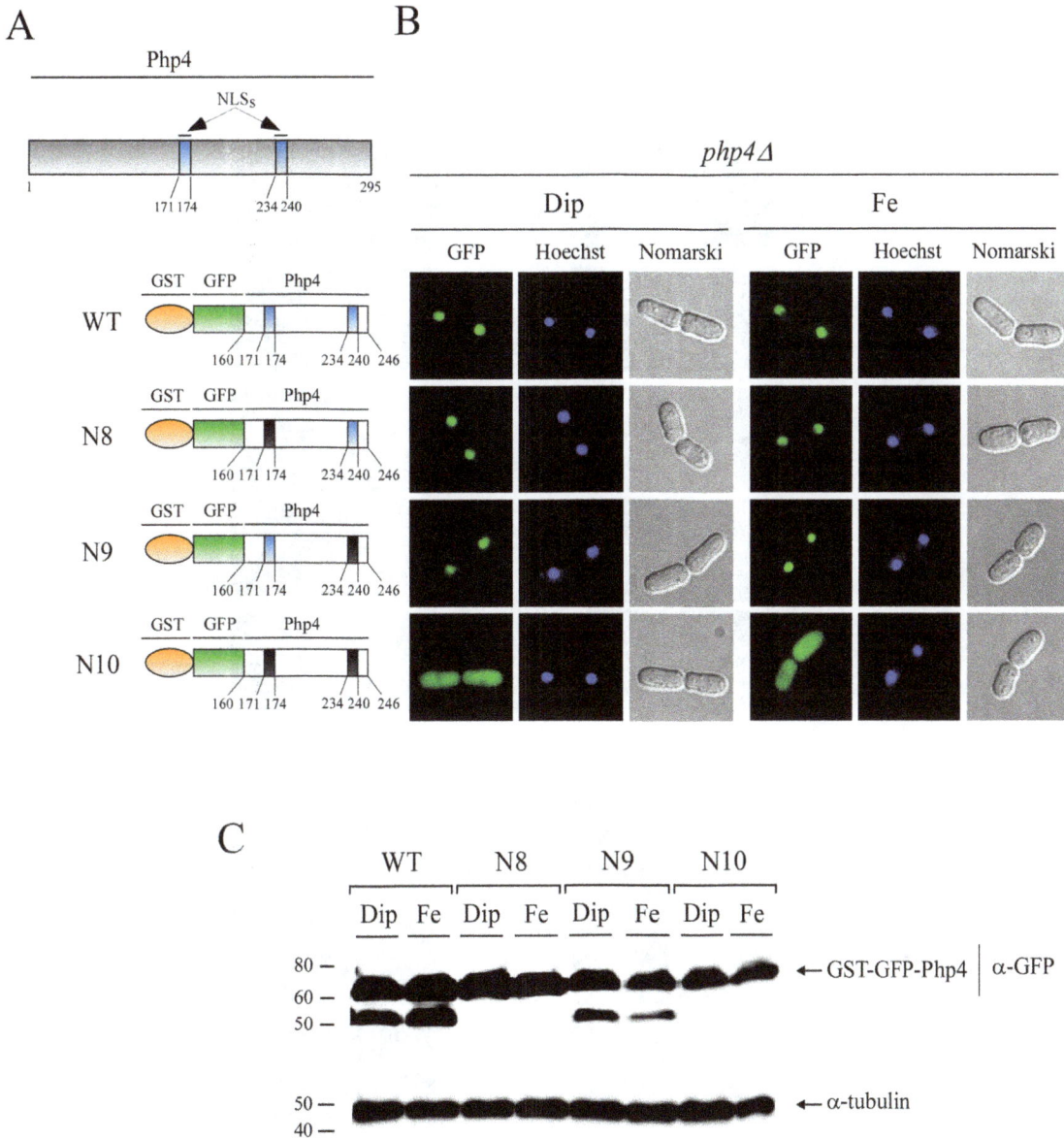

Figure 6. Identification of two functional *Php4* NLSs. *A*, Schematic representation of Php4 that shows relative locations of NLSs (blue boxes). The left bottom panel shows GST-GFP fusion proteins containing the amino acid fragment 160–246 of Php4, including wild-type (WT) and mutant (N8 to N10) versions. Color codes are, orange (GST), green (GFP), blue (NLS) and black (mutated NLS). Amino acid sequence numbers refer to the position relative to the first amino acid of Php4. *B*, Cells harboring a *php4Δ* deletion were transformed with the indicated integrative constructs. Cells were grown to early-logarithmic phase and then thiamine was withdrawn from cell cultures. Thiamine-free cultures were grown for 12 h, and then incubated in the presence of Dip (250 µM) or FeCl$_3$ (Fe) (100 µM) for 3 h. Subsequently, cells were subjected to fluorescence microscopy for GFP detection. Cell morphology was examined through Nomarski optics (Nomarski) and nuclear DNA was detected by Hoechst staining. The results shown are representative of five independent experiments. *C*, Cell extracts were prepared from strains observed in panel B, and analyzed by immunoblotting. GST-GFP-^{160}Php4^{246} (WT) and its mutant (N8 to N10) versions were detected using anti-GFP antibody. As an internal control, extracts preparations were probed with anti-α-tubulin antibody. The positions of the molecular weight standards are indicated to the left.

import (Fig. 7A). As expected, when *sal3Δ* deletion cells were treated with iron, GFP-Php4 was primarily distributed in the cytoplasm (Fig. 7A). Nuclear accumulation of GFP-Php4 in response to iron starvation appeared to rely on more than one karyopherins. Thus, we investigated whether a *imp1Δ sal3Δ* double deletion would favor increased mislocalization of Php4 under low-iron conditions. Results showed that a double deletion of *imp1+* and *sal3+* exhibited a greater cytoplasmic accumulation of GFP-Php4, suggesting that Imp1 and Sal3 may use distinct nuclear import mechanisms for targeting Php4 to the nucleus

(Fig. 7A). As a control, we tested whether the absence of Kap104 influenced Php4 localization. Kap104 is a Kapβ2 that specifically binds proline-tyrosine-NLS (PY-NLS) rather than classical NLS [12]. In this case, GFP-Php4 was properly localized in the nucleus in iron-starved *kap104Δ* cells, supporting the interpretation that the negative effect of the absence of Imp1, Cut15, or Sal3 on Php4 nuclear import was specific (Fig. 7A). When wild-type and mutant karyopherin strains were incubated in the presence of exogenous iron, GFP-Php4 was distributed in the cytoplasm of cells (Fig. 7). Taken together, the results revealed

A

B

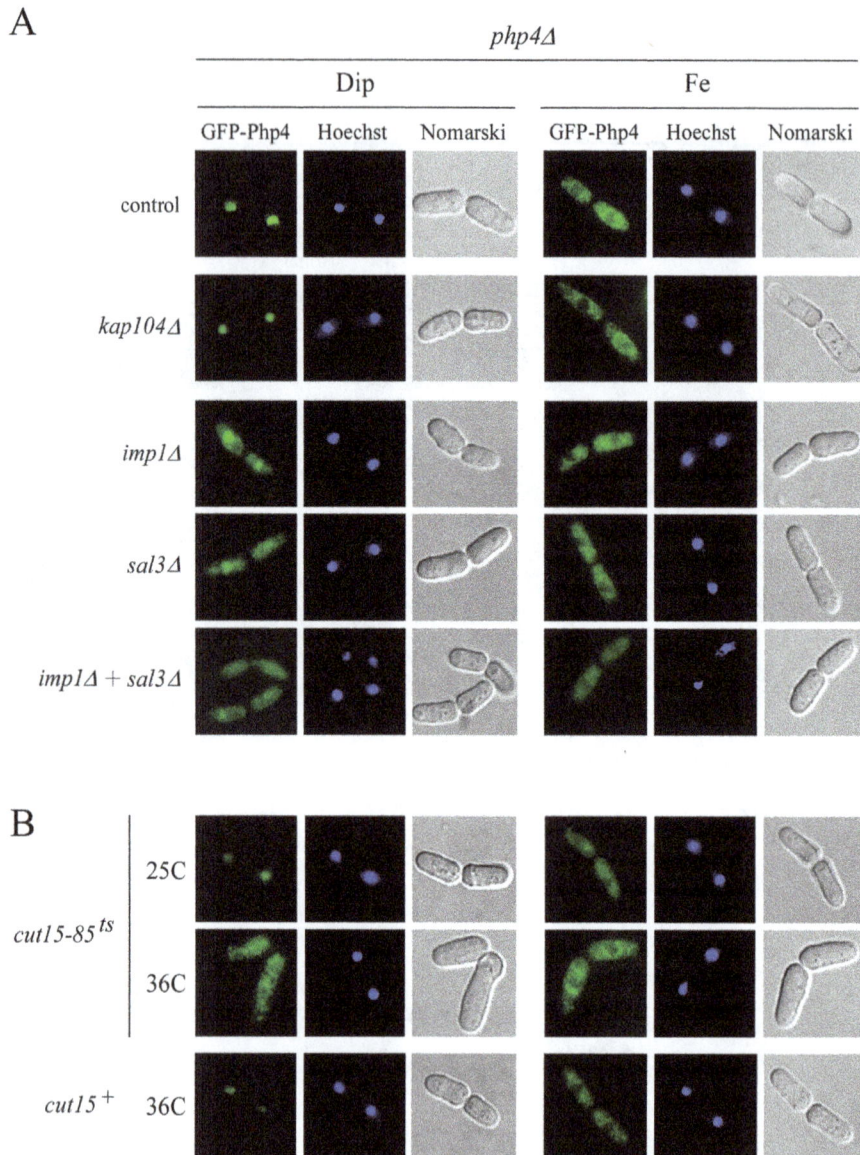

Figure 7. Inactivation of imp1⁺, cut15⁺ and sal3⁺ produced defect in nuclear import of GFP-Php4. *A,* An integrative plasmid expressing a functional GFP-tagged *php4⁺* allele was transformed into *php4Δ, php4Δ kap104Δ, php4Δ imp1Δ, php4Δ sal3Δ,* and *php4Δ sal3Δ imp1Δ* mutant strains. Mid-logarithmic phase cultures were treated with Dip (250 µM) or FeCl₃ (Fe, 100 µM) for 3 h. Fluorescence microscopy was used to visualize cellular location of GFP-Php4. Cells were treated with Hoechst dye for nuclear DNA staining. Cell morphology was examined using Nomarski optics. *B,* Mid-logarithmic phase cultures of the indicated strains were grown at either the permissive (25°C) or nonpermissive (36°C) temperature for 1 h. Cultures were subsequently divided into four separate aliquots which were treated with Dip (250 µM) or FeCl₃ (Fe, 100 µM) at permissive (25°C) or non-permissive (36°C) temperature. After 3 h treatment, cells were analyzed by fluorescence microscopy for GFP detection. The results shown are representative of five independent experiments.

that Imp1, Cut15 or Sal3 could participate in nuclear accumulation of Php4 when cells are grown under low iron conditions.

Given the involvement of Imp1, Cut15 and Sal3 in nuclear import of Php4, we tested whether the repression of *isa1⁺* expression was affected in *imp1Δ, cut15-85* and *sal3Δ* mutant cells. Deletion of *imp1⁺* (*imp1Δ*) resulted in steady-state levels of *isa1⁺* that were increased (~30%) in cells treated with Dip in comparison with iron-starved control cells (Fig. 8). In the case of disruption of *sal3⁺* (*sal3Δ*) that resulted in a modest upregulation (~10%) of *isa1⁺* transcription under low iron conditions. Similarly to *imp1Δ* cells, mRNA levels of *isa1⁺* were

upregulated (~40%) in *imp1Δ sal3Δ* cells, especially in the case of iron-starved control cells (Fig. 8). Similar increases in *isa1⁺* mRNA levels were observed in *imp1Δ, sal3Δ* and *imp1Δ sal3Δ* cells expressing an endogenous Php4 protein (Figure S2 in File S1). We also examined steady-state mRNA levels of *isa1⁺* in *cut15–85* cells expressing a thermolabile Cut15. *php4Δ* and *php4Δ cut15–85* cells expressing GFP-Php4 were grown at the permissive temperature (25°C). At mid-logarithmic phase, cells were divided in aliquots which were then incubated at permissive (25°C) or nonpermissive (36°C) temperature in the presence of Dip (250 µM), FeCl₃ (100 µM), or left without treatment. At 25°C, a temperature where Cut15 was functional, cells displayed very low $isa1^+$ transcript levels under low iron conditions (Dip). In contrast, $isa1^+$ mRNA levels were up-regulated under basal and iron-replete conditions (Fig. 8). At nonpermissive temperature (36°C), inactivation of Cut15 resulted in a 34% increase in $isa1^+$ transcription under low iron conditions compared to levels of $isa1^+$ observed in a $cut15^+$ strain under the same conditions (Fig. 8). In $cut15–85$ cells expressing an endogenous Php4 protein, inactivation of Cut15 resulted in a 20% increase in $isa1^+$ transcription under iron starvation conditions (Figure S2 in File S1). As expected, $isa1^+$ mRNA levels in both untreated and iron-treated $php4\Delta$ $cut15–85$ GFP-$php4^+$ or $cut15–85$ cells were induced as compared to iron-starved cells (Fig. 8 and Figure S2 in File S1). Collectively, these results indicated that Php4 is less competent to repress gene expression under low iron conditions in the absence of Imp1, Cut15 or Sal3.

Imp1, Cut15 and Sal3 are interacting partners of Php4

Given the fact that inactivation of $imp1^+$, $cut15^+$ or $sal3^+$ negatively altered import of Php4 to a different extent, we examined whether Php4 could form complexes with Imp1, Cut15 or Sal3 *in vivo*. To address this possibility, we investigated Php4 capacity to interact with these proteins using TAP pull-down experiments. In these assays, we used iron-starved cells co-expressing distinct pairs of fusion proteins, including GFP-Php4 and Imp1-TAP, GFP-Php4 and Cut15-TAP, GFP-Php4 and Sal3-TAP or GFP-Php4 and TAP (Fig. 9). Total cell extracts were incubated in the presence of IgG-Sepharose beads that selectively bound unfused TAP or TAP-tagged proteins. This strategy allowed an enrichment of Imp1, Cut15 or Sal3 and detection of their potential interacting partners. Western blot analysis of proteins retained by the beads using an anti-GFP antibody revealed that GFP-Php4 was present in the immunoprecipitate fraction of cells expressing Imp1-TAP, Cut15-TAP or Sal3-TAP (Fig. 9). In contrast, GFP-Php4 was absent in the bound fraction of cells expressing TAP alone (Fig. 9). Whole-cell extract fractionation was confirmed using an antibody directed against α-tubulin. Results showed that α-tubulin was present in total cell extracts but not in the retained protein fractions (Fig. 9). To ascertain the steady-state protein levels of Imp1-TAP, Cut15-TAP, or Sal3-TAP, Western blot analyses of both whole cell protein preparations and bound fractions were performed using an anti-IgG antibody (Fig. 9). Taken together, these results showed the existence of Php4-Imp1, Php4-Cut15 and Php4-Sal3 interactions in *S. pombe*.

Discussion

Php4-like proteins are widely distributed among fungal species [20,39]. These proteins include Hap43 (from *Candida albicans*), AnHapX (from *Aspergillus nidulans*), AfHapX (from *Aspergillus fumigatus*) and CnHapX (from *Cryptococcus neoformans*) [40–43] [44]. Of note, *Saccharomyces* species are one of the rare groups that lack Php4 orthologs. Although Php4-like proteins are key nuclear regulators for preventing futile expression of genes encoding iron-using proteins under low-iron conditions, the nature of their NLSs and the mechanisms responsible for triggering their nuclear import have remained poorly characterized. In this study, we have identified two functionally independent and redundant NLSs that are responsible for delivery of Php4 into the nucleus. The first NLS (^{171}KRIR^{174}) possessed a sequence that matched the degenerate consensus K(K/R)X(K/R) motif, which represents one of the two types of conventional monopartite NLSs. Furthermore, classical monopartite NLSs are known to specifically

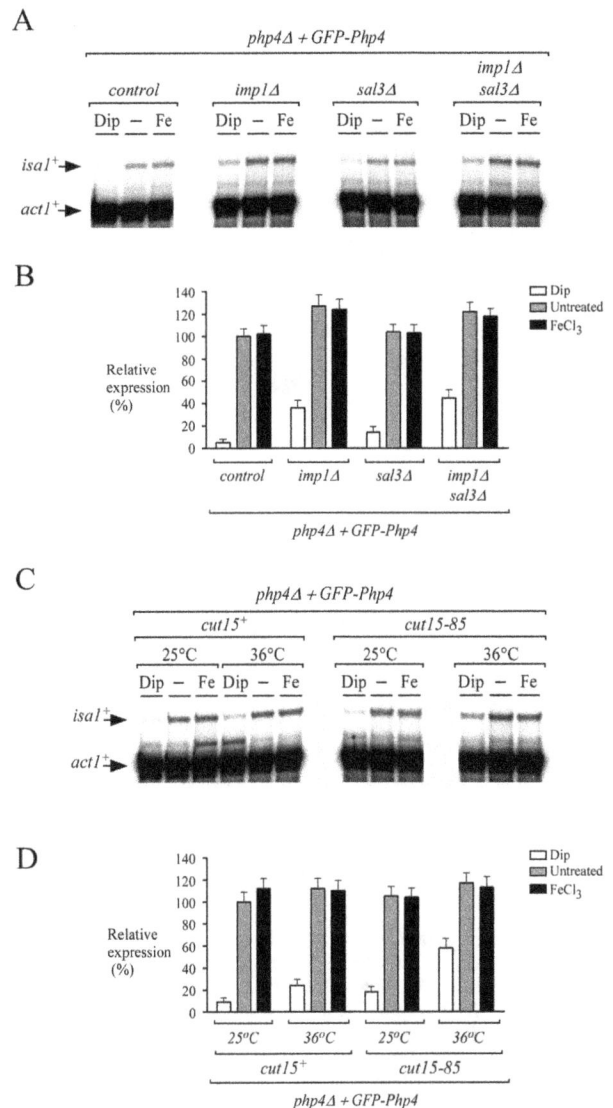

Figure 8. Loss of Imp1, Cut15 or Sal3 resulted in increased expression of isa1+ under low iron conditions. A, Strains harboring insertionally inactivated $php4\Delta$, $php4\Delta$ $imp1\Delta$, $php4\Delta$ $sal3\Delta$, or $php4\Delta$ $imp1\Delta$ $sal3\Delta$ genes were transformed with the GFP-tagged $php4^+$ allele. The indicated strains were assessed for their ability to repress $isa1^+$ gene expression in the presence of Dip (250 µM) versus basal (−) or iron-replete (Fe, 100 µM) conditions. After 3 h of treatment, total RNA was prepared and then analyzed by RNase protection assays. Steady-state levels of $isa1^+$ and $act1^+$ mRNAs are shown with arrows. B, Quantification of three independent RNase protection assays, including the experiment shown in panel A. C, $php4\Delta$ and $php4\Delta$ $cut15–85$ strains were transformed with an integrative plasmid expressing a functional GFP-Php4 protein. Mid-logarithmic phase cultures were divided into four aliquots which were treated with Dip (250 µM) or FeCl3 (100 µM) at permissive (25°C) or nonpermissive (36°C) temperature. After 3 h, total RNA was extracted and used in RNase protection protocol to determine $isa1^+$ and $act1^+$ mRNA levels. D, Quantification of $isa1^+$ transcript levels after treatments. Data are shown as the mean values of triplicate ± standard deviations.

bind Kap α proteins. The second NLS (^{234}KSVKRVR^{240}) is a modified version of the first one. It has ^{237}KRVR^{240} [K(K/R)X(K/R)] as a basic core motif and few flanking residues (^{234}KSV^{236}) immediately upstream of the core basic residues.

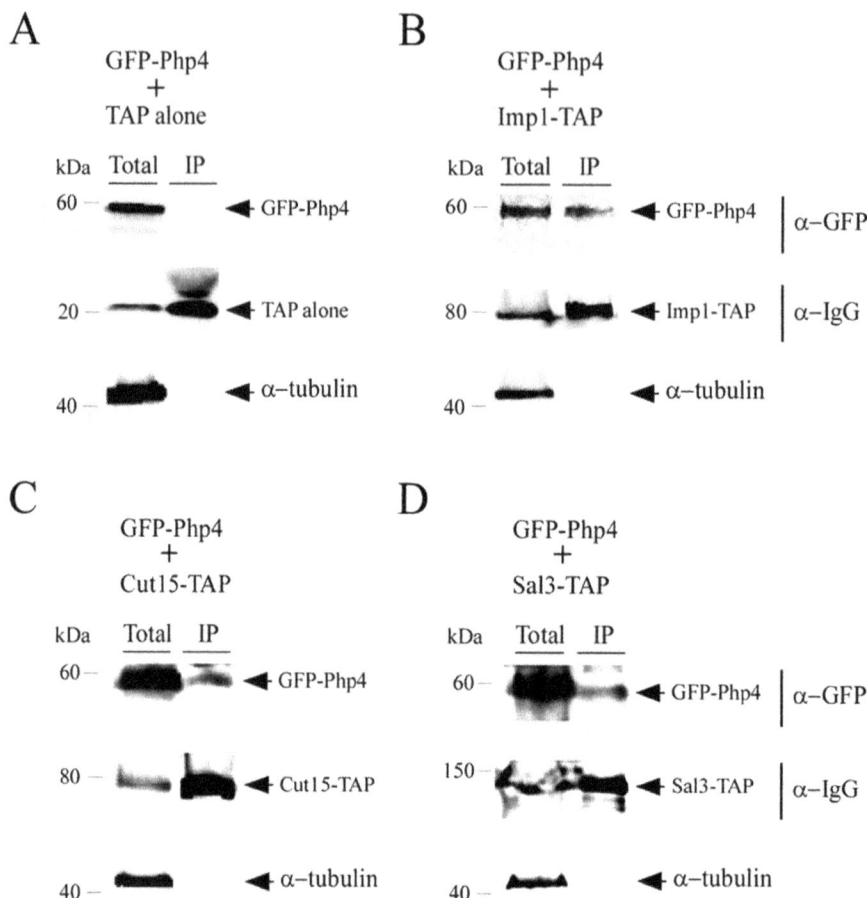

Figure 9. *Php4* **interacts with Imp1, Cut15, and Sal3 in S. pombe.** *php4Δ* cells expressing GFP-tagged Php4 and TAP alone (*A*), GFP-tagged Php4 and TAP-tagged Imp1 (*B*), GFP-tagged Php4 and TAP-tagged Sal3 (*C*), or GFP-tagged Php4 and TAP-tagged Cut15 (*D*) were grown to mid-logarithmic phase in EMM without thiamine in the presence of Dip (250 μM). Extracts (Total) were subjected to immunoprecipitation (IP) using IgG-Sepharose beads. The bound proteins were eluted and analyzed by immunoblot assays using a mouse anti-GFP antibody (α-GFP). A portion of the total cell extracts (~2%) was included to ascertain the presence of proteins prior to chromatography. As additional controls, aliquots of whole-cell extracts and bound fractions were probed with an anti-mouse IgG antibody (α-IgG) and an anti-tubulin antibody (α-tubulin). The positions of the molecular weight of protein standards (in kDa) are indicated on the left-hand side.

These properties represent a modified pattern of classical monopartite NLS that has been previously shown to be competent for binding with Kap α proteins [27]. Indeed, a previous study has shown that the RVSKRPR motif, which is highly reminiscent to KSVKRVR found in Php4, is specifically recognized by Kap α [27]. When we examined the effect of mutating ^{234}K to Ala on the ability of GST-GFP-^{219}Php4^{246} protein to localize to the nucleus, we observed only a weak mislocalization of the protein to the cytoplasm (in comparison with an unmodified GST-GFP-^{219}Php4^{246}). Yet, most GST-GFP-^{219}Php4^{246} ^{234}K→A signal was detected in the nucleus in response to iron starvation (unpublished data). When ^{237}KRVR^{240} were mutated to Ala residues in GST-GFP-^{219}Php4^{246}, the mutant exhibited a pancellular distribution pattern, revealing that the basic core amino acid residues were essential for nuclear import (unpublished data).

Consistent with the amino acid composition of the two Php4 NLSs, we found that the two *S. pombe* Kap α proteins, Imp1 and Cut15, were involved in nuclear import of Php4. This observation meant that Php4 is a common cargo for Imp1 and Cut15. This situation has been reported before. SV40 NLS is functional in *S. pombe* and has been used to assess nuclear protein import

competence. As observed in the case of Php4, both *cut15-85* and *imp1Δ* mutant cells were less efficient at accumulating a SV40 NLS fusion protein in the nucleus than wild type cells [9], revealing that Imp1 and Cut15 have overlapping functions for the import of an SV40 NLS-containing protein. Similarly to Php4, it has been reported that *S. pombe* transcription factor Pap1 interacts with both Imp1 and Cut15 [9]. Neither *imp1Δ* nor *cut15-85* mutant cells were competent to efficiently import Pap1 into the nucleus as compared to wild-type cells. This observation suggested an overlapping function of Imp1 and Cut15 for nuclear import of Pap1. In *S. cerevisiae*, Kap95 is a Kapβ1 involved in the nuclear import of proteins with classical NLSs. One pathway by which Kap95 mediates nuclear import of cargo proteins involves its association with a Kap α protein. One could envision that *S. pombe* Kap95, which is essential for cell viability, is required for the Imp1- or Cut15-mediated nuclear import of Php4. However, the potential involvement of Kap95 remains speculative at this time and needs further investigation.

In general, protein containing NLSs that are recognized by Kap α proteins are transported as a trimeric complex with Kapβ1 proteins. However, it has been shown in the case of some proteins that their nuclear import can be mediated by distinct

Kaps or groups of Kaps. These proteins include histones, ribosomal proteins and stress-responsive transcription factors such as Asr1 and AlcR [45,46]. These findings led us to examine whether some nonessential members of the Kap β family could be required for nuclear import of Php4. Results showed that the inactivation of Sal3 caused a mislocalization of Php4 to the cytoplasm (although a significant proportion of Php4 could still be seen into the nucleus). In *S. pombe*, Sal3 is the ortholog of *S. cerevisiae* Pse1. Interestingly, Pse1 is required for the nuclear localization of the iron-responsive transcription factor Aft1 in *S. cerevisiae*. Although Aft1 is a transcriptional activator and in contrast, Php4 is a transcriptional repressor, both are active and accumulate in the nucleus under conditions of iron starvation. Similarly to Php4, Aft1 possesses two functionally independent NLSs. Although their amino acid composition (KPKKKR and RKPK) is different than those of Php4 (KRIR and KSVKRVR), each of these NLSs is monopartite and is closely related to the consensus sequence K(K/R)X(K/R). However, as opposed to Kap α proteins that are required for nuclear import of Php4, *S. cerevisiae* Kap α (Srp1) is not involved in nuclear import of Aft1. Furthermore, it has been shown that nuclear translocation of Aft1 is exclusively dependent on Pse1 in *S. cerevisiae* and does not depend on other Kap β family members [38].

In contrast, some proteins in *S. cerevisiae* are import substrates of more than one Kaps. For instance, Kap114, Kap95, Kap123, Pse1, and Kap104 recognize NLSs present in histones H2A and H2B, whereas these Kaps mediate nuclear transport of Asr1 [46]. Based on these data, it is likely that *S. pombe* Php4 interacts with more than one type of nucleo-cytoplasmic factors, thereby leaving more options for its nuclear import when iron levels are low. However, the question whether one Php4 NLS is more specific than the others in being recognized by either Kaps α (Imp1 and Cut15) or Kap β1 (Sal3) awaits further studies.

In *A. nidulans*, the CCAAT-binding factor is composed of the HapB, HapC, HapE and HapX subunits [40,47]. Whereas HapC and HapE lack NLSs, HapB contains one functional NLS. In the case of HapX, the presence of functional NLS has not been reported. Under iron sufficient conditions, while the *HAPX* gene is repressed, HapB, HapC and HapE are expressed and assembled as a heterotrimeric complex. To enable cells to provide equimolar concentrations of HapB/C/E subunits to the nucleus, HapB subunit acts as a primary cargo for nuclear import of HapC and HapE. According to a proposed model, HapC and HapE have first to form a heterodimer that is transported into the nucleus only in complex with HapB by way of a piggy-back mechanism [47]. Although the nuclear import mechanism of *S. pombe* CCAAT-binding Php2/3/5 subunits is unknown, we investigated whether nuclear import of Php4 was dependent on the presence of these subunits. Under iron-limiting conditions, disruption of *php2*$^+$, *php3*$^+$ and *php5*$^+$ had no effect on the nuclear import of Php4. Results showed that Php4 accumulated within the nucleus of iron-starved *php2Δ php3Δ php5Δ* triple mutant cells. In the presence of iron, Php4 exhibited a steady-state distribution in the cytoplasm of both *php2*$^+$*/3*$^+$*/5*$^+$ and *php2Δ/3Δ/5Δ* strains. We concluded that nucleocytoplasmic trafficking of Php4 was Php2/3/5-independent. This mechanism is different in comparison with the piggy-back nuclear import mechanism that occurs for the heterotrimeric CCAAT-binding complex in *A. nidulans*.

Our findings suggest that NLS-mediated import of Php4 is not iron-regulated, as we found that the presence of iron did not affect the nuclear localization of the three GST-GFP-Php4 NLS fusion proteins (GST-GFP-^{160}Php4^{190}, GST-GFP-^{219}Php4^{246}, and GST-GFP-^{160}Php4^{246}) (Figs 5 and 6). Furthermore, in the context of full-length protein, when nuclear export sequence (NES) of Php4 was mutated, Php4 exhibited a constitutive nuclear localization under both iron-depleted and iron-replete conditions [21]. This observation suggested that the recognition of Php4 NLSs by Imp1, Cut15 or Sal3 occurred regardless of iron conditions. Nevertheless, it is intriguing to note that Php4 NLSs (positions 171 to 174 and 234 to 240) are included in a region of Php4 from residues 152 to 254 that is known to be required for interaction with the GRX domain of Grx4 [22]. As opposed to the TRX domain, the GRX domain of Grx4 interacts in an iron-dependent manner with Php4. Under high-iron conditions, the GRX domain interacts with the region 152 to 254 of Php4, which may induce conformational changes that negatively affect interactions between NLSs and their import receptors. This may contribute in cytoplasmic accumulation of Php4 under iron-replete conditions. In contrast, under iron-limiting conditions, the GRX domain is no longer able to interact with Php4, which may favor associations between Php4 NLSs and Kaps, therefore contributing in nuclear accumulation of Php4. Although this dynamic interplay may occur in the context of the full-length Php4 protein, further investigation is needed to address this possibility.

Supporting Information

File S1   Figure S1, Detection of intact GST-GFP and GST-GFP fusion proteins. Figure S2, Inactivation of *imp1Δ*, *cut15–5* or *sal3Δ* resulted in increased expression of isa1$^+$ under iron starvation conditions.

Acknowledgments

We are grateful to Dr. Gilles Dupuis for critically reading the manuscript and for his valuable comments. We are indebted to Alexandre Mercier, Pierre-Luc Mallet, François Bachand and the Japanese Yeast Genetic Resource Center (NRBP/YGRC) for yeast strains.

Author Contributions

Conceived and designed the experiments: MGMK J-FJ JB SL. Performed the experiments: MGMK J-FJ JB. Analyzed the data: MGMK J-FJ JB SL. Contributed reagents/materials/analysis tools: MGMK J-FJ JB SL. Wrote the paper: MGMK J-FJ SL.

References

1. Hoelz A, Debler EW, Blobel G (2011) The structure of the nuclear pore complex. Annu Rev Biochem 80: 613–643.
2. Chook YM, Suel KE (2011) Nuclear import by karyopherin-βs: Recognition and inhibition. Biochim Biophys Acta 1813: 1593–1606.
3. Chook YM, Blobel G (2001) Karyopherins and nuclear import. Curr Opin Struct Biol 11: 703–715.
4. Hoelz A, Blobel G (2004) Cell biology: Popping out of the nucleus. Nature 432: 815–816.
5. Pemberton LF, Paschal BM (2005) Mechanisms of receptor-mediated nuclear import and nuclear export. Traffic 6: 187–198.
6. Lange A, Mills RE, Lange CJ, Stewart M, Devine SE, et al. (2007) Classical nuclear localization signals: Definition, function, and interaction with importin alpha. J Biol Chem 282: 5101–5105.
7. Xu D, Farmer A, Chook YM (2010) Recognition of nuclear targeting signals by karyopherin-β proteins. Curr Opin Struct Biol 20: 782–790.
8. Kuersten S, Ohno M, Mattaj IW (2001) Nucleocytoplasmic transport: Ran, beta and beyond. Trends Cell Biol 11: 497–503.
9. Umeda M, Izaddoost S, Cushman I, Moore MS, Sazer S (2005) The fission yeast *Schizosaccharomyces pombe* has two importin-alpha proteins, Imp1p and Cut15p, which have common and unique functions in nucleocytoplasmic transport and cell cycle progression. Genetics 171: 7–21.

10. Wood V, Gwilliam R, Rajandream MA, Lyne M, Lyne R, et al. (2002) The genome sequence of *Schizosaccharomyces pombe*. Nature 415: 871–880.

11. Chen XQ, Du X, Liu J, Balasubramanian MK, Balasundaram D (2004) Identification of genes encoding putative nucleoporins and transport factors in the fission yeast *Schizosaccharomyces pombe*: A deletion analysis. Yeast 21: 495–509.

12. Mallet PL, Bachand F (2013) A proline-tyrosine nuclear localization signal (PY-NLS) is required for the nuclear import of fission yeast PAB2, but not of human PABPN1. Traffic 14: 282–294.

13. Lee BJ, Cansizoglu AE, Suel KE, Louis TH, Zhang Z, et al. (2006) Rules for nuclear localization sequence recognition by karyopherin β 2. Cell 126: 543–558.

14. Labbé S, Khan MG, Jacques JF (2013) Iron uptake and regulation in *Schizosaccharomyces pombe*. Curr Opin Microbiol 16: 669–676.

15. Rustici G, van Bakel H, Lackner DH, Holstege FC, Wijmenga C, et al. (2007) Global transcriptional responses of fission and budding yeast to changes in copper and iron levels: A comparative study. Genome Biol 8: R73.

16. Labbé S, Pelletier B, Mercier A (2007) Iron homeostasis in the fission yeast *Schizosaccharomyces pombe*. Biometals 20: 523–537.

17. Jbel M, Mercier A, Pelletier B, Beaudoin J, Labbé S (2009) Iron activates in vivo DNA binding of *Schizosaccharomyces pombe* transcription factor Fep1 through its amino-terminal region. Eukaryot Cell 8: 649–664.

18. Mercier A, Pelletier B, Labbé S (2006) A transcription factor cascade involving Fep1 and the CCAAT-binding factor Php4 regulates gene expression in response to iron deficiency in the fission yeast *Schizosaccharomyces pombe*. Eukaryot Cell 5: 1866–1881.

19. Mercier A, Watt S, Bahler J, Labbé S (2008) Key function for the CCAAT-binding factor Php4 to regulate gene expression in response to iron deficiency in fission yeast. Eukaryot Cell 7: 493–508.

20. Haas H, Eisendle M, Turgeon BG (2008) Siderophores in fungal physiology and virulence. Annu Rev Phytopathol 46: 149–187.

21. Mercier A, Labbé S (2009) Both Php4 function and subcellular localization are regulated by iron via a multistep mechanism involving the glutaredoxin Grx4 and the exportin Crm1. J Biol Chem 284: 20249–20262.

22. Vachon P, Mercier A, Jbel M, Labbé S (2012) The monothiol glutaredoxin Grx4 exerts an iron-dependent inhibitory effect on php4 function. Eukaryot Cell 11: 806–819.

23. Sabatinos SA, Forsburg SL (2010) Molecular genetics of *Schizosaccharomyces pombe*. Methods Enzymol 470: 759–795.

24. Ho SN, Hunt HD, Horton RM, Pullen JK, Pease LR (1989) Site-directed mutagenesis by overlap extension using the polymerase chain reaction. Gene 77: 51–59.

25. Beaudoin J, Labbé S (2007) Crm1-mediated nuclear export of the *Schizosaccharomyces pombe* transcription factor Cuf1 during a shift from low to high copper concentrations. Eukaryot Cell 6: 764–775.

26. Kalderon D, Roberts BL, Richardson WD, Smith AE (1984) A short amino acid sequence able to specify nuclear location. Cell 39: 499–509.

27. Kosugi S, Hasebe M, Matsumura N, Takashima H, Miyamoto-Sato E, et al. (2009) Six classes of nuclear localization signals specific to different binding grooves of importin alpha. J Biol Chem 284: 478–485.

28. Kudo N, Taoka H, Toda T, Yoshida M, Horinouchi S (1999) A novel nuclear export signal sensitive to oxidative stress in the fission yeast transcription factor Pap1. J Biol Chem 274: 15151–15158.

29. Chen D, Toone WM, Mata J, Lyne R, Burns G, et al. (2003) Global transcriptional responses of fission yeast to environmental stress. Mol Biol Cell 14: 214–229.

30. Pelletier B, Trott A, Morano KA, Labbé S (2005) Functional characterization of the iron-regulatory transcription factor Fep1 from *Schizosaccharomyces pombe*. J Biol Chem 280: 25146–25161.

31. Bahler J, Wu JQ, Longtine MS, Shah NG, McKenzie A, et al. (1998) Heterologous modules for efficient and versatile PCR-based gene targeting in *Schizosaccharomyces pombe*. Yeast 14: 943–951.

32. Tasto JJ, Carnahan RH, McDonald WH, Gould KL (2001) Vectors and gene targeting modules for tandem affinity purification in *Schizosaccharomyces pombe*. Yeast 18: 657–662.

33. Jacques JF, Mercier A, Brault A, Mourer T, Labbé S (2014) Fra2 is a co-regulator of Fep1 inhibition in response to iron starvation. PLoS One 9: e98959.

34. Kosugi S, Hasebe M, Tomita M, Yanagawa H (2009) Systematic identification of cell cycle-dependent yeast nucleocytoplasmic shuttling proteins by prediction of composite motifs. Proc Natl Acad Sci USA 106: 10171–10176.

35. Saydam N, Georgiev O, Nakano MY, Greber UF, Schaffner W (2001) Nucleo-cytoplasmic trafficking of metal-regulatory transcription factor 1 is regulated by diverse stress signals. J Biol Chem 276: 25487–25495.

36. Moreno MB, Duran A, Ribas JC (2000) A family of multifunctional thiamine-repressible expression vectors for fission yeast. Yeast 16: 861–872.

37. Yamaguchi-Iwai Y, Ueta R, Fukunaka A, Sasaki R (2002) Subcellular localization of Aft1 transcription factor responds to iron status in *Saccharomyces cerevisiae*. J Biol Chem 277: 18914–18918.

38. Ueta R, Fukunaka A, Yamaguchi-Iwai Y (2003) Pse1p mediates the nuclear import of the iron-responsive transcription factor Aft1p in *Saccharomyces cerevisiae*. J Biol Chem 278: 50120–50127.

39. Kronstad JW (2013) Iron in eukaryotic microbes: Regulation, trafficking and theft. Curr Opin Microbiol 16: 659–661.

40. Hortschansky P, Eisendle M, Al-Abdallah Q, Schmidt AD, Bergmann S, et al. (2007) Interaction of HapX with the CCAAT-binding complex-a novel mechanism of gene regulation by iron. EMBO J 26: 3157–3168.

41. Hsu PC, Yang CY, Lan CY (2011) Candida albicans Hap43 is a repressor induced under low-iron conditions and is essential for iron-responsive transcriptional regulation and virulence. Eukaryot Cell 10: 207–225.

42. Jung WH, Saikia S, Hu G, Wang J, Fung CK, et al. (2010) HapX positively and negatively regulates the transcriptional response to iron deprivation in *Cryptococcus neoformans*. PLoS Pathog 6: e1001209.

43. Schrettl M, Beckmann N, Varga J, Heinekamp T, Jacobsen ID, et al. (2010) HapX-mediated adaption to iron starvation is crucial for virulence of *Aspergillus fumigatus*. PLoS Pathog 6: e1001124.

44. Gsaller F, Hortschansky P, Beattie SR, Klammer V, Tuppatsch K, et al. (2014) The janus transcription factor HapX controls fungal adaptation to both iron starvation and iron excess. EMBO J Aug 4 Epub ahead of print.

45. Nikolaev I, Cochet MF, Felenbok B (2003) Nuclear import of zinc binuclear cluster proteins proceeds through multiple, overlapping transport pathways. Eukaryot Cell 2: 209–221.

46. Fries T, Betz C, Sohn K, Caesar S, Schlenstedt G, et al. (2007) A novel conserved nuclear localization signal is recognized by a group of yeast importins. J Biol Chem 282: 19292–19301.

47. Steidl S, Tuncher A, Goda H, Guder C, Papadopoulou N, et al. (2004) A single subunit of a heterotrimeric CCAAT-binding complex carries a nuclear localization signal: Piggy back transport of the pre-assembled complex to the nucleus. J Mol Biol 342: 515–524.

48. Pelletier B, Beaudoin J, Mukai Y, Labbé S (2002) Fep1, an iron sensor regulating iron transporter gene expression in *Schizosaccharomyces pombe*. J Biol Chem 277: 22950–22958.

9

Identification and Characterization of Novel Renal Sensory Receptors

Premraj Rajkumar, William H. Aisenberg, Omar W. Acres[¤a], Ryan J. Protzko[¤b], Jennifer L. Pluznick*

Department of Physiology, Johns Hopkins University School of Medicine, Baltimore, Maryland, United States of America

Abstract

Recent studies have highlighted the important roles that "sensory" receptors (olfactory receptors, taste receptors, and orphan "GPR" receptors) play in a variety of tissues, including the kidney. Although several studies have identified important roles that individual sensory receptors play in the kidney, there has not been a systematic analysis of the renal repertoire of sensory receptors. In this study, we identify novel renal sensory receptors belonging to the GPR (n = 76), olfactory receptor (n = 6), and taste receptor (n = 11) gene families. A variety of reverse transcriptase (RT)- PCR screening strategies were used to identify novel renal sensory receptors, which were subsequently confirmed using gene-specific primers. The tissue-specific distribution of these receptors was determined, and the novel renal ORs were cloned from whole mouse kidney. Renal ORs that trafficked properly *in vitro* were screened for potential ligands using a dual-luciferase ligand screen, and novel ligands were identified for Olfr691. These studies demonstrate that multiple sensory receptors are expressed in the kidney beyond those previously identified. These results greatly expand the known repertoire of renal sensory receptors. Importantly, the mRNA of many of the receptors identified in this study are expressed highly in the kidney (comparable to well-known and extensively studied renal GPCRs), and in future studies it will be important to elucidate the roles that these novel renal receptors play in renal physiology.

Editor: Johannes Reisert, Monell Chemical Senses Center, United States of America

Funding: This work was supported by funding from the NIH (DK081610 to J.L.P.) and the minigrant from the National Kidney Foundation, Maryland, (to P.R.). Publication of this article was funded in part by the Open Access Promotion Fund of the Johns Hopkins University Libraries. The funders had no role in study design, data collection and analysis, decision to publish, or preparation of the manuscript.

Competing Interests: The authors have declared that no competing interests exist.

* Email: jpluznick@jhmi.edu

¤a Current address: Howard University College of Medicine, Washington, D. C., United States of America
¤b Current address: Department of Molecular and Cell Biology, University of California, Berkeley, California, United States of America

Introduction

A recent paradigm in sensory physiology suggests that several classes of understudied receptors (olfactory receptors (ORs), taste receptors, and orphan G-protein coupled receptors (GPRs)) play key roles in non-sensory tissues, where they serve as selective and sensitive chemoreceptors [1,2,3,4,5,6,7,8,9,10,11,12,13]. For example, ORs are expressed in a variety of non-olfactory tissues (including sperm, muscle, brain, and liver) [3,11,14,15] and it has been shown that an OR in the spermatozoa of both humans and mice functions as a chemosensor to help guide the sperm towards the egg [11]. In addition, sweet taste receptors are found in the bladder [2], sour taste receptors facilitate pH sensing in the cerebrospinal fluid [5], bitter taste receptors mediate both bronchodilation and ciliary beat frequency in airways [1,10], and GPR receptors play important roles in whole-animal physiology as sensors of metabolites [4,6,9,12,13,16]. Ligands for these receptors are often generated by metabolic pathways or other physiological processes [5,12,17], indicating that known metabolites may have additional (and as-of-yet unknown) signaling functions [4,12,17].

We and others have recently demonstrated that the olfactory and GPR signaling pathways play a role in the kidney [7,8,12,13,18]. We also reported, using a degenerate PCR screen

[19,20], that 6 individual ORs are expressed in mouse kidney by RT-PCR [8]. However, the expression of specific sensory receptors in the kidney and their potential roles is understudied, and the full complement of renal sensory receptors is unknown [16]. The OR gene family alone consists of ~1000 genes in the mouse, and despite being the largest gene family in the genome it is largely unexplored in the kidney. In addition, although taste receptors have been shown to play important roles in several tissues outside the tongue [1,2,10,21,22,23], taste receptor expression in the kidney had not been previously examined. In order to categorize and identify which sensory receptors are present in the murine kidney, we employed several strategies. First, to identify novel renal GPRs, we employed a real-time RT-PCR screen for detection of mouse GPCR transcripts within a mouse kidney cDNA and determined their relative levels of expression. In order to identify whether additional renal ORs (beyond the 6 reported previously [8]) are expressed in the kidney, we performed several small-scale directed RT-PCR screens. Finally, we performed a directed RT-PCR screen for all known murine taste receptors. Together, our study identified 76 novel GPRs, 6 novel ORs, and 11 novel taste receptors expressed in the murine kidney. Subsequently, for a subset of these receptors we analyzed the tissue

distribution patterns outside of the kidney, and cloned and studied the receptors *in vitro*.

Materials and Methods

RT-PCR

This study was carried out with mice that were housed and treated in accordance with policies and protocol (M013M109) approved by the Johns Hopkins University Animal Care and Use Committee (ACUC), as well as the National Institutes of Health principles and guidelines for the Care and Use of Laboratory Animals. Mice were asphyxiated with CO2 and the tissues required for RNA isolation were quickly removed and stored until future use. C57Bl/6 (Charles River) male mice were asphyxiated with CO_2 and tissues (tongue, colon, heart, liver, lung, skeletal muscle, small intestine, stomach, kidney and testes) were quickly removed and stored in RNALater (Qiagen) until further use. All efforts were taken to minimize any suffering. RNA was isolated from the tissues using TRIzol reagent (Invitrogen) and samples were further processed using the RNeasy RNA clean-up protocol with on-column DNAase digestion (Qiagen). Tissue specific cDNA was synthesized from 1 μg of purified RNA by reverse transcription (RT; iScript cDNA Synthesis Kit, BioRad). Mock-reverse transcription controls were also prepared from each tissue by omitting the iScript reverse transcriptase enzyme (replaced with an equal volume of water) in reaction mixtures.

PCR was performed using HotStarTaq Master Mix (Qiagen) following standard thermocycling conditions. Murine PCR gene specific primer (GSP) sets were designed using the NCBI Primer Blast PCR primer designer tool for a total of 40 ORs, selected as described in the results section. The nucleotide sequences of the GSP sets along with the expected size band for taste receptors and GPCRs are also listed in Table S1. Prior to screening kidney for novel receptors, we tested our primers and optimized PCR cycling conditions by using either tongue RT (taste) or genomic DNA (ORs). Mock RT reactions were run in parallel with all RT reactions, and all PCR amplicons were sequenced to confirm identity. All RT-PCR products were sequenced to confirm identity.

Taqman array GPCR screen

To identify novel GPRs and determine their relative expression levels in the kidney, we performed an unbiased screen of whole kidney tissue cDNA using the Taqman array mouse GPCR panel (Applied Biosystems, catalog # 4378703) according to the manufacturers protocol. Briefly, two C57Bl/6 mice (one male & one female) were asphyxiated with CO_2 and their kidneys were quickly removed and stored. RNA was isolated from the left kidney of both mice using TRIzol reagent (Invitrogen) and 2ug of RNA per reaction was used to synthesize cDNA using the High Capacity RNA-to-cDNA Kit (AB). Each reservoir in the Taqman array microfluidic card was filled with 1000 ng of cDNA per reservoir and the array cards were run on the AB 7900HT Fast RT-PCR system and analyzed using the SDS2.4 software. Each mouse kidney cDNA was screened on 2 chips, for a total of 4 chips. The screen targeted 380 GPCRs including retinal receptors, small molecule receptors, and chemokine receptors in addition to other 'classic -endogenous' genes as controls. From the obtained Ct measurements, we calculated ΔCt values of each receptor by normalizing to beta-actin, and further estimated standard deviation (S.D.).

Surface Immunofluorescence

Full-length coding sequences of mouse Olfr 31, 99, 545, 691, 693 and 1426 were cloned by PCR from mouse kidney RT into a mammalian expression vector, pME18S, with N-terminal Flag and Rho sequences (kind gift from Kazushige Touhara, Univ. of Toyko [24] and Stuart Firestein, Columbia University) between EcoRI and XhoI cloning sites. We also cloned another set of constructs for each OR with a Lucy tag [25] at the N-terminus in addition to Flag and Rho tags. OR constructs were transiently expressed in HEK293T cells with and without chaperone RTP1S (Lipofectamine 2000, Invitrogen). The trafficking of Flag-Rho-tagged/Lucy-Flag-Rho-tagged ORs (+/− RTP1S/Ric8b) in transfected cells was assayed using a surface immunocytochemistry staining procedure as previously described [25,26], in which a rabbit polyclonal anti-Flag antibody (Sigma) was used in live cells at 4°C. Subsequently, the cells were fixed with 4% PFA, permeabilized using 0.3% triton-X 100 and then exposed to a mouse monoclonal anti-Flag antibody to label internal receptor (Sigma). Fluorescent secondary antibodies (AlexaFluor, Invitrogen) were used to localize the Flag-tagged ORs to the membrane surface or the cytosol of HEK293T cells.

Luciferase Assay

For ORs that trafficked to the plasma membrane of HEK293T cells (trafficking conditions determined in the surface immunofluorescence assay as described above), we performed an unbiased ligand screen using a dual-luciferase reporter assay (Promega) to identify potential ligands for orphan ORs and to expand the ligand profile of previously deorphanized Olfr691 [26]. Under the conditions (+/− Lucy tag, +/− RTP/Ric8b) that yielded strong surface trafficking for each OR, ORs were transfected into HEK293T cells along with a CREB-dependent luciferase (*Firefly*) and a constitutively expressed luciferase (*Renilla*) [26]. Upon a ligand-OR binding event, a rise in cAMP drives the measurable expression of *Firefly* luciferase, which was normalized to the activity of the *Renilla* luciferase to control for variation in cell number and transfection efficiency. Transfected cells were exposed to potential ligands for 4 hours and their corresponding luciferase values were measured in triplicates, in a semi-automated fashion using a FLUOstar Omega microplate reader (BME Labtech). Cells expressing each OR were tested with a set of odorant mixes (described in [18]) and with an additional mix termed CYCONE (containing cyclopentanone, cyclohexanone, cycloheptanone and cyclooctanone each at a final concentration of 0.3 mM). Any activation to the mixes was further explored by exposing the cells to individual components of the chemical mixture to identify the active ligand of that particular OR. In addition, cells were also exposed to a library of chemicals (listed in Table S2) each tested separately at 500 μM. Following the identification of an active ligand, additional candidate ligands were chosen by varying carbon atomic number (CAN) and functional group type and position. Additionally, a metric for odorant comparison was referenced for identifying multifaceted and structurally diverse analogues of active ligands for testing [27]. EC_{50} values of Olfr691 were calculated based on the response to 10 μM, 50 μM, 100 μM, 0.5 mM, 1 mM, 5 mM and 7.5 mM of active ligands by using Sigmaplot data analysis software. Furthermore, all active ligands for Olfr691 were repeated and confirmed by at least three independent trials.

Results

Identification of novel renal murine sensory receptors

GPRs. GPR is the gene name given to orphaned GPCRs; in recent years, as GPRs have been deorphanized, they have been found to play important sensory roles in a variety of tissues [4,6,9,12,17]. To screen for the expression of GPRs in the kidney, we took advantage of a real time Taqman based mouse GPCR array (Applied Biosystems) which assays the expression of 380 transcripts. This screen is targeted to 380 GPCRs, including 91 GPRs (as well as multiple retinal receptors, small molecule receptors, chemokine receptors and 'classic -endogenous genes' as controls). Our analysis focused on the GPRs on this array, as this is a large family of (primarily orphan) receptors which have been shown to play 'sensory-receptor like' roles in a variety of tissues [4,6,9,12,13]. We screened whole kidney tissue from a male & female C57Bl/6 mouse (each kidney was screened on 2 chips, for a total of 4 chips), and calculated average ΔCt values for each receptor by normalizing to beta-actin (Table S3). Based on the average Ct values, we classified the expression level of each receptor in the mouse kidney into one of four categories: high expression (ΔCt $\leq$7.5), medium expression (7.5$\geq$ ΔCt $\leq$12.5); low expression (12.5$\geq$ ΔCt $\leq$20) and null expression (ΔCt $\geq$20). Among 380 receptors assayed, a total of 30 receptors were highly expressed (average ΔCt $\leq$7.5) in all four chips, out of which six were GPRs. In addition, a total of 95 receptors had medium levels of expression (7.5$\geq$ ΔCt $\leq$12.5), out of which 23 were GPRs, and 175 receptors were present with low levels of expression (12.5$\geq$ ΔCt $\leq$20) of which 51 were GPRs. 80 receptors were found to be not expressed (ΔCt $\geq$20), of which 11 were GPRs. As a point of reference, GPCRs that are well known to be in the kidney, such as the Angiotensin II 1a receptor (Agtr1a), Arginine vasopressin 2 receptor (Avpr2) and Parathyroid hormone 1 receptor (PTHR1) are present at a high expression level in our array and had an average ΔCt of 6.25$\pm$0.24, 7.27$\pm$0.17 and 4.38$\pm$0.30 respectively. We validated the expression of the top 25 receptors identified through our Taqman array (which includes the top six GPRs: Gpr137, Gpr137b, Gpr56, Gpr48, Gprc5c and Gpr116) by performing 'conventional' RT-PCR on the mouse whole kidney cDNA with a separate set of GSP primers, followed by sequencing to confirm identity (Figure 1A, Figure 1B). We did not observe any noticeable differences in the GPR expression levels among male and female kidney.

Olfactory receptors. In order to identify novel renal olfactory receptors (ORs) in mouse kidney, we undertook an RT-PCR approach. Although most ORs are orphan receptors with no known ligands, a minority of ORs do have identified ligands, and we reasoned that this group of receptors may be advantageous to study if they are expressed ectopically since (at least one) ligand(s) are already known. Therefore, we first performed an RT-PCR screen using primers for murine olfactory receptors that already have reported ligands [4,6,27,28,29,30,31,32]. The primers (Table S1) were first verified using mouse gDNA (taking advantage of the fact that ORs do not contain introns) as a positive control and to optimize cycling conditions, and then were used on reverse-transcribed kidney cDNA using identical cycling parameters. Of the twenty-nine OR primer sets used, two detected novel renal ORs: Olfr545 (MOR42-1, S50) and Olfr691 (MOR31-6) (Figure 2).

A second RT-PCR screen was performed on mouse kidney cDNA using gene specific primer (GSP) sets directed against ORs which had been previously reported in the literature to be present in renal tissues or cells. First, we assayed for the presence of nine mouse ORs listed by NCBI- Homologene as the corresponding orthologs of rat ORs identified in native rat inner medullary collecting duct (IMCD) cells by a proteomic screen [29] (additional ORs were identified in the original study for which murine orthologs had not been identified, and thus these ORs were not pursued in our study). Bands of the correct size were obtained for two OR primer sets: the murine ortholog of rat Olr1739 (mouse Olfr99), (Figure 2), and the murine ortholog of rat Olr217 (mouse Olfr705). Sequencing confirmed the presence of Olfr99 (MOR156-1), but revealed that the Olfr705 primers had actually amplified the closely related murine OR, Olfr693. We did not observe any chimeric olfactory receptor products in our sequencing results, and subsequent PCR using Olfr693-specific primers confirmed that Olfr693 (aka MOR283-8) is expressed in the kidney (Figure 2). In addition, we were also able to successfully amplify and clone full length Olfr693 receptor from the mouse kidney cDNA. We also screened for the murine homolog of human renal olfactory receptor, OR2T1 [33], and identified that the murine ortholog Olfr31 is present in the whole kidney cDNA (Figure 2). Finally, we also identified that murine Olfr1426, ortholog of a rat OR in the collecting duct and thick ascending limb (M. Knepper, NIH, personal communication), is expressed in the whole kidney cDNA.

Taste receptors. We designed thirty-five GSPs to identify known taste receptors expressed in the kidney using an RT-PCR approach. We used mouse tongue cDNA as the positive control to validate primers (Table S1) and to optimize PCR cycling conditions, and subsequently used the exact cycling conditions on reverse-transcribed kidney cDNA. We identified expression of the three Tas1r receptors, which together mediate both sweet and umami taste (Tas1r2+ Tas1r3 mediate sweet taste, whereas Tas1r1+ Tas1r3 mediate umami taste) [34]. In addition, seven bitter taste receptors (Tas2r108, Tas2r119, Tas2r135, Tas2r137, Tas2r138, Tas2r140 and Tas2r143) and a sour taste receptor, PKD1L3 [35], were identified in the kidney (Figure 3). The salt receptor (ENaC) is already known to be expressed in the kidney where it plays an important role in sodium handling [36,37,38]; therefore, we did not include it in our screen. In addition, we also identified expression of G_{NAT3} (the G-protein that mediates taste perception in the tongue) [39] in whole mouse kidney as well (Figure 3L).

Tissue distribution of renal Sensory Receptors

Ultimately, we are interested in understanding the physiological roles played by these receptors. We were curious, therefore, whether these receptors are expressed ectopically only in the kidney, or if they have wider tissue distributions. To that end, we used an RT-PCR and sequencing approach to assay whether the novel renal sensory receptors we had identified were also found in other tissues. As summarized in Table 1, we found that the expression of these receptors was not limited to the kidneys, and that the tissue distribution profile was unique to each individual receptor. Of the six ORs assayed, Olfr99 had the widest tissue expression profile, present in every tissue that we screened except for skeletal muscle. In addition to the kidneys, Olfr31 and Olfr1426 were expressed only in one another tissue (testes), whereas the remaining ORs were found in at least 3 additional tissues. Every tissue screened except skeletal muscle expressed at least one of the renal ORs, and intriguingly, each of the novel renal ORs was also expressed in the testes (a tissue where ORs have previously been shown to play an important role) [11]. It should be noted that the cDNA from all tissues yielded bands for β-actin.

Of the eleven taste receptors that we assayed by the RT-PCR approach, Tas2r135 had the widest tissue distribution profile, with

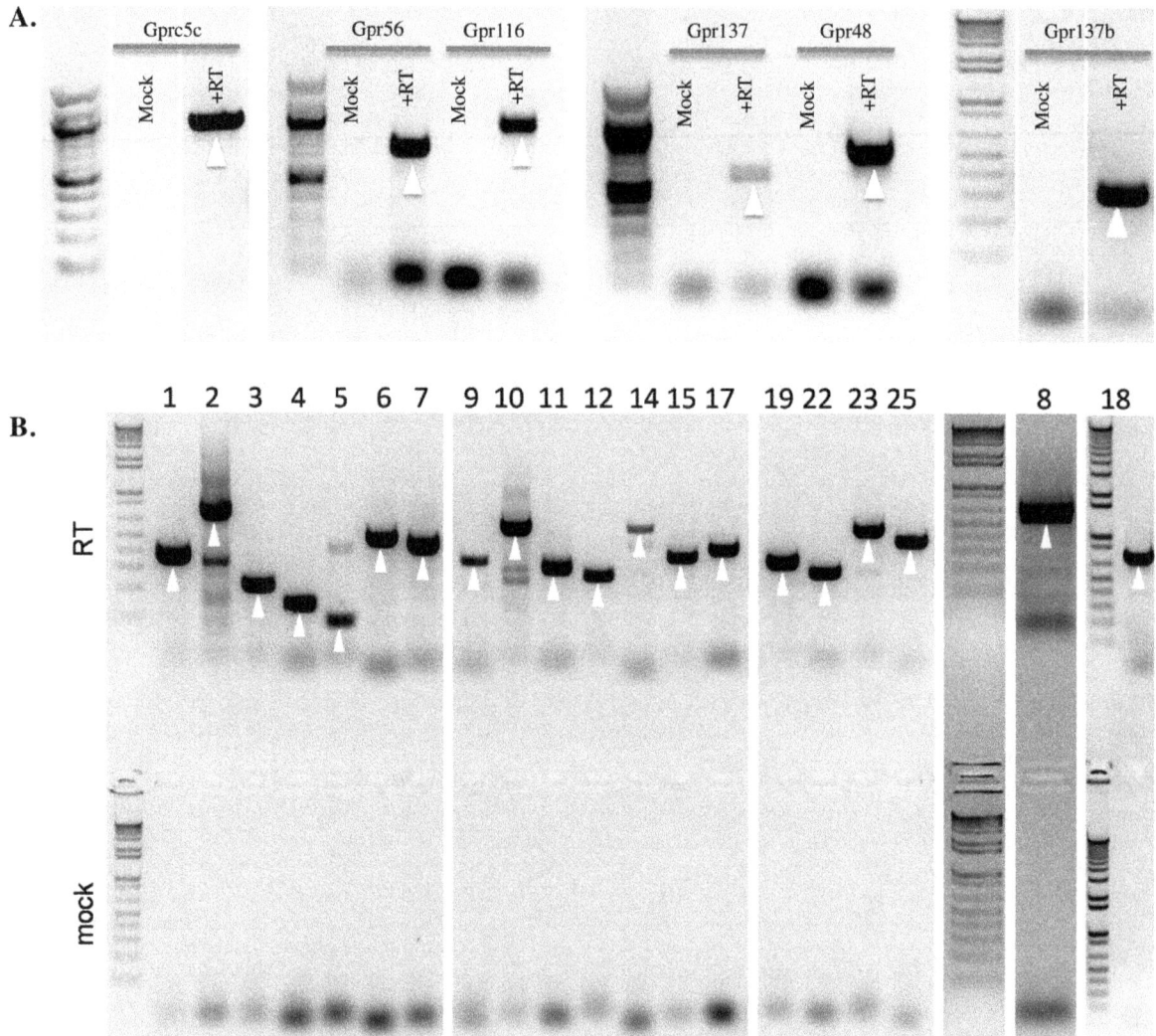

Figure 1. Conventional RT-PCR confirms expression of the six most highly expressed GPRs and top 25 highly expressed transcripts from the TaqMan screen. (**A**) Gprc5c, Gpr56, Gpr116, Gpr137, Gpr48 and Gpr137b were identified as the top six highly expressed GPRs in the mouse kidney based on the TaqMan screen. Mock lanes without RT are negative for all GPRs. The white arrow indicates the expected sized band for each GPR. (**B**) Whole kidney RT and mock RT reaction mixture were screened to validate expression of top 25 highly expressed targets identified from our Taqman array screen: (1)Actb (2) Gapdh (3) Ppia (4) Pgk1 (5) Ubc (6) Calm1 (7) B2 m (8) Pth1r (9) Ywhaz (10) Calm2 (11) Gpr137b (12) Tm7sf3 (14) Agtr1a (15) Sfrp1 (17) Ptger3 (18)Tfrc (19) Hprt (22) Fzd4 (23) Avpr2 (25) Polr2a. All products were sequenced to confirm their identities. The white arrow indicates the expected size band for each receptor.

expression identified in the heart, lung and testes. In contrast, among the tissues we screened, the expression of Tas2r119, Tas2r137 and Tas2r140 was seen only in kidney. The remaining receptors were found in at least one tissue in addition to kidney (Table 1).

We also assayed the tissue distribution of the five most highly-expressed renal GPRs from the TaqMan array (Gprc5c, Gpr48, Gpr56, Gpr116 and Gpr137). These five novel renal GPRs were all found to be expressed in the testes, as well as 1–3 additional tissues (Table 1). All PCR reactions were run along with mock RT controls and amplicons were sequenced to confirm identity.

Trafficking of newly identified murine renal ORs in HEK293T

In order to understand the function of these receptors in physiology, it is necessary to understand their ligand profiles. Unfortunately, the majority of ORs are orphan receptors with no known ligands. Therefore, using RT-PCR we cloned Olfr99, 545, 691, 693, 31 and 1426 from kidney into expression vectors with N-terminal Flag and Rho tags (+/− Lucy tags; clones were sequenced to confirm identity).

In order to screen an OR for potential ligands, it must be expressed on the cell surface and unfortunately, trafficking of ORs to the cell surface has historically been a problem in the field [30]. Surface expression can sometimes be achieved or enhanced by the concurrent expression of chaperones, most notably receptor transport protein 1 short (RTP1S) [32,33] or by the use of N-terminal tags (such as Rho [33] or Lucy [25]). We have previously tested and published the conditions under which Olfr99, 545, 691 and 693 reach the cell surface [25]. To determine whether the other novel renal ORs identified here are able to traffic to the cell surface, we used surface immunofluorescence to assay the ability of Flag and Rho (+/− Lucy) tagged ORs to traffic to the surface of HEK293T cells (+/− RTP1S). The optimized condition that

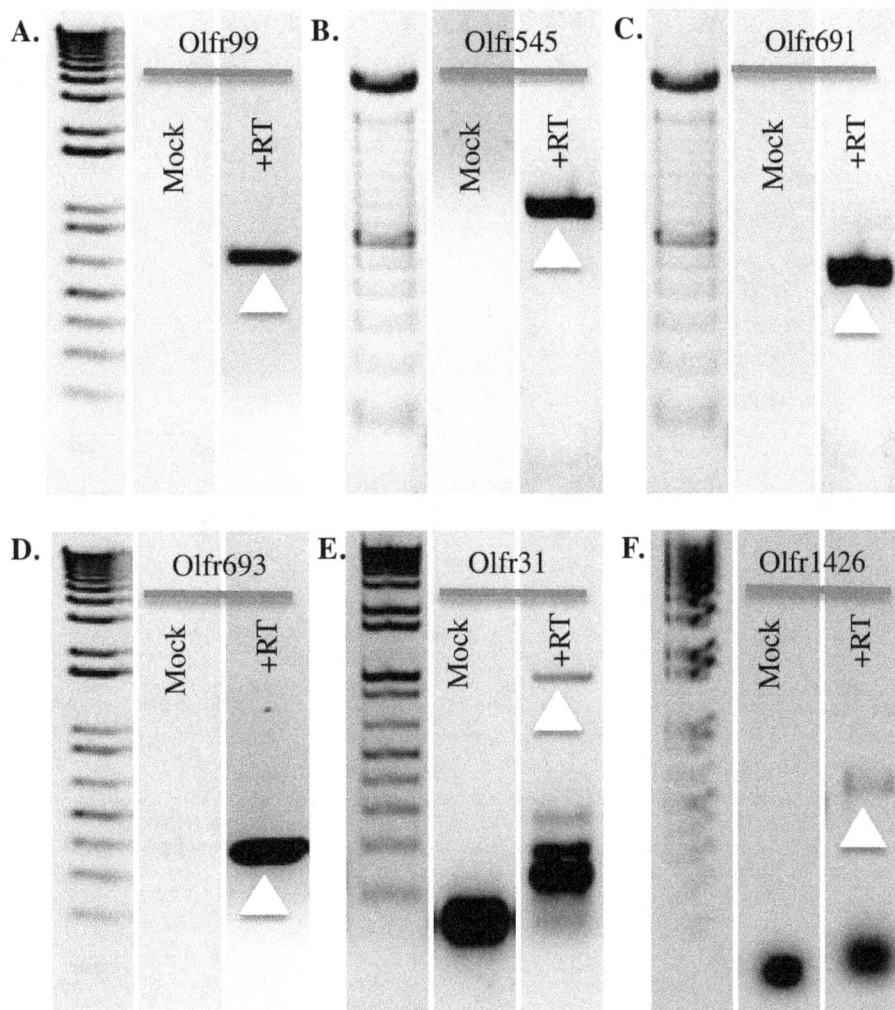

Figure 2. RT-PCR with mouse whole kidney cDNA as template to identify novel renal olfactory receptors. Olfr99 (A), Olfr545 (B), Olfr691 (C), Olfr693 (D), Olfr31(E) and Olfr1426(F) expression is detectible in mouse whole kidney cDNA by PCR and sequencing confirms the identity of amplified products. Mock RT template controls are negative for OR GSP sets and β-actin (not shown). The white arrow indicates the band of the expected size for each olfactory receptor.

facilitates membrane surface trafficking varies for each OR. Briefly, Olfr31 requires co-expression of RTP1S; Olfr691 & Olfr693 require presence of N-terminal Lucy tag along with co-expression of RTP1S; Olfr99 and Olfr545 requires presence of N-terminal Lucy tag along with co-expression of RTP1S and Ric8b (Figure 4). As seen in Figure 4, we observe surface expression for every OR tested with the exception of Olfr1426, which failed to reach the cell surface.

Ligand profiles

Because Olfr31, 99, 545, 691 and 693 trafficked to the cell surface of HEK 293T cells, we proceeded to examine the ligands of these ORs. Using a cAMP-luciferase reporter assay [26], we tested the response of ORs to an unbiased library of odorant mixes that cover a wide range of odorant space [18], as well as a library of diverse chemicals not biased to a particular olfactory receptor (listed in Table S2). The ligand mixes, the chemical library, and mouse urine all failed to evoke any response from Olfr31, 99, 545 and 693. However, we confirmed previous reports [32] that Olfr691 responds to carboxylic acids valerate and isovalerate in a

dose dependent manner (Figure 5A). Previously reported ligands such as pentanal for Olfr691 [32] and sebacic acid for Olfr545 [40] did not induce a response in our luciferase assay.

To determine if we could expand upon the known ligands for Olfr691, we then tested Olfr691 using compounds similar to valerate and isovalerate (Figure 5B). We selected ligands using a multidimensional physiochemical metric for odorant prediction which takes into account a variety of molecular characteristics in addition to the traditional values of carbon number and functional group [27]. As summarized in Figure 6, we found that Olfr691 senses a wide range of both short and medium chain fatty acids, binding to carboxylic acids with carbon lengths of three (propionate) to eight (octanoate). Chemical structures for ligands tested are shown in Figure 6 (in their carboxylate form, for simplicity and to reduce space). Olfr691 was not responsive to dicarboxylic acids, amino acids or aldehydes with similar carbon lengths and structures to identified ligands (A complete list of compounds tested for Olfr31, 99, 545, 691 and 693 is shown in Table S2). In this study, we identified thirteen new ligands for Olfr691 in addition to its previously published ligands [32]. Response values in Figure 6 (0.5 mM) have been normalized to

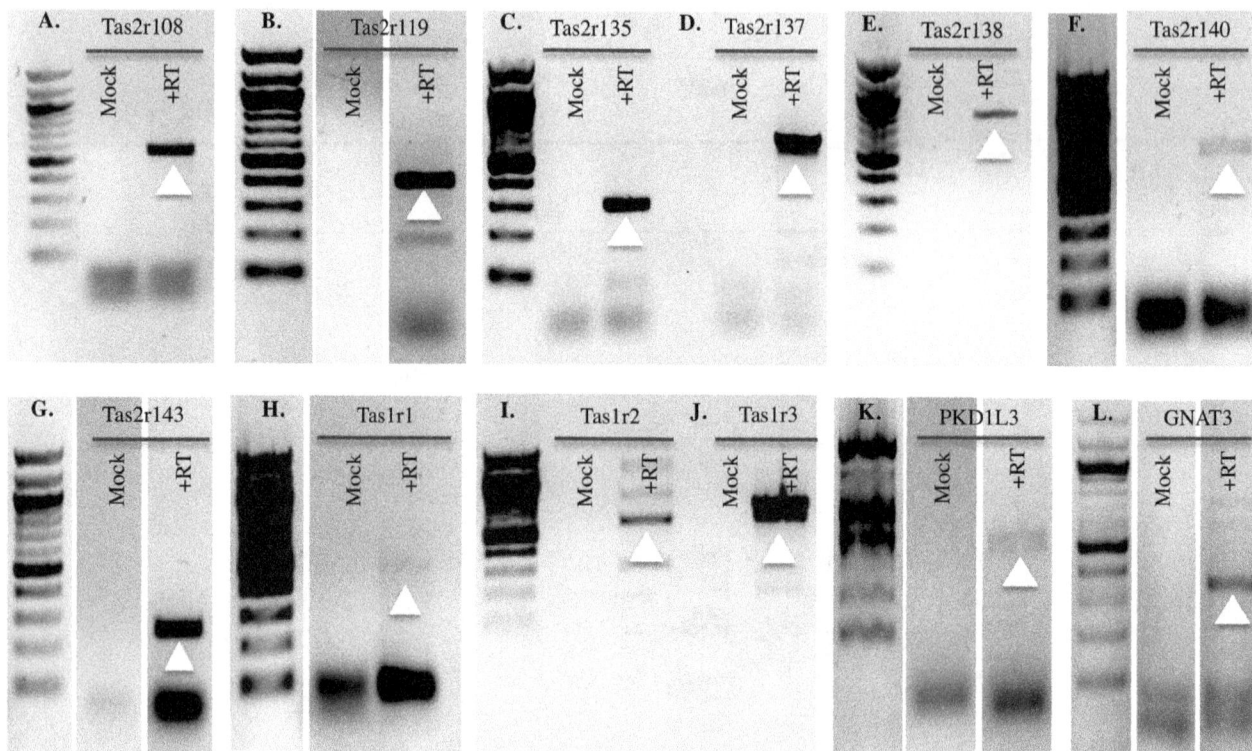

Figure 3. RT-PCR with mouse whole kidney cDNA as template to identify novel renal taste receptors. Tas2r108 (A), Tas2r119 (B), Tas2r135 (C), Tas2r137 (D), Tas2r138 (E), Tas2r140 (F), Tas2r143 (G), Tas1r1 (H), Tas1r2 (I) and Tas1r3 (J) PKD1L3 (K) and G_{NAT3} (L) expression detected in the mouse whole kidney cDNA by RT-PCR and confirmed by sequencing. Mock controls without RT are negative in all the lanes. The white arrow indicates the band of the expected size for each olfactory receptor.

the response of Olfr691 to the strongest ligand, 4-pentenoate. Branched chain and alkene analogues of short chain fatty acids and the aromatic carboxylic acid, benzoic acid, as suggested by the physiochemical metric for odorants [12], also induced Olfr691 responses. Detailed dose response curves and EC_{50} values were calculated for four ligands inducing the strongest response at 0.5 mM (Figure 7), shown in bold in Figure 6. An allylic analogue of valerate, 4-Pentenoate, induced the strongest response (Figure 6), whereas valproate had the lowest EC_{50} (0.4778 mM).

Discussion

Recent studies in the literature have highlighted the important roles that sensory receptors, including ORs, taste receptors and novel GPRs, play in a variety of different tissues [1,2,3,4,5,6,7,8,9,10,11,12,13]. To better understand the roles that sensory receptors play in the kidney, it is first necessary to identify and categorize the full complement of such receptors. To this end, in this study we aimed to identify expression of novel ORs, taste receptors and GPRs in the kidney by using a variety of approaches.

Identification of Novel Receptors

We previously identified 6 renal ORs using a degenerate OR (dOR) primer screen [19,20]; however, this approach is biased towards those ORs with the highest levels of expression in a given tissue. In this study, we wanted to determine whether there are any renal ORs beyond the 6 identified previously. Others have successfully used microarrays to detect the expression of ORs in the olfactory epithelium (OE) and elsewhere [41,42], but ORs in

the kidney are expressed in a lower level than in the OE. This low level of expression increases the probability that a microarray may result in false negatives. Therefore, in this study we employed a PCR-based approach using GSPs in order to assay whether additional renal ORs, not detected in our original degenerate OR primer screen, may also be expressed in the kidney. Although not a comprehensive screen, our results clearly show that additional ORs are found in the kidney. In addition to novel renal ORs which have published ligands, we also assayed for renal ORs which had been reported by others to be present in the renal tissues. However, in some cases we found some but not all of the previously reported ORs: for example, previous work on freshly isolated native rat IMCD cells conducted by the Knepper Laboratory at the NIH had identified 19 novel renal ORs [29]. We generated GSP sets against the mouse homologs of these renal rat ORs and detected only Olfr99 and Olfr693 in murine kidney. The discrepancy between our findings is very likely due to the difficulty of identifying OR homologues across species, especially when there are a large number of highly homologous ORs in both species (~1000 OR genes in mice, and ~1400 in rats (13; 37)). Therefore, although we only confirmed 2 out of the 19 ORs reported by the Knepper Laboratory, this may represent the limitations of the ability to correctly assign homologues based on sequence similarities. However, the fact that 19 novel ORs were identified by looking at the IMCD alone indicates that a more thorough screen for ORs within the whole kidney is necessary and justified in order to identify the full complement of renal ORs.

It is worth noting that Olfr691 and Olfr31 do have a human homologue listed in NCBI (OR52B2 and OR2T1 respectively), however, there are no human homologs listed for Olfr99, 545,

Table 1. Summary of the tissue expression profile of all the novel sensory receptors identified in the mouse whole kidney cDNA.

Receptor	Kidney	Testes	Colon	Heart	Liver	Lung	Skeletal	Small Intestine	Stomach
Olfr31	+	+	–	–	–	–	–	–	–
Olfr99	+	+	+	+	+	+	–	+	+
Olfr545	+	+	–	+	+	–	–	+	–
Olfr691	+	+	+	+	–	+	–	–	+
Olfr693	+	+	–	–	–	+	–	–	+
Olfr1426	+	+	–	–	–	–	–	–	–
Tas2r108	+	+	–	–	–	–	–	–	–
Tas2r119	+	–	–	–	–	–	–	–	–
Tas2r135	+	+	–	+	–	+	–	–	–
Tas2r137	+	–	–	–	–	–	–	–	–
Tas2r138	+	+	–	–	–	–	–	–	–
Tas2r140	+	–	–	–	–	–	–	–	–
Tas2r143	+	–	–	+	–	+	–	–	–
PKD1L3	+	+	–	–	–	–	–	–	–
Tas1r1	+	–	+	–	–	–	–	–	–
Tas1r2	+	–	–	–	–	–	+	–	–
Tas1r3	+	+	–	–	–	+	–	–	–
Gpr56	+	+	–	–	–	–	+	–	+
Gprc5c	+	+	–	–	+	+	+	–	+
Gpr116	+	+	–	–	–	+	+	+	–
Gpr137	+	+	+	–	–	–	–	+	+
Gpr48	+	+	+	–	–	–	–	+	+

A '+' sign indicates expression of the corresponding receptor in our RT-PCR screen whereas a '–' sign indicates absence in that particular tissue. All '+' signs in the table were confirmed by sequencing to confirm identity. In each case, the mock sample without reverse transcriptase during cDNA synthesis was negative.

A.

B.

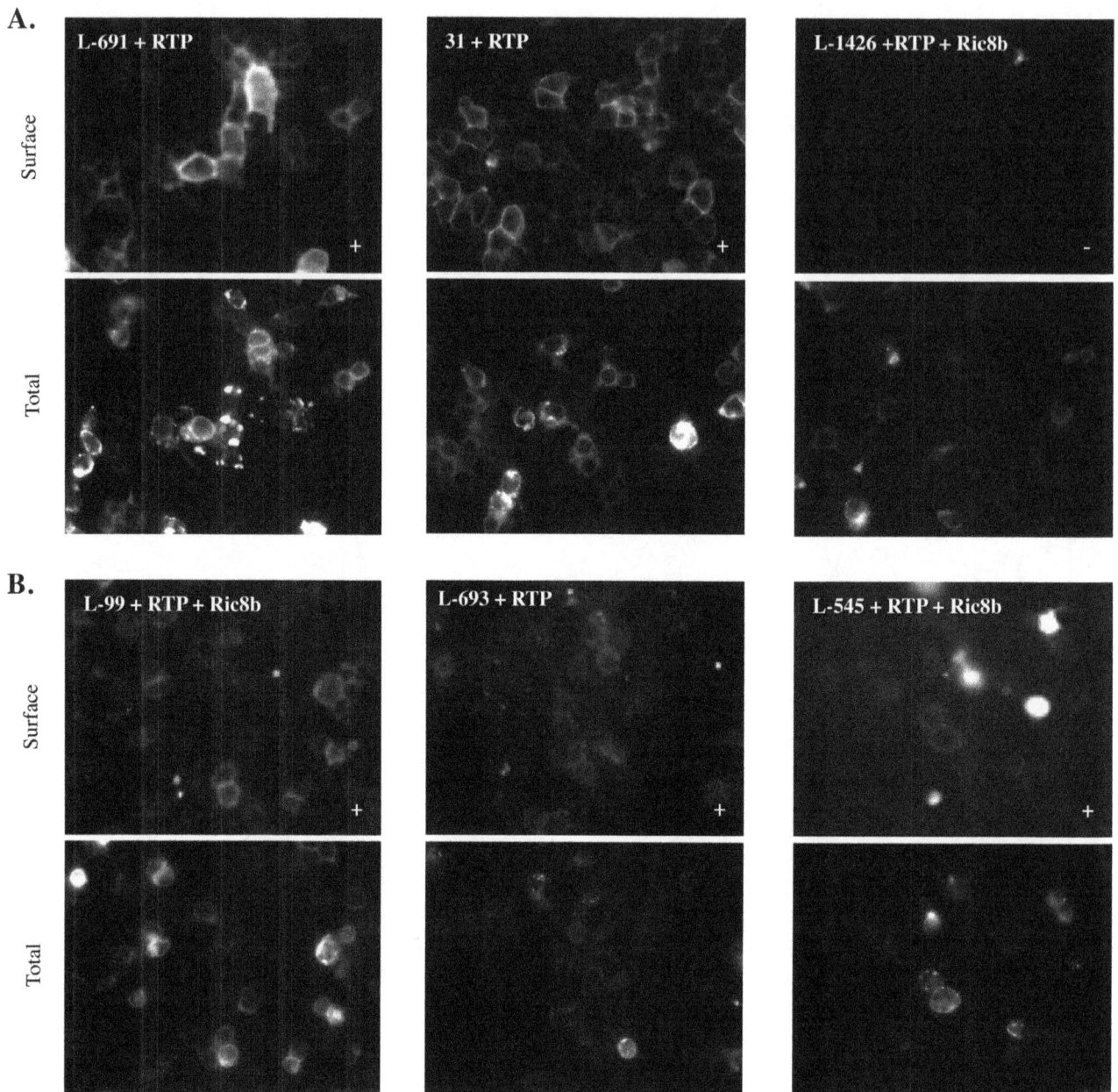

Figure 4. Immunohistochemistry showing surface expression of ORs. Each OR is shown under the experimentally determined condition which allowed for optimized surface expression in HEK293T cells. The surface trafficking conditions vary for each OR and we have published the corresponding conditions for Olfr99, 545, 691 and 693 previously [25]. Briefly, Olfr31 requires co-expression of RTP1S; Olfr691 & Olfr693 require presence of N-terminal Lucy tag along with co-expression of RTP1S; Olfr99, Olfr545 & Olfr1426 requires presence of N-terminal Lucy tag along with co-expression of RTP1S and Ric8b respectively. Olfr31 requires co-expression of RTP1S and Olfr1426 failed to reach the membrane surface at all the tested conditions. HEK293T cells were first stained with a poly-flag antibody (surface) then subsequently permeabilized and stained with a mono-flag antibody (total). The images were taken at equal exposure between all surface and total conditions. Surface images are marked with either a+or – in their lower right-hand corners to indicate the presence or absence of surface expression, respectively. Images in (B) have been enhanced to better display surface expression. Images in (A) are presented as they were taken. Unenhanced images for Figure 4B can be found in Figure S1.

693 or 1426. Therefore, future studies will be necessary to determine if Olfr99, 545, 693 or 1426 may have functional orthologs in human.

As the taste receptor family is relatively small, we screened for the full complement of taste receptors using 35 gene specific primer (GSP) sets. This direct approach is well suited to screen small families of receptors, as it is cost effective and sensitive to low-level expression receptors. Expression of taste receptors

(including the bitter receptors) has been previously identified in non-gustatory tissues [1,2,10,21,22,23,43], however, our study is the first to identify taste receptors in the mouse kidney. From the previous literature on taste receptors, we know that the mouse heteromeric umami (Tas1r1+Tas1r3) and sweet (Tas1r2+Tas1r3) receptors are broadly tuned and that they respond to a variety of L-amino acids and sugars, respectively [34,44,45]. Since we detected expression of all three Tas1r subunits (Tas1r1/Tas1r2/

A.

B.

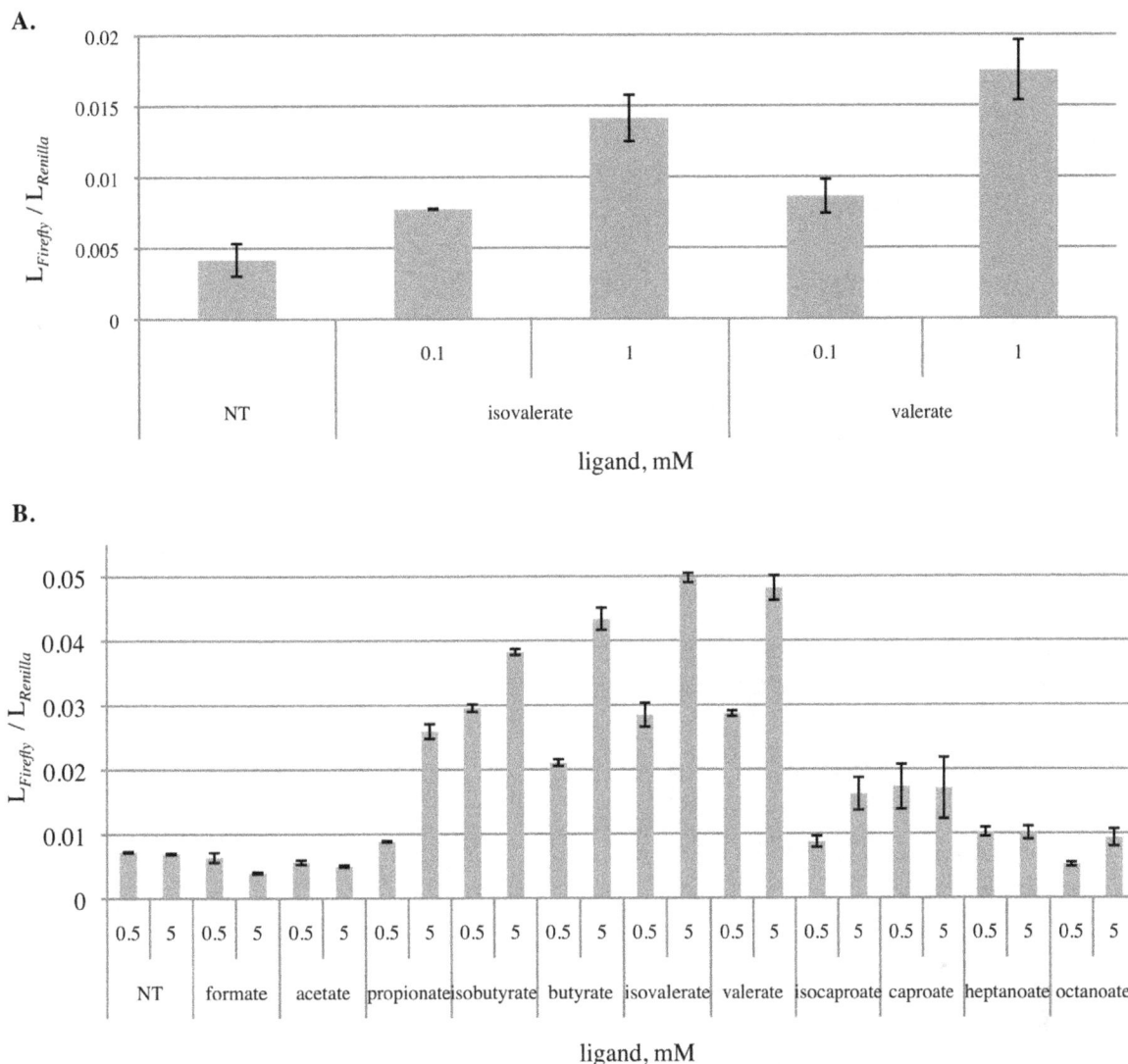

Figure 5. Ligand screening for Olfr691. Olfr691 responds to published short chain fatty acids, isovalerate and valerate, in a dose dependent manner when co-expressed with RTP1S (A). Further ligand screening shows that Olfr691 responds to wide range of saturated short and medium chained fatty acids, from propionate to octanoate, but not including formate and acetate (B). NT represents measurements obtained from non-treated cells (with no ligand) transfected with Olfr691 and RTP1S.

Tas1r3) in the mouse kidney, future work is necessary to understand the dimerization characteristics (Tas1r1+Tas1r3 vs. Tas1r2+Tas1r3) of these receptor subunits in the kidney along with their potential renal role towards mediating amino acid and energy homeostasis. Although there is no previous data in the literature regarding taste receptors in the kidney, Tas2r135 and Tas2r143 were previously reported to be expressed in the heart [43], in agreement with our findings.

In addition, because several novel GPRs have been found to play sensory roles in a variety of non-renal tissues, but have not been well studied in the kidney, we undertook a high-throughput approach to assay for the expression of 91 GPRs in the kidney using real-time PCR. In our screen, we detected expression of previously reported GPRs with known cardiovascular and renal functions (C_t: Gpr30 = 28.67 ± 0.49; Gpr43 = 32.29 ± 0.26; Gpr48 = 23 ± 0.29 and Gpr91 = 23.91 ± 1.04) [6,12,46,47,48].

Finally, it should be noted that we assayed receptor expression in whole kidney tissue. We cannot rule out that receptors we found

to be 'absent' are in fact expressed, but only in a small subset of cells (i.e., macula densa, intercalated cells, etc.). In this case, although these receptors may have significant renal roles, they may appear as null expressers in a screen of whole kidney.

Identification of Novel Ligands

In this study, we screened Olfr31, 99, 545, 691 and 693 in a luciferase assay system to identify their ligands. We identified thirteen novel ligands for Olfr691, but did not identify any ligands for Olfr31, 99, 545 and 693. It is possible that Olfr31, 99, 545, and 693 are narrowly tuned receptors which do not respond to the chemical profiles in our odorant library mixtures [49]. Valerate, isovalerate and pentanal were previously reported [32] as Olfr691 ligands. Of these three, we were able to confirm valerate and isovalerate, but not pentanal. In addition to these previously reported ligands, we now show that Olfr691 is broadly tuned towards carboxylic acid activation, including short and medium chain fatty acids (physiological concentrations are within the range

Ligand	Structure	Relative response at 0.5mM
Propionate	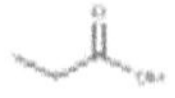	42 ± 3.1
Isobutyrate	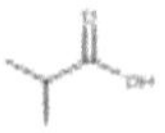	**97 ± 10.8**
Butyrate		55 ± 2.5
2-Methylbutyrate	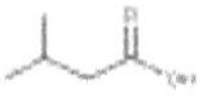	65 ± 3.4
Isovalerate	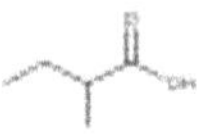	73 ± 4.8
Valerate		**98 ± 5.7**
Valproate	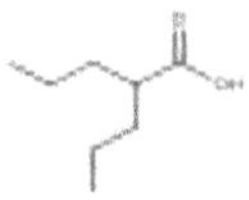	**98 ± 5.0**
Isocaproate	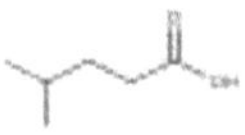	37 ± 2.7
Caproate		80 ± 1.4
2-Ethylhexanoate	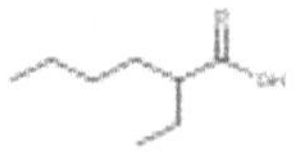	76 ± 3.6
Heptanoate		41 ± 2.1
Octanoate		20 ± 0.9
Crotonoate		39 ± 3.0
4-Pentenoate	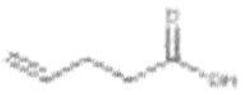	**100 ± 10.6**
Benzoate	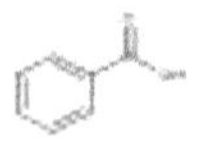	34 ± 1.0

Figure 6. Relative response values at 0.5 mM for Olfr691 ligands. The structures of the ligands are shown in the figure for reference.

of the ligand concentrations assayed for isovalerate (0.89±0.93 uM [50]). Interestingly, gut bacterial metabolism is the primary physiological source of short chain fatty acids in the bloodstream, with the concentrations reported for propionate varying from 0.1–10 mM [6]. In addition, the response of Olfr691 to valproate is quite intriguing. In clinical trials, patients treated with valproate as an antiepileptic drug have been shown to develop Fanconi syndrome [51,52], where the renal proximal tubules are affected resulting in an excessive spillage of amino acids, phosphate, glucose, bicarbonate, and uric acid in their urine. In support of this hypothesis, in preliminary studies we observed successful amplification of Olfr691 in cDNA isolated specifically from the S1 and S3 segment of proximal tubule (n = 3).

Clearly, future work will need to be done to investigate the relevant *in vivo* renal role of Olfr691.

Summary

In this study, we have identified expression of novel olfactory receptors, taste receptors and GPRs in the kidney, thereby extending the list of previously known renal sensory receptors. Despite the fact that we screened only part of the OR gene family, and did not screen the trace amine- associated receptor (TAAR) or vomeronasal receptor (VR) families, we were able to identify 93 novel murine sensory receptors, many of which were expressed at high levels by real-time PCR. These data imply that there is a large and robust complement of sensory receptors in the kidney

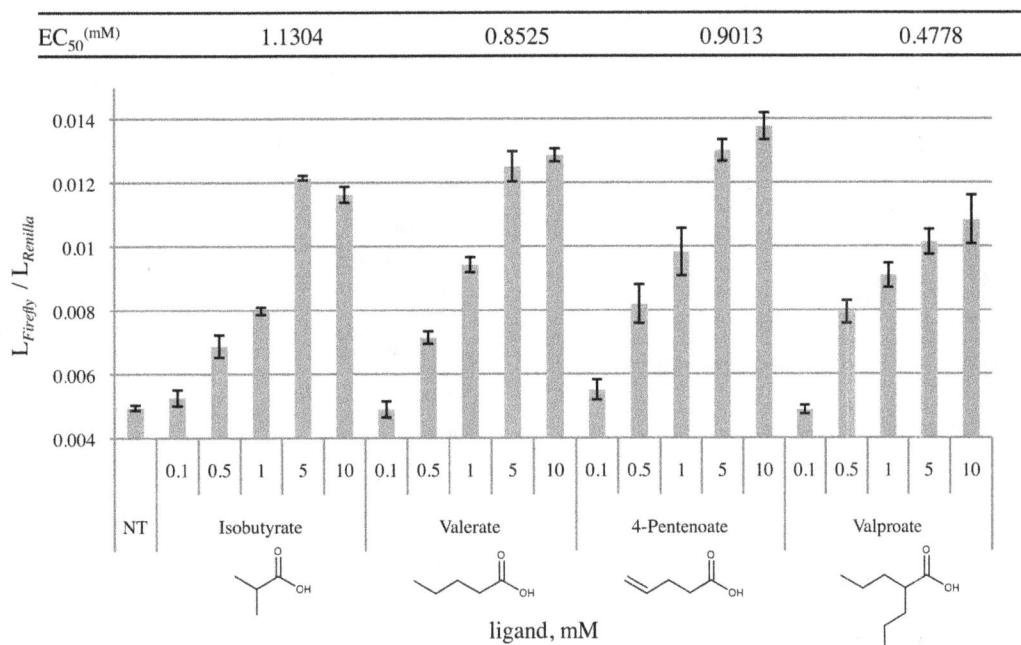

Figure 7. Dose response curves for the novel Olfr691 ligands. Dose response curves show that Olfr691 has the highest affinity for valproate when co-expressed with RTP1S in HEK293T cells, with an EC_{50} value of 0.4778 mM; however 4-pentenoate induced the strongest cAMP responses at all doses when compared to isobutyrate, valerate and valproate. NT represents measurements obtained from non-treated cells (with no ligand) transfected with Olfr691 and RTP1S.

which have not yet been examined in a functional context. Our study is an important first step in identification of novel renal receptors, and future work is now required to localize these receptors within the kidney and to elucidate the physiological role of each receptor.

Supporting Information

Figure S1 Unenhanced surface images from Figure 4B. Unenhanced images of Olfr99, Olfr693 and Olfr545 in their corresponding conditions that facilitate plasma membrane surface trafficking in HEK293T cells.

Table S1 Nucleotide sequences of the primers used to screen cDNA synthesized from mouse whole kidney. Sequences of both the forward and reverse primers used to screen whole kidney cDNA in our RT-PCR approach along with their expected size bands.

Table S2 List of all ligands used to screen Olfr691, Olfr99, Olfr545, Olfr693 and Olfr31 in the dual-lucifer-

ase assay. '+' or '−' in each column indicates if the olfactory receptor responded or had no effect to that specific chemical.

Table S3 Mouse Taqman GPCR array data. List of all the GPCRs screened in this study and their corresponding C_t values are listed. The average $\Delta C_t \pm SD$ values are also listed for each GPCR based on the data obtained from four independent runs.

Acknowledgments

We would like to thank Dr. Mark Knepper (NIH) for sharing unpublished data, Drs. Alain Doucet and Lydie Cheaval (Centre de Recherche des Cordeliers de Jussieu) for providing microdissected nephron segments for preliminary studies, Daniel Gergen and Blythe Shepard for assistance with OR primer design, Wennie Sansing for assistance with taste receptor primer design, and the current members of the Pluznick Lab for helpful discussions.

Author Contributions

Conceived and designed the experiments: RJP JLP. Performed the experiments: PR WHA RJP OWA JLP. Analyzed the data: PR WHA RJP OWA JLP. Contributed to the writing of the manuscript: PR JLP.

References

1. Deshpande DA, Wang WC, McIlmoyle EL, Robinett KS, Schillinger RM, et al. (2010) Bitter taste receptors on airway smooth muscle bronchodilate by localized calcium signaling and reverse obstruction. Nat Med 16: 1299–1304.

2. Elliott RA, Kapoor S, Tincello DG (2011) Expression and distribution of the sweet taste receptor isoforms T1R2 and T1R3 in human and rat bladders. J Urol 186: 2455–2462.

3. Griffin CA, Kafadar KA, Pavlath GK (2009) MOR23 promotes muscle regeneration and regulates cell adhesion and migration. Dev Cell 17: 649–661.

4. He W, Miao FJ, Lin DC, Schwandner RT, Wang Z, et al. (2004) Citric acid cycle intermediates as ligands for orphan G-protein-coupled receptors. Nature 429: 188–193.

5. Huang AL, Chen X, Hoon MA, Chandrashekar J, Guo W, et al. (2006) The cells and logic for mammalian sour taste detection. Nature 442: 934–938.

6. Maslowski KM, Vieira AT, Ng A, Kranich J, Sierro F, et al. (2009) Regulation of inflammatory responses by gut microbiota and chemoattractant receptor GPR43. Nature 461: 1282–1286.

7. Pluznick JL, Caplan MJ (2012) Novel sensory signaling systems in the kidney. Curr Opin Nephrol Hypertens 21: 404–409.

8. Pluznick JL, Zou DJ, Zhang X, Yan Q, Rodriguez-Gil DJ, et al. (2009) Functional expression of the olfactory signaling system in the kidney. Proc Natl Acad Sci U S A 106: 2059–2064.

9. Samuel BS, Shaito A, Motoike T, Rey FE, Backhed F, et al. (2008) Effects of the gut microbiota on host adiposity are modulated by the short-chain fatty-acid binding G protein-coupled receptor, Gpr41. Proc Natl Acad Sci U S A 105: 16767–16772.

10. Shah AS, Ben-Shahar Y, Moninger TO, Kline JN, Welsh MJ (2009) Motile cilia of human airway epithelia are chemosensory. Science 325: 1131–1134.

11. Spehr M, Gisselmann G, Poplawski A, Riffell JA, Wetzel CH, et al. (2003) Identification of a testicular odorant receptor mediating human sperm chemotaxis. Science 299: 2054–2058.

12. Vargas SL, Toma I, Kang JJ, Meer EJ, Peti-Peterdi J (2009) Activation of the succinate receptor GPR91 in macula densa cells causes renin release. J Am Soc Nephrol 20: 1002–1011.

13. Wang J, Li X, Ke Y, Lu Y, Wang F, et al. (2012) GPR48 increases mineralocorticoid receptor gene expression. J Am Soc Nephrol 23: 281–293.

14. Feldmesser E, Olender T, Khen M, Yanai I, Ophir R, et al. (2006) Widespread ectopic expression of olfactory receptor genes. BMC Genomics 7: 121.

15. Fukuda N, Yomogida K, Okabe M, Touhara K (2004) Functional characterization of a mouse testicular olfactory receptor and its role in chemosensing and in regulation of sperm motility. J Cell Sci 117: 5835–5845.

16. Pluznick JL (2014) Extra sensory perception: the role of Gpr receptors in the kidney. Curr Opin Nephrol Hypertens 23: 507–512.

17. Kimura I, Inoue D, Maeda T, Hara T, Ichimura A, et al. (2011) Short-chain fatty acids and ketones directly regulate sympathetic nervous system via G protein-coupled receptor 41 (GPR41). Proc Natl Acad Sci U S A 108: 8030–8035.

18. Pluznick JL, Protzko RJ, Gevorgyan H, Peterlin Z, Sipos A, et al. (2013) Olfactory receptor responding to gut microbiota-derived signals plays a role in renin secretion and blood pressure regulation. Proc Natl Acad Sci U S A 110: 4410–4415.

19. Malnic B, Hirono J, Sato T, Buck LB (1999) Combinatorial receptor codes for odors. Cell 96: 713–723.

20. Otaki JM, Yamamoto H, Firestein S (2004) Odorant receptor expression in the mouse cerebral cortex. J Neurobiol 58: 315–327.

21. Bezencon C, le Coutre J, Damak S (2007) Taste-signaling proteins are coexpressed in solitary intestinal epithelial cells. Chem Senses 32: 41–49.

22. Dyer J, Salmon KS, Zibrik L, Shirazi-Beechey SP (2005) Expression of sweet taste receptors of the T1R family in the intestinal tract and enteroendocrine cells. Biochem Soc Trans 33: 302–305.

23. Wu SV, Rozengurt N, Yang M, Young SH, Sinnett-Smith J, et al. (2002) Expression of bitter taste receptors of the T2R family in the gastrointestinal tract and enteroendocrine STC-1 cells. Proc Natl Acad Sci U S A 99: 2392–2397.

24. Kajiya K, Inaki K, Tanaka M, Haga T, Kataoka H, et al. (2001) Molecular bases of odor discrimination: Reconstitution of olfactory receptors that recognize overlapping sets of odorants. J Neurosci 21: 6018–6025.

25. Shepard BD, Natarajan N, Protzko RJ, Acres OW, Pluznick JL (2013) A cleavable N-terminal signal peptide promotes widespread olfactory receptor surface expression in HEK293T cells. PLoS One 8: e68758.

26. Zhuang H, Matsunami H (2008) Evaluating cell-surface expression and measuring activation of mammalian odorant receptors in heterologous cells. Nat Protoc 3: 1402–1413.

27. Haddad R, Khan R, Takahashi YK, Mori K, Harel D, et al. (2008) A metric for odorant comparison. Nat Methods 5: 425–429.

28. Godfrey PA, Malnic B, Buck LB (2004) The mouse olfactory receptor gene family. Proc Natl Acad Sci U S A 101: 2156–2161.

29. Huling JC, Pisitkun T, Song JH, Yu MJ, Hoffert JD, et al. (2012) Gene expression databases for kidney epithelial cells. Am J Physiol Renal Physiol 302: F401–407.

30. Lu M, Echeverri F, Moyer BD (2003) Endoplasmic reticulum retention, degradation, and aggregation of olfactory G-protein coupled receptors. Traffic 4: 416–433.

31. Ma M, Shepherd GM (2000) Functional mosaic organization of mouse olfactory receptor neurons. Proc Natl Acad Sci U S A 97: 12869–12874.

32. Saito H, Kubota M, Roberts RW, Chi Q, Matsunami H (2004) RTP family members induce functional expression of mammalian odorant receptors. Cell 119: 679–691.

33. Zhuang H, Matsunami H (2007) Synergism of accessory factors in functional expression of mammalian odorant receptors. J Biol Chem 282: 15284–15293.

34. Zhao GQ, Zhang Y, Hoon MA, Chandrashekar J, Erlenbach I, et al. (2003) The receptors for mammalian sweet and umami taste. Cell 115: 255–266.

35. Ishimaru Y, Inada H, Kubota M, Zhuang H, Tominaga M, et al. (2006) Transient receptor potential family members PKD1L3 and PKD2L1 form a candidate sour taste receptor. Proc Natl Acad Sci U S A 103: 12569–12574.

36. Heck GL, Mierson S, DeSimone JA (1984) Salt taste transduction occurs through an amiloride-sensitive sodium transport pathway. Science 223: 403–405.

37. Chandrashekar J, Kuhn C, Oka Y, Yarmolinsky DA, Hummler E, et al. (2010) The cells and peripheral representation of sodium taste in mice. Nature 464: 297–301.

38. Hummler E (1999) Implication of ENaC in salt-sensitive hypertension. J Steroid Biochem Mol Biol 69: 385–390.

39. McLaughlin SK, McKinnon PJ, Margolskee RF (1992) Gustducin is a taste-cell-specific G protein closely related to the transducins. Nature 357: 563–569.

40. Abaffy T, Matsunami H, Luetje CW (2006) Functional analysis of a mammalian odorant receptor subfamily. J Neurochem 97: 1506–1518.

41. Zhang X, De la Cruz O, Pinto JM, Nicolae D, Firestein S, et al. (2007) Characterizing the expression of the human olfactory receptor gene family using a novel DNA microarray. Genome Biol 8: R86.

42. Zhang X, Rogers M, Tian H, Zou DJ, Liu J, et al. (2004) High-throughput microarray detection of olfactory receptor gene expression in the mouse. Proc Natl Acad Sci U S A 101: 14168–14173.

43. Foster SR, Porrello ER, Purdue B, Chan HW, Voigt A, et al. (2013) Expression, regulation and putative nutrient-sensing function of taste GPCRs in the heart. PLoS One 8: e64579.

44. Nelson G, Hoon MA, Chandrashekar J, Zhang Y, Ryba NJ, et al. (2001) Mammalian sweet taste receptors. Cell 106: 381–390.

45. Toda Y, Nakagita T, Hayakawa T, Okada S, Narukawa M, et al. (2013) Two distinct determinants of ligand specificity in T1R1/T1R3 (the umami taste receptor). J Biol Chem 288: 36863–36877.

46. Dang Y, Liu B, Xu P, Zhu P, Zhai Y, et al. (2014) Gpr48 deficiency induces polycystic kidney lesions and renal fibrosis in mice by activating wnt signal pathway. PLoS One 9: e89835.

47. Pluznick JL (2013) Renal and cardiovascular sensory receptors and blood pressure regulation. Am J Physiol Renal Physiol 305: F439–444.

48. Hofmeister MV, Damkier HH, Christensen BM, Olde B, Fredrik Leeb-Lundberg LM, et al. (2012) 17beta-Estradiol induces nongenomic effects in renal intercalated cells through G protein-coupled estrogen receptor 1. Am J Physiol Renal Physiol 302: F358–368.

49. Nara K, Saraiva LR, Ye X, Buck LB (2011) A large-scale analysis of odor coding in the olfactory epithelium. J Neurosci 31: 9179–9191.

50. Arthur K, Hommes FA (1995) Simple isotope dilution assay for propionic acid and isovaleric acid. J Chromatogr B Biomed Appl 673: 132–135.

51. Endo A, Fujita Y, Fuchigami T, Takahashi S, Mugishima H (2010) Fanconi syndrome caused by valproic acid. Pediatr Neurol 42: 287–290.

52. Knorr M, Schaper J, Harjes M, Mayatepek E, Rosenbaum T (2004) Fanconi syndrome caused by antiepileptic therapy with valproic Acid. Epilepsia 45: 868–871.

10

Characteristic and Functional Analysis of a Newly Established Porcine Small Intestinal Epithelial Cell Line

Jing Wang, Guangdong Hu, Zhi Lin, Lei He, Lei Xu, Yanming Zhang*

College of Veterinary Medicine, Northwest A&F University, Yangling, Shaanxi, China

Abstract

The mucosal surface of intestine is continuously exposed to both potential pathogens and beneficial commensal microorganisms. Recent findings suggest that intestinal epithelial cells, which once considered as a simple physical barrier, are a crucial cell lineage necessary for maintaining intestinal immune homeostasis. Therefore, establishing a stable and reliable intestinal epithelial cell line for future research on the mucosal immune system is necessary. In the present study, we established a porcine intestinal epithelial cell line (ZYM-SIEC02) by introducing the human telomerase reverse transcriptase (hTERT) gene into small intestinal epithelial cells derived from a neonatal, unsuckled piglet. Morphological analysis revealed a homogeneous cobblestone-like morphology of the epithelial cell sheets. Ultrastructural indicated the presence of microvilli, tight junctions, and a glandular configuration typical of the small intestine. Furthermore, ZYM-SIEC02 cells expressed epithelial cell-specific markers including cytokeratin 18, pan-cytokeratin, sucrase-isomaltase, E-cadherin and ZO-1. Immortalized ZYM-SIEC02 cells remained diploid and were not transformed. In addition, we also examined the host cell response to *Salmonella* and LPS and verified the enhanced expression of mRNAs encoding IL-8 and TNF-α by infection with *Salmonella enterica serovars Typhimurium* (*S. Typhimurium*). Results showed that IL-8 protein expression were upregulated following Salmonella invasion. TLR4, TLR6 and IL-6 mRNA expression were upregulated following stimulation with LPS, ZYM-SIEC02 cells were hyporeponsive to LPS with respect to IL-8 mRNA expression and secretion. TNFα mRNA levels were significantly decreased after LPS stimulation and TNF-α secretion were not detected challenged with *S. Typhimurium* neither nor LPS. Taken together, these findings demonstrate that ZYM-SIEC02 cells retained the morphological and functional characteristics typical of primary swine intestinal epithelial cells and thus provide a relevant *in vitro* model system for future studies on porcine small intestinal pathogen-host cell interactions.

Editor: Michael Hensel, University of Osnabrueck, Germany

Funding: This work was supported by Basic Research grants from the Doctoral Scientific Fund Project of the Ministry of Education of China (No.20110204110015). The funders had no role in study design, data collection and analysis, decision to publish, or preparation of the manuscript.

Competing Interests: The authors have declared that no competing interests exist.

* Email: zhangym@nwsuaf.edu.cn

Introduction

Pigs of all ages are susceptible to intestinal diseases, which most commonly present as diarrhea [1]. However, piglets are especially vulnerable to infection by bacteria, viruses, parasites and other etiologic agents that cause primary intestinal diseases. Intestinal diseases in piglets have both high morbidity and mortality, which results in large losses in the livestock industry each year. Previous studies have been largely performed in animal infection models [2], however, the study of molecular mechanisms of enteropathogen infections is limited by the availability of reliable and relevant established porcine cell lines.

The intestinal epithelial monolayer acts not only as a physical barrier but also plays a critical role in preventing macromolecules and pathogenic microorganisms in the gut lumen from penetrating to the underlining mucosa [2]. The mucosal surface is continuously exposed to commensal microorganisms and/or innocuous environmental antigens, and the intestinal mucosal immune system is exquisitely sensitive to the challenge of constant immunological stimulation [3]. Many studies have described the host-pathogen interaction in short-term intestinal epithelial cell cultures derived from humans [4–7] and from a variety of animals [8], including mice [9–11], rats [12–15], rabbits [16], and cattle

[17–21]. Non-transformed long-term swine epithelial cell lines from intestinal sections are available so far, e.g. IPEC-1 from pig ileum and jejunum [22] and IPEC-J2 from pig jejunum [2]. The majority of studies have been carried out on IPEC-J2, which generated in 1989 by Berschneider [23] and is considered a useful model for ion transport research. However, except for an abstract form the annual meeting of the American Gastroenterological Association, few studies have documented the generation of a stable, non-transformed porcine intestinal epithelial cell line.

Immortalized cell lines have numerous advantages over primary cultures, particularly the retention of reasonably constant characteristics for following numerous passages [24]. Human telomerase reverse transcriptase (hTERT) is the catalytic subunit of the telomerase enzyme, which together with the telomerase RNA component (TERC), comprise the telomerase ribonucleoprotein complex. Telomerase activation is a critical step in cellular immortalization and tumorigenesis [25,26], and hTERT alone has been found to be necessary and sufficient for inducing the telomerase activity [27]. Overexpression of hTERT has been previously used as a strategy for immortalization of human retinal pigment epithelial cells [27], swine vascular endothelial cells [28] and the cattle type II alveolar epithelial cell line [29].

In this study, the hTERT gene was successfully introduced into swine small intestinal epithelial cells, resulting in stable hTERT expression. After screening and identification, an immortalized cell line designated ZYM-SIEC02 was established. Immortalized ZYM-SIEC02 cells retained morphological and functional characteristic typical of primary small intestinal epithelial cells, and can be used as an *in vitro* model for mechanistic studies of pathogenic infections.

Materials and Methods

Ethics Statement

All animal experiments were approved by Care and Use of Animals Center, Northwest A & F University. This study was carried out in strict accordance with the Guidelines for the Care and Use of Animals of Northwest A & F University. Every effort was made to minimize animal pain, suffering and distress and to reduce the number of animal used.

Reagents, antibodies and experimental animals

DMEM/F12 and FBS were purchased from Gibco. EGF, ITS-G and Lipofectamine Plus were products of Invitrogen. The WST-1 Cell Proliferation and Cytotoxicity Assay Kit was obtained from Beyotime, Shanghai, China. The Annexin V-FITC Apoptosis Detection Kit was a product of Calbiochem, Darmstadt, Germany. The TRAP-silver staining Telomerase Detection Kit was purchased from KeyGEN Biotech, Nanjing, China. The following primary antibodies were used: Pan-cytokeratin (clone AE1/AE3, AbD Serotec, Oxford, UK), mouse anti-Cytokeratin 18 (clone CY90, Sigma, St. Louis, MO, USA), rabbit anti-Sucrase-isomaltase (clone H-123, Santa Cruz, Heidelberg, Germany), rabbit anti-E-cadherin antibody (Genscript), mouse anti-OCLN (Clones 1G, AbD, Oxford, UK), rabbit anti-villin and anti-ZO-1 (Bioss, Beijing, China).

Healthy unsuckled 1-day-old Landrace piglets were purchased from the pig farm of Northwest Agriculture & Forestry University.

Nude mice at 4 weeks old of age were purchased from the Experimental Animal Center of The Fourth Military Medical University (Xi'an, China) and maintained in pathogen-free conditions.

Isolation and culture of primary intestinal epithelial cells

Small intestines were collected from two 1-day-old, unsukled piglets. The mid-jejunum was dissected from each piglet, cut into 8–10 cm segments, then placed into ice-cold phosphate buffered saline (PBS) containing 200 U/ml of penicillin and 200 g/ml of streptomycin. The intestinal lumen was flushed with PBS using a 20 ml syringe until the liquid ran clear. The lumen was cut into small pieces with eye scissors, washed 3 times with PBS and once with DMEM/F12, and, cell pellets from each intestinal segment were collected by centrifugation at 800 g for 7 min. Cell pellets were resuspended in DMEM/F-12, washed twice with DMEM/F12, and resuspended in DMEM/F-12 containing 10 mM HEPES, 2 mM L-glutamine, 100 U/ml penicillin, 100 µg/ml streptomycin, 10 ng/ml EGF, ITS (insulin 1.0 g/l, transferrin 0.55 g/l and selenium 0.67 mg/l), and 5% heat-inactivated FBS and cultured at 37°C in a humidified incubator with 5% CO_2. Cells reached confluence after 10–12 days and were then used for further experiments. The method of trypsin digestion with Citric acid was used to purify primary intestinal epithelial cells, purified epithelial cells were diluted at a density of 50 cells/ml and then, 100 µl suspensions were cultured to 96 well plates to select epithelial clones.

Transfection and screening

After 2 passages, primary swine intestinal epithelial cells (pSIECs) were seeded onto 24-well plates at a density of 1×10^4 cells/cm². After 24 h, cells at 60–70% confluence were transfected with pCI-neo-hTERT by lipofection. Forty-eight hours following transfection, a selecting dose of G418 (600 µg/ml) was added to the culture medium, and the G418-containing. Medium was replenished every another day. After 12 days, drug-resistant cells were selected and maintained in 300 µg/ml of G418 to ensure a stable selected population, and positive clones were enlarged in further culture.

Scanning and Transmission electron microscopy

Samples for scanning electron microscopy (SEM) and transmission electron microscopy (TEM) were prepared as described previously [30]. Briefly, cells were collected and fixed with 2.5% glutaraldehyde in 0.1 M sodium cacodylate buffer, pH 7.4 for 1 h, then post-fixed with 0.2% uranyl acetate. After dehydration in a

Table 1. List of primers used for Real-time PCR.

Genes	Primer sequences (5′–3′)	Reference/accession
TLR-4	F.GCCATCGCTGCTAACATCATC	1
	R. CTCATACTCAAAGATACACCATCGG	
IL-8	F. CTGGCTGTTGCCTTCTTG	NM_213867.1
	R. TCGTGGAATGCGTATTTATG	
TLR-6	F. AACCTACTGTCATAAGCCTTCATTC	1
	R.GTCTACCACAAATTCACTTTCTTCAG	
TNF-α	F. CGCATCGCCGTCTCCTACCA	NM_214022
	R. CTGCCCAGATTCAGCAAAGTCCA	
IL-6	F. TGGATAAGCTGCAGT CACAG	2
	R. ATTATCCGAATGGCCCTCAG	
β-actin	F. CAAGGACCTCTACGCCAACAC	DQ845171.1
	R. TGGAGGCGCGATGATCTT	

1 [49]; 2 [56].

A

B

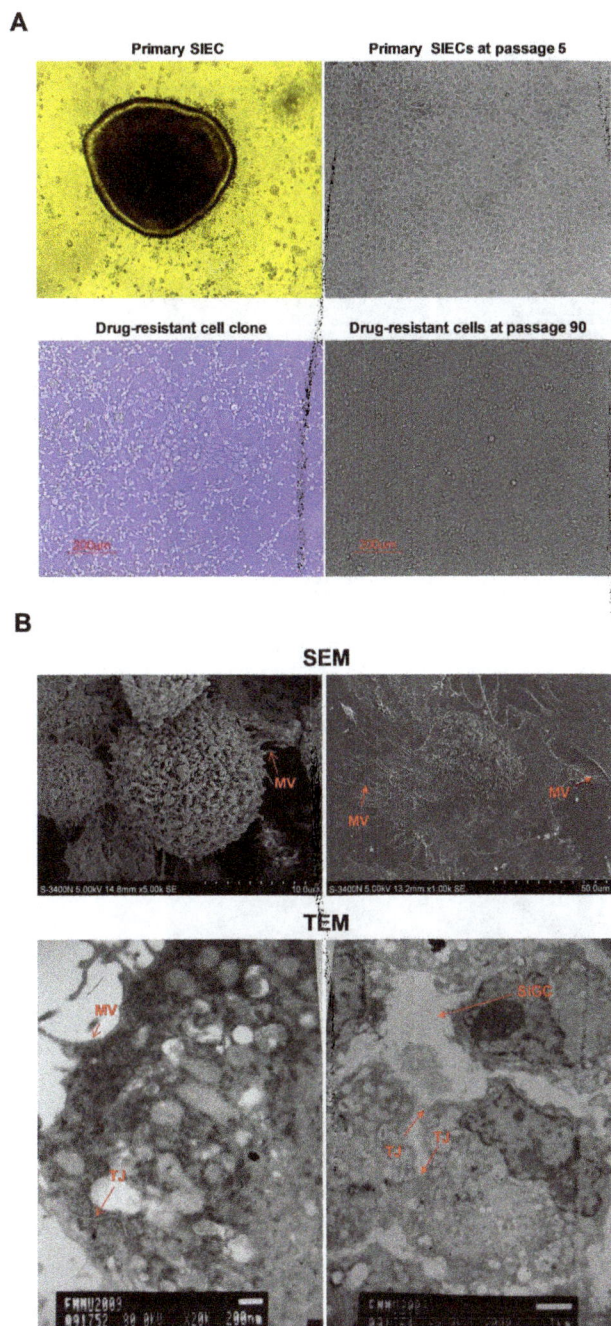

Figure 1. Morphological features of pSIECs and ZYM-SIEC02 cells. A. Cellular morphology of pSIECs at 7 days in culture (100×) and at passage 5 (100×); a drug-resistant cell clone (100×) and ZYM-SIEC02 cells at passage 90 (100×).No obvious morphological differences were observed between ZYM-SIEC02 cells and primary SIECs. **B**. Electron micrographs of monolayer cultures of ZYM-SIEC02 cells indicate microvilli (MV) in 3 dimensions andin monolayer culture; tight junctions (TJ) and a small intestine glandular configuration (SIGC) (Red arrow).

graded series of ethanol, samples were gold-coated for SEM or embedded in EpON 812 resin for TEM. SEM specimens were visualized using a S-3400N scanning electron microscope at a 5 kV potential and TEM specimens were visualized on a JEM-1400 transmission electron microscope.

Immunofluorescent staining

Immunostaining for keratin was performed according to the method of Shierack et al [2]. Briefly, cells were fixed in acetone for 10 min at $-20°C$ and blocked with 1% bovine serum albumin (BSA). After washing 3 times in PBS, samples were immunolabeled using antibodies against pan-cytokeratin (1:400) or cytokeratin 18 (1:400) overnight at 4°C. Subsequently, samples were incubated with secondary antibody (FITC-conjugated goat anti-mouse IgG (H+L)) at room temperature (RT) for 1 h. The samples were mounted and then imaged using fluorescent microscopy.

Western Blot analysis

Confluent cells were rinsed twice with PBS, harvested by scraping into RIPA buffer containing protease inhibitors, and sonicated for 10 s. Equibalent amounts of proteins were separated by 10% SDS-PAGE, and transferred to a PVDF membrane (Millipore, Massachusetts, USA). Membranes were washed 4 times for 5 min each with PBS, blocked with 5% non-fat milk (Yili Industrial Group Co., Ltd, Hollyhock, China) in PBST (PBS containing 0.5% Tween 20) for 1 h, at RT. After washing, the membranes were incubated with an anti-GAPDH antibody or primary antibodies diluted in the blocking solution according the manufacturer's instructions for 1 h at 37°C with shaking (or overnight at 4°C). Secondary antibody (HRP-labeled goat anti-mouse or anti-rabbit IgG (H+L)) (Bioss, Beijing, China) diluted at 1:5,000 in blocking solution was added to the membranes and was incubated for 1 h, at RT. Antigen-antibody complexes were detected using the Western Light kit (Advansta, Menlo Park, USA).

Flow cytometry

Cells were collected, fixed with 70% ethanol, and stored at 4°C until analysis. After removing the supernatants and washing, cells were resuspended in PBS at a concentration of 1×10^6 cells/ml, then 100 µl of cell suspensions was placed into preparatory tubes combined with 200 µl DNA-PREPTMLPR, and incubated for 30 s. Samples were then mixed with 2 ml of DNA-PREPTMLPR staining reagent (propidium iodide solution) and incubated at RT for 30 min. Cell cycle phases were examined by flow cytometry, and data were processed using SYSTEM II software.

Cell apoptosis was detected using an Annexin V-FITC Apoptosis Detection Kit (Calbiochem, Darmstadt, Germany). Cells (5×10^5) were collected in PBS and centrifuged twice at 100 g for 5 min. The supernatants were discarded and 0.5 ml 1× binding buffer and 1.25 µl Annexin V-FITC were added to the cells, followed by incubation at RT for 15 min, protected from light. Cells were washed twice with 1× binding buffer, then resuspended in binding buffer to which 10 µl of propidium iodide was added; cells were then incubated for 10 min. The percentage of apoptotic cells was measured by flow cytometry, and the data were processed using WinMDI2.9 software.

Telomerase activity analysis

pSIECs (passage 3), ZYM-SIEC02 cells (passage 50) and A549 cells (a lung adenocarcinoma cell line) were trypsinized, washed twice with PBS, and pelleted by centrifugation at 2,000 rpm for 5 min. Telomerase activity was analyzed using a TRAP-silver staining Telomerase Detection Kit (KeyGEN Biotech, Nanjing, China) according to the manufacturer's instructions.

Soft agar assay

Growth in soft agar is the minimum requirement to demonstrate in *vitro* transformation. A lower layer of 0.5% agar gel was

A

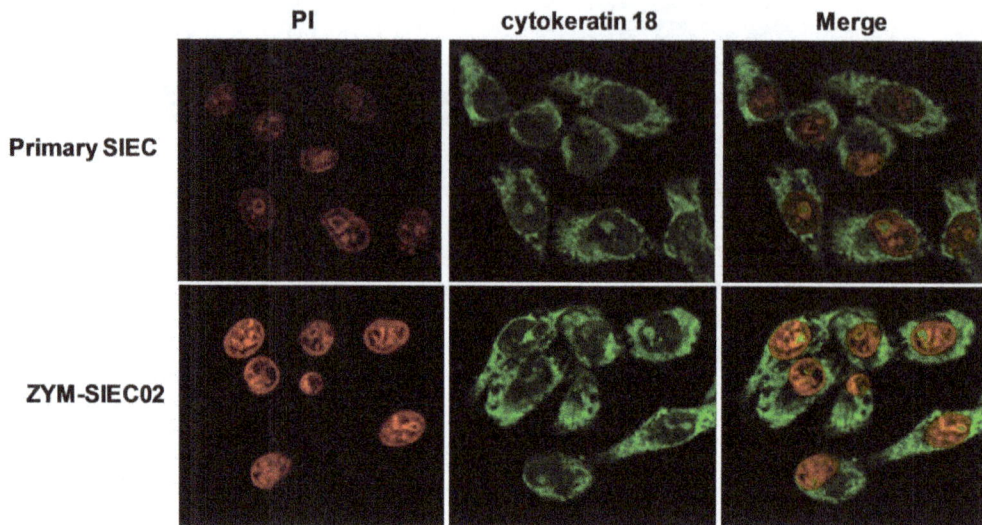

B

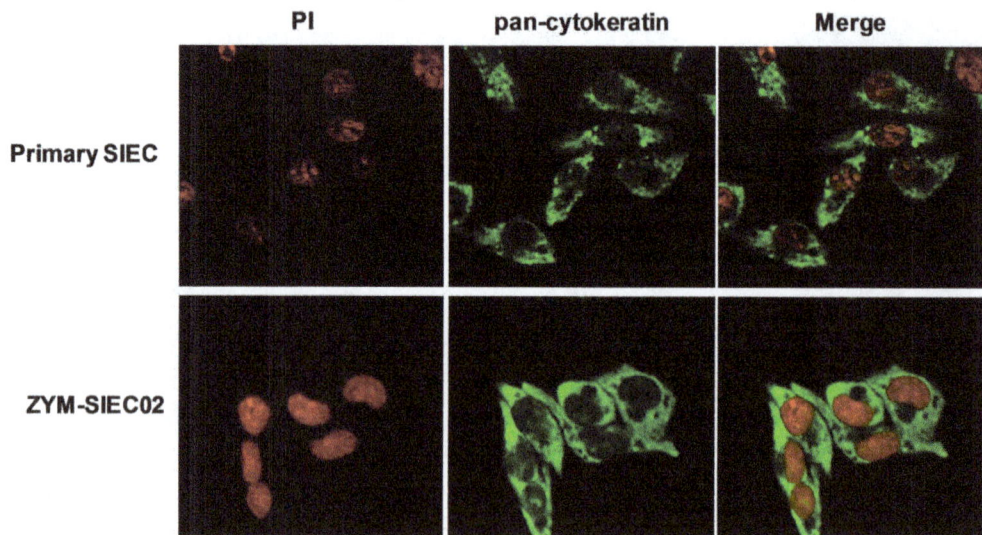

C

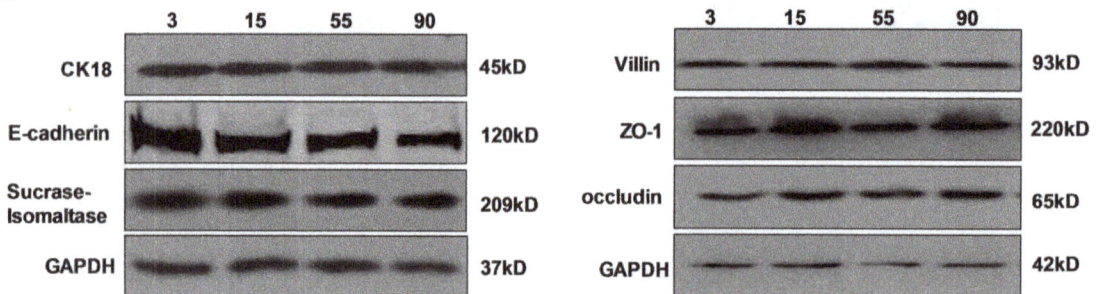

Figure 2. Immunostaining and western blot analysis of ZYM-SIEC02 cell cultures. A. pSIECs and ZYM-SIEC02 cells were stained with cytokeratin 18 antibodies (green) and propidium iodide (red). **B**. pSIECs and ZYM-SIEC02 cells were stained with pan-cytokeratin antibodies (green) and propidium iodide (red). **C**. expression of cytokeratin 18, E-cadherin, SI, villin, ZO-1 and occludin were determined by western blot analysis. Cytokeratin 18, pan-cytokeratin, E-cadherin, SI, villin, ZO-1 and occludin were expressed in both pSIECs (at passage 3) and ZYM-SIEC02 cells (at passages 15, 55, and 90).

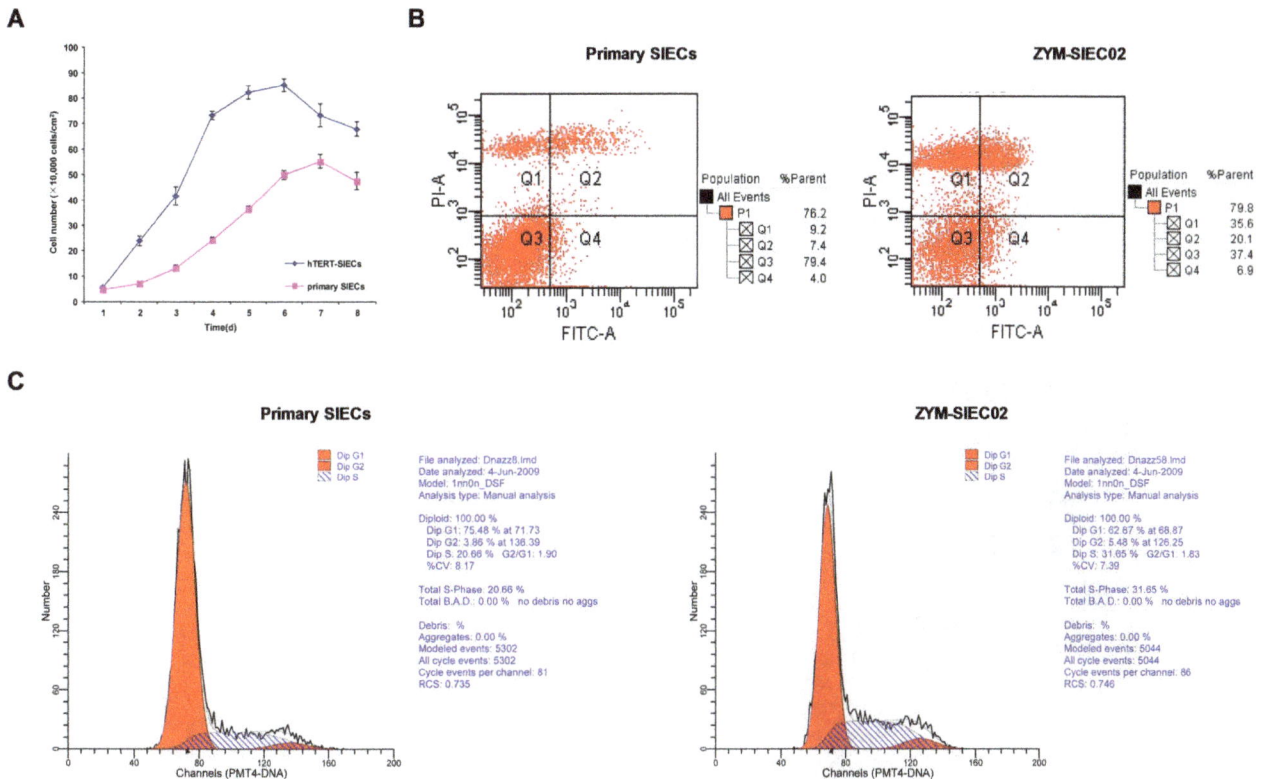

Figure 3. Growth curves, apoptosis and cell cycle analysis of SIECs before and after hTERT transfection. A. Growth of pSIECs and ZYM-SIEC02 cells. Data are represented as the mean ± SD of 3 independent experiments. **B**. Apoptosis analysis of control pSIECs and ZYM-SIEC02 cells, The percentage of apoptosis in pSIECs and ZYM-SIEC02 cells under basal growth conditions was 26.9% and 11.4% (Q2+Q4), respectively. **C**. Cell cycle distributions of control SIECs and ZYM-SIEC02 cells, the percentages of pSIECs and ZYM-SIEC02 cells in S phase were 20.66% and 31.65%, respectively.

prepared in 24-well plates at 4°C. Subsequently, ZYM-SIEC02 cells were trypsinized and resuspended at concentrations of 5×10^3, 1×10^4 and 2×10^4 cells/ml in DMEM/F-12 containing 20% FBS. Thereafter, 2 ml of 0.5% agar was added to 1 ml of cell suspension (final concentration: 0.33% agar) and cell suspensions (1 ml) were overlaid onto the solidified lower layer and incubated at 37°C in a humidified atmosphere with 5% CO_2 for 2 weeks. Colonies were then analyzed under a microscope. We considered the presence of even a single colony an indication that cells were capable of anchorage-independent growth.

Karyotype analysis

The number of chromosomes of transfected cells were determined by karyotype analysis performed as previously described by Hong et al [28].

Tumorigenicity assay

To evaluate *in vivo* tumorigenic potential, primary cells and ZYM-SIEC02 cells were trypsinized and resuspended in DMEM/F-12 at a concentration of 1×10^6 cells/ml, and 0.2 ml of the cell suspension was injected subcutaneously into the flanks of 6 nude mice (4 weeks old). As a positive control, equivalent number of EMF6 cells (a breast cancer cell lines derived from BALB/c mice) were injected into the flanks of 6 nude mice, both group of mice were monitored for 1 month to observe tumor formation.

Paraffin sections

After sacrificing the mice, tissue samples were removed from both control animals and from injected nude mice, then fixed in neutral buffered 10% formalin and processed into paraffin-embedded blocks for haematoxylin eosin staining. The study of histological structures was performed using light microscopy.

Proliferation assay

The growth curves were determined as previously described [28]. ZYM-SIEC (passage 50) and pSIECs (passage 3) were seeded in 24-well plates at 1×10^4 cells/well. Cells were digested with trypsin and counted each day for 8 days. Three independent experiments were performed in triplicates. The growth curves of the cells were plotted.

The proliferation of ZYM-SIEC02 cells in response to medium supplements was evaluated as previously described [2]. Briefly, ZYM-SIEC-02 cells were seeded into 96-well plates at 5×10^3 cells per well and cultured for 24 h. Cell culture medium and non-adherent cells were removed and replaced with fresh medium containing either FBS (2%, 5%, 10%, 15%, 20%), EGF (5 ng/ml), or ITS for an additional 24 h. Cell growth determined using the WST-1 Cell Proliferation and Cytotoxicity Assay Kit (Beyotime, shanghai, China) during the last 24 h of incubation. Cell proliferation was quantified using a Multiskan FC Microplate Photometer (Thermo Scientific, USA). Three independent experiments were performed in triplicates.

Bacterial invasion

Invasion assays were performed essentially as previously described [31]. *Salmonella enterica serovars Typhimurium(S. Typhimurium)* was grown in L-broth to an optical density at 600 nm (OD_{600}) of approximately 1. After centrifugation and

A

B

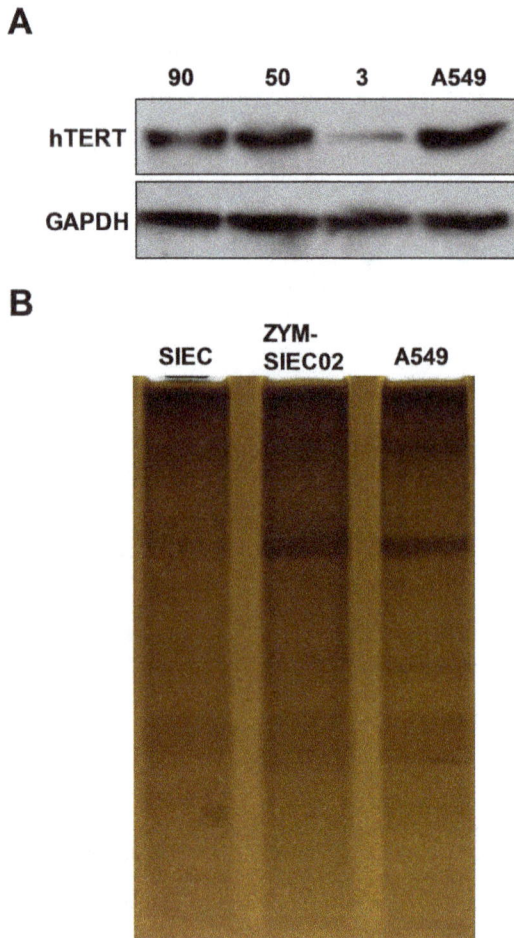

Figure 4. Detection of Telomerase activity. A. Expression of hTERT protein was detected by western blot analysis. Expression of hTERT was readily detected in ZYM-SIEC02 cells at passage 50 and 90, but was barely detectable in pSIECs at passage 3. The A549 cell line was used as a positive control. **B.** Silver staining method was used to detect telomerase activity. The laddering patterns indicated that ZYM-SIEC02 cells display higher telomerase activity as compared to pSIECs and A549 cell line.

washing, bacteria were resuspended in cell culture medium and diluted to a multiplicity of infection (MOI) of 10:1 (*Salmonella*: host cells) in a 12-well plate. Confluent monolayers were infected at 37°C, 5% CO_2 for 1 h to allow bacterial entry. After removal of the extracellular bacteria, cultures were incubated for an additional hour in medium containing 50 μg/ml gentamincin to kill any remaining extracellular bacteria. Infected cells were washed twice with PBS, lysed with 0.1% Triton ×−100 in deionized water. Dilutions of cell lysates were plated onto L-broth agar plates for quantification of intracellular bacteria. Infections were carried out in triplicate.

Exposure of ZYM-SIEC02 cells to inflammatory stimuli

Treatments included control (uninfected cells), LPS (L-2880, from *E.coli* 055:B5; Sigma Chemical Co., St. Louis, MO; 1 μg/ml), or *Salmonella*. Bacteria were grown as described for invasion assays. After the bacterial population was estimated by spectrophotometry at OD600, bacteria were pelleted and resuspended in DMEM/F-12 growth media lacking FBS and antibiotics. Confluent ZYM-SIEC02 cells were washed twice with PBS and 1 ml of

media alone (control), bacteria was added to the wells and plates were further incubated for 1 h. Cell culture medium from control treatments or cultures exposed to bacteria was replaced with fresh media containing 50 μg/ml gentamicin. Medium was removed from wells 4 h after LPS or bacteria exposure for determination of IL-8 and TNF-α secretion, and adherent cells were washed 3 times with PBS and lysed with Trizol reagent (Invitrogen, Carlsbad, USA).

Inhibition of LPS-mediated response

E.coli LPS 055:B5 (0.0125, 0.25 or 5 μg/mL) were preincubated with or without 10 μg/mL of polymyxin B sulfate (Sigma, Shanghai, China) for 3 hours. ZYM-SIEC02 cells were then washed 3 times with PBS and lysed with Trizol reagent (Invitrogen, Carlsbad, USA).

Real time PCR

Total RNA was isolated according to the manufacturer's instructions. The quality and purity of total RNA was determined by optical density (OD) at 260 nm and 280 nm wavelengths using a spectrophotometer. Total cDNA was synthesized using M-MLV Reverse Transcriptase (Takara, Dalian, China) according to the manufacturer's instructions. Relative gene expression was determined by qPCR using an iQ5 Thermo Cycler (Bio-Rad). 20 μl real-time PCR reactions were carried out using 2 μl of diluted cDNA as template and iQTM SYBR Green Supermix (Takara, Dalian, China) according to the manufacturer' instructions. Thermal cycling parameters were utilized according to manufacturer recommendations by 40 cycles of amplification with alternating 5 s 95°C denaturation and 30 s 56°C anneal/extension. Specific genes were amplified with DNA polymerase using the primers listed in Table 1.

ELISA assay

Infection of cell monolayers with Salmonella typhimurium was performed as described for invasion assays. Supernatants from infected and uninfected cells were harvested after 4 h. *ELISA* were performed according to the manufacturer's protocols (Swine IL-8 or TNF- *ELISA* Kit, Invitrogen, Camarillo, CA, USA).

Statistical analysis

Relative gene expression of chemokines and cytokines in ZYM-SIEC02 cells was determined from real time PCR data using the $\Delta\Delta CT$ method as previously described [32,33] Data were expressed as mean ± (SEM). Differences between groups were examined for statistical significance using the Student's t-test. A P value <0.05 was considered as statistically significant.

Results

Immortalized intestinal epithelial cells maintained the morphological features of the primary cells

Cell material obtained from porcine mid-jejunum consisted of cell aggregates or organoids that adhered to collagen-coated culture flasks and formed circular proliferating foci (Fig. 1A, upper left). Initial expansion of the foci resulted in fusion of the cell plaques, forming a confluent cell layer within 10–12 days due to the presence of residual multicellular organoids dispersed in the newly formed monolayer. During the first passage, epithelial cell sheets exhibited a homogeneous cobblestone-like morphology. No obvious morphological changes were observed during prolonged passage in culture, defined as 8–10 passages at a split ratio of 1:2 over 4–5 weeks (Fig. 1A, upright). Cells reached confluence within

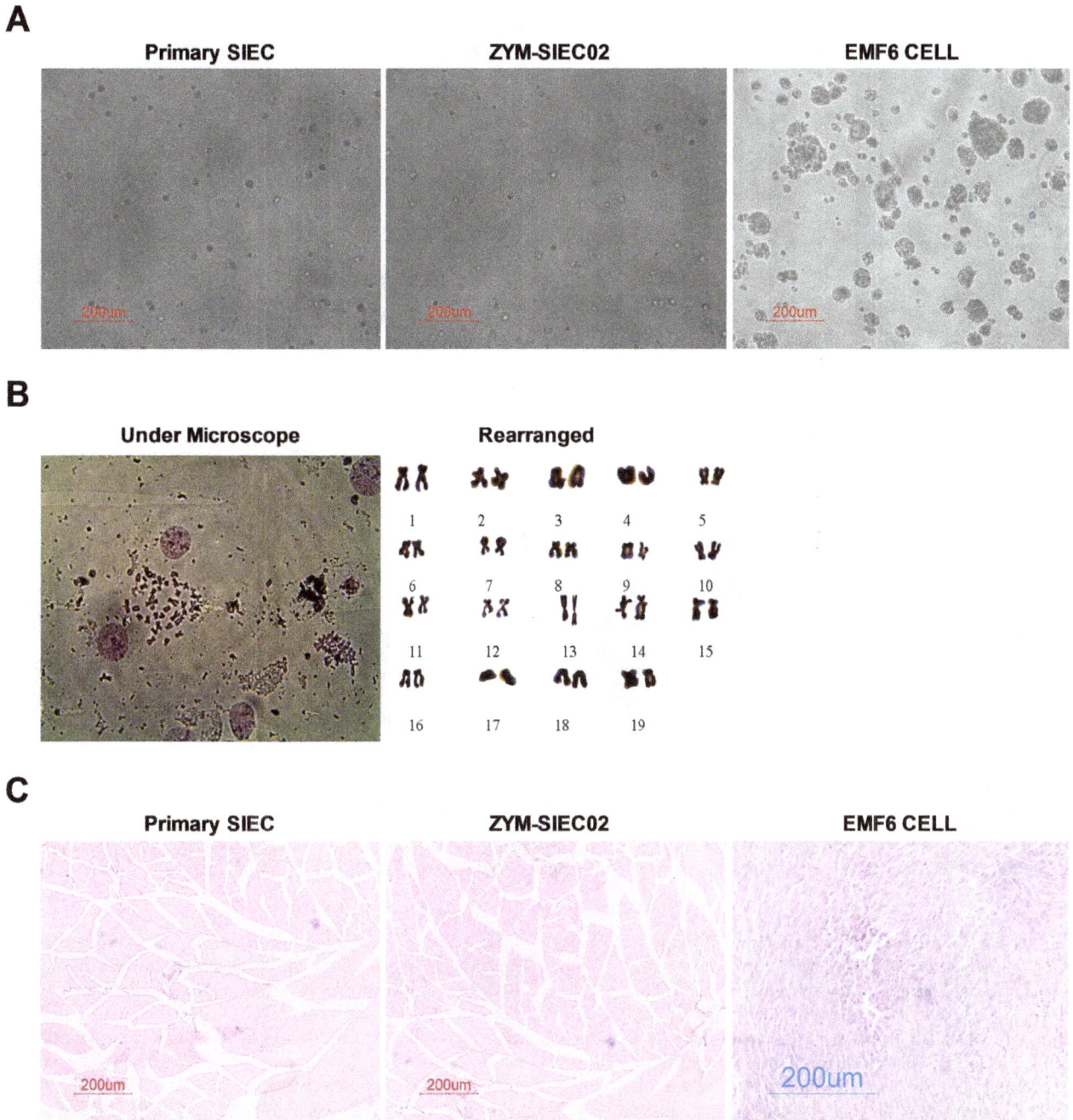

Figure 5. Tumorigenicity assay of ZYM-SIEC02 cells. A. Representative images of pSIECs and ZYM-SIEC02 cells growing in soft agar. Neither cell line formed colonies after 2 weeks in culture, indicating that the cells did not exhibit anchorage-independent growth (100×). B. Karyotype analysis of ZYM-SIEC02 cells at passage 55, indicated that the cells contained 38 chromosomes, consistent with a diploid karyotype. C. Tumor formation assay. pSIECs and ZYM-SIEC02 cells were injected subcutaneously into nude mice. After one month, neither group of mice had developed tumors, in contrast, mice injected with EMF6 cells rapidly developed tumors within 8 days. Histological examination indicated morphologically normal tissue below the injection site of pSIECs (passage 3) and ZYM-SIEC02 cells (passage 55), but a dense cellular mass below the EMF6 cell injection site.

2–3 days during the early passages (p1–p6). However, a progressive slowdown in the proliferation rate was observed until cells reached senescence. Eleven passages later, cells became enlarged and the cobblestone-like morphology was lost, even though the expression of cytokeratin 18 was maintained (data not shown).

In order to establish an immortalized porcine intestinal epithelial cell line, pSIECs at passage 2 were transfected with the plasmid of pCI-hTERT-neo, and selected with G418 for 2 weeks, at which time we obtained a drug-resistant clone, which we designated ZYM-SIEC02 (Fig. 1A, lower left). ZYM-SIEC02 cell exhibited similar morphological features as compared to control pSIEC, even at passage 90 pSIECs (Fig. 1A lower right). To date,

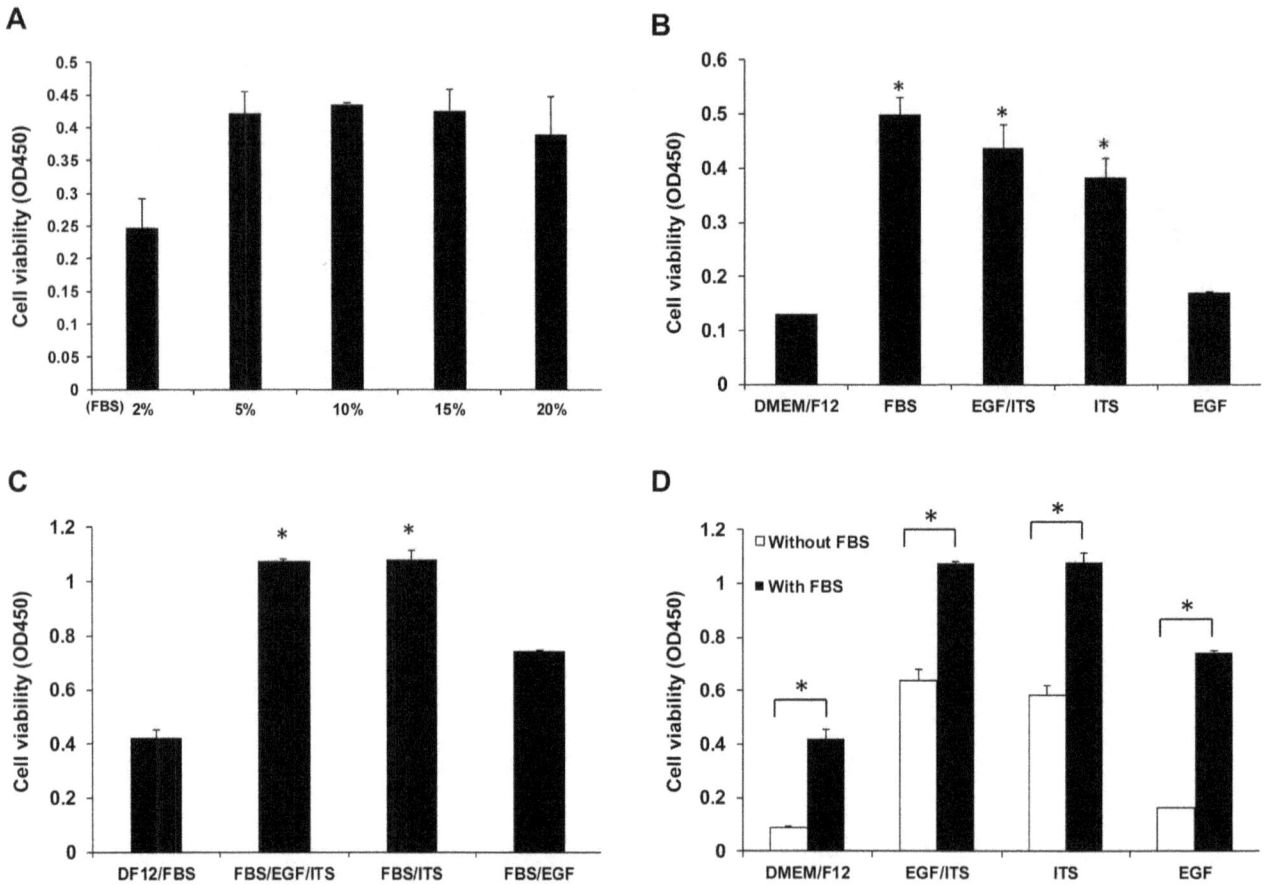

Figure 6. Optimization of culture medium for ZYM-SIEC02 cells. A. ZYM-SIEC02 cells were cultured for 24 h with 2%, 5%, 10%, 15%, or 20% FBS. 5% FBS was found to be the optimal concentration. **B**. Addition of 5% FBS or ITS, and/or EGF to the cell medium individually, compared with DMEM/F12 alone; Addition of ITS or ITS/EGF enhanced cell growthsimilar to the addition of 5% FBS, while addition of EGF alone did not have a significant effect. **C**. ITS/EGF or ITS significantly enhanced cell growth with the use of cell culture medium containing 5% FBS. **D**. Addition of ITS, EGF, or EGF/ITS to DMEM/F12 had little effect on the proliferation rate compared to the addition to 5% FBS. * $P \square 0.05$ versus basal conditions.

the established ZYM-SIEC02 cell line has been cultured for more than 100 passages at split ratios of 1:2 or 1:3.

Ultrastructural analyses indicated that ZYM-SIEC02 cells in culture maintained the structures of microvilli and apical tight junctions (Fig. 1B). Interestingly, a small intestine glandular configuration (Fig. 1B, lower right) was also observed. In addition, the majority of cells possessed sponge-like nucleoli with a high nucleus: cytoplasm ratio, suggests that the cells have a robust proliferative activity.

Epithelial cell-specific markers are stably expressed in immortalized ZYM-SIEC02 cells

ZYM-SIEC02 cells were characterized using immunofluorescence and western blot analysis. Immunostaining for cytokeratin indicated characteristic disposition of intermediate filaments, and intracellular expression of cytokeratin 18 (Fig. 2A) and pan-cytokeratin (Fig. 2B).

Western blot analysis confirmed the expression of epithelial cell markers in ZYM-SIEC02 cells. A cytokeratin 18 antibody recognized a 45 kD protein in homogenates from primary porcine enterocytes, as well as in those from early (passage 5), intermediate (passage 55) and late passages (passage 90) of ZYM-SIEC02 cells. ZYM-SIEC02 cells also expressed the epithelial cell marker E-cadherin (~120 kD), which plays an important role in the

maintenance of tight junctions. Moreover, the expression of tight junction proteins ZO-1 and occludin confirmed that ZYM-SIEC02 cells formed intercellular tight junctions. In order to determine whether these cells were differentiated, sucrase-isomaltase (SI, 209 kD), a marker of differentiation expressed in the intestinal brush border, was detected as described previously [34] (Fig. 2C). In addition, both pSIECs and ZYM-SIEC02 cells expressed detectable villin protein (93 kD), confirming the ultrastructural analysis (Fig. 1B).

Enhanced proliferation and decreased apoptosis in immortalized cells

The growth curves of ZYM-SIEC02 cells and control pSIEC are shown in Figure 3A. In order to ascertain whether hTERT expression affected cell proliferation or cell cycle progression, cell cycle distribution and basal levels of apoptosis in pSIECs and ZYM-SIEC02 cells were determined by flow cytometry. The percentages of apoptotic cells in pSIECs and ZYM-SIEC02 cultures were 26.9% and 11.4% (Q2+Q4), respectively (Fig. 3B). The percentages of pSIECs and ZYM-SIEC02 cells in S phase were 20.7% and 31.7%, respectively (Fig. 3C), indicating that the immortalized ZYM-SIEC02 cell line exhibits increased proliferative activity and a lower incidence of apoptosis compared to that of primary swine intestinal epithelial cells.

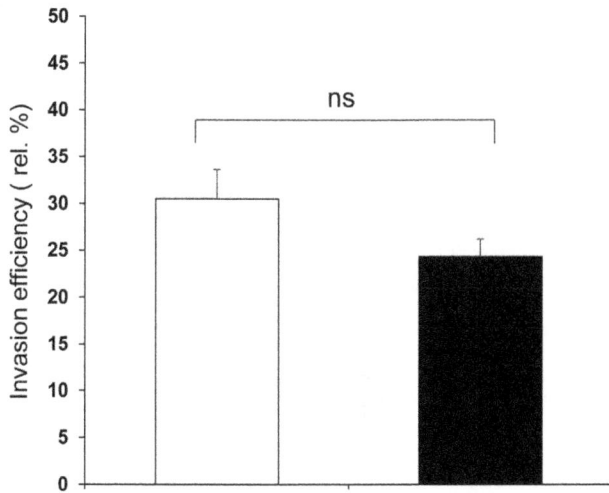

Figure 7. Invasion efficiency of *Samonlla Thpyimurium* in the ZYM-SIEC02 cell line. Confluent monolayers of ZYM-SIEC02 cells and pSIECs were infected with a MOI of 10:1 (*S. Typhimurium* to host cells) for 1 h, and incubated in media containing gentamicin for an additional hour. Cells used were pSIECs (open bars) and ZYM-SIEC02 cells (filled bars). Data shown are normalized to the efficiency (relative percent of input bacteria) and representative of 5 independent experiments performed in triplicate.

Increased telomerase activity in transfected cells

Telomerase activity in both pSIECs and ZYM-SIEC02 cells was detected by Western blot and a TRAP-silver staining Telomerase Detection Kit. There were no significant differences observed in the level of hTERT protein in ZYM-SIEC02 cells at early or late passage, but ZYM-SIEC02 cells exhibited higher hTERT expression and activity than pSIECs (Fig. 4A, B).

The newly established ZYM-SIEC02 cell line is immortalized but not transformed

Anchorage-independent growth was tested using a clonogenic soft agar assay. While EMF6 cells formed numerous colonies, ZYM-SIEC02 cells and pSIECs did not form colonies after 2 weeks in soft agar (Fig. 5A), indicating that ZYM-SIEC02 cells did not exhibit anchorage-independent growth.

Karyotype analysis of ZYM-SIEC02 cells at passage 60 was performed to determine whether transduction with hTERT affected the chromosomal status. We observed a normal complement of chromosomes in ZYM-SIEC02s with a modal chromosome number of 38 (Fig. 5B).

A definitive functional assay for determining tumorigenicity was carried out by injecting either ZYM-SIEC02 cells or tumorigenic EMF6 cells subcutaneously into nude mice. Tumors were observed with 7–8 days in all mice injected with EMF 6 cells, while no tumors developed in mice injected with ZYM-SIEC02 cells even after 1 month. Further histological examination revealed a normal tissue structure at the site of ZYM-SIEC02 cells or pSIEC cell injection, whereas dense cellular masses were observed at the EMF6-injection site (Fig. 5C).

ITS is beneficial to cell growth

To optimize the culture conditions of our newly immortalized epithelial cells, medium supplements for growth of the cell line were tested. Based on cell growth responses, a concentration of 5% fetal bovine serum (FBS) was found to be optimal (Fig. 6A). To further confirm whether medium additives promote cell prolifer-

ation, 5% FBS or ITS (1.0 g/l insulin, 0.55 g/l transferrin, and 0.67 mg/l selenium) and/or epidermal growth factor (EGF, 10 ng/ml) were added to the culture medium. We found that, ITS alone or a combination of ITS and EGF enhanced cell growth equivalent to that induced by 5% FBS alone (Fig. 6B). Moreover, the addition of ITS to the culture medium containing 5% FBS further promoted cell proliferation while EGF alone had no effect (Fig. 6C, D).

Invasion efficiency of Salmonella typhimurium in the ZYM-SIEC02 cell line

Salmonellae are facultative intracellular pathogens. An essential step in Salmonella pathogenicity is the formation of an intracellular *Samonella*-containing vacuole in which the pathogen replicates [2]. To determine whether ZYM-SIEC02 cells are susceptible to *Samonella typhimurium* invasion in *vitro*, intracellular bacteria were quantified in order to calculate invasion efficiency. As shown in Fig. 7, both pSIECs and ZYM-SIEC02 cells could engulf *Salmonella* during the invasion process.

ZYM-SIE02 cells maintained the immune functions of pSIECs

To further verify the immunomodulatory responses of the ZYM-SIEC02 cell line, we evaluated the expression of the inflammatory factors and chemokines expressed in intestinal epithelial cells that are known to be involved in the immune response to pathogenic infection. ZYM-SIEC02 cells were used to evaluate the effect of *S. Typhimurium* and LPS on IL8 and TNF-α mRNA expression. As shown in Fig. 8, the expression of IL-8, TNF-α was upregulated following *S. Typhimurium* challenge and IL8 expression was upregulated approximately 2.5-fold within 4 h of *S. Typhimurium* challenge relative to control, while TNF-α mRNA expression was a modest increased by *S. Typhimurium* stimulation. Interestingly, cells that stimulated with LPS demonstrated a significant reduction in IL-8 and TNF-α mRNA expression compared to controls (Fig. 8A, B). To determine if this functional immune response were via TLR4 recognition on intestinal epithelial cells, we choose to detect TLR4 and TLR6 mRNA expression following stimulation with LPS. The result showed that TLR4 and TLR6 were upregulated (Fig. 8C).

In order to determine if challenged ZYM-SIEC02 cells exhibit differential protein expression during an inflammatory response, IL-8 and TNF-α secretion were evaluated by ELISA. Exposure to *S. Typhimurium* resulted in an initial trend toward an increase in IL8 secretion (Fig. 8D). However, the lack of response of swine intestinal epithelial cell in *vitro* to stimulate with LPS was found in terms of IL8 production. Moreover, TNF-α secretion in ZYM-SIEC02 cells remained undetectable while challenged with both *S. Typhimurium* and LPS (data not shown).

Interestingly, antagonism of TLR4 signaling with polymyxin B prevented the increase in IL-6 mRNA expression but increased TNF-α mRNA expression when the concentration of LPS was lower than 250 ng/ml (Fig. 9A). As expected, ZYM-SIEC02 cells responded to LPS in a dose-dependent manner and polymyxin B blocked LPS recognition by host cells inhibited this response (Fig. 9B).

Discussion

In the present study, we describe the establishment of the ZYM-SIEC02 cell line and demonstrate that ZTM-SIEC02 cells exhibit normal epithelial cell morphological and functional characteristics similar to pSIECs. Morphological analysis showed a typical cobblestone-like morphology, apical tight junctions, and the

A

IL-8

B

TNF-α

C

D

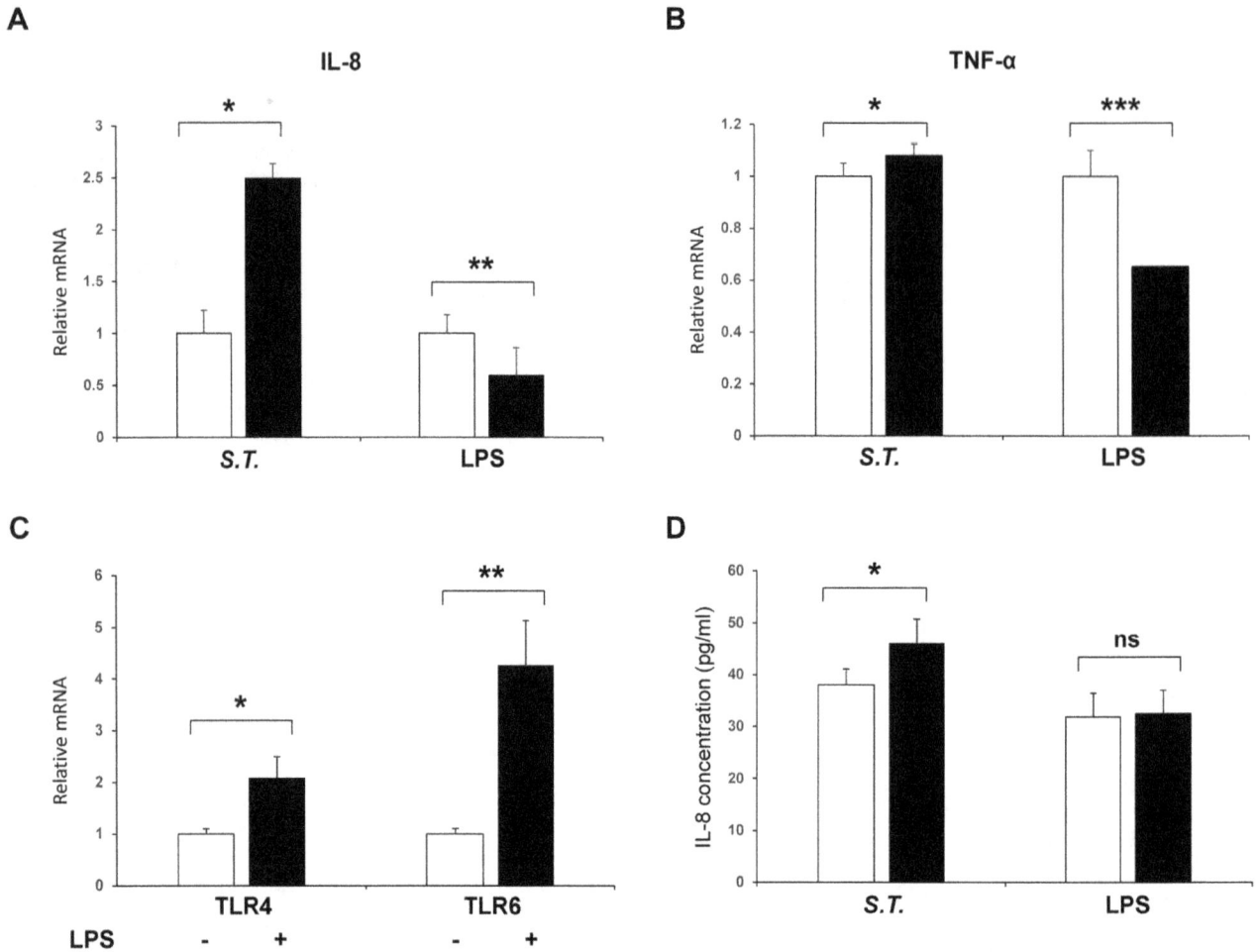

Figure 8. Cytokine expression in ZYM-SIEC02 cells infected with *Salmonella typhimurium* and LPS. Confluent ZYM-SIEC02 cell monolayers were infected for 4 h with *S. Typhimurium* or LPS. **A**. Relative mRNA expression of IL-8 in ZYM-SIEC02 cells were upregulated approximately 2.5-fold within 4 h of *S. Typhimurium* challenge relative to control, while the cells that stimulated with LPS demonstrated a significant reduction; **B**. TNF-α mRNA expression was a modest increased by *S. Typhimurium* stimulation, while the cells that stimulated with LPS demonstrated a significant reduction; **C**. TLR4 and TLR6 mRNA expression were upregulated following stimulation with LPS. D. IL-8 protein expression in ZYM-SIEC02 cells infected with *Salmonella typhimurium* and LPS. ZYM-SIEC02 cells were incubated with *S. Typhimurium* at a MOI of 100:1 followed by an incubation in media containing gentamicin, and then cell culture supernatants were harvested after 4 h. ZYM-SIEC02 cells were stimulated with LPS, and cells were harvested after 4 h. IL-8 concentrations were determined by ELISA. The data shown are means±SEM of 3 independent experiments. * $P<0.05$;** $P<$ 0.01; *** $P<0.001$; ns, not significant.

presence of microvilli. In addition, functional analysis suggested that this cell line was not transformed and was able to respond to *S. Typhimurium* infection by upregulating expression and secretion of IL8. These results demonstrate that introducing the hTERT gene can be used to immortalize porcine intestinal epithelial cells, consistent with previous findings that stable overexpression of hTERT gene is sufficient to immortalize swine umbilical vascular endothelial cells [28], and cattle typealveolar epithelial cell line [29].

Cytokeratins play a pivotal role in cell differentiation and tissue specialization, functioning to maintain the overall structural integrity of epithelial cells. The expression of cytokeratin is considered as the marker of epithelial cells [2]. In this study, expression of cytokeratin 18 and pan-cytokeratin was detected in ZYM-SIEC02 cells, demonstrating that the cells were epithelial in origin. E-cadherin is a membrane-spanning protein found in epithelial cells and plays a crucial role in formation of cell-cell junction formations [35]. Here, we found that E-cadherin was

expressed by both pSIECs and ZYM-SIEC02 cells, indicating that ZYM-SIEC-02 cells preserved this characteristic of epithelial cells. Moreover, the enterocyte marker sucrase-isomaltase (SI) was detected in both pSIECs and ZYM-SIEC02 cells. SI is a small-intestinal microvillus hydrolase that localizes to the apical membrane of adult intestinal enterocytes along the intestinal crypt-villus axis [36]. Deficiency in SI protein results in osmotic diarrhea due to an inability to hydrolyze intestinal disaccharides into component monosaccharides.

Therefore, we conclude that the newly established cell line is comprised of small intestine-derived epithelial cells, which in turn suggests that this cell line can be used in future research on disease inducing porcine diarrhea.

Typically, differentiated cells were taken for representing the intestinal villus tip cells, while the undifferentiated cells were used to mimic the basilar crypt cells of the intestines [37,38]. Thus, the prototypical characteristics of differentiated porcine intestinal epithelial cells are the presence of tight junctions and distinct

A

TNF-α

B

IL-6

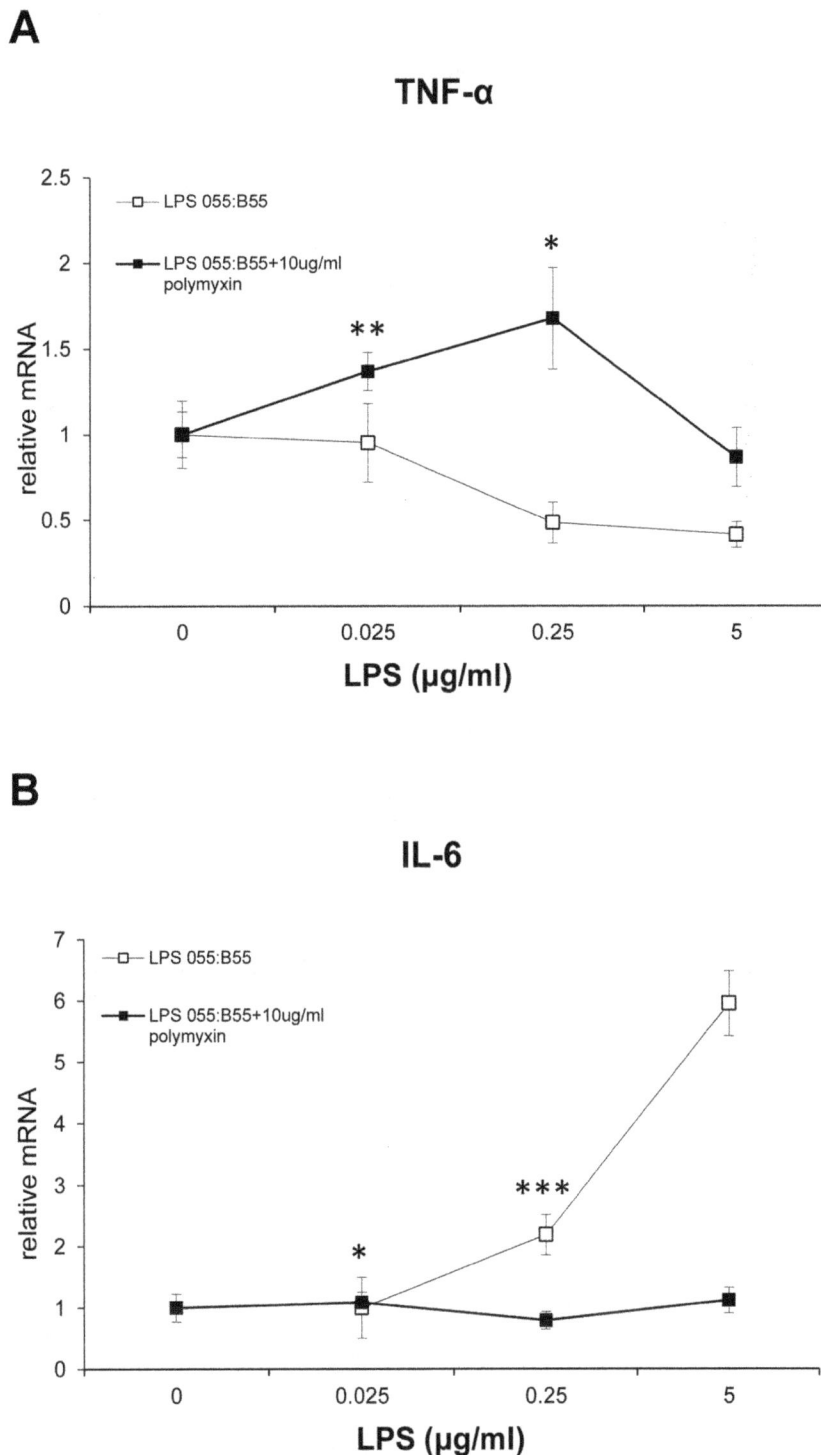

Figure 9. Inhibition of LPS-mediated response. ZYM-SIEC02 cells were stimulated with the indicated dose of *E.coli* 055:B55 LPS in the presence or absence of polymyxin B sulfate(Sigma) for 3 hours. A. Polymyxin B increased TNF-α mRNA expression when the concentration of LPS was lower than 250 ng/ml. B. Polymyxin B prevented the increase in IL-6 mRNA expression. The data shown are means±SEM of 3 independent experiments. * $P<0.05$;** $P<0.01$; *** $P<0.001$.

microvilli on their apical surfaces [38,39]. Here, we demonstrated that porcine intestinal epithelial cells expressed markers typical of differentiated enterocytes, specifically, ZYM-SIEC02 cells were positive for E-cadherin, ZO-1, Occludin, villin, and sucrose isomaltase, which is the most reliable indicator of intestinal cell

differentiation in vitro [40], indicating the presence of differentiated villus cells in the culture. IPEC-J2 cells were cultured for 1 days or 21 days, representing undifferentiated proliferating and highly differentiated IPEC-J2 cells, respectively [41,42]. However, in our experiment, all the ZYM-SIEC02 cells we used were cultured to have a confluent monolayer within 3 or 6 days according to the density of cell cultures. In this regard, the characteristics of differentiated ZYM-SIEC02 cell were evidently due to the time that isolated and purified the primary porcine intestinal epithelial cells was long enough for the differentiation of pSIECs, which prepared for the lipofection. Thus, ZYM-SIEC02 cells remain the characteristic of differentiated pSIECs, and markers typical of differentiated enterocytes were detected. Moreover, we used tissue culture method to isolate epithelial cells from small intestine, it was unavoidable that the growth of fibroblasts over cultured primary intestinal epithelial cells, however, on the other side of the coin, the intestinal epithelial cell differentiation need the heterologous cell-cell contacts for *denovo* synthesis to trigger cell polarity and differentiation [40]. The addition of EGF may be another important inducement for differentiation as it plays a pivotal role in the regulation of intestinal epithelial proliferation and differentiation [43]. Over time in culture, cells maintained their epithelial morphology as well as expression of markers (8–10 passages for pSIECs and more than 90 passages in ZYM-SIEC02 cells). These characteristics were not affected by cryogenic freezing and retrieval (data not shown).

An early study by Hayflick et al demonstrated that normal human somatic cells have a finite replicative potential *in vitro* [44], and will stop dividing after a finite number of population doublings and enter senescence. Moreover, the lifespan of human cells *in vitro* is known to be dependent on telomere length and telomerase activity. Telomere shortening in humans is a prognostic marker of disease risk, progression and premature mortality. However, telomere shortening can be counteracted by activity of the cellular enzyme telomerase [45]. Telomerase adds telomeric repeat sequences to the ends of chromosomal DNA, thus preserving telomere length, cellular function, and long-term immune function [46]. The introduction of the hTERT gene is a widely used strategy to extend the lifespan of many cell types and successful immortalization has been reported in human retinal pigment epithelial cells, porcine umbilical vein endothelial cells and cattle type II alveolar epithelial cells [27–29]. In the present study, primary porcine small intestinal epithelial cells were transfected with the pCI-hTERT-neo plasmid and stable transfectants were selected with G418. The newly established epithelial cell line (ZYM-SIEC02) was detected to have a significantly higher expression of hTERT and higher telomerase activity compared to primary cells, suggesting that the lifespan of pSIECs was successfully extended.

Various synthetic culture media, supplemented with 5–20% FBS, have been used in studies of the mammalian intestinal epithelium [47], And supplementation with ITS or EGF is beneficial for epithelial cell proliferation [2]. In the present study, the addition of 5% FBS was found to be optimal for the growth of ZYM-SIEC02 cells. ITS supplementation promoted cell proliferation, while the combination of ITS and FBS further enhanced proliferation. In contrast, EGF had minimal effects on the proliferation of ZYM-SIEC02 cells. Overall, our data suggests that ITS can partially substitute for the mitogenic effects of FBS and is capable of sustaining proliferation of ZYM-SIEC02 cells, which is beneficial for future studies on viral infection.

The newly established ZYM-SIEC02 cell line was not transformed and was not tumorigenic in nude mice, indicating that this cell line can be used in addition to pSIECs. Furthermore, we found that this cell line closely mimics the porcine intestinal environment *in vivo*, contributing to the establishment of a porcine-derived infection model.

The functionality of the intestinal epithelium is critically important to neonatal swine as this period of time represents a window of significant vulnerability to pathogens including *Salmonella enteria serovar typhimurium (S. Typhimurium)* [48]. Moreover, we detected the invasion efficiency of *S. Typhimurium* in the ZYM-SIEC02 cells and found *S. Typhimurium* could invade host cells effectively and we used MOI of 100:1 (*S. Typhimurium*: cell) to ensure enough number of intracellular bacteria.

Cytokines impact on the functional state of immune cell populations by eliciting altered patterns of gene expression [2]. In response to pathogenic microorganism infection, host cells secrete numerous types of cytokines and cytokine expression patterns are species- and tissue-dependent [2]. Melania etal [49] found that all porcine intestinal sections tested were able to sense *S. Typhimurium* presence as indicated by changes in IL8 and TNF-α gene expression. Our study of IL8 and TNF-α gene expression suggests that this newly established porcine intestinal epithelial cell line could respond similarly to bacterial infection, we also showed that the ZYM-SIEC02 cells increased secretion of IL-8 infection by *S. Typhimurium*. This is in agreement with previous observations showing increased levels of IL-8 production in the serum of *S. Typhimurium* infected piglets [49], and is also in accordance with a marked upregulation of IL-8 secretion observed in human intestinal epithelial cells and IPEC-J2 cells following *S. Typhimurium* infection [2,50]. These results suggest that the ZYM-SIEC02 cell line can effectively respond to pathogen infections and that an inflammatory process has been triggered along after infection with *S. Typhimurium* Interestingly, while previous studies demonstrated that IPEC-J2 was of hyporesponsiveness to LPS with respect to IL-8 mRNA expression and secretion [32]. This is consistent with our results but in contrast to the conclusions drawn from microarray analysis following stimulation with 1 µg/ml LPS [51].

Previous work has revealed conflicting results regarding changes in TNF-α gene and protein expression. Several groups observed increase in TNF-α, IL-8 protein and mRNA expression in intestinal tissues from pigs infected with *S. Typhimurium* [52,53], however, Hyland et al [54] reported that neither TNF-α mRNA nor protein was detected in pigs that infected with Salmonella ssp. In this study, *S. Typhimurium* elicited a modest increase ($P<0.05$) in TNF-α gene expression from ZYM-SIEC02 cells compared to untreated wells. Furthermore, TNF-α mRNA levels were significantly decreased after LPS stimulation, and indicating that TNF-α may not be critical in the early response to LPS as it is in IPEC-J2. We did not detect any TNF-α production from ZYM-SIEC02 cell challenged with *S. Typhimurium* or LPS. TLR4 was upregulated following stimulation with LPS, as the expected values, due to the function of TLR4 as the LPS receptor. The observation that LPS can be the stimulus for IL-6 mRNA production by ZYM-SIEC02 cells suggests that TLR4 were involved in the activation of porcine intestinal epithelial cells in response to LPS. Interestingly, we reported here that ZYM-SIEC02 cells express mRNA for TLR4 and TLR6 and were upregulated significantly after stimulation with LPS.

TLR4 signaling is known that two major signaling pathways, the MyD88-dependent and TIR-domain-containing adapter-inducing interferon-β (TRIF)-dependent signaling pathways, are activated when TLR4 recognizes LPS. Previous studies have demonstrated that the induction of postnatal tolerance in mouse and human small intestine is a mechanism that relies on microRNA (miR)-146a upregulation and subsequent IRAK-1 degradation [55]. Basing on this point, it is also reasonable for TLR4 expression as IRAK-1 is an adapter molecule for MyD88-dependent signaling pathway. The intracellular recognition of LPS initiates the TRIF-dependent pathway, which is important for the induction of adaptive immune responses. It is possible that strategies aimed at opening the TRIF-dependent pathway or others will broaden therapeutic opportunities for controlling TLR trafficking and localization. This may expain why both TLR4 and TLR6 were increased. Be that as it may, the results of our studies establish a physiologically significantly role played by TLR-4 in mediating signaling elicited by LPS. Whether the intracellular recognition of LPS may initiate TRIF-dependent pathway is currently being investigated.

In summary, the newly established ZYM-SIEC02 cell line retains the morphological and functional features of primary swine intestinal epithelial cells. Moreover, this cell line is not transformed and can be safely used for future studies. Therefore, the establishment of a stable ZYM-SIEC02 cell line is of great importance for future studies of the mechanisms of pathogen infection *in vitro*, particularly for swine based infection studies, and also a potential model of zoonotic infections (e.g. *S.*

Typhimurium). Given the high degree of homology between porcine and human intestines, studies performed on the porcine intestinal epithelial ZYM-SIEC02 cell line may provide valuable insight into human intestinal disease.

Acknowledgments

We would like to thank Dr R.A. Weinberg for kindly providing the pCI-neo-hTERT plasmid. We are also grateful to Dr. Hongchao Zhou, GangWang, Xiaoyun Yang, Chen Dai, Tuo Zhang, and Chouzhi Zhao for their expert technical assistance.

Author Contributions

Conceived and designed the experiments: JW YZ. Performed the experiments: JW GH ZL LH LX. Analyzed the data: JW GH YZ. Contributed reagents/materials/analysis tools: JW ZL LX. Wrote the paper: JW YZ.

References

1. Burch DG (1982) Tiamulin feed premix in the prevention and control of swine dysentery under farm conditions in the UK. Vet Rec 110: 244–246.
2. Schierack P, Nordhoff M, Pollmann M, Weyrauch KD, Amasheh S, et al. (2006) Characterization of a porcine intestinal epithelial cell line for in vitro studies of microbial pathogenesis in swine. Histochem Cell Biol 125: 293–305.
3. Sun Z, Huber VC, McCormick K, Kaushik RS, Boon AC, et al. (2012) Characterization of a porcine intestinal epithelial cell line for influenza virus production. J Gen Virol 93: 2008–2016.
4. Gibson-D'Ambrosio RE, Samuel M, D'Ambrosio SM (1986) A method for isolating large numbers of viable disaggregated cells from various human tissues for cell culture establishment. In Vitro Cell Dev Biol 22: 529–534.
5. Panja A (2000) A novel method for the establishment of a pure population of nontransformed human intestinal primary epithelial cell (HIPEC) lines in long term culture. Lab Invest 80: 1473–1475.
6. Perreault N, Beaulieu JF (1998) Primary cultures of fully differentiated and pure human intestinal epithelial cells. Exp Cell Res 245: 34–42.
7. Perreault N, Jean-Francois B (1996) Use of the dissociating enzyme thermolysin to generate viable human normal intestinal epithelial cell cultures. Exp Cell Res 224: 354–364.
8. Kaeffer B (2002) Mammalian intestinal epithelial cells in primary culture: a mini-review. In Vitro Cell Dev Biol Anim 38: 123–134.
9. Booth C, Patel S, Bennion GR, Potten CS (1995) The isolation and culture of adult mouse colonic epithelium. Epithelial Cell Biol 4: 76–86.
10. Macartney KK, Baumgart DC, Carding SR, Brubaker JO, Offit PA (2000) Primary murine small intestinal epithelial cells, maintained in long-term culture, are susceptible to rotavirus infection. J Virol 74: 5597–5603.
11. Whitehead RH, Demmler K, Rockman SP, Watson NK (1999) Clonogenic growth of epithelial cells from normal colonic mucosa from both mice and humans. Gastroenterology 117: 858–865.
12. Booth C, Evans GS, Potten CS (1995) Growth factor regulation of proliferation in primary cultures of small intestinal epithelium. In Vitro Cell Dev Biol Anim 31: 234–243.
13. Evans GS, Flint N, Somers AS, Eyden B, Potten CS (1992) The development of a method for the preparation of rat intestinal epithelial cell primary cultures. J Cell Sci 101 (Pt 1): 219–231.
14. Kaeffer B, Benard C, Blottiere HM, Cherbut C (1997) Treatment of rat proximal and distal colonic cells with sodium orthovanadate enhances their adhesion and survival in primary culture. Cell Biol Int 21: 303–314.
15. Kaeffer B, Briollais I (1998) Primary culture of colonocytes in rotating bioreactor. In Vitro Cell Dev Biol Anim 34: 622–625.
16. Vidrich A, Ravindranath R, Farsi K, Targan S (1988) A method for the rapid establishment of normal adult mammalian colonic epithelial cell cultures. In Vitro Cell Dev Biol 24: 188–194.
17. Birkner S, Weber S, Dohle A, Schmahl G, Follmann W (2004) Growth and characterisation of primary bovine colon epithelial cells in vitro. Altern Lab Anim 32: 555–571.
18. Follmann W, Weber S, Birkner S (2000) Primary cell cultures of bovine colon epithelium: isolation and cell culture of colonocytes. Toxicol In Vitro 14: 435–445.
19. Hoey DE, Sharp L, Currie C, Lingwood CA, Gally DL, et al. (2003) Verotoxin 1 binding to intestinal crypt epithelial cells results in localization to lysosomes and abrogation of toxicity. Cell Microbiol 5: 85–97.
20. Rusu D, Loret S, Peulen O, Mainil J, Dandrifosse G (2005) Immunochemical, biomolecular and biochemical characterization of bovine epithelial intestinal primocultures. BMC Cell Biol 6: 42.
21. Kaushik RS, Begg AA, Wilson HL, Aich P, Abrahamsen MS, et al. (2008) Establishment of fetal bovine intestinal epithelial cell cultures susceptible to bovine rotavirus infection. J Virol Methods 148: 182–196.
22. Lu S, Yao Y, Cheng X, Mitchell S, Leng S, et al. (2006) Overexpression of apolipoprotein A-IV enhances lipid secretion in IPEC-1 cells by increasing chylomicron size. J Biol Chem 281: 3473–3483.
23. Berschneider H (1989) Development of normal cultured small intestinal epithelial cell lines which transport Na and Cl. Gastroenterology 96: A41.
24. Miyazawa K, Hondo T, Kanaya T, Tanaka S, Takakura I, et al. (2010) Characterization of newly established bovine intestinal epithelial cell line. Histochem Cell Biol 133: 125–134.
25. Counter CM, Gupta J, Harley CB, Leber B, Bacchetti S (1995) Telomerase activity in normal leukocytes and in hematologic malignancies. Blood 85: 2315–2320.
26. Tavelin S, Milovic V, Ocklind G, Olsson S, Artursson P (1999) A conditionally immortalized epithelial cell line for studies of intestinal drug transport. J Pharmacol Exp Ther 290: 1212–1221.
27. Bodnar AG, Ouellette M, Frolkis M, Holt SE, Chiu CP, et al. (1998) Extension of life-span by introduction of telomerase into normal human cells. Science 279: 349–352.
28. Hong HX, Zhang YM, Xu H, Su ZY, Sun P (2007) Immortalization of swine umbilical vein endothelial cells with human telomerase reverse transcriptase. Mol Cells 24: 358–363.
29. Su F, Liu X, Liu G, Yu Y, Wang Y, et al. (2013) Establishment and Evaluation of a Stable Cattle Type II Alveolar Epithelial Cell Line. PLoS One 8: e76036.
30. Golaz JL, Vonlaufen N, Hemphill A, Burgener IA (2007) Establishment and characterization of a primary canine duodenal epithelial cell culture. In Vitro Cell Dev Biol Anim 43: 176–185.
31. Lee CA, Falkow S (1990) The ability of Salmonella to enter mammalian cells is affected by bacterial growth state. Proc Natl Acad Sci U S A 87: 4304–4308.
32. Skjolaas KA, Burkey TE, Dritz SS, Minton JE (2006) Effects of Salmonella enterica serovars Typhimurium (ST) and Choleraesuis (SC) on chemokine and cytokine expression in swine ileum and jejunal epithelial cells. Vet Immunol Immunopathol 111: 199–209.
33. Arce C, Ramirez-Boo M, Lucena C, Garrido JJ (2010) Innate immune activation of swine intestinal epithelial cell lines (IPEC-J2 and IPI-2I) in response to LPS from Salmonella typhimurium. Comp Immunol Microbiol Infect Dis 33: 161–174.
34. Jourdan N, Brunet JP, Sapin C, Blais A, Cotte-Laffitte J, et al. (1998) Rotavirus infection reduces sucrase-isomaltase expression in human intestinal epithelial cells by perturbing protein targeting and organization of microvillar cytoskeleton. J Virol 72: 7228–7236.
35. Redmer T, Diecke S, Grigoryan T, Quiroga-Negreira A, Birchmeier W, et al. (2011) E-cadherin is crucial for embryonic stem cell pluripotency and can replace OCT4 during somatic cell reprogramming. EMBO Rep 12: 720–726.
36. Schwitalla S, Fingerle AA, Cammareri P, Nebelsiek T, Goktuna SI, et al. (2013) Intestinal tumorigenesis initiated by dedifferentiation and acquisition of stem-cell-like properties. Cell 152: 25–38.
37. Parthasarathy G, Mansfield LS (2005) Trichuris suis excretory secretory products (ESP) elicit interleukin-6 (IL-6) and IL-10 secretion from intestinal epithelial cells (IPEC-1). Vet Parasitol 131: 317–324.
38. Parthasarathy G, Mansfield LS (2009) Recombinant interleukin-4 enhances Campylobacter jejuni invasion of intestinal pig epithelial cells (IPEC-1). Microb Pathog 47: 38–46.
39. Geens MM, Niewold TA (2011) Optimizing culture conditions of a porcine epithelial cell line IPEC-J2 through a histological and physiological characterization. Cytotechnology 63: 415–423.
40. Simon-Assmann P, Turck N, Sidhoum-Jenny M, Gradwohl G, Kedinger M (2007) In vitro models of intestinal epithelial cell differentiation. Cell Biol Toxicol 23: 241–256.

41. Vandenbroucke V, Croubels S, Martel A, Verbrugghe E, Goossens J, et al. (2011) The mycotoxin deoxynivalenol potentiates intestinal inflammation by Salmonella typhimurium in porcine ileal loops. PLoS One 6: e23871.

42. Verbrugghe E, Vandenbroucke V, Dhaenens M, Shearer N, Goossens J, et al. (2012) T-2 toxin induced Salmonella Typhimurium intoxication results in decreased Salmonella numbers in the cecum contents of pigs, despite marked effects on Salmonella-host cell interactions. Vet Res 43: 22.

43. Costello CM, Hongpeng J, Shaffiey S, Yu J, Jain NK, et al. (2014) Synthetic small intestinal scaffolds for improved studies of intestinal differentiation. Biotechnol Bioeng.

44. Hayflick L, Moorhead PS (1961) The serial cultivation of human diploid cell strains. Exp Cell Res 25: 585–621.

45. Ornish D, Lin J, Daubenmier J, Weidner G, Epel E, et al. (2008) Increased telomerase activity and comprehensive lifestyle changes: a pilot study. Lancet Oncol 9: 1048–1057.

46. Blackburn EH (2000) Telomere states and cell fates. Nature 408: 53–56.

47. Baten A, Sakamoto K, Shamsuddin AM (1992) Long-term culture of normal human colonic epithelial cells in vitro. FASEB J 6: 2726–2734.

48. Burkey TE, Skjolaas KA, Dritz SS, Minton JE (2009) Expression of porcine Toll-like receptor 2, 4 and 9 gene transcripts in the presence of lipopolysaccharide and Salmonella enterica serovars Typhimurium and Choleraesuis. Vet Immunol Immunopathol 130: 96–101.

49. Collado-Romero M, Arce C, Ramirez-Boo M, Carvajal A, Garrido JJ (2010) Quantitative analysis of the immune response upon Salmonella typhimurium infection along the porcine intestinal gut. Vet Res 41: 23.

50. McCormick BA, Colgan SP, Delp-Archer C, Miller SI, Madara JL (1993) Salmonella typhimurium attachment to human intestinal epithelial monolayers: transcellular signalling to subepithelial neutrophils. J Cell Biol 123: 895–907.

51. Geens MM, Niewold TA (2010) Preliminary Characterization of the Transcriptional Response of the Porcine Intestinal Cell Line IPEC-J2 to Enterotoxigenic Escherichia coli, Escherichia coli, and E. coli Lipopolysaccharide. Comp Funct Genomics 2010: 469583.

52. Cho WS, Chae C (2003) Expression of inflammatory cytokines (TNF-alpha, IL-1, IL-6 and IL-8) in colon of pigs naturally infected with Salmonella typhimurium and S. choleraesuis. J Vet Med A Physiol Pathol Clin Med 50: 484–487.

53. Watson PR, Paulin SM, Jones PW, Wallis TS (2000) Interaction of Salmonella serotypes with porcine macrophages in vitro does not correlate with virulence. Microbiology 146 (Pt 7): 1639–1649.

54. Hyland KA, Kohrt L, Vulchanova L, Murtaugh MP (2006) Mucosal innate immune response to intragastric infection by Salmonella enterica serovar Choleraesuis. Mol Immunol 43: 1890–1899.

55. Chassin C, Hempel C, Stockinger S, Dupont A, Kubler JF, et al. (2012) MicroRNA-146a-mediated downregulation of IRAK1 protects mouse and human small intestine against ischemia/reperfusion injury. EMBO Mol Med 4: 1308–1319.

56. Moue M, Tohno M, Shimazu T, Kido T, Aso H, et al. (2008) Toll-like receptor 4 and cytokine expression involved in functional immune response in an originally established porcine intestinal epitheliocyte cell line. Biochim Biophys Acta 1780: 134–144.

11

Neurospora crassa Female Development Requires the PACC and Other Signal Transduction Pathways, Transcription Factors, Chromatin Remodeling, Cell-To-Cell Fusion, and Autophagy

Jennifer L. Chinnici, Ci Fu, Lauren M. Caccamise, Jason W. Arnold, Stephen J. Free*

Department of Biological Sciences, SUNY University at Buffalo, Buffalo, New York, United States of America

Abstract

Using a screening protocol we have identified 68 genes that are required for female development in the filamentous fungus *Neurospora crassa*. We find that we can divide these genes into five general groups: 1) Genes encoding components of the PACC signal transduction pathway, 2) Other signal transduction pathway genes, including genes from the three *N. crassa* MAP kinase pathways, 3) Transcriptional factor genes, 4) Autophagy genes, and 5) Other miscellaneous genes. Complementation and RIP studies verified that these genes are needed for the formation of the female mating structure, the protoperithecium, and for the maturation of a fertilized protoperithecium into a perithecium. Perithecia grafting experiments demonstrate that the autophagy genes and the cell-to-cell fusion genes (the MAK-1 and MAK-2 pathway genes) are needed for the mobilization and movement of nutrients from an established vegetative hyphal network into the developing protoperithecium. Deletion mutants for the PACC pathway genes *palA*, *palB*, *palC*, *palF*, *palH*, and *pacC* were found to be defective in two aspects of female development. First, they were unable to initiate female development on synthetic crossing medium. However, they could form protoperithecia when grown on cellophane, on corn meal agar, or in response to the presence of nearby perithecia. Second, fertilized perithecia from PACC pathway mutants were unable to produce asci and complete female development. Protein localization experiments with a GFP-tagged PALA construct showed that PALA was localized in a peripheral punctate pattern, consistent with a signaling center associated with the ESCRT complex. The *N. crassa* PACC signal transduction pathway appears to be similar to the PacC/Rim101 pathway previously characterized in *Aspergillus nidulans* and *Saccharomyces cerevisiae*. In *N. crassa* the pathway plays a key role in regulating female development.

Editor: Stefanie Pöggeler, Georg-August-University of Göttingen Institute of Microbiology & Genetics, Germany

Funding: Funding for this study was provided by National Institutes of Health Grant R01 GM078589 and by the University at Buffalo Foundation. The authors acknowledge the Neurospora Genome Project for the creation of the deletion mutants and the Fungal Genetics Stock Center for providing the mutants for their research efforts. Funding for the Neurospora Genome Project was provided by Grant #P01 GM068087 from the National Institutes of Health. The funders had no role in study design, data collection and analysis, decision to publish, or preparation of the manuscript.

Competing Interests: The authors have declared that no competing interests exist.

* Email: free@buffalo.edu

Introduction

From the viewpoint of a developmental biologist, the study of perithecial development in *Neurospora crassa*, and in the related ascomycetes *Sordaria macrospora*, *Podospora anserina*, and *Gibberella zeae* (anamorph *Fusarium graminearum*) has several advantages [1]. The morphological stages of perithecial development have been carefully cataloged by Lord and Read [2]. The perithecia is a reasonably simple structure with a limited number of different cell types [3]. Large numbers of perithecia can be generated and studied as they go through six days of development in a somewhat synchronous manner [4,5]. The genomes of these ascomycetes have been sequenced [6,7]. The expression pattern of the entire *N. crassa* genome has been examined by RNAseq during perithecia maturation [4]. Similarly, genome-wide expression patterns have been extensively analyzed during the develop-ment of *S. macrospora* and *Gibberella zeae* [5,8,9]. The organisms are haploid except during mating when a dikaryotic ascogenous tissue is generated and a diploid cell, which immediately under-goes meiosis, is formed. The haploid nature of the organisms facilitates the isolation and characterization of mutants affected in female development [1,10–12]. Most importantly, the creation of the *N. crassa* single gene deletion library provides a unique opportunity to carry out a comprehensive genetic analysis of female development in a filamentous fungus [6,13,14].

The morphological events that occur during female develop-ment have been well documented in *N. crassa*, *S. macrospora*, *P. anserina*, and *G. zeae* [1,2,11,15–21]. Female development can be initiated in *N. crassa* by nitrogen deprivation [22]. At the onset of female development a specialized coiled hyphal structure, called an ascogonium, is generated from a vegetative hypha. The ascogonium grows into a tightly woven spherically-shaped structure called a protoperithecium. The protoperithecium contains two types of cells, the ascogenous hyphae in the middle of the structure, and an outer layer of protective hyphae called the peridium. The ascogenous hyphae are generated from the ascogonium, but some of the hyphal elements of the peridium are thought to be generated by hyphae from the vegetative tissue that join with the ascogonium hyphae to create the protective outer layer. A long hypha, termed a trichogyne, is generated from the ascogenous hyphae and grows out of the protoperithecium. The trichogyne is attracted to a pheromone released by conidia and hyphae of the opposite mating type, and is able to undergo a cell fusion event with a cell of the opposite mating type to generate a dikaryon (a cell with two different types of nuclei) [23]. The pheromones and pheromone receptors that function in the chemotrophic growth of the trichogyne to a cell of the opposite mating type have been identified and characterized [24–27]. The movement of the male nucleus from the trichogyne into the ascogenous tissue, and the nuclear events within the ascogenous hyphae have been carefully characterized in *N. crassa* by Raju [23]. The male nucleus travels down the trichogyne into the ascogenous hyphae, where it undergoes several rounds of nuclear division. Hyphae having the shape of a "shepherd's crook" (an upside down J) called crosiers, are generated within the ascogenous tissue. A single ascus cell containing a female and a male nucleus is generated at the top of the "crook" within each of these crosiers. Nuclear fusion occurs in the ascus cell, followed immediately by a meiotic division and a single mitotic division to generate a linear array of four pairs of nuclei. These then mature into eight ascospores. While the male nuclei are being replicated and ascospore formation occurs, the other cell types in the perithecium undergo development. The fertilized perithecium dramatically increases in size and becomes flask shaped. The outer layers of the peridium become highly melanized [10]. Specialized cells called paraphyses, which are thought to help support the development of the ascospores, are generated from the inner layer of the peridium and grow between the developing asci [2]. As perithecium development nears completion, an ostiolar pore (opening) is generated at the top of the flask-shaped perithecium. At the end of female development the mature ascospores are ejected through the ostiolar pore.

During the 1970 s, studies demonstrated that *N. crassa* female development was amenable to genetic analysis. Johnson [10] isolated a large number of mutants affected in female development and ordered these mutants based on the size of the developing protoperithecia. Vigfusson and Weiljer [28], Tan and Ho [29], and Mylyk and Thelkeld [30] also isolated female sterile mutants. Some of the mutations were mapped onto the *N. crassa* genetic map, while others were unmapped. Although these female developmental mutants were deposited in the Fungal Genetics Stock Center, only recently have any of the genes defined by these mutants been identified. Four of these "classical" female developmental genes, *ff-1* (female fertile-1), *fs-n* (female sterile-n), *ty-1/ste-50* (tyrosinaseless-1 or sterile-50), and *per-1* (perithecial-1) were recently identified by whole genome sequencing of the mutant strains [31]. The *ff-1* gene (NCU01543) encodes a LIM domain-containing protein with homology to Pat1p, a topoisomerase II-associated protein. The mutation in the *fs-n* strain was shown to be in the *so/ham-1* gene (NCU02794), a WW domain-containing protein that has been shown to be needed for cell-to-cell fusion as well as for female development [32]. The *per-1* mutation affects the melanization of the perithecia, and the mutant was initially identified by the presence of an unmelanized perithecia [33]. The *per-1* gene (NCU03584) encodes a polyketide synthase, and probably functions in a melanin biosynthetic

pathway. The *ty-1/ste-50* mutant has a complex phenotype. It produces short aerial hyphae (flat conidiation), is female infertile, and is "tyrosinaseless" [34]. The *ty-1/ste-50* gene (NCU00455) encodes a homolog of the *S. cerevisiae* Ste50 protein, a scaffold protein for the MAP kinase pathways.

In addition to a large number of *N. crassa* genes that have been identified as being needed for female development, many additional female development genes have been identified and characterized in *S. macrospora*, *P. anserina*, *Magnaporthe grisea*, and *Aspergillus nidulans* (see review by Pöggeler et al. [1]). These additional genes include transcription factors, signal transduction pathway proteins, and autophagy proteins, and provide a wealth of information on female development in filamentous fungi.

In this report, a morphological screening of the *N. crassa* single gene deletion library was used to identify the genes that are needed for the development of protoperithecia and their subsequent maturation into perithecia. In characterizing the mutants identified in this screening, we identified 68 genes that are needed for female development. One of the more interesting findings from the screening experiments was that the PACC signal transduction pathway is required for female development in *N. crassa*.

Results and Discussion

Isolation of protoperithecia-defective mutants and co-segregation experiments

To identify genes that are required for female development in *N. crassa*, a large-scale mutant screening experiment was carried out. As described in Materials and Methods, each of the 10,000 isolates in plates 1 to 119 of the *N. crassa* single gene deletion library was tested for the ability to produce protoperithecia and perithecia on a 3 ml slant of synthetic crossing medium (Figure 1). This screening was done in addition to our previous screening of the library to identify cell-to-cell fusion mutants, which were shown to be defective in protoperithecium development [35]. The purpose of the screening experiments described in this report was to identify other types of genes that are needed for female development. In these screening experiments we identified 649 mutants (representing 508 genes) that were either protoperithecia-defective or perithecia-defective. Co-segregation experiments were done for 443 of these genes. Co-segregation experiments were not done for most of the genes that encoded a known function not directly related to female development (genes important for mitochondrial ATP synthesis, ribosome functions, vesicular trafficking, etc.; see Table S1) and genes where the library contained two deletion isolates and only one of the isolates displayed the mutant phenotype. These co-segregation experiments showed that for 123 of the genes, the mutant phenotype co-segregated with the gene deletion, suggesting that the deletion might be responsible for the female-defective phenotype. Included in these genes were 31 genes that had been previously identified as being required for *N. crassa* female development (see Tables 1–3). This demonstrates that the screening and co-segregation experiments effectively identified genes that function in the process of female development. For 320 of the genes identified in the screening experiments, the co-segregation experiments showed that the female developmental defect did not co-segregate with the gene deletion, indicating that these mutants had additional mutations that were responsible for the mutant phenotype. Excel files showing the results of our screening, co-segregation, complementation, and RIP experiments can be accessed at our website   (http://www.biology.buffalo.edu/Faculty/Free/ KO_list_2014/KO_List.html).

Figure 1. Screening and complementation analysis. Wild type, a Δ*palF* isolate, and a Δ*palF* isolate that has been transformed with a wild type copy of the *palF* gene were grown on 3 ml slants of SCM agar medium for 10 days to allow them to form protoperithecia. The three panels show images of the hyphae and protoperithecia on the test tube glass adjacent to the agar slant. An abundance of protoperithecia are produced by the wild type isolate on the glass at the edge of the agar (first panel/A) while the Δ*palF* mutant is unable to produce protoperithecia (second panel/B). Transformation of the Δ*palF* mutant with a wild type copy of the *palF* gene restores the ability to generate protoperithecia (third panel/C). Arrows point to examples of protoperithecia.

Among the mutants we identified in the screening experiment were several mutants that grew very poorly and had mutations in mitochondrial proteins, in general transcription proteins, in protein translation functions, in vesicular trafficking, and other "general cellular health" functions. Co-segregation experiments on a few of these mutants showed that the "general health" mutation did co-segregate with the female-defective phenotype. A list of these genes is found in Table S1, and, for the most part, these genes were not further characterized in our study. These mutants demonstrate that the hyphae must be "healthy" in order to participate in female development. We also noted that our experiments identified 18 genes that function in chromatin assembly and remodeling. The co-segregation experiments, along with some complementation experiments, clearly demonstrated that chromatin assembly and remodeling are required for female development. Others have previously shown chromatin assembly and remodeling mutants are defective in female development [36–39]. We have listed these chromatin organization genes in Table S2. Although a study of how chromatin remodeling is involved in directing female development is a very interesting topic, we did not focus on these mutants in our current study.

Complementation and RIP experiments

Complementation or RIP experiments were carried out on a majority of the putative female developmental genes defined by the co-segregation experiments. However, for some of the genes, pre-existing definitive information showing that the gene was required for *N. crassa* female development had been previously published, and complementation experiments were not carried out on these mutants. In these situations, the publication citation showing that the gene was needed for the protoperithecium development or maturation is provided in Tables 2 through 5. As described in Materials and Methods, the complementation experiments used wild type copies of the putative female developmental genes to transform the deletion mutants. The ability of the wild type gene to restore a wild type phenotype to a deletion mutant was taken as definite proof that the gene was needed for female development.

In cases where the complementation experiment was difficult because the mutant did not produce conidia, the cell type used in the transformation experiments, RIP experiments were used to generate additional mutated copies of the gene. In these cases, one or more of the RIP alleles were PCR amplified and sequenced to verify that the mutant allele(s) had multiple RIP mutations. As a result of our co-segregation, complementation, and RIP experiments, we identified 68 genes that are needed for female development. While many of these genes had been previously reported as being needed for female development, our studies identified 32 genes that had not been previously identified as being required for female development in *N. crassa* (denoted as being defined by this report in Tables 1 through 5). The results of our screening, cosegregation, and complementation experiments demonstrate the value of a careful screening of the *N. crassa* single gene deletion library in characterizing the biology of the model filamentous fungus. Our research also highlights the importance of using co-segregation and complementation exper-

Table 1. Genes from the PacC pathway are required for female development.

Genes	NCU#	Co-segregation	Complementation	Reference information
pacC	00090	Yes	Yes	This report
palA	05876	Yes	Yes	This report
palB	00317	Yes	Yes	This report
palC	03316	Yes	Yes	This report
palH	00007	Yes	Yes	This report
palF	03021	Yes	Yes	This report

A notation of "This report" in the reference information column indicates that the gene was identified as being needed for *N. crassa* development by our experiments.

Table 2. Signal transduction pathway genes are required for female development.

Gene name	NCU #	Co-segregation	Complementation	Reference information
		MAP kinases		
mik-1	02234*	Yes	PP	Maerz et al. [110]; Park et al. [105]
mek-1	06419*	Yes	PP	Maerz et al. [110]; Park et al. [105]
mak-1	09842*	Yes	PP	Maerz et al. [110]; Park et al. [105]
nrc-1	06182*	Yes	PP	Kothe and Free [64]; Maerz et al. [110];Pandey et al. [63]
mek-2	04612*	Yes	PP	Maerz et al. [110]; Pandey et al. [63]
mak-2	02393*	Yes	PP	Maerz et al. [110]; Pandey et al. [63]; Li et al. [111]
os-2	07024	Yes	PP	Lichius et al. [45]
os-4	03071	Yes	PP	Lichius et al. [45]
os-5	00587	Yes	PP	Lichius et al. [45]
		Genes encoding MAPK pathway components		
ste-50 or ty-1	00455	Yes	PP	McCluskey et al. [31]
hym-1	03576	Yes	PP	Dettmann et al. [112]
rac-1	02160*	Yes	PP	Araujo-Palomares et al. [113]; Fu et al. [35]
rho-1	01484	NA	PP	Richthammer et al. [114]
rgf-1	00668	NA(het)	PP	Richthammer et al. [114]
lrg-1	02689	Yes	PP	Vogt and Seiler [115]
pp-2A	06563*	Yes	PP	Fu et al. [35]
PP2A activator	03269*	Yes	–	This report
pp-1	00340*	NA(het)	PP	Leeder et al. [91]; Li et al. [111]
so, fs-n or ham-1	02794*	Yes	PP	Fleissner et al. [32]; Engh et al. [116]
ham-2	03727*	Yes	PP	Xiang et al. [42]; Bloemendal et al. [75]
ham-3	08741*	Yes	PP	Dettmann et al. [43]; Simonin et al. [117]; Bloemendal et al. [118]
ham-4	00528*	Yes	PP	Dettmann et al. [43]; Simonin et al. [117]
mob-3	07674*	Yes	PP	Dettmann et al. [43]; Maerz et al. [41]
ham-5	01789*	Yes	PP	Aldabbous et al. [86]
ham-6	02767*	Yes	PP	Fu et al. [35]; Nowrousian [119]
ham-7	00881*	Yes	PP	Fu et al. [35]; Maddi et al. [120]
ham-8	02811*	Yes	PP	Fu et al. [35]
ham-9	07389*	Yes	PP	Fu et al. [35]
amph-1	01069*	Yes	PP	Fu et al. [35]
whi-2	10518*	Yes	Yes	This report
prs-1	08380*	Yes	Yes	This report
rrg-1	01895	–	PP	Jones et al. [121]
		Nox pathway genes		
nox-1	02110*	Yes	PP	Cano-Dominguez et al. [69]
nor-1	07850*	NA(het)	PP	Cano-Dominguez et al. [69]
		Pheromone signaling genes		
mfa-1	16992	NA	PP	Kim and Borkovich [24]
pre-1	00138	–	PP	Kim and Borkovich [24]; Poggeler et al. [122]
ccg-4	02500	–	PP	Kim and Borkovich [24]
pre-2	05758	–	PP	Kim and Borkovich [24]; Poggeler et al. [122]
		Septation initiation network		
cdc-7	01335	Yes	PP	Park et al. [48]; Heilig et al. [76]
sid-1	04096	–	PP	Heilig et al. [76]
dbf-2	09071	–	PP	Heilig et al. [76]
		Calcium signaling		

Table 2. Cont.

Gene name	NCU #	Co-segregation	Complementation	Reference information
		MAP kinases		
cnb-1	03833	NA	PP	Kothe and Free [64]
camk-1	09123	–	–	Park et al. [48]
ham-10	02833*	Yes	PP	Fu et al. [35]
		Other signaling pathways		
gna-1	06493	NA	PP	Ivey et al. [123]; Kamerewerd et al. [124]
gnb-1	00440	Yes	PP	Yang et al. [125]; Kamerewerd et al. [124]
gng-1	00041	NA	PP	Krystofova et al. [126]; Kamerewerd et al. [124]
cpc-2	05810	NA(het)	PP	Müller et al. [127]
fl	04990	Yes	PP	McCluskey et al. [31]
stk-22	03523	Yes	Yes	Park et al. [48]; This report
stk-16	00914	Yes	–	Park et al. [48]
div-4	04426	Yes	Yes	Park et al. [48]; This report
stk-47	06685	Yes	–	Park et al. [48]

PP – Previously published data demonstrated that the gene was needed for female development.
NA – a deletion strain is not available in the single gene deletion library.
NA(het) – the deletion strain in the single gene deletion library is a heterokaryon and a homokaryon isolate was not available during the screening experiments.
RIP – a RIP experiment was used to verify that the gene is required for female development.
An * by the NCU number indicates that the gene is needed for CAT (conidia anastomosis tube) formation (a cell fusion phenotype) and is likely to be a component of either the MAK-1 or MAK-2 signal pathway.
A notation of "This report" in the reference information column indicates that the gene was either newly identified or verified by co-segregation and complementation analysis as being needed for N. crassa development by our experiments.

iments in verifying that the deletions identified in screening the deletion library give rise to the observed mutant phenotypes. As demonstrated by our study, the presence of secondary mutations in some isolates in the library does not detract from the value and importance of the single gene deletion library in analyzing *N. crassa* gene functions.

Cell-to-cell fusion assays

To further characterize our mutants, the ability of the mutants to participate in cell-to-cell fusion was assessed with the CAT fusion assay [40]. We and others have previously shown that cell-to-cell fusion is needed for female development [32,35,41,42]. The genes required for cell-to-cell fusion are designated with an asterisk in Table 2. Interestingly, we found that all of the cell-to-cell fusion genes that we identified encode proteins that are either confirmed or likely participants in one of three closely related signaling pathways: 1) the STRIPAK signaling complex, which is required for the movement of MAK-1 into the nucleus in a MAK-2-dependent manner; 2) the MAK-1 cell wall integrity signal transduction pathway; and 3) the MAK-2 hyphal growth signal transduction pathway used to direct the growth of fusion hyphae towards each other [35,43,44].

Table 3. Transcription factors needed for female development.

Gene	NCU#	Co-segregation	Complementation	Reference information
pacC	00090	Yes	Yes	This report
asm-1	01414	NA (het)	PP	Aramayo et al. [85]
rco-1	06205	Yes	PP	Yamashiro et al. [128]; Aldabbous et al. [86]
rcm-1	06842	NA (het)	PP	Kim and Lee [129]; Aldabbous et al. [86]
ada-1	00499	Yes	Yes	Colot et al. [84]; This report
adv-1	07392	Yes	Yes	Colot et al. [84]; Fu et al. [35]
fsd-1	09915	Yes	PP	Hutchinson and Glass [90]
fmf-1	09387	–	PP	Iyer et al. [89]
mcm-1	07430	NA	PP	Nolting and Poeggeler [93,94]
pp-1	00340	NA (het)	PP	Leeder et al. [91]

PP – Previously published data demonstrated that the gene was needed for female development.
NA – a deletion strain is not available in the single gene deletion library.
NA (het) – the deletion strain in the single gene deletion library is a heterokaryon and a homokaryon isolate was not available during the screening experiments.

Ascogonia formation

Female development begins with the formation of the ascogonium, a specialized coiled hyphae structure. To determine whether the mutants were affected in ascogonia formation, mutant conidia were used to inoculate either synthetic crossing medium with sucrose or cellophane filters placed on synthetic crossing medium (without an additional added carbon/energy source), and the formation of ascogonia was assessed under a compound microscope. The PACC pathway mutants (palA, palB, palC, palF, palH, and pacC) had a rather interesting ascogonia formation phenotype. We found that they were unable to produce ascogonia on the synthetic crossing medium with sucrose but were able to form ascogonia on cellophane.

In the ascogonia formation on cellophane experiments, we found that 17 of the mutants were affected in the formation of ascogonia. These included the mutants for the OS MAP kinase pathway gene os-2, os-4, and os-5 genes, which have been previously shown to be defective in ascogonia formation by Lichius et al. [45]. We also found that deletion mutants for the transcription factor rco-1 (NCU06205) and 3 calcium signaling genes, cnb-1 (NCU03833), camk-1 (NCU09123), and ham-10 (NCU02833) were affected in ascogonia formation. This suggests that Ca++ signaling may be required for the initiation of female development in N. crassa. Previous work on each of these genes has shown that they are needed for female development [46–48]. The other 10 genes affecting ascogonia formation were rac-1 (NCU02160), div-4 (NCU04426), stk-16 (NCU00914), stk-22 (NCU03523), stk-47 (NCU006685), fem-1 (NCU03589), fem-4 (NCU06243), fem-5 (NCU02073), fem-6 (NCU09052), and fem-7 (NCU03985). Most of these mutants produced an abundance of conidia instead of making ascogonia on the cellophane filter. These mutants also shared a second phenotypic characteristic, a hyper-production of conidia when grown on slants containing synthetic crossing medium with sucrose. We hypothesize that these genes may function in allowing the fungus to choose between two alternative developmental pathways, a sexual pathway leading to female development and the asexual pathway leading to conidiation. These genes may well define a signal transduction pathway needed for initiating female development.

Perithecium grafting experiments

Female development on synthetic crossing medium occurs in response to nitrogen limitation, a condition in which the fungus may be restricted in the synthesis of new amino acids, and need to rely on pre-existing amino acids for the synthesis of new proteins. We developed a perithecia grafting assay (see Materials and Methods) that allowed us to ask whether the vegetative hyphae from the mutants we identified in our screening experiments were able to support the development of newly fertilized wild type perithecia (Figure 2). The results of the perithecia grafting experiments were quite instructive. We found that deletion mutants for virtually all of the genes known to be needed for autophagy and deletion mutants for all of the genes known to be components of the MAK-1 and MAK-2 signal transduction pathways needed for cell-to-cell fusion were defective in supporting the development of wild type perithecia in these grafting experiments (Figure 2). This demonstrates that both autophagy and cell-to-cell fusion are needed within the vegetative hyphal network. We conclude that the release of amino acids, and perhaps other nutrients, from the vegetative hyphal network and the transferring of these nutrients into the developing perithecium are needed to support female development.

In doing the perithecia grafting experiments, we found that the mating types of the graft and host vegetative network had to be the same in order for the host to be able to support graft development, suggesting that the vegetative incompatibility system is operating between the host and graft tissues. In addition to providing an assessment of whether a mutant "host" supported a wild type "graft", we noted a second unexpected phenomenon in the grafting experiments. When testing the PACC pathway mutants (palA, palB, palC, palF, palH, and pacC) as hosts, we found that the presence of wild type perithecia grafts of either mating type were able to induce protoperithecia formation in the host hyphal network surrounding the graft (Figure 2). We conclude that the PACC pathway operates to allow the fungus to choose the female developmental pathway in response to environmental/nitrogen limitation cues present in synthetic crossing medium. However, the fungus apparently has other pathways for inducing female development, and one of these pathways is responsive to the presence of nearby perithecia.

Functional grouping of the genes required for female development

An examination of the genes defined by the deletion mutants showed that we could assign the genes into five different categories. These were: 1) Genes encoding proteins which have been found to function in the PACC signal transduction pathway; 2) Genes encoding proteins that are likely to function in other signal transduction pathways, including the three N. crassa MAP kinase pathways, the STRIPAK pathway, the pheromone-responsive pathway, the NOX pathway, the heterotrimeric G protein signaling pathway, the septation initiation network (SIN), a Ca++ signaling pathway, and perhaps other signaling pathways; 3) Genes encoding probable transcription factors; 4) Genes that are required for autophagy; and 5) a few miscellaneous genes that don't fit into the other categories. These genes are listed in Tables 1 through 5 respectively, along with details about the mutant phenotypes. To provide a somewhat comprehensive gene list, we have included a number of additional genes that have been shown by others to be involved in female development. In most of these cases, the deletion mutants for these genes weren't identified in our screening experiments because they were not available in the library or they were found as heterokaryons in the library. These "added" genes include mfa-1 (NCU06992), pre-1 (NCU00138), ccg-4 (NCU02500), pre-2 (NCU05758), asm-1 (NCU01414), fmf-1 (NCU09387), pp-1 (NCU00340), rho-1 (NCU01484), nor-1 (NCU07850), gna-1 (NCU06493), gng-1 (NCU00041), cnb-1 (NCU03833), cpc-2 (NCU05810), mcm-1 (NCU07430), per-1 (NCU03584), rgf-1 (NCU00668), rrg-1 (NCU01895), rcm-1 (NCU06842) and tyrosinase (NCU00776). Citations for the publications that demonstrated that these genes play roles in female development are included in Tables 2, 3, and 5. We will examine each category of mutants one at a time and describe their phenotypes and how they might function in supporting or directing female development.

Category #1: The PACC signal transduction pathway

The PacC/rim101 signal transduction pathway has been well characterized in A. nidulans and S. cerevisiae, where the pathway is regulated by the pH of the medium [49–53]. A representation of the pathway with the components we have identified in N. crassa is shown in Figure 3. In S. cerevisiae and A. nidulans, the pathway is activated at neutral-to-alkaline pHs, and the PACC protein functions as a transcription factor to activate expression of a number of genes needed for growth under neutral-to-alkaline pH conditions. The PALH protein, a seven transmembrane protein found in the plasma membrane, functions as the receptor or pH sensor for the pathway. PALH is found in a complex with the

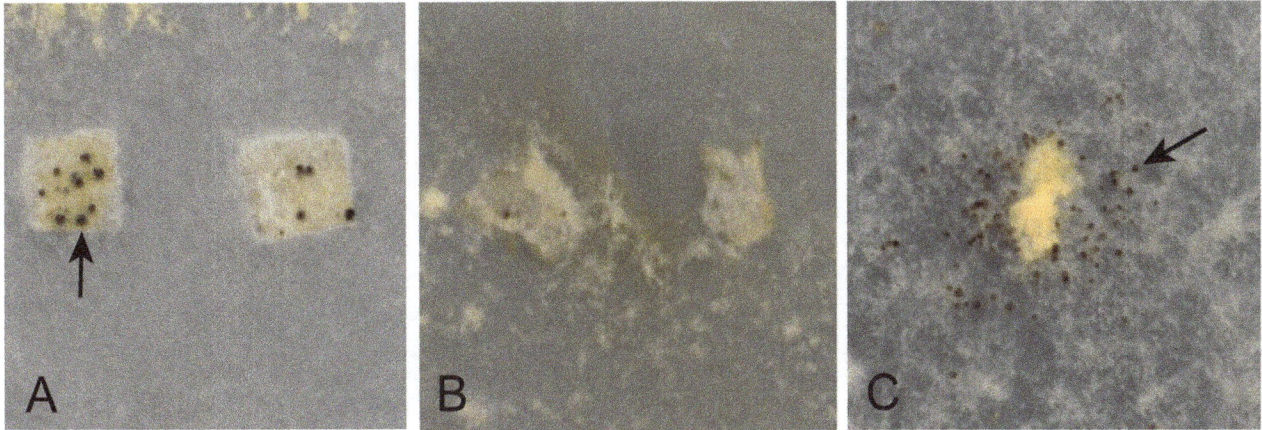

Figure 2. Perithecia grafting experiments. Small pieces of cellophane containing "graft" fertilized wild type protoperithecia were placed on mutant "host" vegetative hyphal networks grown on synthetic crossing medium. The "host" shown are: 1) Δ*fmf-1* (left panel/A) which shows a host supporting the development of the graft perithecia. 2) Δ*ada-1* (middle panel/B) which shows a host not supporting the development of the graft perithecia. 3) Δ*palA* (right panel/C) which shows the graft inducing protoperithecia in the Δ*palA* host. Arrows point to examples of perithecia on the cellophane (left panel/A) and protoperithecia induced in the host vegetative hyphal network (right panel/C).

PALF protein, which has homology to arrestins. Under activating pH conditions, the PALF protein is phosphorylated and ubiquitinated. These modifications to PALF lead to the endocytosis of PALH protein, which joins with the PALA, PALB, PALC, and PACC proteins to create a multimeric signaling center within the ESCRT (endosomal sorting complexes required for transport) complex. Within the signaling center, PALB functions as a processing protease and cleaves an inhibitory C terminal domain from the PACC transcription factor. The activated PACC is released from the ESCRT complex and enters the nucleus. Since the PACC pathway has been characterized as a pH-dependent pathway, finding that it was required for female development in *N. crassa*, which occurs during nitrogen deprivation, was unexpected.

We found that the Δ*palA*, Δ*palB*, Δ*palC*, Δ*palF*, Δ*palH*, and Δ*pacC* mutants are all defective in protoperithecium development. We were unable to identify any ascogonia, the initial stage of protoperithecial development, when examining mutant hyphae growing on synthetic crossing medium, while ascogonia and protoperithecia were observed in wild type hyphae. Complementation analysis for these PACC pathway mutants verified that each of the genes was needed for protoperithecium formation (Figure 1, Table 1). The PacC pathway has been well-defined in *A. nidulans*, *S. cerevisiae*, and *C. albicans* as being required to regulate genes involved in growth in neutral-to-alkaline media [49–52,54–60]. To the best of our knowledge, this is the first report demonstrating that the pathway is required for protoperithecial development. However, the PacC mutant of *Sclerotinia sclerotium* has been found to be defective in the formation of sclerotia, melanized structures that can remain dormant for many years and give rise to apothecia, the female structure for this species [61].

As noted in the ascogonia formation and perithecia graft experiments, the PacC pathway deletion mutants were unable to generate ascogonia and protoperithecia on synthetic crossing medium, but could produce protoperithecia in response to a signal from nearby fertilized perithecia from either mating type (Figure 2) or when grown of cellophane. We also tested for protoperithecia production on corn meal agar, a medium containing complex carbohydrates which can be used for *N. crassa* matings, and found that the PACC mutants produced protoperithecia on this medium. Our data shows that the PACC

pathway is needed for the induction of female development (ascogonial development) only when female development is induced by growth on synthetic crossing medium, a medium that is generally ascribed as inducing female development in response to nitrogen deprivation, but that other environmental cues can induce female development in PACC pathway mutants. It is interesting to note that the environmental conditions inducing female development differ for different fungal species [1], which may reflect differences in their life cycles. We hypothesize that PACC induction of ascogonia production may be restricted to those fungi that induce female development in response to nitrogen limitation.

To determine whether the protoperithecia produced by PacC pathway deletion mutants were fully functional, we fertilized protoperithecia that had been induced by the presence of nearby perithecia or growth on corn meal agar. Upon fertilization, the mutant perithecia increased in size and began to melanize. However, the perithecia did not complete development and eject ascospores. Several of these mutant perithecia were examined, and we found that the perithecia did not grow as large as wild type perithecia. They also did not become as melanized as wild type perithecia. Microscopic examination of squashed mutant perithecia shows that they do not generate asci (Figure 4). Thus, we find that the PACC pathway is used at two different stages of female development, first during the initial induction of protoperithecial development in response to environmental cues, and then later in the maturing perithecia during the development of asci from the ascogenous tissue.

To determine whether the removal of the C terminal inhibitory domain of PACC was sufficient to direct *N. crassa* cells into the female developmental pathway, we prepared a version of *pacC* in which a stop codon was inserted into the gene at amino acid 492. The truncated protein made from this construct would lack the predicted inhibitory C terminus domain, and would be predicted to give rise to a constitutively active form of PACC. Transformation of wild type isolates with the plasmid encoding this constitutively active PACC resulted in the abundant formation of protoperithecia, even in the Vogel's medium where female development is normally repressed. Transformants of the Δ*palA*, Δ*palB*, Δ*palC*, Δ*palF*, and Δ*palH* with plasmid encoding the

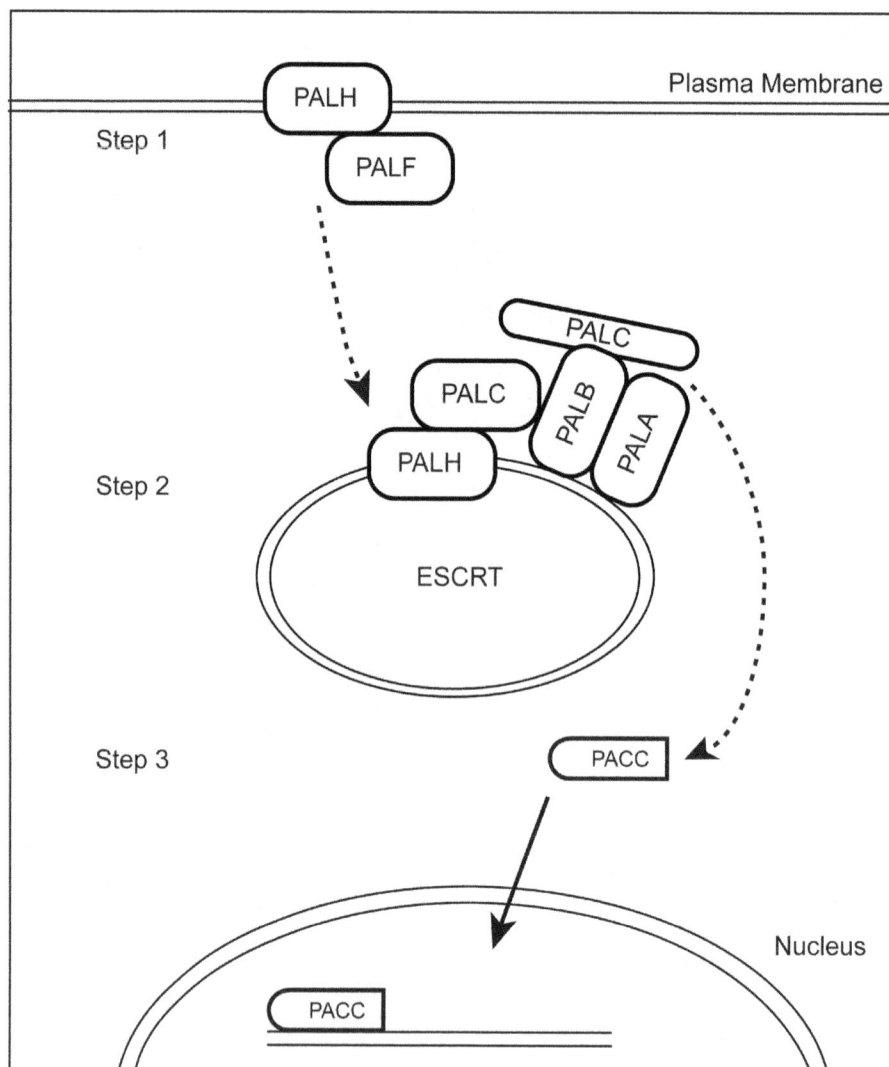

Figure 3. Schematic representation of the _N. crassa_ PACC pathway. The PACC signal transduction pathway elements found in _N. crassa_, and the model for how the pathway might function are depicted. The PALH and PALF proteins are thought to be found at the plasma membrane. PALH is a seven transmembrane receptor which is sensitive to environmental cues. PALF is an arrestin type protein that associates with PALH. PALF is phosphorylated and ubiquitinated in response to the environmental cues. These events lead to the endocytosis of the PALH/PALF complex. Following endocytosis, the PALH is directed into an ESCRT compartment, where it enters into a signaling complex containing PALA, PALB, PALC, and PACC. Within the signaling complex, PALB functions as a protease which cleaves PACC. This cleavage event removes a C-terminal inhibitory domain from the PACC transcription factor, and the processed PACC is released from the signaling complex. The activated PACC then enters the nucleus and directs transcriptional activity leading to the formation of the protoperithecium.

constitutive active form of PACC, also produced an abundance of protoperithecia (Figure 5). This demonstrates that PALA, PALB, PALC, PALF, and PALH are all upstream of PACC within the signaling pathway. We conclude that the activation of PACC is a major event in triggering female development in _N. crassa_, and that activation of PACC is sufficient to direct cells to undergo female development, even in the absence of the normal nitrogen limitation cue.

To examine whether the PACC pathway has the same characteristics in _N. crassa_ that have been previously identified in the _A. nidulans_ and _S. cerevisiae_ systems, we decided to examine the intracellular location of PACC pathway components in wild type and mutant isolates. Endogenous promoter-driven GFP-tagged versions of PALA, PALC, and PALH were prepared using the pMF272 vector [62]. The GFP-tagged versions of PALC

and PALH did not complement the deletion mutants and we were unable to detect a GFP signal, suggesting that the GFP-tagged proteins were rapidly degraded. However, the GFP-tagged PALA fully complemented the deletion mutation and gave a faint, but detectable intracellular signal in conidia and germlings. The faint signal was localized in a punctate pattern near the periphery of the conidia and germlings. We were unable to detect a signal from the GFP-tagged PALA within early developing protoperithecia (ascogenous coils), suggesting that the signal was weak. We also prepared a _ccg-1_ promoter-driven GFP-tagged PALA vector as described in Materials and Methods. Transformation with the _ccg-1_ promoter-driven GFP-tagged _palA_ provided for a stronger signal in wild type and _ΔpalA_ isolates (Figure 6). Examination of the GFP-tagged PALA in germinating conidia showed that the protein was localized in a punctate pattern near the cell surface. The data

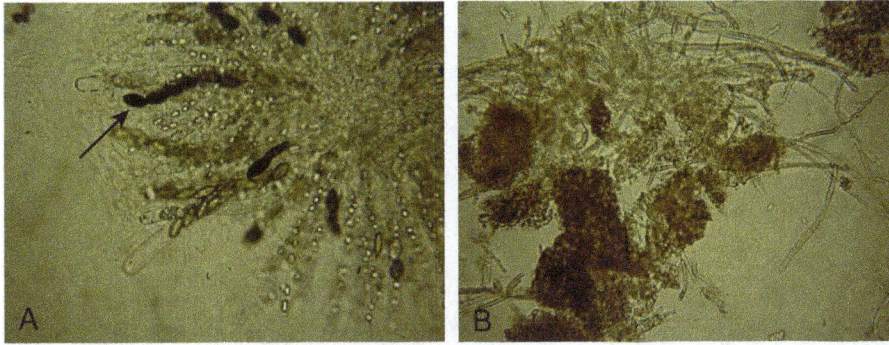

Figure 4. The PACC pathway is required for the development of asci. Fertilized perithecia from wild type (left panel/A) and ΔpalA (right panel/B) were allowed to develop for 7 days. The perithecia were squashed between a glass slide and a glass coverslip and examined with a transmitted light microscope. The wild type perithecia has generated ascospores (arrow) while the ΔpalA peirthecia is defective in ascospore formation. The arrow in the left panel points to an ascospore.

suggests that the *N. crassa* PACC pathway is localized to the ESCRT complex, just as was previously shown to occur for the *A. nidulans* and *S. cerevisiae* pathways [51,52]. Our findings are consistent with the *N. crassa* PACC pathway functioning like the canonical PACC pathway defined in *A. nidulans* and *S. cerevisiae*. In *S. cerevesiae*, the PACC homlog, Rim101p, has been shown to be required for meiosis [55,57], and the inability of the PACC pathway mutants to produce asci would suggest that the *N. crassa* PACC pathway may function in an analogous manner during the later stages of female development.

Category #2: Genes encoding other signal transduction pathway elements

Table 2 lists 53 genes that are likely components of signal transduction pathways. An extensive analysis of the serine-threonine kinases encoded in the *N. crassa* genome was recently published, and many of the kinases we identify in Table 2 were also found to be defective in female development in that study [48]. Fourteen of these signal transduction pathway genes were previously defined as being required for cell-to-cell fusion in our previous study [35]. All three of the MAP kinase pathways encoded in *N. crassa* genome have been previously shown to be needed for female development [45]. Table 2 contains 25 genes required for the production of CATs (conidial anastomosis tubes),

the cell type needed for cell-to-cell fusion. Most of these genes are known to be components of either the MIK-1/MEK-1/MAK-1 cell wall integrity pathway, the NRC-1/MEK-2/MAK-2 hyphal growth pathway, or the NOX pathway. These pathways have been identified as being needed for cell-to-cell fusion [32,35,45,63,64]. Table 2 contains three newly identified genes, *pp-2A activator* (NCU03269), *whi-2* (NCU10518) and *prs-1* (NCU08380), which are required for the CAT formation and are therefore likely components or regulators of the MAK-1 or MAK-2 pathways. The PP-2A activator (NCU03269) is likely associated with PP-2A (NCU06563), which has been shown to be a member of the STRIPAK complex that regulates the movement of MAK-1 into the nucleus [43]. WHI-2 is a homolog of a yeast general stress regulator [65–68]. Recent work in our laboratory has shown that MAK-1 and MAK-2 pathways are affected in the *N. crassa whi-2* mutant [47]. PRS-1 is a putative membrane-associated protein phosphatase and forms a complex with WHI-2 [65]. In yeast, Whi2p and Psr1p have roles in autophagy and in responding to mitochondrial dysfunction [68]. Given the probable association between the PP2A activator and the STRIPAK complex and between PRS-1 and WHI-2, it seems likely that PP2A activator and PRS-1 are also involved in regulating the MAK-1 and MAK-2 pathways in *N. crassa*. Those signal transduction pathway components found in Table 2 that do not

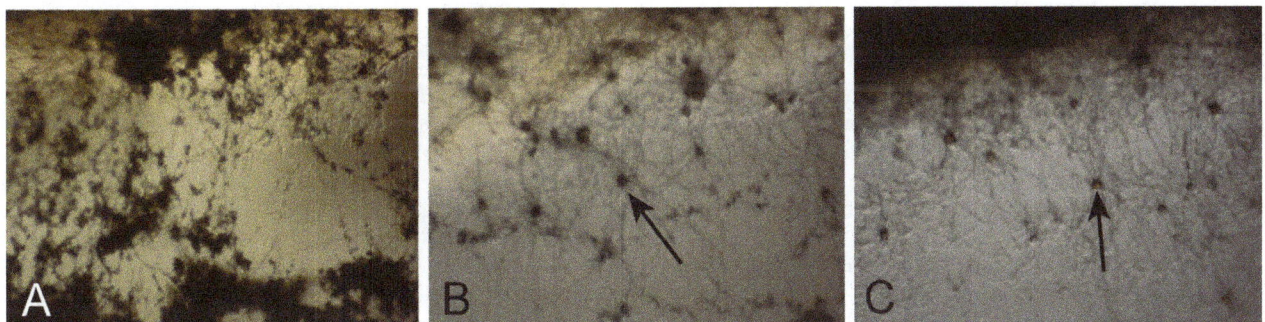

Figure 5. Constitutively active PACC activates female development. Cells were inoculated onto agar slants and allowed to grow for 10 days at room temperature. The left panel (A) shows a ΔpalC isolate on Vogel's sucrose medium. The middle panel (B) shows a ΔpalC isolate that has been transformed with the constitutively activated PACC construct growing on synthetic crossing medium. The right panel (C) shows a ΔpalC isolate that has been transformed with the constitutively activated PACC construct growing on Vogel's sucrose medium, a medium that represses female development. Note that the constitutively activated PACC caused protoperithecia production in the absence of PALC on both media. The arrows in the middle and right panels point to protoperithecia.

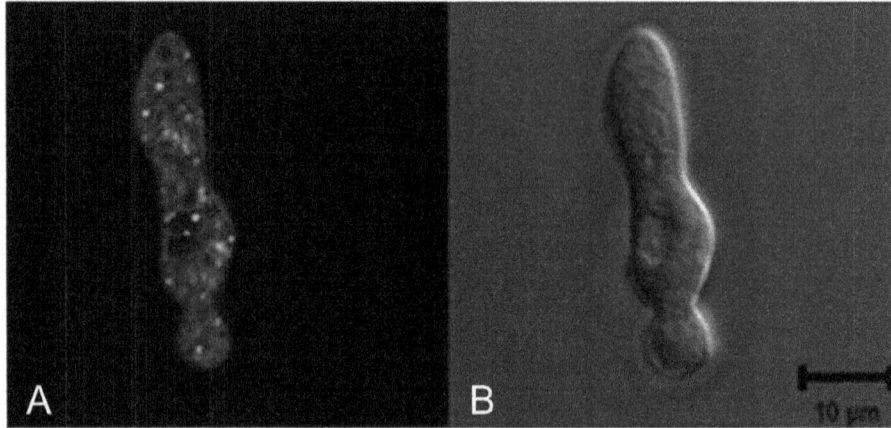

Figure 6. PALA is localized to the small intracellular vesicles. GFP-tagged PALA was expressed under the regulation of the *ccg-1* promoter in a Δ*palA* isolate. The GFP fluorescent image of a germling is shown in the panel on the left (A). The DIC image of the same germling is shown in the right panel (B). The bar in the DIC image is 10 μm in length.

affect CAT formation are unlikely to be part of the MAK-1 and MAK-2 pathways, and are more likely to be functioning in some other pathway.

We noted that our list of signal transduction genes included the *nox-1* (NCU02110) and *nor-1* (NCU07850) genes, which function in superoxide production and signaling [69–71]. These genes have been previously identified by others as being needed for protoperithecia formation, and may play a key role in cell-to-cell signaling during female development [69,72,73].

The *mfa-1* (NCU016992) and *ccg-4* (NCU02500) pheromone genes and the genes encoding their receptors, *pre-1* (NCU000138) and *pre-2* (NCU005758), have been previously identified as playing vital functions during mating and perithecium maturation in *N. crassa* and *S. macrospora* [24,27,74]. The PRE-1 and PRE-2 receptors have been previously shown to function in directing growth of the trichogyne toward conidia of the opposite mating type to facilitate the fertilization event [27]. The signal transduction pathway(s) through which the PRE-1 and PRE-2 receptors function has not been characterized. Further analysis will be needed to characterize how this pathway functions.

Mutants that are affected in the process of septation formation have been previously shown to be unable to generate protoperithecia in *S. macrospora* and *N. crassa* [41,75–77]. Our screening experiments corroborate the need for septation during female development. We identified *cdc-7* (NCU01335), *sid-1* (NCU04096), and *dbf-2* (NCU09071) as being needed for protoperithecia formation (Table 2). Several other genes encoding components of the *N. crassa* sepatation initiation network (SIN) have been shown to needed for female development [76–80], but the deletion mutants were either absent from the single gene deletion library or found as heterokaryons, and these genes are not listed in Table 2.

Table 2 lists three genes involved in calcium signaling that were identified in our experiments. These are the gene for the calcineurin subunit b, *cnb-1* (NCU03833), the calcium/calmodulin-dependent protein kinase *camk-1* (NCU09123), and *ham-10* (NCU02833), which encodes a C2 domain-containing protein. C2 domains are thought to function as calcium-dependent lipid-binding domains involved in vesicular trafficking. All three of these genes have been previously identified as affecting female development [46–48].

In addition to the components of the three MAP kinase pathways, the PACC pathway, the NOX-1 pathway, the septation initiation network, the pheromone pathway, and the calcium signaling pathway, Table 2 also contains five additional kinases. These kinases are *fi* (NCU04990), *div-4* (NCU04426), *stk-16* (NCU00914), *stk-22* (NCU03523), and *stk-47* (NCU06685). Four of these kinases, *div-4*, *stk-16*, *stk-22* and *stk-47* are needed for ascogonia formation, and may be part of a signaling pathway regulating entry into female development. However, further work is needed to characterize how these kinases function during protoperithecia formation and maturation.

Category #3: Transcription factors

Table 3 lists the 10 transcription factor genes that we have identified as being needed for female development. Although not listed in Table 3, the genes at the two mating alleles encode transcription factors that have been well characterized and are known to play critical roles in regulating transcriptional activity during female development [81]. In *P. anserina*, a group of ten HMG-box proteins, including the mating type proteins, have been shown to function in directing perithecial development [82,83]. Except for *pacC*, all of the transcription factors in Table 3 have been previously identified as being involved in female development by others [84]. The *asm-1* (NCU01414) transcription factor was previously identified by Aramayo et al. [85] as being needed for protoperithecium development. The *rco-1* (NCU06205), *rcm-1* (NCU06842) and *adv-1* (NCU07392) genes had been previously identified as encoding transcription factors needed for cell-to-cell fusion [35]. The *rco-1* (NCU06205) and *rcm-1* (NCU06842) genes are homologs of the *S. cerevisiae* SSN6 and TUP1genes, which encode the subunits of a general dimeric transcription factor. RCO-1 and RCM-1 have been previously identified as being needed for female development [86]. ADV-1 is a homolog of the *S. macrospora* PRO-1 protein, which has been shown to be needed for female development [87,88]. The *ada-1* (NCU00499) and *adv-1* (NCU07392) genes were identified as transcription factors needed for both asexual and sexual development [84], and we verified their importance for female development by complementation experiments. The *fmf-1* (NCU09387) gene was previously characterized by Iyer et al. [89] as a homolog of the *S. pombe* Ste11p, a transcription factor involved in the expression of genes involved in pheromone signaling. Hutchinson and Glass [90] previously reported that the *fsd-1* (NCU09915) gene was needed for the transition from protoperithecium to a mature perithecium. The *pp-1* (NCU00340) and *pacC* (NCU00090) genes encode transcription factors that are known to function in the MAK-2 and PACC signal transcription pathways [51,52,91,92]. In addition to these transcription factors that had been previously identified and characterized in *N. crassa*, one additional transcription factor for female development, MCM-1 (NCU07430) is listed in Table 3. The deletion mutant for *mcm-1* is not found in the first 119 plates of the library, but a *mcm-1* mutant has been characterized in the closely related *S. macrospora* and shown to be unable to produce perithecia [93,94]. The *S. macrospora* MCM-1 has been shown to associate with a protein encoded by the mating type locus gene, and to regulate gene expression during perithecium development [93].

Category #4: Autophagy genes

The identification of 10 genes in our screening and co-segregation experiments that are required for autophagy (Table 4), clearly demonstrates that autophagy is a required activity during female development. Previous work from the Pöggeler laboratory has demonstrated the importance of autophagy during *S. macrospora* female development [95–98]. The *atg-4*, *atg-7*, *atg-8*, *vps-34* and *vps15* genes have each been shown to be required for perithecial development in *S. macrospora*. Autophagy has been studied in *Magnaporthe oryzae*, where it has been shown to be important for asexual development, perithecial development, and appressorium formation [99,100]. In *Aspergillus oryzae*, the deletion of autophagy genes was shown to affect the conidiation process [101]. Complementation experiments with 2 of our autophagy genes, *atg-3* and *atg-8*, verified that autophagy was required for the development of *N. crassa* protoperithecia. Some of our autophagy mutants were initially identified as being protoperithecia-defective while others were identified as being perithecia-defective. However, careful examination of the autophagy mutants shows that these mutants can initiate female development (produce ascogonia) and make a few small protoperithecia. Thus, the autophagy mutants might be best described as being able to initiate protoperithecium development, but unable to fully support subsequent female development. The protoperithecia grafting experiments demonstrate that autophagy is required within the vegetative hyphal network. We found that fertilized autophagy mutant perithecia were able to complete female development and produced ascospores when grafted onto a wild type host. We conclude that the autophagy mutants are affected in female development because the vegetative hyphal network is unable to provide an adequate supply of nutrients to the developing female.

Category #5: Miscellaneous genes

There were 10 genes that are required for female development which didn't fall into one of the 4 categories discussed above (Table 5). Among the miscellaneous genes we identified as being needed for female development were 3 genes that were annotated as encoding enzymes that are likely to function in melanin biosynthesis. Tyrosinase, which is encoded by the T gene (NCU00776), catalyzes the formation of melanin from dihydroxyphenylalanine (DOPA), and has been previously shown to be needed for protoperithecium formation [102]. In addition to tyrosinase, the *per-1* gene (NCU03584), which encodes a polyketide synthase that is probably needed for the production of melanin from dihydroxynapthalene (DHN), is needed for perithecium melanization. The Δ*per-1* mutant has an unmelanized perithecia phenotype [31]. We also found that an aldo-keto

reductase gene (NCU01703) was needed for female development. The aldo-keto reductase may function in the pathway with the polyketide synthase for the production of DHN melanin. A polyketide synthase gene cluster has also been shown to be important for *S. macrospora* perithecial development [103,104].

Table 5 lists 7 genes that were annotated as encoding "hypothetical proteins" or "conserved hypothetical proteins" in the *N. crassa* genome site at the Broad Institute website. These genes have been verified by complementation as being needed for female development (Table 5). We have named these genes as *fem* genes to designate that they are required for female development. Some of the encoded hypothetical proteins are highly conserved in the genomes of the filamentous fungi, suggesting that the genes play important roles in the life cycles of the filamentous fungi. As mentioned in the ascogonia formation section, *fem-1*, *fem-4*, *fem-5*, *fem-6*, and *fem-7* may function with the *div-4*, *stk-16*, *stk-22* and *stk-47* kinases in a signal transduction pathway regulating the choice between asexual and sexual developmental programs. As the annotation of the *N. crassa* genome improves and further research is done, we hope that we will be able to identify functions for many of these conserved hypothetical proteins.

Conclusions

Our research demonstrates that a rather large number of signal transduction pathways and transcription factors are required to regulate female development (Tables 1, 2 and 3). Our analysis, in conjunction with other previously published studies, points to the involvement of at least nine different signal transduction pathways regulating female development. These are: 1) the MIK-1/MEK-1/MAK-1 cell wall integrity MAP kinase pathway; 2) the HAM-2/HAM-3/HAM-4/MOB-3 striatin pathway involved in making a complex on the nuclear envelope and in directing the movement of MAK-1 into the nucleus in a MAK-2 dependent manner; 3) the NRC-1/MEK-2/MAK-2 hyphal growth MAP kinase pathway; 4) the OS-2/OS-4/OS-5 osmotic stress MAP kinase pathway; 5) the NOX-1/NOR-1 superoxide pathway; 6) the PRE-1/PRE-2 pheromone signal transduction pathway; 7) the septation initation network (SIN), 8) a calcium signaling pathway, and 9) the PACC signal transduction pathway, which functions during the induction of protoperithecium development and during the maturation of the perithecium. Our experiments suggest that there is a signaling pathway that regulates the choice between sexual development and asexual development, and points to several genes that are likely to function in the pathway. We also found that chromatin remodeling was needed for female developments. Our experiments highlighted the role of autophagy and cell-to-cell fusion in the neighboring vegetative hyphae, which functions to provide nutrients to the developing female. What is clear from the analysis is that female development requires the coordinated activities of a number of signaling pathways.

Materials and Methods

Strains and growth conditions

Strains were routinely grown on Vogel's minimal medium with 2% sucrose or on synthetic crossing medium with 0.5% sucrose [22]. The single gene deletion library was obtained from the Fungal Genetics Stock Center (Kansas City, MO). The Δ*pacC* mutant was a kind gift from Dr. Maria Bertolini (Sao Paulo, Brazil) [92]. To provide the strains needed for the complementation experiments with the pBM60/pBM61 vector system, deletion mutants were mated with a *his-3* mutant of the opposite mating type (*his-3*, *mat*-A FGSC#6103 or a *his-3*, *mat*-a isolate obtained

Table 4. Autophagy genes required for female development.

Gene	NCU#	Co-segregation	Complementation	Reference information
atg-3	01955	Yes	Yes	This report
atg-8	01545	Yes	Yes (RIP)	Fu et al. [35]; Voigt et al. [97]; Liu et al. [100]
atg-12	10049	Yes	–	This report
atg-7	06672	Yes	–	Nolting et al. [98]; This report
atg-9	02422	Yes	–	This report
atg-10	02779	Yes	–	This report
atg-1	00188	Yes	–	This report
atg-5	04662	Yes	–	This report
atg-13	04840	Yes	–	Kikuma and Kitamoto [101]; This report
atg-18	03441	Yes	–	This report

RIP – a RIP experiment was used to verify that the gene is required for female development.
A notation of "This report" in the reference information column indicates that the gene was identified as being needed for *N. crassa* development by our experiments.

by mating FGSC#6103 with wild-type *mat-a*) and *his-3* isolates containing the deletion mutation were isolated from among the progeny. These mating experiments were carried out as described in Davis and DeSerres [22].

To determine whether mutant isolates could form ascogonia, cellophane filters overlaid on synthetic crossing medium (without carbon/energy supplementation) were inoculated with conidia. Two to three days post-inoculum, pieces of cellophane were cut from the filters and the presence of ascogonia was assessed by observation under a compound microscope at 200 X magnification. Ascogonia formation was also assessed by inoculating synthetic crossing medium containing 0.1% sucrose with conidia and allows the cells to grow for one to three days. Agar samples were then removed from the culture and examined for ascogonia under the compound microscope.

Screening the library

A screening procedure was used to identify mutants in the *Neurospora* deletion library that were defective in the formation of protoperithecia or for their subsequent development into mature perithecia. Plates 1 through 119 of the library contain approximately 10,000 haploid deletion mutants. Each of these mutants was individually inoculated into a glass test tube (16 × 100 mm) containing a 3 ml synthetic crossing medium agar slant. The isolates were allowed to grow for 10 days at room temperature, and examined for the presence of protoperithecia under the dissecting microscope. Protoperithecia were readily observed on the glass test tube near the edge of the synthetic crossing medium (Figure 1). After screening for protoperithecia formation, we added approximately 500 µl of water containing conidia of both *mat*-A and *mat*-a mating types to each of the slants to fertilize the protoperithecia and induce perithecium development. The slants were visually screened for the development of melanized perithecia three to four days after adding the conidial suspension. The haploid mutants isolated by the screening procedure would be considered as maternal or female developmental mutants. Most of the deletion mutants were tested for complementation in heterokaryons, and the deletion mutations found to be recessive.

Co-segregation analysis

Many of the isolates in the *N. crassa* single gene deletion library have mutations in addition to the targeted deletion [35]. To help

Table 5. Miscellaneous genes needed for female development.

Gene	NCU#	Co-segregation	Complementation	Reference information
Tyrosinase	00776	NA	PP	Fuentes et al. [102]
per-1 polyketide synthase	03584	NA(het)	PP	McCluskey et al. [31]
Aldo-keto reductase	01703	Yes	Yes	This report
fem-1 Hypothetical	03589	Yes	Yes	This report
fem-2 Hypothetical	07135	Yes	Yes	This report
fem-3 Hypothetical	03588	Yes	Yes	This report
fem-4 Hypothetical	06243	Yes	Yes	This report
fem-5 Hypothetical	02073	Yes	Yes	This report
fem-6 Hypothetical	09052	Yes	Yes	This report
fem-7 Hypothetical	03985	Yes	Yes	This report

PP – Previously published data demonstrated that the gene was needed for female development.
NA – a deletion strain is not available in the single gene deletion library.
NA(het) – the deletion strain in the single gene deletion library is a heterokaryon and a homokaryon isolate was not available during the screening experiments.
A notation of "This report" in the reference information column indicates that the gene was identified as being needed for *N. crassa* development by our experiments.

determine if the deletion mutations were responsible for the protoperithecium-defective and perithecium-defective phenotypes observed in the screening procedure, the mutants were subjected to a co-segregation analysis. The mutants were mated with a wild type isolate of the opposite mating type and ascospore progeny from these matings were collected [22]. Because all of the mutants were defective in female development, the mutant isolates were used as the conidial (male) partners in these matings. Ascospores from each of the matings were activated and 24 single ascospore progeny were isolated using standard procedures [22]. Each of the progeny were tested for hygromycin resistance and for the ability to form protoperithecia or perithecia with the same procedure used in the initial screening. The deletion mutations are marked by the insertion of the hygromycin resistance cassette in replacement of the deleted gene [84]. The co-segregation of hygromycin resistance with the mutant phenotype was taken as preliminary evidence that the deletion mutation was responsible for the mutant phenotype. The mutant phenotype was ascribed to a mutation other than the deletion in those cases where the mutant phenotype did not co-segregate with hygromycin resistance.

Many of the deletion mutations are represented by two isolates in the single gene deletion library, one of each mating type. In screening the library for protoperithecia-defective and perithecia-defective mutants, we found several cases where only one of the two deletion isolates in the library was defective in female development. Prior experience doing co-segregation analysis with the library showed that in such cases, the mutant phenotype invariably does not co-segregate with the deletion mutation [35]. Thus, the phenotype was assumed to be due to a mutation other than the deletion in those cases where the library contained two isolates and only one of the isolates had a mutant phenotype.

Cell-to-cell fusion assays

A number of the previously characterized cell-to-cell fusion (anastomosis) mutants have been previously shown to be affected in female development [32,35,42,45,105]. To identify cell-to-cell fusion mutants among the mutants affected in female development, the mutants were individually tested for the ability to participate in cell-to-cell fusion with the conidial anastomosis tube fusion (CAT fusion) assay as previously described by Fu et al. [35]. The CAT fusion assay was originally described and developed by Roca et al. [40].

Perithicia grafting experiments

The primary purpose of the protoperithecia grafting experiments was to determine whether a mutant vegetative hyphal network could support the development of transplanted wild type perithecia. A P. anserina perithecia grafting technique, which consists of transferring perithecia directly onto a host hyphal network has been previously described [106,107]. In using this technique with N. crassa, we found that complementation between a wild type graft and a mutant host resulted in the production of multiple perithecia at the graft site. It was very difficult to differentiate between the grafted perithecia and perithecia produced by complementation within the mutant host hyphae. We therefore used a modification of the Podospora grafting techniques to evaluate N. crassa female development. To perform the grafting experiments, mutant isolates were grown on an agar plates containing SCM with 0.5% sucrose for ten days to provide "hosts" for the grafted perithecia. Wild type isolates were grown for 10 days at room temperature on a cellophane filters that had been overlaid on Petri dishes containing SCM with 0.5% sucrose agar medium. An abundance of wild type protoperithecia developed on the cellophane. A conidial suspension of the opposite

mating type was then used to fertilize these wild type protoperithecia. Twenty four hours after fertilization, small pieces of cellophane (approximately 0.5 cm squares) with fertilized "graft" perithecia were cut from the filter with a sterilized new razor blade and transferred onto the Petri dishes containing mutant "host" vegetative hyphae. Three or four small pieces of cellophane containing fertilized wild type perithecia were transferred to each "host" Petri dish (Figure 2). The ability of the mutant vegetative hyphae to support development of the "graft' wild type fertilized perithecium was determined by whether or not the perithecia on the cellophane filter ripened and shot ascospores.

In doing these grafting experiments, we determined that the host and graft had to be of the same mating type in order for the host to support perithecial development in the graft, suggesting that the vegetative incompatibility phenomenon was operative during the grafting experiments. This is in contrast with the P. anserina grafting experiments, in which hosts of either mating type could sustain the development of the grafted perithecia [106].

The perithecia grafting experiments allowed us to evaluate two different aspects of female development. First and foremost, they allowed us to determine whether a mutant host vegetative hyphal network could support the development of fertilized graft perithecia of the same mating type. Second, we found that grafted wild type perithecia could induce protoperithecia formation in a vegetative hyphal network of some of our mutants.

Cloning, complementation, and RIP analysis

Complementation experiments were used to definitely demonstrate that a gene identified in the deletion library was required for female development. These complementation experiments were carried out with the pBM60/pBM61 cloning system as previously described [35]. To test for complementation, PCR primers were used to amplify the putative female development genes along with approximately 1500 base pairs of upstream sequence and 500 base pairs of downstream sequence, and the genes were cloned into the pBM60 and pBM61 vectors [108]. The primers were designed so that they contained restriction enzyme sites that allowed for the cloning of the genes into a multicloning site in the plasmids. The primers used in cloning the PACC pathway genes are given in Table S3. The pBM60 and pBM61 vectors are designed for the targeted insertion of plasmid sequences into the intergenic region downstream of the his-3 locus. pBM60 or pBM61-derived plasmids for each of the cloned genes were used to transform an isolate having the gene deletion in a his-3 background. Insertion of the plasmid sequences by homologous recombination generated a wild-type copy of the his-3 gene, and allowed for the isolation of the transformant. Several transformants for each of the cloned genes were isolated and tested for the ability to make protoperithecia or perithecia on 3 ml slants of synthetic crossing medium. The ability of the cloned wild type gene to complement the deletion mutation was taken as definite proof that the gene was needed for female development.

For those mutants that did not produce conidia, the cell type used in the transformation experiments for the complementation analysis, we carried out RIP (repeat induced point mutation) analyses to verify that the mutant phenotype was due to the mutation of the targeted gene. RIP is a phenomenon in which genes that are present in two or more copies in a haploid genome are extensively mutated during the N. crassa sexual cycle [109]. Both copies of a duplicated gene receive multiple C to T (G to A) mutations during the RIP process. The RIP experiments were performed by cloning genes into the pBM60 or pBM61 as described above. The plasmids were used to transform his-3 conidia to generate a strain having two copies of the cloned gene.

These transformants were then mated with a *his-3* strain of the opposite mating type to activate the RIP process. Individual *his-3* progeny with the mutant phenotype were then isolated. Being *his-3* isolates, these progeny will have a single copy of the gene in question (they don't have the "second copy" of the gene which was targeted into the *his-3* locus during the transformation). The gene was then PCR amplified and sequenced to determine if the mutant phenotype was the result of RIP mutations within the gene.

Constitutively activated version of PACC

A constitutively activated version of PACC was created by introducing a stop codon into the *pacC* gene. The stop codon was placed such that the encoded protein lacked the C terminal 129 amino acids. This construct was generated by using the Gibson Assembly Master Mix kit (New England BioLabs, Ipswich, MA). Primers pacC-F and pacC-activated-R (Table S3) were used to amplify the 5′UTR and *pacC* coding region through the added stop codon region. Primers pacC-activated-F and pacC-R (Table S3) were used to amplify the region beginning with the stop codon and containing 3′ UTR sequences. The two PCR products were mixed with *XbaI* and *EcoRI* digested pBM60 and the Gibson Assembly Master Mix to generate a full length *pacC* containing the early stop codon.

Intracellular localization of GFP-tagged PALA

Two versions of GFP-tagged PALA were generated to examine the localization of PALA in wild type and mutant backgrounds. Both versions were created in the pMF272 vector [62] and contain the complete PALA coding region followed by the GFP coding sequence. The two versions differed in the promoter region used to drive expression of the protein. One of the versions contained the *ccg-1* promoter found in pMF272 to drive high level expression of PALA, while the other version contained the normal *palA* promoter. For the *palA* promoter version, PCR primers palA-GFP-F and palA-GFP-R (Table S3) were used to amplify the region from 1374 base pairs upstream of the coding region to the amino acid preceding the stop codon for PALA. The PCR primers contained restriction enzyme sites to facilitate the cloning of the amplified gene and its upstream regulatory sequence immediately preceding, and in frame with, the GFP sequences in pMF272. To construct the *ccg-1* promoter version of *pal-A*, primers pal-GFP-ccg-1-F and palA-GFP-ccg-1-R (Table S3) were used to amplify and clone the coding region of *palA* into pMF272. The pMF272 vector is designed to facilitate the insertion of the cloned sequences into the intergenic region downstream of the *his-3* gene by homologous recombination [62]. The GFP-tagged *palA* plasmid constructs were used to transform the *ΔpalA, his-3* mutant to demonstrate by complementation that the GFP-tagged PALA was fully functional. The location of the GFP-tagged PALA was assessed by fluorescence confocal microscopy. Similar GFP tagging experiments were carried out to produce GPF versions of PALC and PALH, but these GFP-tagged proteins failed to complement their deletion mutants.

Confocal microscopy

Confocal laser scanning microscopy was performed using a Zeiss LSM 710 Confocal Microscope (Carl Zeiss, Inc., USA).

Plan-Apochromat 63X/1.40 oil DIC M27 objective lens or Plan-Apochromat 40X/1.3 Oil DIC M27 objective lens were used for imaging GFP expression.

Supporting Information

Table S1 List of protoperithecia defective/deficient mutants where the gene deletion is likely to be causing the phenotype because of general cellular health problems. Genes that function in general metabolic pathways needed for female development are listed along with their NCU numbers. For those genes where co-segregation and complementation experiments were done to verify that the genes were needed for female development, the information is provided in the co-segregation and complementation column. Information about the type of encoded protein and the general metabolic functions it is involved in are given in the notations and protein function columns.

Table S2 Chromatin organization genes required for female development. The genes identified in the screening procedure as being involved in generating and remodeling chromatin are listed. The deletion mutations that were shown to co-segregate with the female developmental phenotype are noted with a "yes" in the co-segregation column. Those genes that we verified as being required for female development by complementation are noted with a "yes" in the complementation column. The designation of PP in the complementation column indicates that previously published information demonstrates that the gene is needed for female development, and the reference for the information is given in the reference information column.

Table S3 Primers used in cloning PACC pathway genes. The primers used for PCR amplification and cloning of the PACC pathway genes are listed. The added restriction sites used for inserting the genes into pMB60 and pMB61 are underlined. The pacC-F, pacC-R, pacC-activated-F, and pacC-activated-R were used with the Gibson cloning kit to introduce a stop codon into the pacC gene.

Acknowledgments

We acknowledge the Neurospora Genome Project for the creation of the deletion mutants and the Fungal Genetics Stock Center for providing the mutants for our research efforts. We thank James Stamos for help in the preparation of the manuscript and Alan Siegel for help with fluorescence microscopy. We also thank Zhao Na, Amanda Lickfield, Yichen Li, and Genessis Capellan for their assistance with screening, co-segregation, complementation, RIP, and grafting experiments.

Author Contributions

Conceived and designed the experiments: JLC CF LMC JWA SJF. Performed the experiments: JLC CF LMC JWA SJF. Analyzed the data: JLC CF LMC JWA SJF. Contributed reagents/materials/analysis tools: JLC CF LMC JWA SJF. Contributed to the writing of the manuscript: JLC CF SJF.

References

1. Poggeler S, Nowrousian M, Kuck U (2006) Fruiting-Body Development in Ascomycetes; Kues U, Fischer R, editors: Springer Berlin Heidelberg.
2. Lord KM, Read ND (2011) Perithecium morphogenesis in Sordaria macrospora. Fungal Genet Biol 48: 388–399.
3. Bistis GN, Perkins DD, Read ND (2003) Different cell types in Neurospora crassa. Fungal Genet Newsl 50: 17–19.
4. Wang Z, Lopez-Giraldez F, Lehr N, Farre M, Common R, et al. (2014) Global gene expression and focused knockout analysis reveals genes associated with

fungal fruiting body development in Neurospora crassa. Eukaryot Cell 13: 154–169.

5. Hallen HE, Huebner M, Shiu SH, Guldener U, Trail F (2007) Gene expression shifts during perithecium development in Gibberella zeae (anamorph Fusarium graminearum), with particular emphasis on ion transport proteins. Fungal Genet Biol 44: 1146–1156.

6. Galagan JE, Calvo SE, Borkovich KA, Selker EU, Read ND, et al. (2003) The genome sequence of the filamentous fungus Neurospora crassa. Nature 422: 859–868.

7. Nowrousian M, Stajich JE, Chu M, Engh I, Espagne E, et al. (2010) De novo assembly of a 40 Mb eukaryotic genome from short sequence reads: Sordaria macrospora, a model organism for fungal morphogenesis. PLoS Genet 6: e1000891.

8. Teichert I, Wolff G, Kuck U, Nowrousian M (2012) Combining laser microdissection and RNA-seq to chart the transcriptional landscape of fungal development. BMC Genomics 13: 511.

9. Nowrousian M, Ringelberg C, Dunlap JC, Loros JJ, Kuck U (2005) Cross-species microarray hybridization to identify developmentally regulated genes in the filamentous fungus Sordaria macrospora. Mol Genet Genomics 273: 137–149.

10. Johnson TE (1978) Isolation and characterization of perithecial development mutants in neurospora. Genetics 88: 27–47.

11. Engh I, Nowrousian M, Kuck U (2010) Sordaria macrospora, a model organism to study fungal cellular development. Eur J Cell Biol 89: 864–872.

12. Nowrousian M, Teichert I, Masloff S, Kuck U (2012) Whole-Genome Sequencing of Sordaria macrospora Mutants Identifies Developmental Genes. G3 (Bethesda) 2: 261–270.

13. Collopy PD, Colot HV, Park G, Ringelberg C, Crew CM, et al. (2010) High-throughput construction of gene deletion cassettes for generation of Neurospora crassa knockout strains. Methods Mol Biol 638: 33–40.

14. Borkovich KA, Alex LA, Yarden O, Freitag M, Turner GE, et al. (2004) Lessons from the genome sequence of Neurospora crassa: tracing the path from genomic blueprint to multicellular organism. Microbiol Mol Biol Rev 68: 1–108.

15. Debuchey R, Bertiaux-Lecellier V, Silar P (2010) Mating systems and sexual morphogenesis in Ascomycetes. In: Borkovich KA, Ebbole DJ, editors. Cellular and Molecular Biology of Filamentous Fungi. Washington, D.C.: ASM Press. 501–535.

16. Searle T (1973) Life cycle of Neurospora crassa viewed by scanning electron microscopy. J Bacteriol 113: 1015–1025.

17. Harris JL, Howe HB, Roth IL (1975) Scanning electron microscopy of surface and internal features of developing perithecia of Neurospora crassa. J Bacteriol 122: 1239–1246.

18. Read ND (1983) A scanning electron microscopic study of the external features of perithecium development in Sordaria macrospora. Canad J Bot 61: 3217–3229.

19. Mai SH (1976) Morphological studies in Podospora anserina. Amer J Bot 63: 821–825.

20. Guenther JC, Trail F (2005) The development and differentiation of Gibberella zeae (anamorph: Fusarium graminearum) during colonization of wheat. Mycologia 97: 229–237.

21. Trail F, Common R (2000) Perithecial development by Gibberella zeae: a light microscopy study. Mycologia 92: 130–138.

22. Davis RH, DeSerres FJ (1970) Genetic and microbiological research techniques for Neurospora crassa. Meth Enzymol 27: 79–143.

23. Raju NB (1992) Genetic control of the sexual cycle of Neurospora. Mycol Res 96: 241–262.

24. Kim H, Borkovich KA (2006) Pheromones are essential for male fertility and sufficient to direct chemotropic polarized growth of trichogynes during mating in Neurospora crassa. Eukaryot Cell 5: 544–554.

25. Kim H, Borkovich KA (2004) A pheromone receptor gene, pre-1, is essential for mating type-specific directional growth and fusion of trichogynes and female fertility in Neurospora crassa. Mol Microbiol 52: 1781–1798.

26. Kim H, Metzenberg RL, Nelson MA (2002) Multiple functions of mfa-1, a putative pheromone precursor gene of Neurospora crassa. Eukaryot Cell 1: 987–999.

27. Kim H, Wright SJ, Park G, Ouyang S, Krystofova S, et al. (2012) Roles for receptors, pheromones, G proteins, and mating type genes during sexual reproduction in Neurospora crassa. Genetics 190: 1389–1404.

28. Vigfusson NV, Weijer J (1972) Sexuality in Neurospora crassa. II. Genes affecting the sexual development cycle. Genet Res 19: 205–211.

29. Tan ST, Ho CC (1970) A gene controlling the early development of protoperithecium in Neurospora crassa. Mol Gen Genet 107: 158–161.

30. Mylyk OM, Threlkeld SF (1974) A genetic study of female sterility in Neurospora crassa. Genet Res 24: 91–102.

31. McCluskey K, Wiest AE, Grigoriev IV, Lipzen A, Martin J, et al. (2011) Rediscovery by Whole Genome Sequencing: Classical Mutations and Genome Polymorphisms in Neurospora crassa. G3 (Bethesda) 1: 303–316.

32. Fleissner A, Sarkar S, Jacobson DJ, Roca MG, Read ND, et al. (2005) The so locus is required for vegetative cell fusion and postfertilization events in Neurospora crassa. Eukaryot Cell 4: 920–930.

33. Howe HB, Bensen EW (1974) A perithecial color mutant of Neurospora crassa. Mol Gen Genet 131: 79–83.

34. Horowitz NH, Fling M, MacLeod H, Sueoka N (1961) A genetic study of two structural forms of tyrosinase in Neurospora. Genetics 46: 1015–1024.

35. Fu C, Iyer P, Herkal A, Abdullah J, Stout A, et al. (2011) Identification and characterization of genes required for cell-to-cell fusion in Neurospora crassa. Eukaryot Cell 10: 1100–1109.

36. Adhvaryu KK, Morris SA, Strahl BD, Selker EU (2005) Methylation of histone H3 lysine 36 is required for normal development in Neurospora crassa. Eukaryot Cell 4: 1455–1464.

37. Brenna A, Grimaldi B, Filetici P, Ballario P (2012) Physical association of the WC-1 photoreceptor and the histone acetyltransferase NGF-1 is required for blue light signal transduction in Neurospora crassa. Mol Biol Cell 23: 3863–3872.

38. Lewis ZA, Adhvaryu KK, Honda S, Shiver AL, Knip M, et al. (2010) DNA methylation and normal chromosome behavior in Neurospora depend on five components of a histone methyltransferase complex, DCDC. PLoS Genet 6: e1001196.

39. Gesing S, Schindler D, Franzel B, Wolters D, Nowrousian M (2012) The histone chaperone ASF1 is essential for sexual development in the filamentous fungus Sordaria macrospora. Mol Microbiol 84: 748–765.

40. Roca MG, Arlt J, Jeffree CE, Read ND (2005) Cell biology of conidial anastomosis tubes in Neurospora crassa. Eukaryot Cell 4: 911–919.

41. Maerz S, Dettmann A, Ziv C, Liu Y, Valerius O, et al. (2009) Two NDR kinase-MOB complexes function as distinct modules during septum formation and tip extension in Neurospora crassa. Mol Microbiol 74: 707–723.

42. Xiang Q, Rasmussen C, Glass NL (2002) The ham-2 locus, encoding a putative transmembrane protein, is required for hyphal fusion in Neurospora crassa. Genetics 160: 169–180.

43. Dettmann A, Heilig Y, Ludwig S, Schmitt K, Illgen J, et al. (2013) HAM-2 and HAM-3 are central for the assembly of the Neurospora STRIPAK complex at the nuclear envelope and regulate nuclear accumulation of the MAP kinase MAK-1 in a MAK-2-dependent manner. Mol Microbiol 90: 796–812.

44. Fleissner A, Leeder AC, Roca MG, Read ND, Glass NL (2009) Oscillatory recruitment of signaling proteins to cell tips promotes coordinated behavior during cell fusion. Proc Natl Acad Sci U S A 106: 19387–19392.

45. Lichius A, Lord KM, Jeffree CE, Oborny R, Boonyarungsrit P, et al. (2012) Importance of MAP kinases during protoperithecial morphogenesis in Neurospora crassa. PLoS One 7: e42565.

46. Kothe GO, Free SJ (1998) Calcineurin subunit B is required for normal vegetative growth in Neurospora crassa. Fungal Genet Biol 23: 248–258.

47. Fu C, Ao J, Dettmann A, Seiler S, Free SJ (2014) Characterization of the Neurospora crassa cell fusion proteins, HAM-6, HAM-7, HAM-8, HAM-9, HAM-10, AMPH-1, and WHI-2. PLoS One in press.

48. Park G, Servin JA, Turner GE, Altamirano L, Colot HV, et al. (2011) Global analysis of serine-threonine protein kinase genes in Neurospora crassa. Eukaryot Cell 10: 1553–1564.

49. Arst HN, Penalva MA (2003) pH regulation in Aspergillus and parallels with higher eukaryotic regulatory systems. Trends Genet 19: 224–231.

50. Penalva MA, Arst HN Jr (2002) Regulation of gene expression by ambient pH in filamentous fungi and yeasts. Microbiol Mol Biol Rev 66: 426–446, table of contents.

51. Penalva MA, Tilburn J, Bignell E, Arst HN Jr (2008) Ambient pH gene regulation in fungi: making connections. Trends Microbiol 16: 291–300.

52. Mitchell AP (2008) A VAST staging area for regulatory proteins. Proc Natl Acad Sci U S A 105: 7111–7112.

53. Lamb TM, Xu W, Diamond A, Mitchell AP (2001) Alkaline response genes of Saccharomyces cerevisiae and their relationship to the RIM101 pathway. J Biol Chem 276: 1850–1856.

54. Diez E, Alvaro J, Espeso EA, Rainbow L, Suarez T, et al. (2002) Activation of the Aspergillus PacC zinc finger transcription factor requires two proteolytic steps. EMBO J 21: 1350–1359.

55. Li W, Mitchell AP (1997) Proteolytic activation of Rim1p, a positive regulator of yeast sporulation and invasive growth. Genetics 145: 63–73.

56. Ramon AM, Porta A, Fonzi WA (1999) Effect of environmental pH on morphological development of Candida albicans is mediated via the PacC-related transcription factor encoded by PRR2. J Bacteriol 181: 7524–7530.

57. Su SS, Mitchell AP (1993) Identification of functionally related genes that stimulate early meiotic gene expression in yeast. Genetics 133: 67–77.

58. Xu W, Smith FJ Jr, Subaran R, Mitchell AP (2004) Multivesicular body-ESCRT components function in pH response regulation in Saccharomyces cerevisiae and Candida albicans. Mol Biol Cell 15: 5528–5537.

59. Davis D, Wilson RB, Mitchell AP (2000) RIM101-dependent and-independent pathways govern pH responses in Candida albicans. Mol Cell Biol 20: 971–978.

60. Li M, Martin SJ, Bruno VM, Mitchell AP, Davis DA (2004) Candida albicans Rim13p, a protease required for Rim101p processing at acidic and alkaline pHs. Eukaryot Cell 3: 741–751.

61. Rollins JA (2003) The Sclerotinia sclerotiorum pac1 gene is required for sclerotial development and virulence. Mol Plant Microbe Interact 16: 785–795.

62. Bowman BJ, Draskovic M, Freitag M, Bowman EJ (2009) Structure and distribution of organelles and cellular location of calcium transporters in Neurospora crassa. Eukaryot Cell 8: 1845–1855.

63. Pandey A, Roca MG, Read ND, Glass NL (2004) Role of a mitogen-activated protein kinase pathway during conidial germination and hyphal fusion in Neurospora crassa. Eukaryot Cell 3: 348–358.

64. Kothe GO, Free SJ (1998) The isolation and characterization of nrc-1 and nrc-2, two genes encoding protein kinases that control growth and development in Neurospora crassa. Genetics 149: 117–130.

65. Kaida D, Yashiroda H, Toh-e A, Kikuchi Y (2002) Yeast Whi2 and Psr1-phosphatase form a complex and regulate STRE-mediated gene expression. Genes Cells 7: 543–552.

66. Leadsham JE, Miller K, Ayscough KR, Colombo S, Martegani E, et al. (2009) Whi2p links nutritional sensing to actin-dependent Ras-cAMP-PKA regulation and apoptosis in yeast. J Cell Sci 122: 706–715.

67. Mendl N, Occhipinti A, Muller M, Wild P, Dikic I, et al. (2011) Mitophagy in yeast is independent of mitochondrial fission and requires the stress response gene WHI2. J Cell Sci 124: 1339–1350.

68. Muller M, Reichert AS (2011) Mitophagy, mitochondrial dynamics and the general stress response in yeast. Biochem Soc Trans 39: 1514–1519.

69. Cano-Dominguez N, Alvarez-Delfin K, Hansberg W, Aguirre J (2008) NADPH oxidases NOX-1 and NOX-2 require the regulatory subunit NOR-1 to control cell differentiation and growth in Neurospora crassa. Eukaryot Cell 7: 1352–1361.

70. Aguirre J, Rios-Momberg M, Hewitt D, Hansberg W (2005) Reactive oxygen species and development in microbial eukaryotes. Trends Microbiol 13: 111–118.

71. Takemoto D, Tanaka A, Scott B (2007) NADPH oxidases in fungi: diverse roles of reactive oxygen species in fungal cellular differentiation. Fungal Genet Biol 44: 1065–1076.

72. Dirschnabel DE, Nowrousian M, Cano-Dominguez N, Aguirre J, Teichert I, et al. (2014) New Insights Into the Roles of NADPH Oxidases in Sexual Development and Ascospore Germination in Sordaria macrospora. Genetics 196: 729–744.

73. Malagnac F, Lalucque H, Lepere G, Silar P (2004) Two NADPH oxidase isoforms are required for sexual reproduction and ascospore germination in the filamentous fungus Podospora anserina. Fungal Genet Biol 41: 982–997.

74. Mayrhofer S, Weber JM, Poggeler S (2006) Pheromones and pheromone receptors are required for proper sexual development in the homothallic ascomycete Sordaria macrospora. Genetics 172: 1521–1533.

75. Bloemendal S, Lord KM, Rech C, Hoff B, Engh I, et al. (2010) A mutant defective in sexual development produces aseptate ascogonia. Eukaryot Cell 9: 1856–1866.

76. Heilig Y, Schmitt K, Seiler S (2013) Phospho-regulation of the Neurospora crassa septation initiation network. PLoS One 8: e79464.

77. Justa-Schuch D, Heilig Y, Richthammer C, Seiler S (2010) Septum formation is regulated by the RHO4-specific exchange factors BUD3 and RGF3 and by the landmark protein BUD4 in Neurospora crassa. Mol Microbiol 76: 220–235.

78. Heilig Y, Dettmann A, Mourino-Perez RR, Schmitt K, Valerius O, et al. (2014) Proper actin ring formation and septum constriction requires coordinated regulation of SIN and MOR pathways through the germinal centre kinase MST-1. PLoS Genet 10: e1004306.

79. Rasmussen CG, Glass NL (2005) A Rho-type GTPase, rho-4, is required for septation in Neurospora crassa. Eukaryot Cell 4: 1913–1925.

80. Riquelme M, Yarden O, Bartnicki-Garcia S, Bowman B, Castro-Longoria E, et al. (2011) Architecture and development of the Neurospora crassa hypha – a model cell for polarized growth. Fungal Biol 115: 446–474.

81. Debuchy R, Berteaux-Lecellier V, Silar P (2010) Mating Systems and Sexual Morphogenesis in Ascomycetes. In: Borkovich KA, Ebbole DJ, editors. Cellular and Molecular Biology of Filamentous Fungi. Washington DC: ASM Press. 501–536.

82. Benkhali JA, Coppin E, Brun S, Peraza-Reyes L, Martin T, et al. (2013) A Network of HMG-box Transcription Factors Regulates Sexual Cycle in the Fungus Podospora anserina. Plos Genetics 9.

83. Coppin E, Berteaux-Lecellier V, Bidard F, Brun S, Ruprich-Robert G, et al. (2012) Systematic deletion of homeobox genes in Podospora anserina uncovers their roles in shaping the fruiting body. PLoS One 7: e37488.

84. Colot HV, Park G, Turner GE, Ringelberg C, Crew CM, et al. (2006) A high-throughput gene knockout procedure for Neurospora reveals functions for multiple transcription factors. Proc Natl Acad Sci U S A 103: 10352–10357.

85. Aramayo R, Peleg Y, Addison R, Metzenberg R (1996) Asm-1+, a Neurospora crassa gene related to transcriptional regulators of fungal development. Genetics 144: 991–1003.

86. Aldabbous MS, Roca MG, Stout A, Huang IC, Read ND, et al. (2010) The ham-5, rcm-1 and rco-1 genes regulate hyphal fusion in Neurospora crassa. Microbiology 156: 2621–2629.

87. Masloff S, Poggeler S, Kuck U (1999) The pro1(+) gene from Sordaria macrospora encodes a C6 zinc finger transcription factor required for fruiting body development. Genetics 152: 191–199.

88. Masloff S, Jacobsen S, Poggeler S, Kuck U (2002) Functional analysis of the C6 zinc finger gene pro1 involved in fungal sexual development. Fungal Genet Biol 36: 107–116.

89. Iyer SV, Ramakrishnan M, Kasbekar DP (2009) Neurospora crassa fmf-1 encodes the homologue of the Schizosaccharomyces pombe Ste11p regulator of sexual development. J Genet 88: 33–39.

90. Hutchison EA, Glass NL (2010) Meiotic regulators Ndt80 and ime2 have different roles in Saccharomyces and Neurospora. Genetics 185: 1271–1282.

91. Leeder AC, Jonkers W, Li J, Glass NL (2013) Early colony establishment in Neurospora crassa requires a MAP kinase regulatory network. Genetics 195: 883–898.

92. Cupertino FB, Freitas FZ, de Paula RM, Bertolini MC (2012) Ambient pH controls glycogen levels by regulating glycogen synthase gene expression in Neurospora crassa. New insights into the pH signaling pathway. PLoS One 7: e44258.

93. Nolting N, Poggeler S (2006) A MADS box protein interacts with a mating-type protein and is required for fruiting body development in the homothallic ascomycete Sordaria macrospora. Eukaryot Cell 5: 1043–1056.

94. Nolting N, Poggeler S (2006) A STE12 homologue of the homothallic ascomycete Sordaria macrospora interacts with the MADS box protein MCM1 and is required for ascosporogenesis. Mol Microbiol 62: 853–868.

95. Voigt O, Herzog B, Jakobshagen A, Poggeler S (2013) bZIP transcription factor SmJLB1 regulates autophagy-related genes Smatg8 and Smatg4 and is required for fruiting-body development and vegetative growth in Sordaria macrospora. Fungal Genet Biol 61: 50–60.

96. Voigt O, Herzog B, Jakobshagen A, Poggeler S (2014) Autophagic kinases SmVPS34 and SmVPS15 are required for viability in the filamentous ascomycete Sordaria macrospora. Microbiol Res 169: 128–138.

97. Voigt O, Poggeler S (2013) Autophagy genes Smatg8 and Smatg4 are required for fruiting-body development, vegetative growth and ascospore germination in the filamentous ascomycete Sordaria macrospora. Autophagy 9: 33–49.

98. Nolting N, Bernhards Y, Poggeler S (2009) SmATG7 is required for viability in the homothallic ascomycete Sordaria macrospora. Fungal Genet Biol 46: 531–542.

99. Lin F-C, Liu X-H, Lu J-P, Liu T-B (2009) Studies on Autophagy Machinery in Magnaporthe oryzae. In: Wang G-L, Valent B, editors. Advances in Genetics, Genomics, and Control of Rice Blast Disease: Springer. 33–40.

100. Liu TB, Liu XH, Lu JP, Zhang L, Min H, et al. (2010) The cysteine protease MoAtg4 interacts with MoAtg8 and is required for differentiation and pathogenesis in Magnaporthe oryzae. Autophagy 6: 74–85.

101. Kikuma T, Kitamoto K (2011) Analysis of autophagy in Aspergillus oryzae by disruption of Aoatg13, Aoatg4, and Aoatg15 genes. FEMS Microbiol Lett 316: 61–69.

102. Fuentes AM, Connerton I, Free SJ (1994) Production of tyrosinase defective mutants in Neurospora crassa. Fungal Genet Newsl 41: 38–39.

103. Nowrousian M (2009) A novel polyketide biosynthesis gene cluster is involved in fruiting body morphogenesis in the filamentous fungi Sordaria macrospora and Neurospora crassa. Curr Genet 55: 185–198.

104. Schindler D, Nowrousian M (2014) The polyketide synthase gene pks4 is essential for sexual development and regulates fruiting body morphology in Sordaria macrospora. Fungal Genet Biol 68: 48–59.

105. Park G, Pan S, Borkovich KA (2008) Mitogen-activated protein kinase cascade required for regulation of development and secondary metabolism in Neurospora crassa. Eukaryot Cell 7: 2113–2122.

106. Silar P (2011) Grafting as a method for studying development in the filamentous fungus Podospora anserina. Fungal Biol 115: 793–802.

107. Silar P (2014) Simple genetic tools to study fruiting body development in fungi. Open Mycol J 8: 148–155.

108. Margolin BS, Frietag M, Selker EU (1997) Improved plasmids for gene targeting at the his-3 locus of Neurospora crassa. Fungal Genet Newsl 44: 34–36.

109. Selker EU (1999) Gene silencing: repeats that count. Cell 97: 157–160.

110. Maerz S, Ziv C, Vogt N, Helmstaedt K, Cohen N, et al. (2008) The nuclear Dbf2-related kinase COT1 and the mitogen-activated protein kinases MAK1 and MAK2 genetically interact to regulate filamentous growth, hyphal fusion and sexual development in Neurospora crassa. Genetics 179: 1313–1325.

111. Li D, Bobrowicz P, Wilkinson HH, Ebbole DJ (2005) A mitogen-activated protein kinase pathway essential for mating and contributing to vegetative growth in Neurospora crassa. Genetics 170: 1091–1104.

112. Dettmann A, Illgen J, Marz S, Schurg T, Fleissner A, et al. (2012) The NDR kinase scaffold HYM1/MO25 is essential for MAK2 map kinase signaling in Neurospora crassa. PLoS Genet 8: e1002950.

113. Araujo-Palomares CL, Richthammer C, Seiler S, Castro-Longoria E (2011) Functional characterization and cellular dynamics of the CDC-42 - RAC - CDC-24 module in Neurospora crassa. PLoS One 6: e27148.

114. Richthammer C, Enseleit M, Sanchez-Leon E, Marz S, Heilig Y, et al. (2012) RHO1 and RHO2 share partially overlapping functions in the regulation of cell wall integrity and hyphal polarity in Neurospora crassa. Mol Microbiol 85: 716–733.

115. Vogt N, Seiler S (2008) The RHO1-specific GTPase-activating protein LRG1 regulates polar tip growth in parallel to Ndr kinase signaling in Neurospora. Mol Biol Cell 19: 4554–4569.

116. Engh I, Wurtz C, Witzel-Schlomp K, Zhang HY, Hoff B, et al. (2007) The WW domain protein PRO40 is required for fungal fertility and associates with Woronin bodies. Eukaryot Cell 6: 831–843.

117. Simonin AR, Rasmussen CG, Yang M, Glass NL (2010) Genes encoding a striatin-like protein (ham-3) and a forkhead associated protein (ham-4) are required for hyphal fusion in Neurospora crassa. Fungal Genet Biol 47: 855–868.

118. Bloemendal S, Bernhards Y, Bartho K, Dettmann A, Voigt O, et al. (2012) A homologue of the human STRIPAK complex controls sexual development in fungi. Mol Microbiol 84: 310–323.

119. Nowrousian M, Frank S, Koers S, Strauch P, Weitner T, et al. (2007) The novel ER membrane protein PRO41 is essential for sexual development in the filamentous fungus Sordaria macrospora. Mol Microbiol 64: 923–937.

120. Maddi A, Dettman A, Fu C, Seiler S, Free SJ (2012) WSC-1 and HAM-7 are MAK-1 MAP kinase pathway sensors required for cell wall integrity and hyphal fusion in Neurospora crassa. PLoS One 7: e42374.

121. Jones CA, Greer-Phillips SE, Borkovich KA (2007) The response regulator RRG-1 functions upstream of a mitogen-activated protein kinase pathway impacting asexual development, female fertility, osmotic stress, and fungicide resistance in Neurospora crassa. Mol Biol Cell 18: 2123–2136.

122. Poggeler S (2000) Two pheromone precursor genes are transcriptionally expressed in the homothallic ascomycete Sordaria macrospora. Curr Genet 37: 403–411.

123. Ivey FD, Kays AM, Borkovich KA (2002) Shared and independent roles for a Galpha(i) protein and adenylyl cyclase in regulating development and stress responses in Neurospora crassa. Eukaryot Cell 1: 634–642.

124. Kamerewerd J, Jansson M, Nowrousian M, Poggeler S, Kuck U (2008) Three alpha-subunits of heterotrimeric G proteins and an adenylyl cyclase have distinct roles in fruiting body development in the homothallic fungus Sordaria macrospora. Genetics 180: 191–206.

125. Yang Q, Poole SI, Borkovich KA (2002) A G-protein beta subunit required for sexual and vegetative development and maintenance of normal G alpha protein levels in Neurospora crassa. Eukaryot Cell 1: 378–390.

126. Krystofova S, Borkovich KA (2005) The heterotrimeric G-protein subunits GNG-1 and GNB-1 form a Gbetagamma dimer required for normal female fertility, asexual development, and galpha protein levels in Neurospora crassa. Eukaryot Cell 4: 365–378.

127. Muller F, Kruger D, Sattlegger E, Hoffmann B, Ballario P, et al. (1995) The cpc-2 gene of Neurospora crassa encodes a protein entirely composed of WD-repeat segments that is involved in general amino acid control and female fertility. Mol Gen Genet 248: 162–173.

128. Yamashiro CT, Ebbole DJ, Lee BU, Brown RE, Bourland C, et al. (1996) Characterization of rco-1 of Neurospora crassa, a pleiotropic gene affecting growth and development that encodes a homolog of Tup1 of Saccharomyces cerevisiae. Mol Cell Biol 16: 6218–6228.

129. Kim SR, Lee B-U (2005) Characterization of the *Neurospora crassa rcm-1* mutant. Kor J Microbiol 41: 246–254.

12

Understanding Strategy of Nitrate and Urea Assimilation in a Chinese Strain of *Aureococcus anophagefferens* through RNA-Seq Analysis

Hong-Po Dong, Kai-Xuan Huang, Hua-Long Wang, Song-Hui Lu*, Jing-Yi Cen, Yue-Lei Dong

Research Center for Harmful Algae and Marine Biology, Key Laboratory of Eutrophication and Red Tide Prevention of Guangdong Higher Education Institutes, Jinan University, Guangzhou, China

Abstract

Aureococcus anophagefferens is a harmful alga that dominates plankton communities during brown tides in North America, Africa, and Asia. Here, RNA-seq technology was used to profile the transcriptome of a Chinese strain of *A. anophagefferens* that was grown on urea, nitrate, and a mixture of urea and nitrate, and that was under N-replete, limited and recovery conditions to understand the molecular mechanisms that underlie nitrate and urea utilization. The number of differentially expressed genes between urea-grown and mixture N-grown cells were much less than those between urea-grown and nitrate-grown cells. Compared with nitrate-grown cells, mixture N-grown cells contained much lower levels of transcripts encoding proteins that are involved in nitrate transport and assimilation. Together with profiles of nutrient changes in media, these results suggest that *A. anophagefferens* primarily feeds on urea instead of nitrate when urea and nitrate co-exist. Furthermore, we noted that transcripts upregulated by nitrate and N-limitation included those encoding proteins involved in amino acid and nucleotide transport, degradation of amides and cyanates, and nitrate assimilation pathway. The data suggest that *A. anophagefferens* possesses an ability to utilize a variety of dissolved organic nitrogen. Moreover, transcripts for synthesis of proteins, glutamate-derived amino acids, spermines and sterols were upregulated by urea. Transcripts encoding key enzymes that are involved in the ornithine-urea and TCA cycles were differentially regulated by urea and nitrogen concentration, which suggests that the OUC may be linked to the TCA cycle and involved in reallocation of intracellular carbon and nitrogen. These genes regulated by urea may be crucial for the rapid proliferation of *A. anophagefferens* when urea is provided as the N source.

Editor: Senjie Lin, University of Connecticut, United States of America

Funding: This study was funded by the National Natural Science Foundation of China through grant 41206126 and 41176087 (http://www.nsfc.gov.cn/), the Strategic Priority Research Program of the Chinese Academy of Sciences (XDA11020304) (http://www.cas.ac.cn/), and the National Basic Research Program of China (973 Program) through grant 2010CB428702 (http://www.most.gov.cn/). The funders had no role in study design, data collection and analysis, decision to publish, or preparation of the manuscript.

Competing Interests: The authors have declared that no competing interests exist.

* Email: lusonghui1963@163.com

Introduction

Brown tides are caused by the pelagophyte *Aureococcus anophagefferens*, which is a small (~2–3 µm) eukaryotic phytoplankton. This harmful algal bloom (HAB) has plagued many coastal ecosystems in the Eastern United States and South Africa since its discovery in 1985. Although brown tides do not produce harmful toxins, these tides still decimate fisheries and seagrass beds because of toxicity to bivalves and extreme light attenuation, respectively [1]. Recently, large-scale brown tides have been reported in China, which occurred in early summer for three consecutive years from 2009 to 2011 in the coastal waters of Qinhuangdao, China [2]. This report shows that brown tides are expanding and spreading to other oceanic regions because of anthropogenic activities. It is important to determine the causes of brown tides.

A. anophagefferens often bloom in periods when levels of dissolved inorganic nitrogen (DIN) are low and dissolved organic nitrogen (DON) concentrations are elevated [3]. *A. anophagefferens* is able to utilize a variety of DON compounds, which may facilitate its growth as both carbon and nitrogen sources [4,5,6]. In addition, it has been shown that *A. anophagefferens* has a significantly greater uptake capacity for urea than for other N sources that were tested, including nitrate, glutamic acid, and ammonium [4], and its growth increases as the DON:DIN ratio increases. These results suggest that blooms of *A. anophagefferens* may be related to the preferred utilization and high uptake of DON by *A. anophagefferens*. However, Pustizzi et al. (2004) found that although low light cultures of *A. anophagefferens* with urea have higher growth rates than those cultures without urea, the growth on urea is not significantly faster than the growth on nitrate [7]. It is assumed that uptake rates of urea by *A. anophagefferens* may be separate from the actual assimilation of urea. These results indicate that the utilization of urea by *A. anophagefferens* is complicated and probably affected by other environmental factors, such as nutrient levels and light intensity. Based on the above studies, it is proposed that the rapid growth of *A. anophagefferens* on urea may be associated with the fixation of urea-C. In-depth studies are required to reveal the real mechanism underlying urea utilization by *A. anophagefferens*.

Organic N sources in the ocean are diverse and presumably include urea, amines, peptides, proteins, nucleic acids, amino sugars, and amides. *A. anophagefferens* has order of priority in options of different N sources. These options are supported by the observation that the fastest growth is observed in cultures that are grown on urea, followed by acetamide, nitrate, ammonium, and formamide [8]. The characterization of a cDNA library and gene expression suggests that *A. anophagefferens* can assimilate eight different forms of N, and growth on different N sources elicits an increase in the relative expression of corresponding N transporters [8]. Recently, genome analysis found that, relative to competing phytoplankton, *A. anophagefferens* is enriched in genes encoding enzymes that degrade organic nitrogen compounds and transporters that are specific for a diverse set of organic nitrogen compounds [9,10]. These studies suggest that *A. anophagefferens* has a greater capacity to use organic nitrogenous compounds compared with its competitors. More recently, transcriptome analysis found that *A. anophagefferens* cells express and regulate a suite of genes that are related to organic nitrogen acquisition/metabolism under nitrogen depletion, which further supported the conclusion that *A. anophagefferens* can metabolize reduced organic forms of N [11].

The recently developed RNA-seq technology [12] has made genome-wide transcript analyses both sensitive and quantitative. Additionally, it has been demonstrated that RNA-seq is an excellent genome-scale platform for analyzing transcript levels [13]. In this study, we report the use of RNA-seq technology to examine transcriptomic differences in a Chinese strain of *A. anophagefferens* that was grown on urea, nitrate, or a mixture of urea and nitrate, and that was under N-replete, limited and recovery conditions. These RNA-seq studies have produced large, quantitative data sets for transcript abundance in *A. anophagefferens* by mapping RNA-seq reads to its gene models. The data strongly suggest significant differences in key cellular metabolic pathways, such as N transport and metabolism, the ornithine-urea cycle (OUC), and the tricarboxylic acid (TCA) cycle, among the different experimental groups. To our knowledge, this study is the first to find OUC activity in *A. anophagefferens*.

Materials and Methods

Algal strain

A. anophagefferens was collected from the coastal water of Qinhuangdao in the Bohai Sea, China on June 20, 2012 at station X01 (119°37.911′ E, 39°54.111′ N). The station X01 was located in a region that was experiencing a brown tide on that date. The oceanic region is open to the public and no specific permissions are required for sampling. *A. anophagefferens* cells were isolated using capillary pipettes under an inverted microscope and subsequently cultures from a single cell were established. The culture strains were maintained in sterilized natural seawater. Here, it should be pointed out that the field sampling did not involve endangered or protected species.

Culture conditions

The cultures were grown in 2-L flasks with 1 L of artificial seawater medium [14], which was enriched with f/2 nutrients, with nitrate as the N source (882 μmol L^{-1} NO$_3^-$ and 36.3 μmol L^{-1} PO$_4^{3-}$). Vitamins (thiamine, biotin and B$_{12}$) were sterile filtered and added to the media after autoclaving. The cultures

were grown at 18°C on a 12:12 h light: dark cycle under cool-white fluorescence lights (100 μmol photon m^{-2} s^{-1}), and the rate of growth was measured by monitoring in vivo fluorescence using a Turner Designs Model 10 Fluorometer (Turner Designs, CA, USA). These cultures were harvested during the late exponential growth phase and then inoculated into three different artificial seawater media with three different sources of N, including nitrate (882 μmol L^{-1} final concentration), urea (441 μmol L^{-1} final concentration), and nitrate + urea (Mixture N, 441 μmol L^{-1} nitrate, 220 μmol L^{-1} urea, final concentration). All cultures were grown in triplicate. The cells were harvested at the onset of the stationary phase by centrifugation (8000 × g for 10 min), covered with RNAlater solution (Sigma) and stored at −80°C until the RNA was extracted.

For nitrogen-limited and recovery experiments, cells from an exponential culture grown in f/2 media with urea as the N source (882 μmol N. L^{-1} final concentration) were collected by centrifugation (6000 × g for 5 min), washed once with nitrogen-free media and then inoculated in N-replete (441 μmol L^{-1} urea) and N-limited (20 μmol L^{-1} urea) media, respectively. The growth of the cultures was monitored by measuring in vivo fluorescence. The cells were harvested by centrifugation (8000 × g for 10 min) after 6 days in the stationary phase when nitrogen was depleted in the N-limited media. The remaining nitrogen-starved cultures were divided into two parts and then received additions of either 882 μmol L^{-1} NO$_3^-$ or 441 μmol L^{-1} urea. RNA samples were collected at 24 h.

Chlorophyll fluorescence measurements

The parameters *Fv/Fm* and the rapid light curves for ETR (electron transport rate) were determined using a Phyto-PAM Phytoplankton Analyzer (Walz, Germany). The culture was dark-acclimated for 15–20 min before determining *Fv/Fm*. The rapid light curves for ETR were measured under different PAR levels. Light-saturated ETR, ETRmax and the efficiency of the electron transport were analyzed from light curves of ETR [15].

Analysis of nutrient

Culture media were collected every other day and filtered through a GF/F filter. All filtrate samples were stored at −20°C until analysis. Urea concentration was colorimetrically determined using the diacetyl monoxime method by Rahmatullah and Boyde [16]. Nitrate were measured using flow injection analyzer LACHAT QC 8500 (HACH, USA) following spectrophotometric method [17].

Total RNA extraction and Illumina sequencing

Total RNA was extracted from frozen cell pellets using Trizol reagent (Invitrogen, CA, USA) according to the manufacturer's instructions. RNA concentrations were determined from A$_{260}$ nm, and its purity was evaluated by the A$_{260}$ to A$_{280}$ nm ratio. The integrity of the total RNA was assessed using an Agilent 2100 Bioanalyzer. RNA-seq libraries were constructed following an Illumina gene expression sample preparation kit. Briefly, total RNA (5–10 μg) was treated with RNase-free DNase I. Poly(A) mRNA was isolated using oligo(dT) magnetic beads and then fragmented into short fragments (approximately 200 bp). The first-strand synthesis of cDNA was performed using random hexamer-primed reverse transcription. The second-strand synthesis was performed by adding the first strand cDNA synthesis reaction to a second strand reaction mix consisting of first strand buffer, second strand buffer, a dNTP mix, RNase H (Invitrogen) and DNA polymerase I (Invitrogen). The double stranded cDNA was subsequently purified using magnetic beads. End reparation and

3'-end single nucleotide A addition was performed. Then, the cDNA fragments were connected with sequencing adaptors and were enriched by PCR amplification. Finally, the library was sequenced in BGI-tech (Shenzhen) using an Illumina HiSeq 2000 sequencer. RNA-seq raw data have been deposited in the NCBI Gene Expression Omnibus (GEO) database with experiment series accession number GSE60576.

Analysis of differentially expressed genes

The raw image data were converted into sequence data by base calling which are defined as raw reads. These raw reads had a sequencing length of 50 bp. To obtain high-quality reads, the raw reads were filtered to remove reads with adaptor sequence, low-quality reads and reads with high percentage of unknown bases using BGI-tech's in-house software SOAPnuk. All processed clean reads were mapped to the reference genome and transcript of *A. anophagefferens* using the program SOAPaligner/SOAP2 (version 2.21)[18], respectively, which were downloaded from http://genome.jgi.doe.gov/Auran1/Auran1. This alignment allowed no more than two mismatches. In addition, considering that *A. anophagefferens* genome just contains 1185 scaffolds and lacks prediction of gene models, the reads that mapped to the genome were not used for quantification analysis. In contrast, the reference transcript includes a total of 11501 gene models built by homology to known proteins from other model organisms and ab initio gene predictions as well as from available *A. anophagefferens* EST and cDNA data, so the differential gene expression analysis was performed on the reads that mapped to the transcript. The number of clean reads for each gene was calculated and then normalized to RPKM (number of transcripts per million clean reads), which is related to the read number with gene expression levels [19]. Fold changes in the differential gene expression between conditions were calculated using the $\log_2$ ratio of RPKM.

The significance of differentially expressed genes between two experimental groups (p-value) were performed following a published method ([20]. A false discovery rate (FDR) $\leqq$ 0.001 and an absolute value of $\log_2$ ratio ≥ 1 were used as cutoffs to judge the significance of gene expression differences [21].

Annotation

No function annotation in the transcript which was used as reference genes is provided. In order to annotate these mapped genes, we performed a BLAST search against the non-redundant (NR) database in NCBI with an e-value cut-off of $1e^{-5}$. Those best hits with specific function whose score is the highest and e-value > $1e^{-5}$ were chosen. The Blast2GO program was used to obtain the Gene ontology (GO) annotation of the genes.

Results

Physiological responses to different N sources, N-limitation and recovery

For *A. anophagefferens* cells that were grown on different N sources, the cell density increased quickly during the first 5 days (exponential phase growth) (Fig. 1A). Maximum growth rates (calculated for days 1 to 5) were observed in cultures grown on urea (0.30 d^{-1}), followed by the mixture of urea and nitrate (0.26 d^{-1}), and nitrate (0.21 d^{-1}) (Fig. 1C). However, there were no significant differences between the maximum growth rates. For cultures with nitrate, the cell density continued to increase after day 5, whereas the cell density did not increase in cultures with urea and mixture N. From day 7 to day 10, the cell density remained at the same level, and the density in all cultures tended to be consistent. The Fv/Fm (Maximum photochemical efficiency

of PSII) of cultures from days 6, 9 and 10 was determined, and no significant difference was observed among the three N sources (Fig. 2A). Interestingly, the light-saturated electron transport rate (ETRmax) and electron transport efficiency (ETE) tended to increase gradually from day 6 to day 10 in cultures with urea, whereas these values appeared to drop from day 6 to day 10 in cultures with nitrate and mixture N (Fig. 3). On day 10, ETRmax and ETE of cultures with urea were higher than those values of cultures with the other N sources. In addition, it is noted that the profile of urea concentration as a function of day in cultures with mixture N was extremely similar to that in cultures with urea, whereas no significant decrease was observed in NO_3^- concentration in cultures with mixture N (Fig. 4), which suggested that *A. anophagefferens* may primarily utilize N from urea instead of nitrate in the mixture N media.

For N-limited and recovery experiments, the cell density of N-limited cells started decreasing after day 3, whereas that of N-replete cells continued increasing (Fig. 1B). On day 5, the Fv/Fm of N-limited cells (0.52) was lower than that of N-replete cells (0.60) (Fig. 2B). The Fv/Fm was recovered at 24 hr after the addition of either urea or nitrate to N-limited cultures, and this recovery was independent of the nitrogen sources (Fig. 2B).

RNA-seq analysis

Deconvolution and filtering of raw reads yielded a mean of 6,938,798 reads (range: 6,539,842 to 7,242,083 reads) per individual RNA-seq library (Table 1). Subsequent alignment of the clean reads to the *A. anophagefferens* reference genome yielded a mean of 5,852,408 reads (84.3%) for each sample that mapped to at least one location in the *A. anophagefferens* genome (Table 1). However, of these mapped reads, only 47.4 to 50.9% of the total reads were mapped to the reference transcript for each sample (Table 2). This result indicated that approximately 30% of the total reads mapped to non-coding regions in the genome.

In *A. anophagefferens* grown on three different N sources, 9148 to 9526 genes were detected for each sample. The summary of the gene information is shown in Tables S1, S2 and S3, including unique read numbers that match each gene, its coverage, the expression level of each gene (represented by RPKM), and a putative function annotation. The data will be valuable for contributing to future genome annotation efforts and to the discovery of novel genes. A comparison of gene expression among the three N sources was performed. 322 differentially expressed genes were detected between nitrate-grown and urea-grown cells, 237 between nitrate-grown and mixture N-grown cells and 29 between urea-grown and mixture N-grown cells (Fig. 5, Table S4, S5 and S6). Fewer differentially expressed genes between urea-grown and mixture N-grown cells were identified, which further suggested the preferred utilization of cells for urea in media with mixture N.

For nitrogen-limited and recovery experiments, 707 genes were up-regulated significantly, and 766 were down-regulated significantly in N-depleted cells relative to N-replete cells. N-depleted cells exhibited a broad transcriptional response to nitrogen re-addition, with 681 genes up-regulated and 874 genes down-regulated in urea recovery cells, and 312 genes up-regulated and 688 genes down-regulated in nitrate recovery cells. The summary of those differentially expressed genes is listed in Table S7, S8 and S9.

Nitrate versus urea

Compared with cells that were grown on urea, 142 transcripts were upregulated 2-fold or greater in cells that were grown on nitrate, and 180 transcripts were downregulated 2-fold or greater

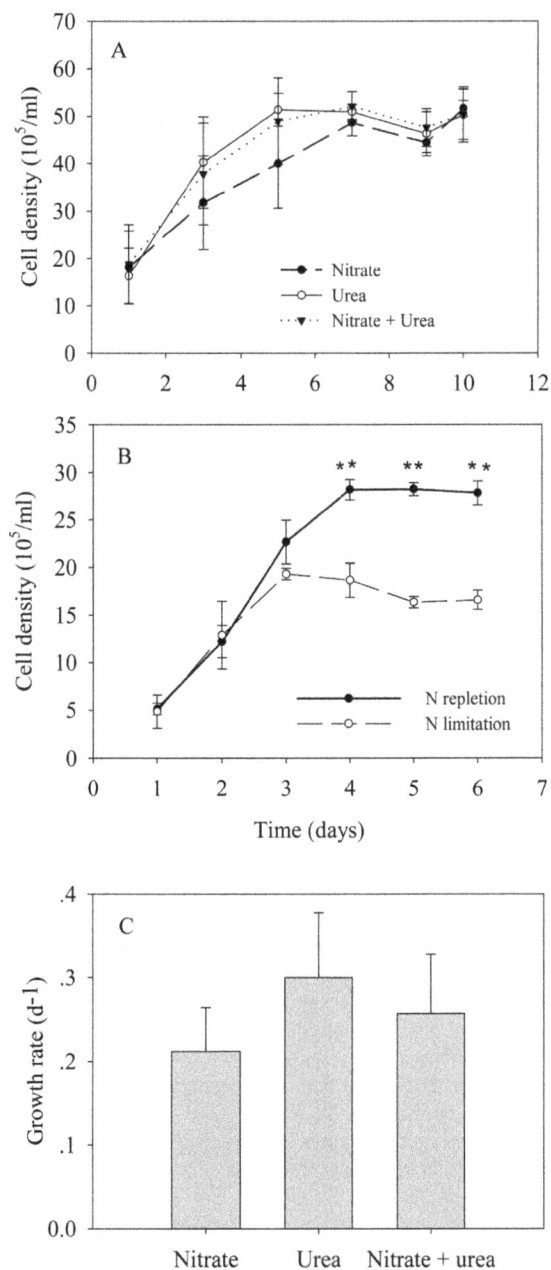

Figure 1. Cell density as a function of culture time. A) Growth of *A. anophagefferens* grown on urea, nitrate, and a mixture of urea and nitrate. B) Growth of *A. anophagefferens* under nitrogen-replete and limited conditions. Cell density under nitrogen-limited condition was compared with that under nitrogen-replete condition. Significance values were expressed as follows: * *P* <0.05, ** *P* <0.001. C) Growth rates of *A. anophagefferens* grown on urea, nitrate, and a mixture of urea and nitrate. Error bars represent standard deviation of the mean for the three biological replicates.

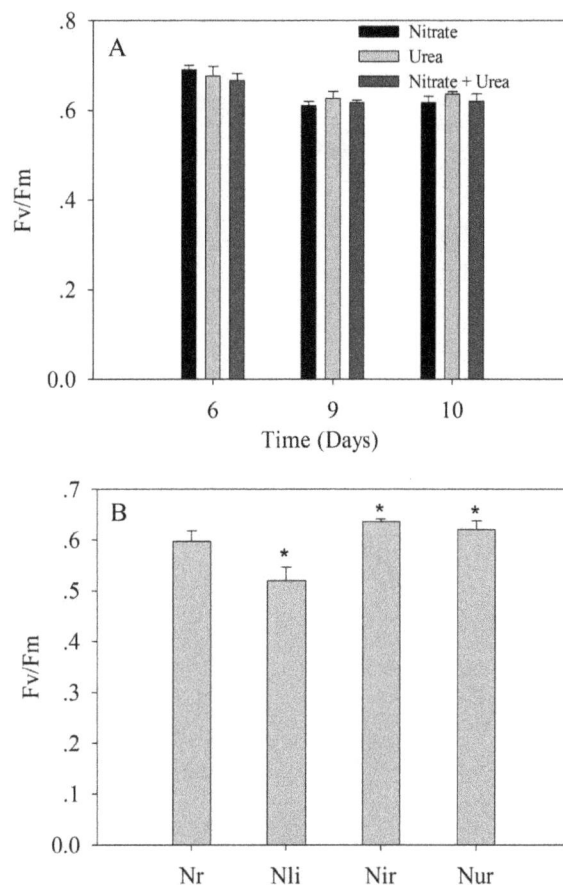

Figure 2. Photosystem II efficiency (Fv/Fm). A) *A. anophagefferens* cultures grown on urea, nitrate, and a mixture of urea and nitrate at the 6-, 9- and 10-day sampling points; B) *A. anophagefferens* cultures grown under nitrogen-replete (Nr), limited (Nli) and recovery conditions. Nir and Nur represent nitrate and urea addition, respectively. Nli was compared with Nr while Nir and Nur were compared with Nli. Significance values were expressed as follows: * *P* <0.05, ** P <0.001. Error bars represent standard deviation of the mean for the three biological replicates.

in cells that were grown on nitrate (Table S4). Approximately 24% of differentially expressed genes could not be assigned a function because the reference genes represented hypothetical or predicted protein or showed no database homology. Transcripts encoding nitrate high affinity transporter, nitrite transporter NAR1, formate/nitrite transporter, and ammonium transporter increased when *A. anophagefferens* cells were grown on nitrate compared

with urea. Transcripts encoding a putative nitrate reductase, NADPH nitrite reductase and glutamine synthetase increased in cells that were grown on nitrate compared with cells that were grown on urea (Fig. 6). Notably, 4 transcripts that were involved in OUC showed 2.0- to 2.6-fold upregulation in urea-grown cells relative to nitrate-grown cells (Table S4, Fig. 6). Thirty nine transcripts encoding proteins that were involved in protein synthesis showed 2.0- to 3.6-fold upregulation in urea-grown cells relative to nitrate-grown cells (Table S4). There were 11 transcripts encoding enzymes that were involved in amino acid synthesis that were upregulated in urea-grown cells relative to nitrate-grown cells. Levels of transcripts encoding proteins that were involved in photosynthesis and central carbon metabolism displayed difference in cells that were grown on urea and nitrate (Table S4, Fig. 7).

Specific genes transcriptionally regulated by urea

The comparison of results for the mixture N and urea groups with the nitrate reference allowed us to find genes for which transcription was specifically regulated by urea. We found 124 common genes differentially regulated between the mixture N and urea groups compared with the reference nitrate group

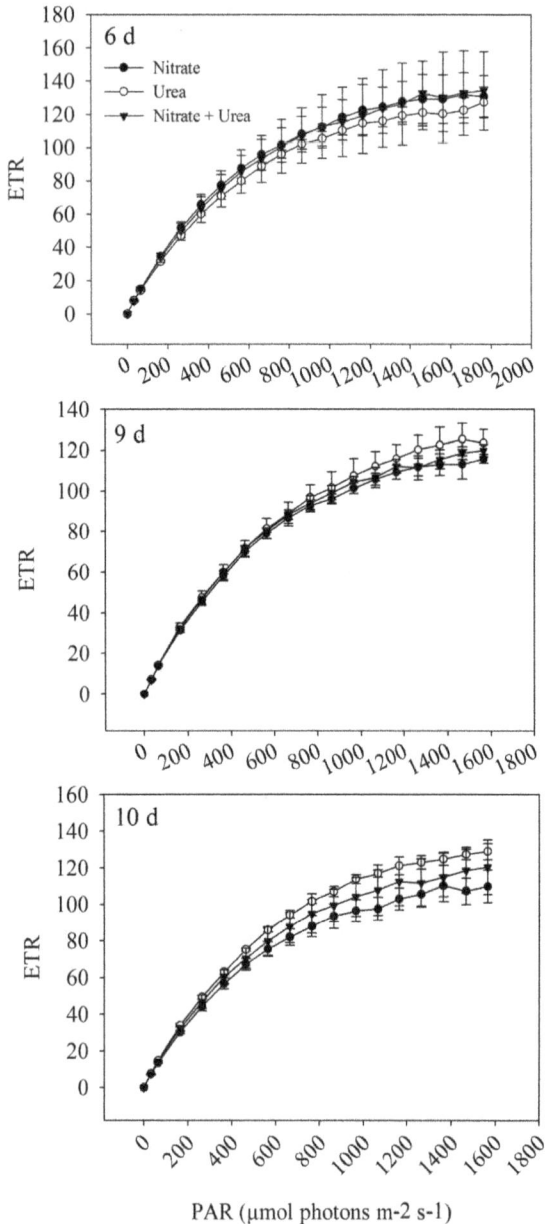

Figure 3. Relative electron transport rate (ETR) as a function of PAR. *A. anophagefferens* was grown on urea, nitrate, and a mixture of urea and nitrate, and samples were meaused at the 6-, 9- and 10-day. Error bars represent standard deviation of the mean for the three biological replicates.

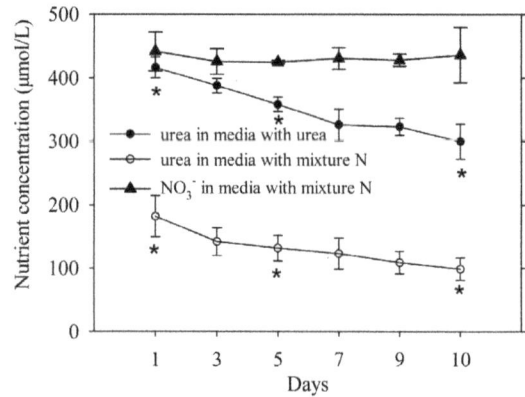

Figure 4. Concentrations of urea and nitrate as a function of culture time in media. Solid triangle represents nitrate concentration in medium with mixture N while solid and open circles represent urea concentrations in media with urea and mixture N, respectively. The difference between time points with asterisk was significant. Significance values were expressed as follows: * $P < 0.05$, ** $P < 0.001$. Error bars represent standard deviation of the mean for the three biological replicates.

(Figure 8A). These genes are listed in Table S10. Interestingly, the pattern of regulation of gene expression for urea and mixture N-grown cells appeared to be perfectly consistent (Figure 8B). Among these 124 genes, those involved in protein synthesis were well represented (Table 3). The next represented gene categories were nitrogen compound metabolism, protein modification and degradation, DNA and RNA binding, transport, signaling, photosynthesis, glycolysis/gluconeogenesis, and stress. Transcripts encoding argininosuccinate synthase, tryptophan synthase and spermine synthase increased in urea-and mixture-grown cells compared with nitrate-grown cells. Twenty two transcripts

encoding proteins that are involved in protein synthesis were induced in urea-and mixture-grown cells.

Nitrogen limitation and recovery

To evaluate the role of the *A. anophagefferens* OUC, RNA-seq was also used to investigate the pattern of gene expression of the OUC, TCA cycle and nitrogen assimilation pathway in response to nitrogen limitation and the addition of different nitrogen substrates to nitrogen-limited *A. anophagefferens* cultures. Under nitrogen-limited condition, two OUC-related transcripts and nine transcripts involved in nitrogen compound transport and assimilation showed upregulation, whereas three key transcripts for the TCA cycle were down-regulated (Table 4). In 24 h nitrogen-recovery experiments, all of the differentially expressed genes involved in these pathways that were significant were down-regulated (fold change >2.0 and $P < 0.05$). No significant difference was observed in expression pattern of these transcripts between the addition of nitrate and urea.

Discussion

In this study, RNA-Seq was used to profile the transcriptome of the Chinese strain of *A. anophagefferens* which was grown on urea, nitrate, and mixture N. Transcripts for the OUC, nitrogen assimilation, and TCA cycle in *A. anophagefferens* under nitrogen-limited and recovery conditions were also analyzed. Our goal is to gain a better understanding of the molecular mechanisms that underlie urea and nitrate metabolism in *A. anophagefferens* and to determine why *A. anophagefferens* prefers organic nitrogen to inorganic nitrogen on a molecular level.

Nitrogen acquisition and assimilation

The levels of transcripts for several N transporters displayed significant differences between urea-grown and nitrate-grown cells (Table S4). Transcripts encoding nitrate high affinity transporter, nitrite transporter NAR1, formate/nitrite transporter, and ammonium transporter increased when *A. anophagefferens* cells were grown on nitrate compared with urea. The result suggested that these transporter genes are inducible by nitrate. In other phytoplankton, such as Chlorophyceae, Haptophyceae, and

Table 1. Summary of RNA-seq sequencing data (mapping to the reference genome).

Samples	Total reads	Total mapped reads	Perfect match	<=2bp mismatch	Unique match	Multi-position match	Total unmapped reads
Mixture N	7,242,083	6,107,412(84.33%)	4,739,972(65.45%)	1,367,440(18.88%)	4,687,001(64.72%)	1,420,411(19.61%)	1,134,671(15.67%)
Nitrate	7,216,883	6,062,505(84.00%)	4,638,395(64.27%)	1,424,110(19.73%)	4,778,646(66.21%)	1,283,859(17.79%)	1,154,378(16.00%)
Urea	7,050,236	5,927,783(84.08%)	4,585,450(65.04%)	1,342,333(19.04%)	4,688,413(66.50%)	1,239,370(17.58%)	1,122,453(15.92%)
Nrep	6,539,842	5,487,229(83.90%)	3,987,158(60.97%)	1,500,071(22.94%)	4,646,256(71.05%)	840,973(12.86%)	1,052,613(16.10%)
Ndep	6,581,716	5,577,020(84.74%)	4,101,357(62.31%)	1,475,663(22.42%)	4,779,171(72.61%)	797,849(12.12%)	1,004,696(15.26%)
Urecov	6,871,619	5,812,055(84.58%)	4,290,560(62.44%)	1,521,495(22.14%)	4,838,377(70.41%)	973,678(14.17%)	1,059,564(15.42%)
Nrecov	7,069,212	5,992,854(84.77%)	4,377,598(61.92%)	1,615,256(22.85%)	4,948,338(70.00%)	1,044,516(14.78%)	1,076,358(15.23%)

Nrep, Ndep, Urecov, and Nrecov refer to the nitrogen-replete, limited, urea recovery, and nitrate recovery samples, respectively.

Table 2. Summary of RNA-seq sequencing data (mapping to the reference transcript).

Samples	Total reads	Total mapped reads	Perfect match	<=2bp mismatch	Unique match	Multi-position match	Total unmapped reads
Mixture N	7,242,083	3,631,779(50.15%)	2,847,747(39.32%)	784,032(10.83%)	2,875,798(39.71%)	755,981(10.44%)	3,610,304(49.85%)
Nitrate	7,216,883	3,639,577(50.43%)	2,804,173(38.86%)	835,404(11.58%)	2,959,722(41.01%)	679,855(9.42%)	3,577,306(49.57%)
Urea	7,050,236	3,590,932(50.93%)	2,797,158(39.67%)	793,774(11.26%)	2,936,461(41.65%)	654,471(9.28%)	3,459,304(49.07%)
Nrep	6,539,842	3,126,200(47.80%)	2,313,610(35.38%)	812,590(12.43%)	2,708,821(41.42%)	417,379(6.38%)	3413642(52.20%)
Ndep	6,581,716	3,118,690(47.38%)	2,355,066(35.78%)	763,624(11.60%)	2,740,361(41.64%)	378,329(5.75%)	3,463,026(52.62%)
Urecov	6,871,619	3,457,608(50.32%)	2,613,980(38.04%)	843,628(12.28%)	2,953,008(42.97%)	504,600(7.34%)	3,414,011(49.68%)
Nrecov	7,069,212	3,445,784(48.74%)	2,565,769(36.29%)	880,015(12.45%)	2,894,006(40.94%)	551,778(7.81%)	3,623,428(51.26%)

Nrep, Ndep, Urecov, and Nrecov refer to the nitrogen-replete, limited, urea recovery, and nitrate recovery samples, respectively.

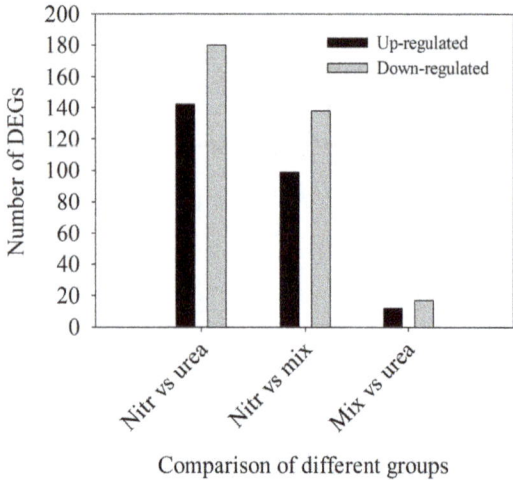

Figure 5. The number of differentially expressed genes among cells that were grown on the three N sources. Nitr, urea and mix represent cells grown on nitrate, urea and mixture N, respectively.

Bacillariophyceae, genes encoding high affinity nitrate transporters were highly induced when the cells were incubated with NO_3^- or N starvation [22]. In addition, the relative expression of a putative nitrate transporter gene is higher in the American strain of *A. anophagefferens* that is grown on nitrate than in cells that are grown on other N sources [8]. Nitrate assimilation involves two membrane barriers, the plasma and the chloroplast membranes. Thus, once nitrate is reduced to nitrite in the cytosol, nitrite has to cross the chloroplast membranes for its subsequent reduction to ammonium [23]. In this study, transcripts for ID 53005 and 15503 had conservative domains of a formate/nitrite transporter and were homologous to the nitrite transporter NAR1 from *Ectocarpus siliculosus* and to a formate/nitrite transporter, respectively. The upregulation of these two transcripts suggested that the two genes might be involved in nitrite transport to chloroplasts. In *Chlamydomonas reinhardtii,* a Nar1 gene that encodes putative formate and nitrite transporters, has been found to play an important role in the regulation of nitrite transport to chloroplasts [24]. In *A. anophagefferens*, a higher fold change was observed in the abundance of the nitrite transporter NAR1, suggesting that the transporter (ID 53005) may be the main contributor to the regulation and transport of nitrite to the chloroplast. Interesting, transcript for the NAR1 was down-regulated under nitrogen

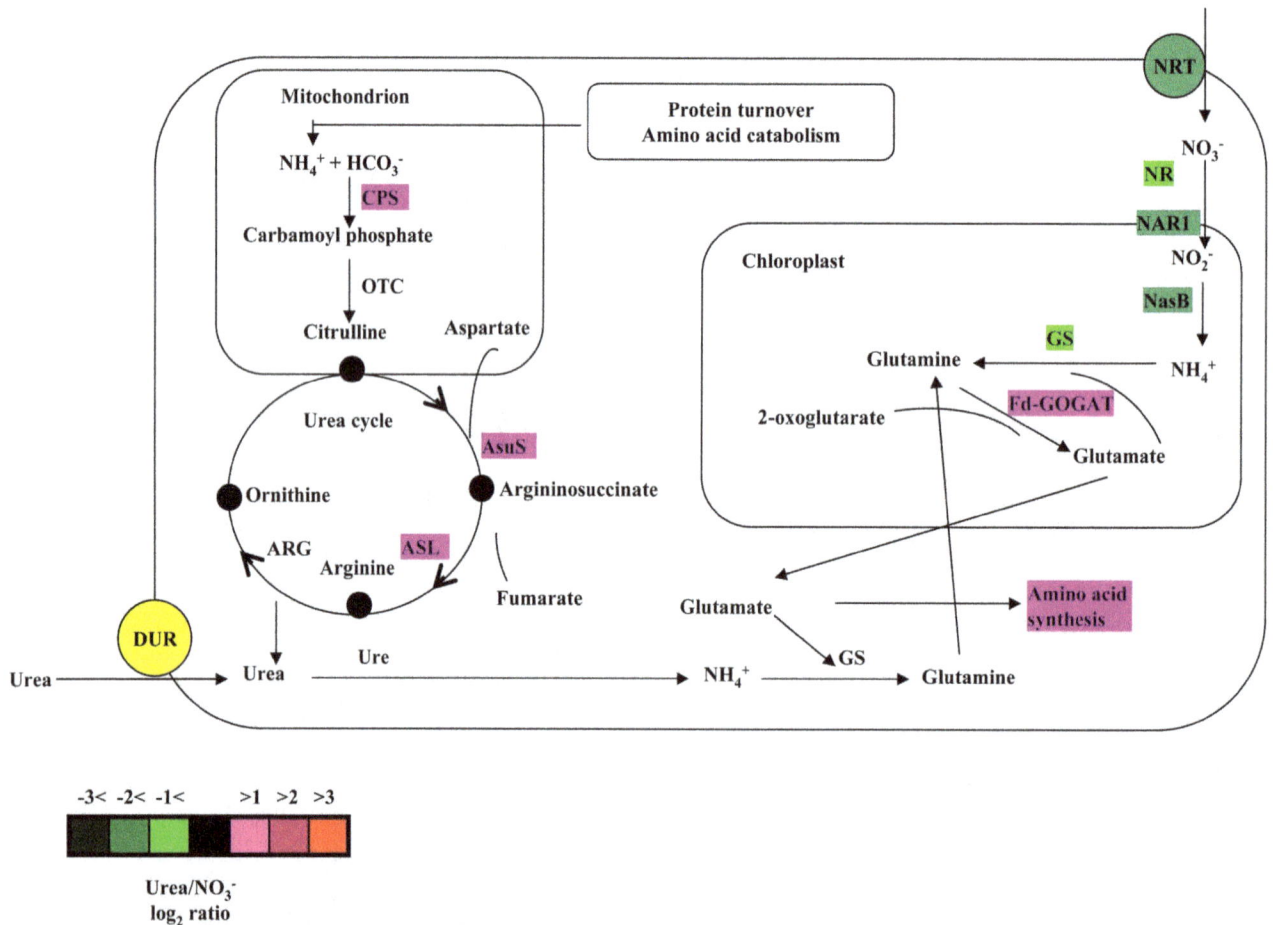

Figure 6. Proposed model showing the urea cycle, nitrate transport and assimilation, and the glutamine-glutamate cycle. Enzymes that are involved in these pathways are labeled with colors that indicate the fold change (log₂) in their transcript levels in urea-grown cells relative to nitrate-grown cells (color code is provided in the figure). NRT, nitrate transporter; NR nitrate reductase; NAR1, nitrite transporter; NasB, NADPH nitrite reductase; GS, glutamine synthetase; Fd-GOGAT, ferredoxin-dependent glutamate synthase; CPS, carbamoyl phosphate synthase; OTC, ornithine transcarboxylase; AsuS, argininosuccinate synthase; ASL, argininosuccinate lyase; ARG, arginase; Ure, urease; DUR, urea transporter.

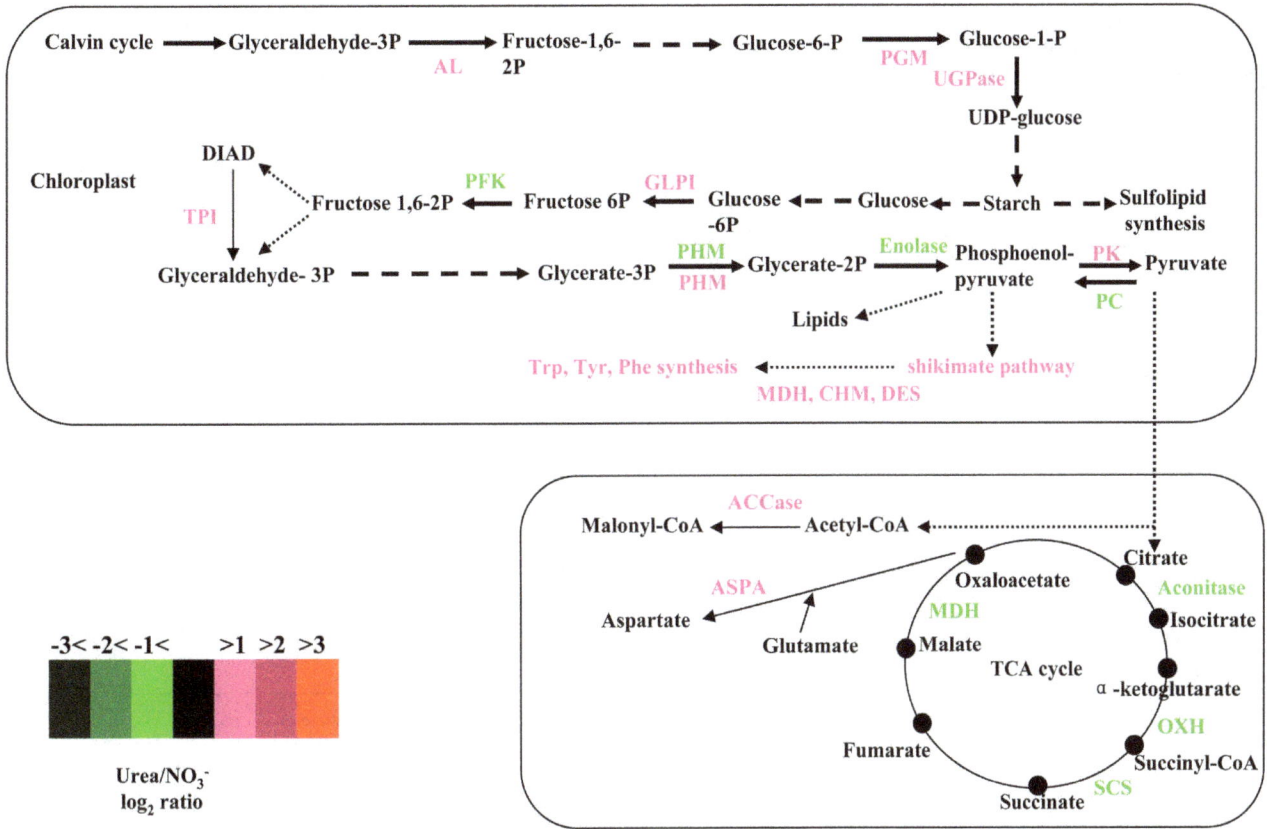

Figure 7. Pathways for starch synthesis, glycolysis, aromatic amino acid synthesis and the TCA cycle. Enzymes that are involved in these pathways are labeled with colors that indicate the fold change (log$_2$) in their transcript levels in urea-grown cells relative to nitrate-grown cells (color code is provided in the figure). Dashed lines mean that no differentially expressed genes were detected in these pathways. AL, aldolase; PGM, phosphoglucomutase; UGPase, UDP-glucose-pyrophosphorylase; PFK, phosphofructokinase; TPI, triose-phosphate isomerase; PHM, phosphoglycerate mutase; PK, pyruvate kinase, PC, pyruvate carboxylase; OXH, oxoglutarate dehydrogenase; SCS, succinyl-CoA ligase; MDH, malate dehydrogenase; EPSPS, 3-phosphoshikimate 1-carboxyvinyltransferase; CHM, chorismate mutase; DES, dehydroquinate synthase; ASPA, aspartate aminotransferase; ACCase, Acetyl-CoA carboxylase; GLPI, glucose-6-phosphate isomerase.

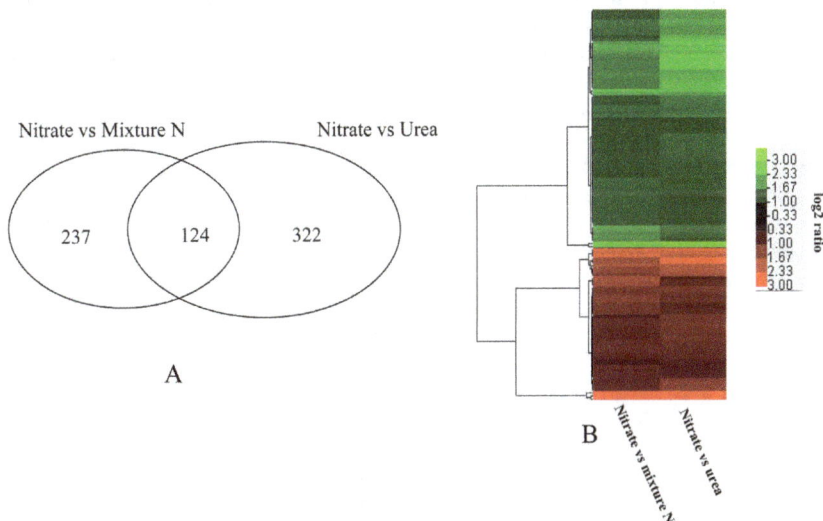

Figure 8. Specific genes transcriptionally regulated by urea. A) Number of common genes differentially regulated between the mixture N and urea groups compared with the reference nitrate group; B) the pattern of regulation of these common genes.

Table 3. GO function class of the genes differentially regulated that are common between the urea and mixture N groups compared with the reference nitrate group.

Function class	Number of genes		
	Up-regulated[a]	Down-regulated[b]	Differentially expressed
Protein synthesis	22	0	22
Nitrogen compound metabolism	7	6	13
Transport	2	5	7
Glycolysis/gluconeogenesis	5	0	5
DNA and RNA (synthesis processing, transcription, regulation)	5	3	8
Photosynthesis	5	0	5
One-carbon compound metabolism	3	0	3
Protein (targeting, modification, degradation)	5	4	9
Signalling	2	4	6
Stress	2	3	5
Urea cycle	1	0	1
Sulfate assimilation	1	0	1
Citric acid cycle	0	1	1
Vanillin synthesis	0	1	1
Cell cycle regulation	0	2	2

Note: a and b represent number of genes that are differentially expressed in urea-and mixture-N cells compared with nitrate-grown cells.

limitation and recovery, whereas two transcripts for formate/nitrite transporters showed a strong upregulation under nitrogen limitation, and then sudden downregulation under short-term nitrogen recovery (Table 4). These results indicated that these nitrite transporter genes were regulated by nitrogen concentration. The rapid responses of these transporter genes to change of nitrate concentration imply that the Chinese strain of *A. anophagefferens* can utilize nitrate or nitrite effectively. In addition, it is expected that some of the genes such as the NAR1 (ID 53005) may be used for molecular biomarkers which are indicative of changes of the DIN during brown tides.

Nitrate also induced the upregulation of transcripts encoding a tryptophan/tyrosine permease and a xanthine/uracil/vitamin C permease (Table S4) other than ammonium transporter, which suggests that amino acids and nucleotides may be superior to nitrate as N sources for this strain. This trait is similar to that of the American strain of *A. anophagefferens* [8]. Notably, transcripts for xanthine uracil permease were upregulated in *A. anophagefferens* cells under N-limited condition (Table 4). The regulation of transcript for the xanthine uracil permease is consistent with a previous study on the American strain of *A. anophagefferens* [11]. These data suggest that the Chinese strain has the ability to utilize amino acids and nucleotides as N sources. This is consistent with other studies on the American strain of *A. anophagefferens* [5,25].

Nitrate elicited an increase in the abundance of transcripts encoding proteins that are involved in $NO3^-$ assimilation and in the synthesis of N metabolites (Table S4). Transcripts encoding a putative nitrate reductase, NADPH nitrite reductase and glutamine synthetase increased in cells that were grown on nitrate compared with cells that were grown on urea. Nitrate is not only an essential nutrient that activates the expression of genes for its assimilation pathway but also a signaling molecule that regulates cellular metabolism [26]. Our data describe how nitrate was transported and then reduced to ammonium in *A. anophageffe-*

rens cells that were grown on nitrate (Fig. 6), and NADPH nitrite reductase may be the main contributor to the conversion of NO_2^- into NH_4^+, instead of ferrdoxin nitrite reductase. Glutamine synthetase (GS) is one of the two enzymes that catalyze the glutamine-glutamate (GS-GOGAT) cycle. In diatoms, two GS isoenzymes have been characterized; GSII is localized to the chloroplast, whereas GSIII is cytosolic [27]. The gene encoding GSII is upregulated in diatom cells assimilating NO_3^- but not in cells assimilating NH_4^+ taken up directly from the environment, where the gene encoding GSIII is constitutively expressed regardless of the presence or absence of nitrogen [28]. Based on these studies, we hypothesized that in *A. anophagefferens*, the GS that is encoded by the transcript (ID 20700) may be in chloroplasts, similar to the GSII of diatom, and can be induced by nitrate in media. Interestingly, no significant change was observed in the abundance of the GS transcript between cells that were grown on mixture N and that were grown on urea, which suggested that the presence of urea might inhibit the induction of the GS gene by the reduction of nitrate. Additionally, the results support the conclusion that *A. anophagefferens* primarily utilizes N from urea instead of nitrate in media with mixture N. Notably, the same three genes involved in nitrogen assimilation (ID 53391, 37238, 20700) induced by nitrate also were upregulated under nitrogen limitation and downregulated under nitrogen recovery (Table 4), suggesting that they were also regulated by nitrogen concentration.

Surprisingly, one transcript encoding a putative ferredoxin-dependent glutamate synthase (Fd-GOGAT) was shown to be upregulated in cells that were grown on urea relative to nitrate (Table S4). GOGAT catalyzes the transfer of the amide group from glutamine to 2-oxoglutarate to yield two molecules of glutamate (Fig. 6). One important fate of glutamate and glutamine is the synthesis of aspartate and asparagine. Evidence from higher plants demonstrates that the abundance of the Fd-GOGAT

Table 4. List of the genes differentially regulated among the nitrogen-replete (Nrep), limited (Ndep) and 24 h recovery (urea recovery, Urecov; nitrate recovery, Nrecov) libraries.

Metabolic pathway	Protein ID	Putative annotation	Ndep/Nrep		Urecov/Ndep		Nrecov/Ndep	
			$\log_2$ ratio	P value	$\log_2$ ratio	P value	$\log_2$ ratio	P value
Urea cycle	jgi|Auran1|20552	Carbamoyl-phosphate synthase	0.77	1.12E-164	-1.11	0	-1.82	0
	jgi|Auran1|33293	Ornithine carbamoyltransferase	0.70	3.33E-06	-1.00	1.26E-10	-0.84	3.35E-08
	jgi|Auran1|26092	Argininosuccinate synthase	1.06	8.24E-36	-0.26	3.55E-14	-0.87	8.00E-27
	jgi|Auran1|28200	Argininosuccinate lyase	0.61	8.78E-09	-0.17	0.08	-1.01	2.32E-20
	jgi|Auran1|72133	Urease	0.66	7.19E-91	-1.17	5.49E-245	-0.20	4.97E-11
	jgi|Auran1|30203	Urease accessory protein UreG	1.04	2.03E-08	-0.97	5.44E-08	-0.63	8.24E-4
Nitrogen assimilation pathway	jgi|Auran1|60332	Nitrate high affinity transporter	0.92	1.51E-101	-1.23	3.32E-171	-0.24	8.16E-11
	jgi|Auran1|53005	Nitrite transporter NAR1	-2.12	5.53E-112	-1.59	1.41E-19	-1.20	4.23E-13
	jgi|Auran1|17695	Formate/nitrite transporter	3.20	1.00E-29	-1.99	8.46E-19	-5.80	3.24E-45
	jgi|Auran1|15503	Formate/nitrite transporter	2.47	6.68E-61	-2.10	8.38E-53	-3.36	9.89E-87
	jgi|Auran1|55502	Xanthine uracil permease	1.52	0	-0.80	9.79E-116	-0.60	1.20E-69
	jgi|Auran1|53391	Nitrate reductase	1.33	4.40E-144	-1.18	1.05E-126	-1.13	2.77E-117
	jgi|Auran1|37238	NADPH nitrite reductase	1.41	3.41E-153	-4.22	0	-3.66	0
	jgi|Auran1|52709	NADH-dependent glutamate synthase	1.10	5.19E-57	-1.27	9.29E-75	-1.58	1.11E-101
	jgi|Auran1|38538	Ferredoxin-dependent glutamate synthase	0.62	2.52E-56	-0.82	7.50E-95	-1.39	1.86E-219
	jgi|Auran1|20700	Glutamine synthetase	1.25	4.90E-41	-1.20	1.40E-40	-0.57	3.68E-12
	jgi|Auran1|55217	Glutamine synthetase, type III	4.07	1.24E-08	-4.20	2.90E-09	-4.16	4.10E-09
	jgi|Auran1|60068	Formamidase	1.52	2.79E-05	-2.72	3.34E-10	-2.69	5.50E-10
	jgi|Auran1|21408	Cyanate lyase	3.64	5.79E-14	-3.09	6.70E-13	-3.06	1.21E-12
Citric acid cycle	jgi|Auran1|37323	Oxoglutarate dehydrogenase	-1.06	2.95E-57	-1.19	9.76E-36	-1.90	9.40E-69
	jgi|Auran1|28269	Succinyl-CoA ligase (SCS)	-0.74	5.08E-08	-0.66	1.15E-4	-1.37	1.35E-12
	jgi|Auran1|60302	Malate dehydrogenase (MDH)	-1.57	9.44E-77	-1.29	2.00E-21	-2.13	9.34E-42
	jgi|Auran1|59240	Isocitrate dehydrogenase (IDH)	-1.28	5.90E-226	0.74	7.43E-67	0.03	0.58

$\log_2$ ratio >1 indicates that the gene is upregulated in nitrogen-limited and recovery conditions relative to the control. $\log_2$ ratio <-1 indicates that the gene is downregulated in nitrogen-limited and recovery conditions relative to the control. Genes were considered differentially regulated at $\log_2$ ratio >1 and $P <0.05$.

transcript is regulated by light but not by nitrogen sources [29]. However, to date, the regulation of the Fd-GOGAT gene in algae is largely unknown. This study is the first to find an increase in the Fd-GOGAT transcript in *A. anophagefferens* cells that were grown on urea relative to nitrate. In addition, no significant change was observed in the abundance of Fd-GOGAT transcript under nitrogen-limited condition (Table 4). The regulation of the gene deserves further study. Interestingly, a NADH-dependent glutamate synthase (NADH-GOGAT) was found to be upregulated under nitrogen limitation (Table 4). The data suggest that compared to Fd-GOGAT, NADH-GOGAT may play more important role in assimilation of ammonia released from intracellular nitrogen compounds in *A. anophagefferens* when N provision is limited.

In this study, two transcripts (ID 37987 and 60068) encoding putative formamidases were increased by 3- to 19-fold in cells that were grown on nitrate relative to urea (Table S4). An InterProScan Sequence Search showed that these transcripts contained a conservative domain of carbon-nitrogen hydrolase. Moreover, one transcript encoding a putative formamidase was shown to be upregulated in *A. anophagefferens* cells under N-limited conditions (Table 4). Increased activities of formamidase were detected in the American strain of *A. anophagefferens* under nitrogen depletion [10]. Our data implicate that this Chinese strain can break down small amides, which is consistent with studies for the American strain of *A. anophagefferens* [8]. Therefore, amides in seawaters may serve as N sources for field populations, especially those experiencing nitrogen depletion. It is noteworthy that one transcript encoding a putative cyanate lyase was increased by16-fold in *A. anophagefferens* cells under N-limited condition (Table 4). A putative cyanase gene has been identified in the American strain of *A. anophagefferens* [8]. The cyanate lyase can hydrolyze cyanate to ammonium and CO_2. The upregulation of the transcript for cyanate lyase implicates that cyanate may also serve as a N source for field populations, especially those experiencing N starvation.

The urea cycle

Interestingly, six genes encoding components of the OUC were found in our dataset including carbamoyl phosphate synthase (CPS), ornithine carbamoyltransferase (OTC), ornithine cyclodeaminase (OCD), argininosuccinate synthase (AsuS), argininosuccinate lyase (ASL), n-acetyl-gamma-glutamyl-phosphate reductase (AggPR). Among them, four transcripts increased in cells that were grown on urea relative to nitrate (Table S4, Fig. 6). Our results confirmed the presence of the OUC in *A. anophagefferens*. In metazoans, the OUC is involved in the catabolism of amino acids and in the generation of urea for export [30]. In diatoms, the OUC serves as a distribution and repackaging hub for inorganic carbon and nitrogen, suggesting that the diatom OUC is a key pathway for anaplerotic carbon fixation into nitrogenous compounds, which are essential for diatom growth [31]. In this study, the utilization of urea elicited an increase in the levels of transcripts encoding enzymes that are involved in the OUC, which contributes to the rapid repackaging and recycling of carbon and nitrogen from urea and protein catabolism. Compared with nitrate, the anaplerotic carbon fixation and rapid use for nitrogen through the OUC may be essential for the elevated growth of *A. anophagefferens* when given urea as a N source. It has been reported that for the American strain of *A. anophagefferens*, the culture with urea had a higher growth rate than the culture with other nitrogen sources tested [8]. However, in our study, growth of this Chinese strain on urea is not significantly faster than growth on nitrate. It is likely that the relative importance of different

nutrient sources is variable, which may depend on availability and other conditions. Light intensity has been shown to be an important condition that regulates utilization of different nutrient sources in *A. anophagefferens* [7]. The influence of organic and inorganic nutrients can change over the course of a brown tide bloom, with varying effects depending on ambient nutrient levels [1]. In addition, the transcripts for the OUC were upregulated under nitrogen limitation and downregulated under nitrogen recovery, similar to the transcripts for nitrogen assimilation pathway (Table 4). The result indicated that the OUC-related genes in *A. anophagefferens* were transcriptionally regulated by nitrogen concentration, implying that they may be involved in anabolic metabolism of nitrogen and carbon in cells. In contrast, for *Arabidopsis thaliana* grown on different N sources, no transcriptional regulation by urea was observed for genes involved in the OUC cycle [32]. In total, it is postulated that for *A. anophagefferens,* the OUC cycle may play important roles in occurrence of blooms, especially when dissolved inorganic and organic nitrogen concentrations in seawater are changed.

Interestingly, we identified four transcripts encoding putative ureases that did not change significantly in abundance between cells that were grown on urea and nitrate (Tables S1 and S2). In the American strain of *A. anophagefferens*, urease activity varied positively with the growth rate, regardless of the N source [33]. Additionally, no significant change was observed in the abundance of the urease transcript in the American strain of *A. anophagefferens* cells under N deficiency (Table 4). Taken together, these results demonstrate that urease genes are constitutively expressed in both Chinese and American strain of *A. anophagefferens*, regardless of N concentration and N substrates. Consistent with its ecogenomic profile, ureases allow *A. anophagefferens* to meet its daily N demand from urea, whereas other phytoplankton do not [33].

Notably, in *A. anophagefferens* cells that are grown on mixture N, profiles of the absolute abundance of transcripts for nitrogen compound transport and metabolism, and the OUC were similar to those profiles of the same transcripts in cells that were grown on urea (Fig. 9). In particular, the absolute abundance of the transcript encoding NADPH nitrite reductase in nitrate-grown cells was approximately 18 times higher than that of the same transcript in urea and mixture N-grown cells (Fig. 9), suggesting that some substances in media with mixture N and urea markedly inhibit the expression of NADPH nitrite reductase gene. Moreover, as discussed above, a similar result was also observed in the GS transcript. Taken together, the results further support the observation that *A. anophagefferens* primarily feeds on urea instead of nitrate when urea and nitrate co-exist, which was obtained from nutrient changes in medium. This conclusion may be supported by the findings of Berg et al. (1997) that urea uptake constituted 58% to 64% of the total N uptake, whereas NO_3^- uptake contributed between 5 and 8% [4]. Therefore, a possible mechanism is that NADPH nitrite reductase is inhibited by urea or its metabolic products in *A. anophagefferens* that is grown on mixture N.

Amino acid and protein metabolism

Many transcripts encoding enzymes that are involved in amino acid synthesis increased in urea-grown cells relative to nitrate-grown cells (Table S4). For example, transcripts encoding two putative aspartate aminotransferases (ASMT) increased. This result was consistent with the observation in *Arabidopsis thaliana* that ASMT and Asp were increased in roots supplied with urea [32]. ASMT catalyzes the interconversion of aspartate and alpha-ketoglutarate to oxaloacetate and glutamate. Considering an

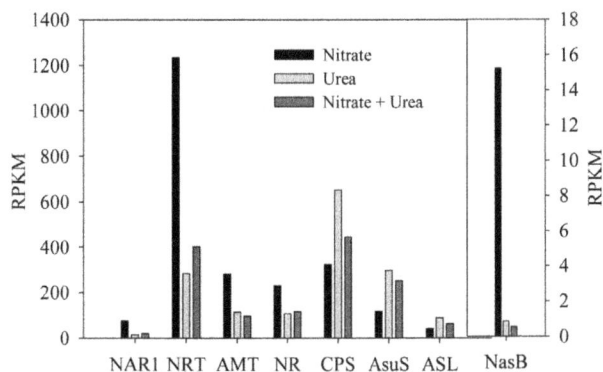

Figure 9. Absolute abundances of transcripts related to N assimilation pathway and the urea cycle. The transcript abundance that was obtained from RNA-seq data is indicated as RPKM (see Methods). The abundance of NasB transcript is shown on the right Y axis, while that of other transcripts is shown on the left Y axis. NAR1, nitrite transporter NAR1; NRT, nitrate high affinity transporter; AMT, ammonium transporter; NasB, NADPH nitrite reductase; NR, nitrate reductase; CPS, CPSase; AsuS, argininosuccinate synthase; ASL, argininosuccinate lyase.

increase in the GOGAT transcript, the elevated level of the ASMT transcript suggests that glutamate may positively regulate the expression of ASMT and direct the reaction to the synthesis of aspartate and alpha-ketoglutarate. In plants and microorganisms, aspartate is the precursor of several amino acids, including methionine, threonine, isoleucine, and lysine. In this study, three transcripts encoding enzymes that are involved in methionine metabolism and three transcripts encoding enzymes that are involved in S-adenosylmethionine (AdoMet) biosynthesis/recycling were shown to be increased in urea-grown cells. The methionine synthase transcript increased 3-fold in urea-grown cells compared with nitrate-grown cells. Methionine synthase not only catalyzes the last reaction in *de novo* methionine synthesis but also serves to regenerate the methyl group of AdoMet after methylation reactions [34]. In addition, two transcripts of methionine S-adenosyl transferase that catalyze the conversion of methionine to S-methyl-methionine also showed 3.9- to 4.7-fold upregulation in urea-grown cells (Table S4). Obviously, these results demonstrated that the synthesis of methionine and the activity of transmethylation were accelerated in urea-grown cells relative to nitrate-grown cells. Furthermore, one transcript encoding a putative ATP-sulfurylase, which is involved in sulfate assimilation, increased by 3.6-fold in urea-grown cells, which presumably indirectly supported the elevated synthesis of methionine and AdoMet. In plants, methionine occupies a central position in cellular metabolism, in which the processes of proteins, methyl-group transfers through AdoMet, and polyamines are interlocked [34]. We also identified two transcripts encoding spermine synthase and spermidine synthase that increased by 3.4- and 4.7-fold in urea-grown cells relative to nitrate-grown cells, respectively (Table S4). Spermine and spermidine are formed from AdoMet and the major polyamines in plants. Spermine and spermidine are involved in various processes, such as cell proliferation, growth, morphogenesis, differentiation, and programmed cell death [35]. Based on these results, it is likely that higher levels of AdoMet, spermine and spermidine promote the rapid growth of *A. anophagefferens* cells that are grown on urea.

Notably, three transcripts encoding enzymes that are involved in the synthesis of tryptophan, tyrosine, and phenylalanine through the shikimate pathway (Table S4, Fig. 7) and one transcript encoding tryptophan synthase increased in urea-grown cells relative to nitrate-grown cells (Table S4). In plants, these aromatic amino acids are not only essential components of protein synthesis but also serve as precursors for a wide range of secondary metabolites that are important for plant growth [36,37]. It has been shown that the exposure of plants to various stresses generally induces the expression of genes that encode the shikimate pathway [37]. Therefore, in this study, the upregulation of transcripts that encode the shikimate pathway in urea-grown cells suggests that the utilization of organic N may boost the defensive ability of *A. anophagefferens* to abiotic and biotic stresses, such as predation of microbes and/or protistan grazers, and contribute to its rapid proliferation and blooms.

Surprisingly, many transcripts encoding ribosomal proteins and translation factors increased in urea-grown cells relative to nitrate-grown cells (Table S4). Meanwhile, genes involved in protein synthesis were the most abundant in the specific genes transcriptionally regulated by urea (Table 3). The results suggest that the sufficient provision of N in urea-grown cells elevated the capability of protein biosynthesis in these cells, which is consistent with an increase in amino acid biosynthesis and in transmethylation. The eukaryotic ribosome is not only responsible for protein synthesis but also plays a major role in controlling cell growth, division, and development [38]. Additionally, a positive correlation was observed between the level of ribosomal protein transcript accumulation and cell division in suspension culture cells [39]. In this study, although it is not clear which signal regulates the expression of ribosomal protein genes, the elevated level of 27 ribosomal protein transcripts may assist in accelerating the division and growth of *A. anophagefferens* cells.

Together, the expression difference of the genes for ribosomal proteins and synthesis of amino acids between urea-grown and nitrate-grown cells may be extended to the field of brown tides. It is hypothesized that the elevated expression levels of these genes may provide *A. anophagefferens* with a greater capacity to exploit organic nitrogenous compounds compared with other phytoplankton when inorganic nitrogen levels are low but organic nitrogen levels are elevated.

Photosynthesis and carbon metabolism

In total, 57 transcripts that encode a putative plastid light harvesting protein were identified in *A. anophagefferens* (Table S1–S3). The genome of *A. anophagefferens* contains 62 genes that encode light-harvesting complex proteins (LHC), which is 1.5–3 times more than other eukaryotic phytoplankton that have been sequenced thus far [9]. LHC proteins bind antenna chlorophyll and carotenoid pigments, which increase the light-capturing capacity of photosynthetic reaction centers. In *Emiliania huxleyi*, LHC genes have been shown to increase under low light [40]. Our data demonstrated that 92% of these genes are useful and can be expressed at 100 μmol photon m^{-2} s^{-1} light, which confers a competitive advantage in absorbing light under low-irradiance conditions. In addition, of these 57 transcripts that encode LHC proteins, six transcripts were found to be upregulated in urea-grown cells relative to nitrate-grown cells and one displayed downregulation (Table S4), suggesting that different N sources may regulate different LHC genes. In combination with the profiles of Fv/Fm for *A. anophagefferens* that was grown on the three N sources (Fig. 2), these data demonstrated that changes in the abundance of a few LHC transcripts may not affect the ability of photosystems to capture light, suggesting that photosynthetic efficiency is independent of N sources when N provision is sufficient. This observation is consistent with the finding of Pustizzi et al. [7] that most of photosynthetic parameters were affected more by light intensity than by nitrogen source.

Interestingly, transcripts encoding two putative FNR (ID 23206 and 31888) increased in urea-grown cells, whereas transcript for another putative FNR (ID 52453) was markedly upregulated in nitrate-grown cells (Table S4). In the chloroplast, FNR catalyzes the interconversion between reduced ferredoxin and oxidized ferredoxin. When cells grow on nitrate, reduced ferredoxin is required for nitrite reductase (NiR) [41]. Thus, it is deduced that the gene encoding FNR (ID 52453) is induced by nitrate and works in reverse to provide reduced ferredoxin with nitrate reduction. Based on rapid light curves on day 10, ETRmax and ETE of cells grown on urea were higher than those values of cells that were grown on other N sources. The genes encoding FNRs (ID 23206 and 31888) may be the main contributors to the transfer of electron from reduced ferredoxin to NADPH during photosynthesis. It is deduced that *A. anophagefferens* cells that were grown on urea were more efficient at maintaining the expression level of genes encoding FNRs (ID 23206 and 31888) in the late phase of growth, whereas cells that were grown on nitrate were not. This observation may be due to the competition for electrons between $NADP^+$ photoreduction and ferredoxin-dependent N assimilation in cells that were grown on nitrate.

Transcripts encoding key proteins of starch synthesis increased in urea-grown cells, whereas transcripts encoding key proteins of the TCA cycle increased in nitrate-grown cells (Fig. 7). For glycolysis, one transcript encoding a putative phosphofructokinase, which is involved in a key regulatory step in the glycolytic pathway, increased by 2.2-fold in nitrate-grown cells, whereas two transcripts encoding putative pyruvate kinase increased by 3.0- to 3.2-fold in urea-grown cells. These enzymes are specific to glycolysis. In addition, other transcripts encoding glucose-6-phosphate isomerase, enolase, triose-phosphate isomerase and pyruvate carboxylase, which are involved in glycolysis and gluconeogenesis, were upregulated in urea-grown or nitrate-grown cells. These results may suggest that glycolytic activity changes little in cells that are grown on different nitrogen sources. However, relative to urea-grown cells, the activity of the TCA cycle was elevated in nitrate-grown cells. The TCA cycle is not only a source of energy and reducing equivalents but also provides carbon skeletons for nitrogen assimilation and for the biosynthesis of compounds. Along with an increase in starch synthesis in urea-grown cells, this result suggests that more carbon from the Calvin cycle may be channeled into the TCA cycle in nitrate-grown cells relative to urea-grown cells. In contrast, more carbon from the Calvin cycle may accumulate as starch in urea-grown cells relative to nitrate-grown cells. This difference may be due to the elevated activity of the OUC in urea-grown cells. In diatoms, correlation analyses of metabolites of the TCA cycle and OUC indicated that the OUC derivatives urea and proline are tightly coupled to TCA cycle intermediates, which suggests that there are important connections between the OUC, the glutamine/glutamate cycle and the TCA cycle in diatoms [31]. In this study, the differential expression of genes for the OUC, the TCA cycle, and the glutamine/glutamate cycle suggested that, similar to animal cells [42], the OUC may be linked to the TCA cycle through the aspartate-argininosuccinate shunt (Fig. 6) in *A. anophagefferens* cells. Different patterns of transcripts for the OUC and TCA cycle in *A. anophagefferens* cells under N-limited and recovery conditions may also support the conclusion. Perhaps, the anaplerotic carbon-fixation for urea-C through the OUC attenuates the activity of the TCA cycle and leads to the accumulation

of starch. This turnover and reallocation of intracellular carbon and nitrogen into key cellular components, such as protein, AdoMet, spermine and aromatic amino acids, may increase competitiveness of *A. anophagefferens* relative to other phytoplankton when concentrations of dissolved organic nitrogen are elevated in the anthropogenically coastal waters.

Signal transduction

Transcripts encoding many putative signal proteins, such as (p)ppGpp synthetase I, G protein, protein kinases, RNA-binding region RNP-1, Nog1 nucleolar GTPase, proteins containing the WD 40 domain, phospholipase D, and RAB family GTPase were shown to be upregulated in urea-grown or nitrate-grown cells (Table S4). These genes may synergistically mediate elaborate cell signaling and density sensing in blooms, which is important for detecting an ambient environment. Based on these data, it is difficult to distinguish the significance of the differential expression of genes that are involved in signal transduction, which are regulated by urea or nitrate. However, as has been reported, the *A. anophagefferens* genome encodes many more proteins that are involved in cell signaling transduction than other phytoplankton [9]. Our study demonstrated the expression of these genes and suggested that these genes may play important roles in the formation of blooms.

Notably, five transcripts encoding enzymes that regulate the biosynthesis of sterols increased in urea-grown cells relative to nitrate-grown cells (Table S4). In higher plants, sterols are precursors of steroid hormones. It is reported that, in *Arabidopsis*, sitosterol, stigmasterol, and some abnormal sterols upregulate the characteristic cell expansion and proliferation of genes [43]. Sterols themselves may act as signaling molecules in plants in a manner that is analogous to the action of cholesterol in mammalian systems [44]. In this study, an increase in transcript for the biosynthesis of sterols may provide cells with sufficient sterols and steroid hormones, which may promote cell expansion and proliferation. Interestingly, one transcript encoding a putative pescadillo-like protein was shown to increase 3.3-fold in urea-grown cells relative to nitrate-grown cells (Table S4). In yeast, pescadillo plays a crucial role in cell proliferation and in the cell cycle [45]. Therefore, in *A. anophagefferens*, the putative pescadillo-like protein may be a key regulatory protein that affects cell proliferation and cell cycle progression.

Comparisons between the Chinese and American strains

This Chinese strain we isolated has been reported to have 99.7–100% similarity to *A. anophagefferens* Hargraves et Sieburth, the causative species of brown tides on the east coast of USA based on the 18S rDNA [2]. In addition, characteristic pigment, 19′-butanoyloxyfucoxanthin in the American strain has also been detected in the Chinese strain of *A. anophagefferens* [46]. For *A. anophagefferens* grown on urea, nitrate, and the mixture of urea and nitrate, some physiological features of the Chinese strain including the growth rate and Fv/Fm were similar to non-axenic cultures of the American strain [7]. However, the maximum growth rate of the Chinese strain was lower than that of the American strain [7].

Our RNA-seq data showed that 83.9–84.7% of the clean reads from the Chinese strain mapped to at least one location in genome of the American strain of *A. anophagefferens* for each sample. Many genes encoding proteins that were involved in nitrogen acquisition and assimilation in the Chinese strain exhibited similar responses with the American strain to nitrate, urea and nitrogen depletion. These data further indicate that the Chinese strain has genetically high similarity to the American strain of *A.*

anophagefferens. In addition, we performed a systems-level analysis of the expression differences of genes regulated by urea or nitrate in the Chinese strain. Changes of some important metabolic pathways including the OUC, TCA cycle, and amino acid synthesis were highlighted. To date, these systemic analyses are not reported in the American strain.

We also noted that the reference transcript we used is not pure transcriptome of *A. anophagefferens*, but a dataset that includes complete gene models predicted from *A. anophagefferens* genome and from available EST and cDNA data. However, only 47.4 to 50.9% of the reads mapped to the gene models in the dataset for each sample. For these reads that did not map to any gene models in the reference transcript, but mapped to the genome sequence, one possible explanation is that they may be associated with exon-3′ UTR and exon-5′ UTR sequences, intergenic locations and intronic regions. As for the reads that did not map to any location in the genome of the American strain of *A. anophagefferens*, this may be due to the gene differences between the Chinese and American strains or errors in gene predictions.

Conclusions

In this study, similar levels of transcripts for nitrate transport and assimilation were detected in mixture N-grown cells and urea-grown cells. Together with changes in nutrient concentrations in media, these results may suggest that *A. anophagefferens* primarily feeds on urea instead of nitrate when urea and nitrate co-exist. A possible mechanism is that NADPH nitrite reductase is inhibited by urea or its metabolic products in urea-grown cells. Transcripts for the OUC, and synthesis of glutamate and aspartate were upregulated by urea, whereas transcripts for the TCA cycle were negatively regulated by urea treatment, suggesting that in *A. anophagefferens* cells, the OUC may be linked to the TCA cycle through the aspartate-argininosuccinate shunt. This speculation was further supported by pattern of transcripts for the OUC and TCA cycles in response to N-limitation and recovery. Transcripts for the biosynthesis of sterols and pescadillo were markedly upregulated in urea-grown cells, presumably regulating the rapid proliferation of *A. anophagefferens* cells and blooms. This study is the first to determine potential roles of the OUC in the reallocation of intracellular carbon and nitrogen in *A. anophagefferens* cells. Our results may provide a partial explanation for blooms of *A. anophagefferens* in estuaries with elevated levels of organic matter.

Acknowledgments

We thank BGI-tech (Shenzhen) for sequencing and Y. Zhang for bioinformatic analysis.

References

1. Gobler CJ, Lonsdale DJ, Boyer GL (2005) A review of the causes, effects, and potential management of harmful brown tide blooms caused by *Aureococcus anophagefferens* (Hargraves et sieburth). Estuaries 28: 726–749.
2. Zhang QC, Qiu LM, Yu RC, Kong FZ, Wang YF, et al. (2012) Emergence of brown tides caused by *Aureococcus anophagefferens* Hargraves et Sieburth in China. Harmful Algae 19: 117–124.
3. Anderson DM, Burkholder JM, Cochlan WP, Glibert PM, Gobler CJ, et al. (2008) Harmful algal blooms and eutrophication: Examining linkages from selected coastal regions of the United States. Harmful Algae 8: 39–53.
4. Berg GM, Glibert PM, Lomas MW, Burford MA (1997) Organic nitrogen uptake and growth by the chrysophyte *Aureococcus anophagefferens* during a brown tide event. Mar Biol 129: 377–387.
5. Mulholland MR, Gobler CJ, Lee C (2002) Peptide hydrolysis, amino acid oxidation, and nitrogen uptake in communities seasonally dominated by *Aureococcus anophagefferens*. Limnol Oceanogr 47: 1094–1108.

Supporting Information

Table S1 The information of genes identified in *A. anophagefferens* grown on urea using RNA-seq technology.

Table S2 The information of genes identified in *A. anophagefferens* grown on nitrate using RNA-seq technology.

Table S3 The information of genes identified in *A. anophagefferens* grown on mixture N using RNA-seq technology.

Table S4 Information of differentially expressed genes in *A. anophagefferens* cells identified when nitrate-grown cells were compared to urea-grown cells.

Table S5 Information of differentially expressed genes in *A. anophagefferens* cells identified when nitrate-grown cells were compared to mixture N-grown cells.

Table S6 Information of differentially expressed genes in *A. anophagefferens* cells identified when mixture N-grown cells were compared to urea-grown cells.

Table S7 Information of differentially expressed genes in *A. anophagefferens* cells identified when N-depleted cells were compared to N-replete cells.

Table S8 Information of differentially expressed genes in *A. anophagefferens* cells identified when urea-recovery cells were compared to N-depleted cells.

Table S9 Information of differentially expressed genes in *A. anophagefferens* cells identified when nitrate-recovery cells were compared to N-depleted cells.

Table S10 Information of common differentially expressed genes in *A. anophagefferens* cells identified when nitrate-grown cells were compared to mixture N-grown and urea-grown cells, respectively.

Author Contributions

Conceived and designed the experiments: HPD SHL. Performed the experiments: KXH HLW JYC YLD. Analyzed the data: HPD. Contributed reagents/materials/analysis tools: SHL KXH HLW JYC YLD. Wrote the paper: HPD.

6. Lomas MW, Glibert PM, Clougherty DA, Huber DR, Jones J, et al. (2001) Elevated organic nutrient ratios associated with brown tide algal blooms of *Aureococcus anophagefferens* (Pelagophyceae). J Plankton Res 23: 1339–1344.
7. Pustizzi F, MacIntyre H, Warner ME, Hutchins DA (2004) Interaction of nitrogen source and light intensity on the growth and photosynthesis of the brown tide alga *Aureococcus anophagefferens*. Harmful Algae 3: 343–360.
8. Berg GM, Shrager J, Glöckner G, Arrigo KR, Grossman AR (2008) understanding nitrogen limitation in *Aureococcus anophagefferens* (Pelagophyceae) through cDNA and qRT-PCR analysis. J Phycol 44: 1235–1249.
9. Gobler CJ, Berry DL, Dyhrman ST, Wilhelm SW, Salamov A, et al. (2011) Niche of harmful alga *Aureococcus anophagefferens* revealed through ecogenomics. P Nat Acad Sci 108: 4352–4357.
10. Gobler CJ, Berry DL, Dyhrman ST, Wilhelm SW, Salamov A, et al. (2011) Niche of harmful alga *Aureococcus anophagefferens* revealed through ecogenomics. P Nat Acad Sci USA 108: 4352–4357.

11. Wurch LL, Haley ST, Orchard ED, Gobler CJ, Dyhrman ST (2011) Nutrient-regulated transcriptional responses in the brown tide-forming alga *Aureococcus anophagefferens*. Environ Microbiol 13: 468–481.

12. Wang Z, Gerstein M, Snyder M (2009) RNA-Seq: a revolutionary tool for transcriptomics. Nat Rev Genet 10: 57–63.

13. Gonzalez-Ballester D, Casero D, Cokus S, Pellegrini M, Merchant SS, et al. (2010) RNA-Seq Analysis of Sulfur-Deprived Chlamydomonas Cells Reveals Aspects of Acclimation Critical for Cell Survival. Plant Cell 22: 2058–2084.

14. Goldman JC, McCarthy JJ (1978) Steady state growth and ammonium uptake of a fast-growing marine diatom. Limnol Oceanogr 23: 695–703.

15. Webb W, Newton M, Starr D (1974) Carbon dioxide exchange of Alnus rubra. Oecologia 17: 281–291.

16. Rahmatullah M, Boyde TRC (1980) Improvements in the determination of urea using diacetyl monoxime; methods with and without deproteinisation. Clinica Chimica Acta 107: 3–9.

17. Anderson L (1979) Simultaneous spectrophotometric determination of nitrite and nitrate by flow injection analysis. Anal Chim Acta 110: 123–128.

18. Li R, Yu C, Li Y, Lam TW, Yiu SM, et al. (2009) SOAP2: an improved ultrafast tool for short read alignment. Bioinformatics 25: 1966–1967.

19. Morrissy AS, Morin RD, Delaney A, Zeng T, McDonald H, et al. (2009) Next-generation tag sequencing for cancer gene expression profiling. Genome Res 19: 1825–1835.

20. Chen S, Yang P, Jiang F, Wei Y, Ma Z, et al. (2010) *De Novo* Analysis of Transcriptome Dynamics in the Migratory Locust during the Development of Phase Traits. PloS one 5: e15633.

21. Audic S, Claverie JM (1997) The significance of digital gene expression profiles. Genome Res 7: 986–995.

22. Song B, Ward BB (2007) Molecular cloning and characterization of high-affinity nitrate transporters in marine phytoplankton. J Phycol 43: 542–552.

23. Crawford NM (1995) Nitrate: nutrient and signal for plant growth. Plant Cell 7: 859–868.

24. Rexach J, Fernandez E, Galvan A (2000) The *Chlamydomonas reinhardtii* Nar1 Gene Encodes a Chloroplast Membrane Protein Involved in Nitrite Transport. Plant Cell 12: 1441–1453.

25. Berg GM, Repeta DJ, Laroche J (2002) Dissolved Organic Nitrogen Hydrolysis Rates in Axenic Cultures of *Aureococcus anophagefferens* (Pelagophyceae): Comparison with Heterotrophic Bacteria. Appl environ microb 68: 401–404.

26. Fernandez E, Galvan A (2008) Nitrate Assimilation in Chlamydomonas. Eukaryotic Cell 7: 555–559.

27. Armbrust EV, Berges JA, Bowler C, Green BR, Martinez D, et al. (2004) The Genome of the Diatom *Thalassiosira Pseudonana*: Ecology, Evolution, and Metabolism. Science 306: 79–86.

28. Takabayashi M, Wilkerson FP, Robertson D (2005) response of glutamine synthetase gene transcription and enzyme to external nitrogen sources in the diatom *Skeletonema Costatum* (Bacillariophyceae). J Phycol 41: 84–94.

29. Coschigano KT, Melo-Oliveira R, Lim J, Coruzzi GM (1998) Arabidopsis gls Mutants and Distinct Fd-GOGAT Genes: Implications for Photorespiration and Primary Nitrogen Assimilation. Plant Cell 10: 741–752.

30. Lee B, Yu H, Jahoor F, O'Brien W, Beaudet AL, et al. (2000) In vivo urea cycle flux distinguishes and correlates with phenotypic severity in disorders of the urea cycle. P Nat Acad of Sci 97: 8021–8026.

31. Allen AE, Dupont CL, Obornik M, Horak A, Nunes-Nesi A, et al. (2011) Evolution and metabolic significance of the urea cycle in photosynthetic diatoms. Nature 473: 203–207.

32. Merigout P, Lelandais M, Bitton F, Renou J-P, Briand X, et al. (2008) Physiological and transcriptomic aspects of urea uptake and assimilation in Arabidopsis Plants. Plant Physiology 147: 1225–1238.

33. Fan C, Glibert PM, Alexander J, Lomas MW (2003) Characterization of urease activity in three marine phytoplankton species, *Aureococcus anophagefferens*, Prorocentrum minimum, and Thalassiosira weissflogii. Mar Biol 142: 949–958.

34. Ravanel S, Gakiere B, Job D, Douce R (1998) The specific features of methionine biosynthesis and metabolism in plants. P Nat Acad Sci USA 95: 7805–7812.

35. Kumar A, Taylor M, Altabella T, Tiburcio AF (1997) Recent advances in polyamine research. Trends Plant Sci 2: 124–130.

36. Dixon RA (2001) Natural products and plant disease resistance. Nature 411: 843–847.

37. Tzin V, Galili G (2010) new insights into the shikimate and aromatic amino acids biosynthesis pathways in plants. Molecular Plant 3: 956–972.

38. Ito T, Kim G-T, Shinozaki K (2000) Disruption of an Arabidopsis cytoplasmic ribosomal protein S13-homologous gene by transposon-mediated mutagenesis causes aberrant growth and development. Plant J 22: 257–264.

39. Gao J, Kim S-R, Chung Y-Y, Lee J, An G (1994) Developmental and environmental regulation of two ribosomal protein genes in tobacco. Plant Mol Biol 25: 761–770.

40. Lefebvre SC, Harris G, Webster R, Leonardos N, Geider RJ, et al. Characterization and expression analysis of the Lhcf gene family in *Emiliania Huxleyi* (Haptophyta) reveals differential responses to light and CO_2 J Phycol 46: 123–134.

41. Brunswick P, Cresswell CF (1988) Nitrite uptake into intact pea chloroplasts: I. kinetics and relationship with nitrite assimilation. Plant Physiol 86: 378–383.

42. Morris SM (2002) Regulation of enzymes of the urea cycle and arginine metabolism. Annu Rev Nutr 22: 87–105.

43. He J-X, Fujioka S, Li T-C, Kang SG, Seto H, et al. (2003) Sterols regulate development and gene expression in Arabidopsis. Plant Physiol 131: 1258–1269.

44. Carland F, Fujioka S, Nelson T (2010) The sterol methyltransferases SMT1, SMT2, and SMT3 influence Arabidopsis development through nonbrassinosteroid products. Plant Physiol 153: 741–756.

45. Kinoshita Y, Jarell AD, Flaman JM, Foltz G, Schuster J, et al. (2001) Pescadillo, a novel cell cycle regulatory protein abnormally expressed in malignant cells. J Biol Chem 276: 6656–6665.

46. Kong F, Yu R, Zhang Q, Yan T, Zhou M (2012) Pigment characterization for the 2011 bloom in Qinhuangdao implicated "brown tide" events in China. Chin J Oceanol Limn 30: 361–370.

13

Computational Analysis of Reciprocal Association of Metabolism and Epigenetics in the Budding Yeast: A Genome-Scale Metabolic Model (GSMM) Approach

Ali Salehzadeh-Yazdi, Yazdan Asgari, Ali Akbar Saboury, Ali Masoudi-Nejad*

Laboratory of Systems Biology and Bioinformatics (LBB), Institute of Biochemistry and Biophysics, University of Tehran, Tehran, Iran

Abstract

Metaboloepigenetics is a newly coined term in biological sciences that investigates the crosstalk between epigenetic modifications and metabolism. The reciprocal relation between biochemical transformations and gene expression regulation has been experimentally demonstrated in cancers and metabolic syndromes. In this study, we explored the metabolism-histone modifications crosstalk by topological analysis and constraint-based modeling approaches in the budding yeast. We constructed nine models through the integration of gene expression data of four mutated histone tails into a genome-scale metabolic model of yeast. Accordingly, we defined the centrality indices of the lowly expressed enzymes in the undirected enzyme-centric network of yeast by CytoHubba plug-in in Cytoscape. To determine the global effects of histone modifications on the yeast metabolism, the growth rate and the range of possible flux values of reactions, we used constraint-based modeling approach. Centrality analysis shows that the lowly expressed enzymes could affect and control the yeast metabolic network. Besides, constraint-based modeling results are in a good agreement with the experimental findings, confirming that the mutations in histone tails lead to non-lethal alterations in the yeast, but have diverse effects on the growth rate and reveal the functional redundancy.

Editor: Julio Vera, University of Erlangen-Nuremberg, Germany

Funding: The authors have no funding or support to report.

Competing Interests: The authors have declared that no competing interests exist.

* Email: amasoudin@ibb.ut.ac.ir

Introduction

Biological systems contain many highly interconnected processes that function in a coordinated fashion to produce cellular behavior. Therefore, understanding the biological networks and their crosstalk properties is important to put an accurate interpretation on the complex nature of biological systems. Metabolism and epigenetics are two central biological processes that are vital to the organisms' survival and reproduction. Metabolism is a complex network of dynamical biochemical reactions empowering organisms to grow, reproduce and maintain their integrity [1]. Epigenetic mechanisms, as a constituent of gene expression regulation machinery, alter transcriptional activities of various genes, independently of changes in their nucleotides sequences [2]. These mechanisms include modifications of DNA and histone proteins, such as different combinations of histone acetylation, methylation and phosphorylation which shape, the so-called 'histone code' and DNA methylation. Ultimately, these modifications lead to transcriptionally active or inactive chromatin [3]. Growing evidence suggests the possible link between different metabolic states of living systems and the epigenetic modifications [4]. For instance, histone acetylation is under control of changes in the intracellular concentration of acetyl-CoA [5]. On the other hand, some of these epigenetic modifications change the metabolic gene expression patterns under varying conditions [6]. Furthermore, recent studies propose that the impairments in the interface between epigenetics and metabolism are strongly associated with development of cancers [7] and metabolic syndromes [8]. Consequently, the cellular metabolism and the epigenetic modifications are not independent entities, and they would better be viewed as an integrated discipline, metaboloepigenetics that focuses on the crosstalk between them [9]. To illustrate the genotype-phenotype relationship of metabolism, a complete map of biochemical reactions and their comprehensive connections in a cell is required [10]. The Genome-Scale Metabolic Model (GSMM) is a mathematical framework, to gain a comprehensive understanding of physiology and the metabolic capacities of the cell, as well as being used for integrative data analysis of genetic, epigenetic and metabolism in combination [11]. Following the introduction of GSMM, the integration of gene expression data into the GSMM, was the new challenge for a better prediction of the metabolic cell fate. The integrative data approach, leads to a deeper understanding of the occurrence of certain changes in different conditions and creates condition- and tissue-specific models [12,13]. For the first time, Covert and Palsson [14] addressed this issue in 2002. In 2004, Akesson et al. [15] used gene expression data as an additional constraint on the metabolic fluxes in yeast. Afterward, different algorithms were developed for tackling this challenge; GIMME [16], E-Flux [17], Moxley [18], MADE [19], RELATCH [20], INIT [21] and mCADRE [22]. Recently, advantages and disadvantages of different approaches of integration of expression data into constraint-based modeling have been evaluated [23]. Due to the availability of curated data [24], we focused on *Saccharomyces cerevisiae*, the budding yeast, as a model for investigating the global influences of epigenetic modifications on metabolism. The first budding yeast GSMM, iFF708, [25] was released in 2003. This model consists of 708 genes and 1175 reactions. Afterward, the improved versions of yeast GSMM were reconstructed; iND750 [26], iLL672 [27], iLN800 [28], iMM904 [29], improved iMM904 [30], and yeast 5 [31] released in the standard format using jamboree approach [32] which is the most up to date version.

In this study, we used the GSMM of *S. cerevisiae*, *Yeast_6.01*, as a scaffold, consisting of 889 genes, 1889 reactions (150 reactions are irreversible), 1456 metabolites and the standard biomass equation. Then we analyzed the nine gene expression profiles of four different states of mutated histones tails (H2A, H2B, H3, and H4) extracted from Gene Expression Omnibus (GEO) [33]. Subsequently, we constructed nine models, by integrating fold change values of the significantly lowly expressed reactions of each gene expression profile, as an additional constraint on metabolic fluxes. Afterward, we have tried to explore possible relation between topological analysis of metabolic network and down-regulated genes with this assumption that down-regulated genes could affect and control the yeast metabolic network. Then, Flux Balance Analysis (FBA), as a constraint-based modeling approach has been used to compute and compare [34], [35] the impact of the mutated histone tails on every reaction flux, the global metabolic fluctuations and the growth rate. The results verifies the prior experimental findings, showing that the histone tails are not essential for the viability of yeast but have a large impact on a vast range of metabolic reactions, which reveals the functional redundancy of the histone tails and their ability to regulate their own metabolite sources [36].

Materials and Methods

2.1. The GSMM of yeast and gene expression data

In this study, we used microarray gene expression profiles and RNA-Seq data of mutated histone tails [37–40] of *S. cerevisiae* extracted from GEO database. The GSE accession numbers and their categorizations are listed below. The GSMM of yeast in SBML format (Systems Biology Markup Language), a representative format for mathematical models of biological processes such as metabolic network, was obtained from http://www.comp-sys-bio.org/yeastnet. The metabolic states of yeast have been studied by integration of gene expression data into the *Yeast_6.01* GSMM as a scaffold model.

GSE1639: Consists of gene expression profiles of H3 and H4 mutated tail. [37].

Group 1: H3 deleted tail.

Group 2: H3 substituted tail, lysine substituted with glutamine residues in histone 3 tail.

Group 3: H4 substituted tail, lysine substituted with glutamine residues in histone 4 tail.

Group 4: H4 deleted tail.

GSE3806: Consists of gene expression profiles of H2B mutated tail. [38].

Group 5: H2B deleted tail.

Group 6: H2B substituted tail, lysine substituted with glutamine residues in histone 2B tail.

GSE7337: Consists of gene expression profile of H2A deleted tail. [39].

Group 7: H2A deleted tail.

GSE7338: Consists of gene expression profile of H2A substituted tail. [39].

Group 8: H2A substituted tail, lysine substituted with glutamine residues in histone 2A tail.

GSE29293: Consists of gene expression profile of H3 depletion. [40].

Group 9: H3 depletion.

2.2. Gene expression analysis

Microarray gene expression data analysis for each given group has been done by GeWorkbench 2.4.0 software [41]. RNA-Seq data analysis has been done according to methodology explained in [42]. Statistical significance of up- and down-regulated genes computed by t-test for the Wild Type (WT) and its corresponding mutated histone tail on log_2-normalized data (using WT and mutated histone tail data sets as case and control, respectively). The differential expression of a gene defined significant, if *p-value* <0.01 and the negative fold change value was indicative of the down-regulated genes. The SBML file of *Yeast_6.01* was converted into COBRA (Constraints Based Reconstruction and Analysis) model structure by COBRA toolbox for subsequent analysis and the lowly expressed reactions determined according to gene-protein-reaction relationship (GPR) in the COBRA model of *Yeast_6.01*.

2.3. Model construction

FBA, a constraint-based modeling approach, calculates the flow of metabolites through a metabolic network. This method allows predicting the rate of production of a metabolite or the growth rate of an organism. FBA includes delineating constraints on the network, based on environmental, physicochemical, regulatory, enzyme capacity and thermodynamics principles for shrinking the solution space. Integration of transcriptomic data into GSMM is a way to generate better predictive computational models through adding an extra biologically meaningful constraint and limiting the solution space of the GSMM. Blazier and Papin in a comprehensive review, summarized the differences, limitations and advantages of all integration algorithms [43]. Considering biological concepts, there are some limitations in the different integration algorithms. For example, the GIMME algorithm reduces gene expression data into binary states (0 and 1 for on and off state of an enzyme, respectively) and the iMAT algorithm [44] into three states (-1, 0, and 1 for lowly expressed, moderately expressed, and highly expressed enzymes, respectively). It is not biologically acceptable that the down-regulated genes and their corresponding fluxes removed from the model, because lowly expressed is not equivalent with the gene silencing. In MADE approach [19], there is no exact threshold to determine which reaction is highly expressed and which one is lowly expressed and in E-Flux [17] there is no function to convert gene expression level into fluxes.

Totally, there are two main classes for creating condition-specific models; switch- and valve-based approaches. In a switch-based approach, the lowly expressed genes removed from constrained model by adjusting their corresponding reactions boundaries (lower and upper bounds) to zero, while in a valve-based approach, the activity of lowly expressed genes decrease by reducing the corresponding reaction boundaries according to expression values [45]. In our integration method, we followed the valve-based approach. First, we used the simulation results of the unconstrained initial model (*Yeast_6.01*) to identify the reaction's fluxes (which matches fluxes observed in vivo) and established the boundaries of the WT model according to the reaction fluxes of the initial model [31]. Then, we imposed the gene expression data of nine groups on these fluxes to construct new models. It means that, the solution space of the constructed GSMMs has been shrunk by adjusting the reaction's fluxes of initial model according to their fold change values of lowly expressed genes. In other words, the upper and lower bounds of our new constructed models

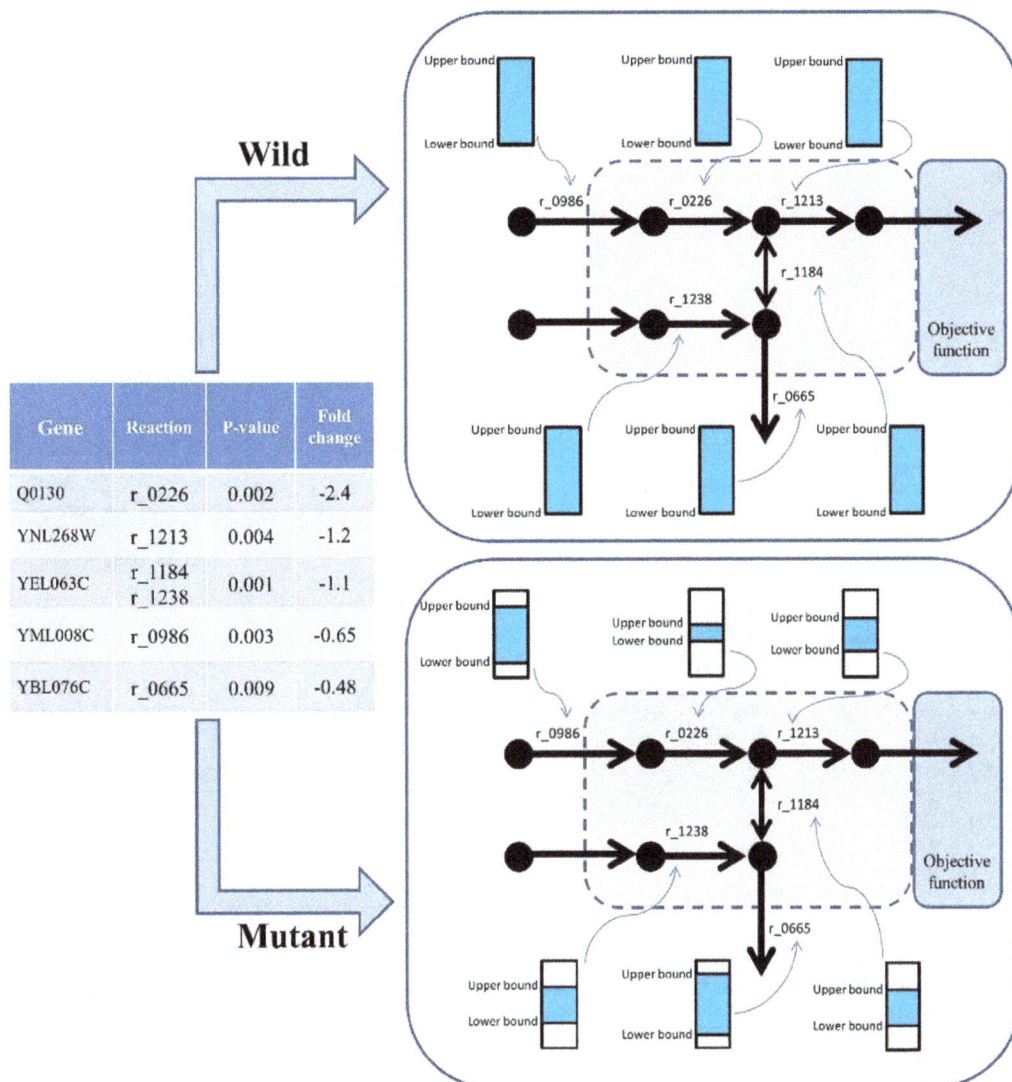

Gene	Reaction	P-value	Fold change
Q0130	r_0226	0.002	-2.4
YNL268W	r_1213	0.004	-1.2
YEL063C	r_1184 r_1238	0.001	-1.1
YML008C	r_0986	0.003	-0.65
YBL076C	r_0665	0.009	-0.48

Figure 1. This figure is a schematic workflow of our method for GSMM construction. The significant down-regulated gene (*p-value* <0.01) and the corresponding reactions (according to GPR) were determined. (In this model, 5 genes and 6 reactions were identified). Afterward, we have restricted the fluxes of the given reactions according to their fold change values. For example if the fold change value is -2, we have restricted the upper bound and lower bound of the corresponding flux to one quarter of the corresponding WT.

are the reduced simulated reaction's fluxes of initial model. However, to represent up-regulated genes, we have to increase the flux ranges of the corresponding reactions which will not have any impact on the FBA-based solution. Therefor we excluded the up-regulated genes from our study. We have used *p-value* as a threshold for identification of highly and lowly expressed genes. Finally, for not reducing the gene expression data into binary or three states, we used fold change values as a quantitative parameter to constraint the fluxes. Figure 1 shows that how expression data integrated into GSMM.

Moreover, we used the standard biomass equation as an objective function of the yeast growth rate. To investigate the comprehensive metabolic properties of the models, several COBRA utilities have used. All GSMMs are available in the File S2.

2.4. Topological analysis

Centrality analysis has been carried out on the undirected enzyme-centric network [46] of *Yeast_601* using CytoHubba plug-in in Cytoscape [47]. We have used twelve centrality indices: Maximal Clique Centrality (MCC), Density of Maximum Neighborhood Component (DMNC), Maximum Neighborhood Component (MNC), Degree, Edge Percolated Component (EPC), Bottleneck, Eccentricity, Closeness, Radiability, Betweenness, Stress and Clustering Coefficient. Then, the lowly expressed enzymes of nine constructed GSMMs, were sorted based on their centrality indices. (For more information, see part A in the File S1).

2.5. COBRA utilities that have been used in this study

Among various software for calculating the FBA, we used the well-known MATLAB toolbox, COBRA, and a standard objective function (maximization of biomass equation) for evaluating the

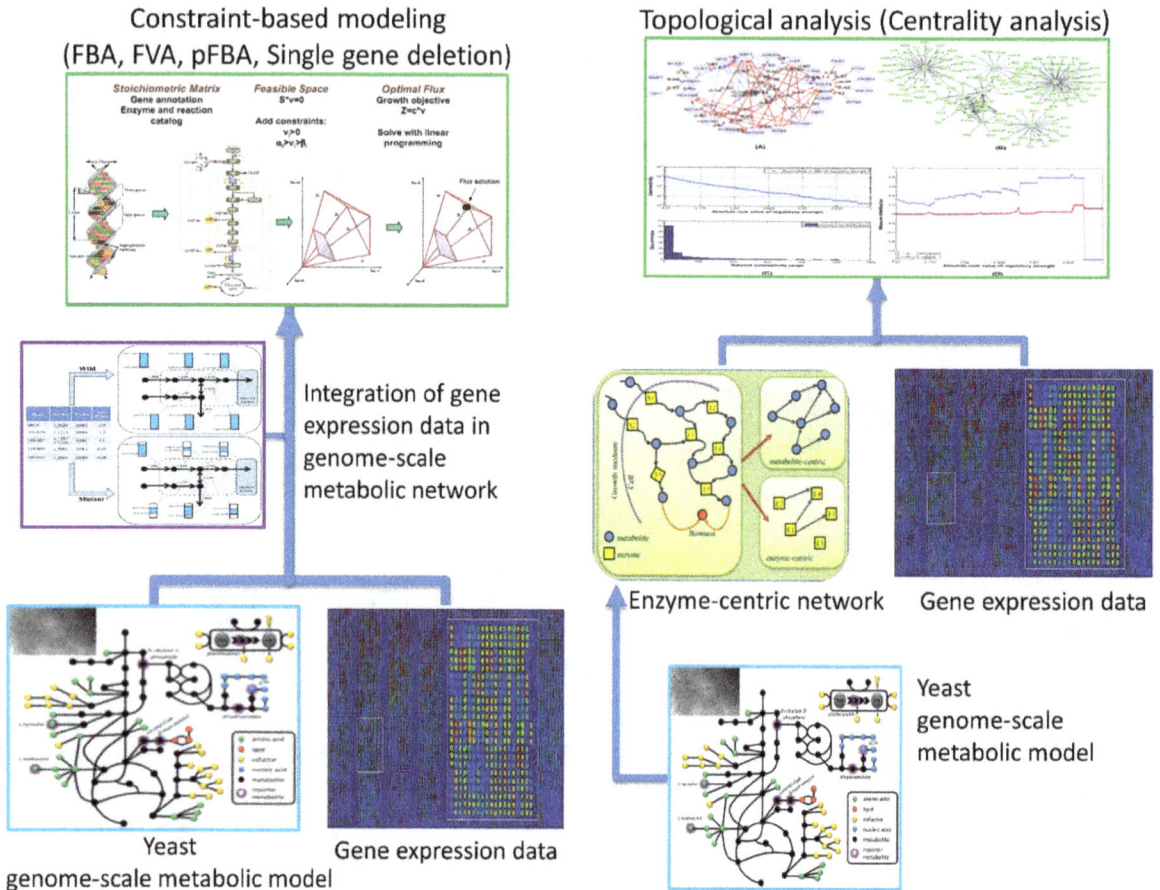

Figure 2. Shows the workflow of our study. This study consists of two parts; topological analysis and constraint-based modeling. In the topological analysis section, we used the twelve centrality indices of the undirected enzyme-centric network of yeast and examined the distribution of the down-regulated genes in the nine mutated histone tails profiles, extracted from geWorkbench software. In the constraint-based modeling section, we integrated gene expression of nine groups of the mutated histone tails profiles to the yeast GSMM and constructed new models. FBA, FVA, pFBA and Single gene deletion analyzed by COBRA toolbox.

metabolic models. After running the *optimizeCbModel* function in the COBRA toolbox, four main output structures are built: f, x, y, and w, where f is optimal objective value, x is containing reaction fluxes of each reaction in the model, y is a vector of shadow prices, and w is a vector of reduced costs.

Then, we calculated the number of carrying-flux reactions (negative and positive carrying-flux reactions) according to the x file of each model extracted after performing FBA.

Flux Variability Analysis (FVA) was carried out on all GSMMs to identify and compare the range of possible flux values (flux

Table 1. Up and down significant metabolic genes in each group.

GEO accession	(Histone Modifications)	Total no. of GSMM Significant genes	No. of Up-Regulated Genes	No. of Down-Regulated genes	Annotation files
1.GSE1639	WT-H3 (Deletion1–28)	36	31	5	YG_S98.na32.annot
2.GSE1639	WT-H3(K4,9,14,18,23,27Q)	96	58	38	YG_S98.na32.annot
3.GSE1639	WT-H4 (K5,8,12,16Q)	35	18	17	YG_S98.na32.annot
4.GSE1639	WT-H4 (Deletion 2–26)	41	14	17	YG_S98.na32.annot
5.GSE3806	WT-H2B (Deletion 3–32)	8	3	5	YG_S98.na32.annot
6.GSE3806	WT-H2B (K-G)	12	3	9	YG_S98.na32.annot
7.GSE7337	WT-H2A (Deletion 4–20)	57	33	24	YG_S98.na32.annot
8.GSE7338	WT-H2A (K4,7G)	9	7	2	YG_S98.na32.annot
9.GSE29293	WT-H3 Depletion	135	130	5	

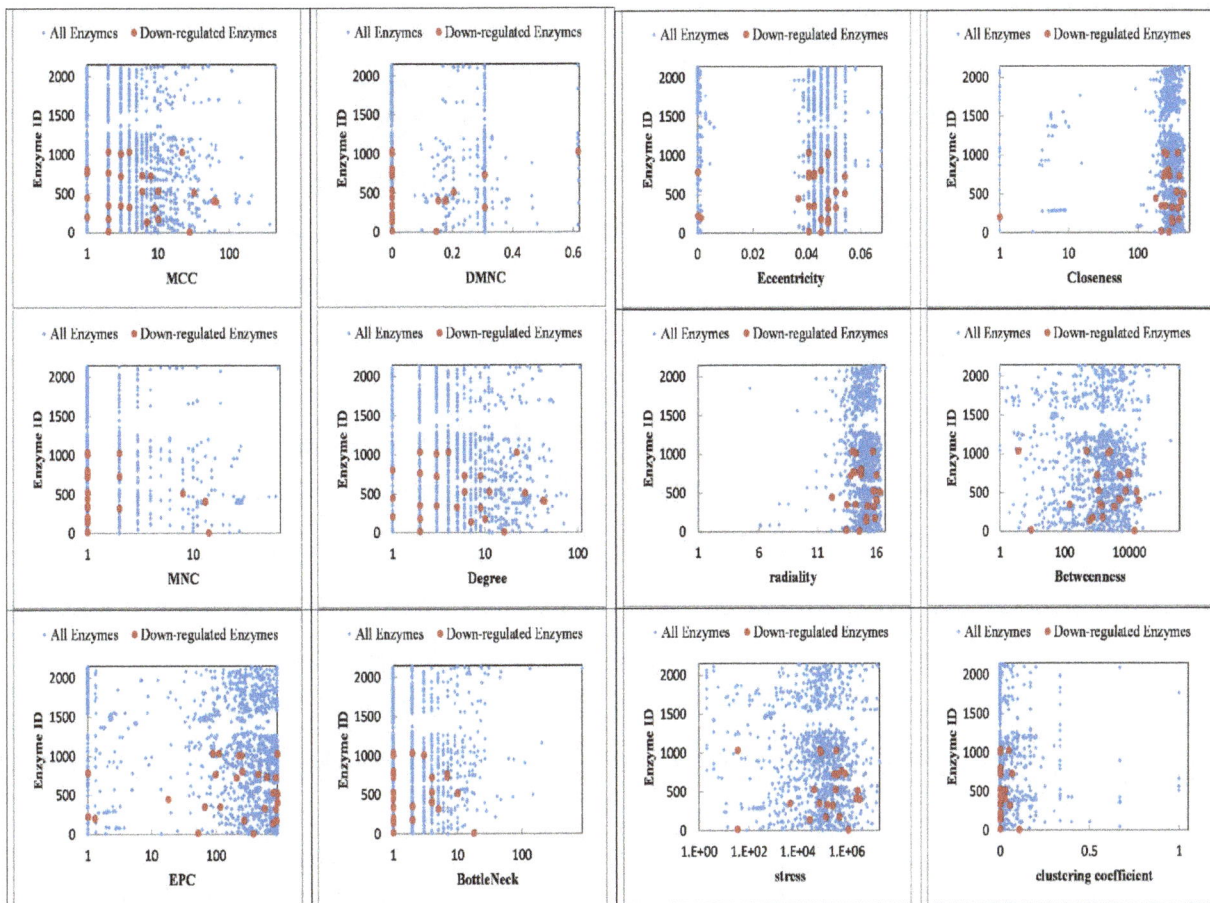

Figure 3. Shows the distribution of down-regulated enzymes in enzyme-centric network of yeast in H4 substituted tail model. The x- and y-axis indicate the centrality score and the enzyme ID of the H4 substituted tail model, respectively.

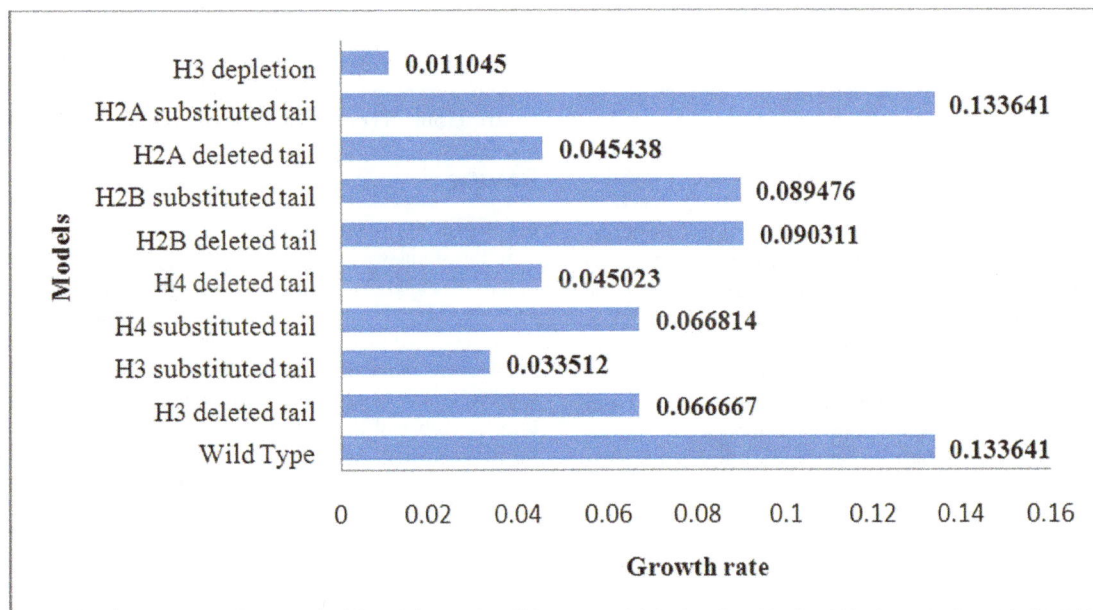

Figure 4. Shows the growth rate (the optimal objective value) of all constructed GSMMs calculated by FBA. The unit of the growth rate is mmol gDW^{-1} hr^{-1} (milimoles per gram dry cell weight per hour).

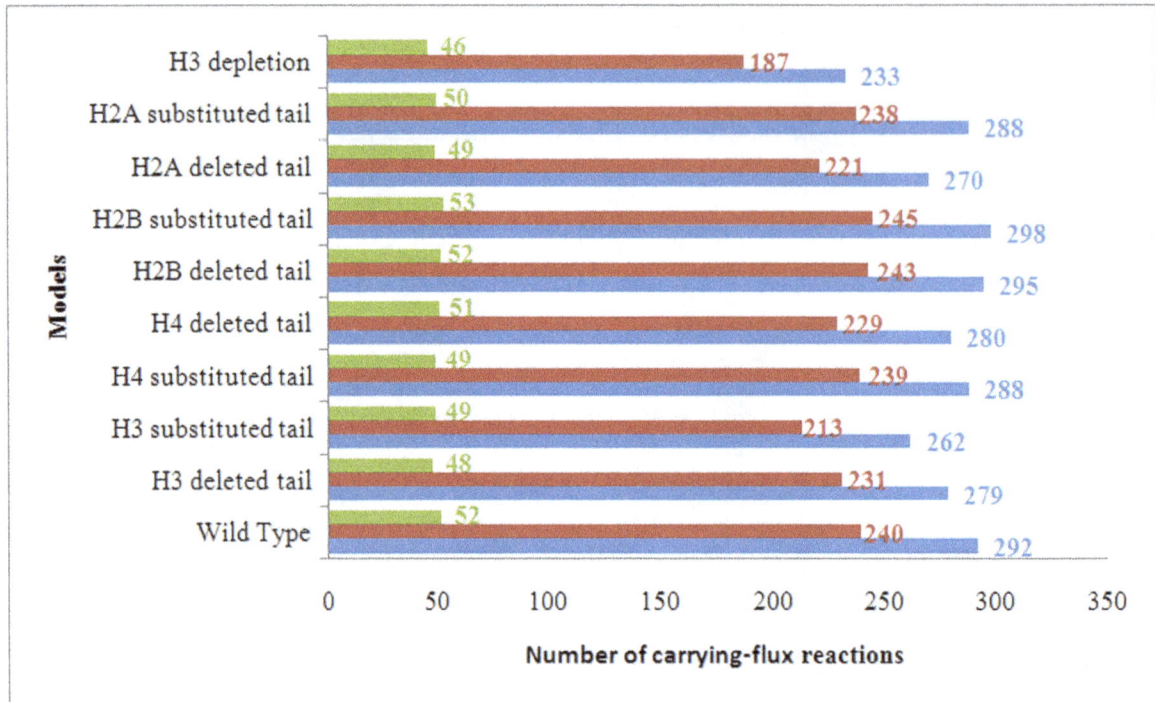

Figure 5. Comparison of the number of all, positive and negative carrying-flux reactions (blue, brown and green bars, respectively) in all constructed GSMMs.

capacity) resulting in the same optimal objective value (maximum and minimum possible fluxes through a particular reaction). We compared the flux capacity of each reaction of a mutated histone tail models with the corresponding reaction of WT model. So, positive values indicate increased flux capacity in mutated histone tail models in comparison to corresponding WT model, whereas negative values indicate decreased flux capacity in mutated histone tail models in comparison to corresponding WT model. Finally, the increased and decreased ranges of flux values categorized in the yeast metabolic subsystems.

Single Gene Deletion used to compute the essential genes, which are important for the growth of the each model.

Parsimonious FBA (pFBA) used to categorize the metabolic reactions to the six groups according to their importance in FBA.

The *SingleGeneDeletion* function is implemented by constraining the flux of deleted gene to zero and then the flux distribution and maximal growth for the new phenotype simply calculated by FBA. If the maximal growth of the new phenotype is reduced, the gene will be essential. The pFBA method is a modified of FBA in which an extra constraint is added. In this approach, after maximizing the growth rate, the net metabolic flux through all gene-associated reactions will be minimized [35,48]. According to the additional constraint (minimization of metabolic adjustment) in pFBA method, the number of non-essential metabolic genes decreases compared with single gene deletion method [48].

Then, we calculated the percentage of the lowly expressed enzymes of the nine constructed GSMMs in each category of pFBA and Single Gene Deletion. (For more information, see part B in the File S1).

Production of cofactors and biomass precursors: We used this capability of FBA for calculating the maximum yield of acetyl-CoA in GSMMs. Acetyl-CoA in nucleocytosolic compartment is the main donor of acetyl group in histone acetylation and is also

used for de novo synthesis of fatty acid. In fact, histone acetylation and synthesis of fatty acids compete for the same acetyl-CoA pool. Acetyl-CoA synthetase is responsible for acetyl-CoA production and acetyl-CoA carboxylase, is an enzyme which carboxylates acetyl-CoA to form malonyl-CoA in de novo synthesis of fatty acids. Actually, acetyl-CoA carboxylase regulates the activity of acetyl-CoA synthetase [49]. For maximizing the production and regulation of acetyl-CoA, we set the objective function of the constructed models to acetyl-CoA carboxylase and acetyl-CoA synthetase reactions respectively, and turning the lower bound of these two reactions to 0 because these two reactions are known to be irreversible. Figure 2 illustrates the workflow of our study.

Results

3.1. Gene expression results

In this study, we just considered the significantly up- and down-regulated metabolic genes extracted from GeWorkbench 2.4.0 software. Table 1 summarizes the basic information about up- and down-regulated metabolic genes. The table shows that the highest number of the significant metabolic genes was found in H3 deletion group, whereas the group 8, H2A substituted tail, had the lowest number of the significant metabolic genes. For constraint-based modeling, we took into account the down-regulated metabolic genes for further analysis. Afterward, the nine mutated histone tail models were constructed. All data are available in the File S3.

3.2. Topological analysis

Centrality indices are a global property of a network that ranks the graph nodes according to their importance in the network. The higher the rank the more important the node is in the network, indicating that it may play key roles in controlling cellular

Table 2. Increased flux range of yeast metablic subnetwork in the 9 mutated histone tail models.

Metabolic subsystem	Models
2-ketoglutarate dehydrogenase complex	1- 3- 4- 5- 9
acetoin biosynthesis	1- 3- 4- 5- 6- 9
arginine biosynthesis	1- 2- 4- 6
arginine degradation	1- 3- 4- 5- 6- 9
beta-alanine biosynthesis	1- 2
butanediol biosynthesis	1- 2- 3- 4- 5- 6- 9
chorismate biosynthesis	1- 2- 3- 4- 5- 9
de novo biosynthesis of purine nucleotides	1- 2- 3- 5- 9
de novo biosynthesis of pyrimidine ribonucleotides	1- 2- 3- 6- 9
ergosterol biosynthesis	1- 5- 6- 9
fatty acid biosynthesis	1- 2- 3- 4- 5- 6- 9
folate biosynthesis	1- 2- 3- 4- 5- 6- 9
formaldehyde oxidation	1- 2- 3- 4- 6- 9
glutamate biosynthesis	1- 2- 3- 4- 5- 6- 9
glutamate degradation	1- 3- 4- 5- 6- 9
glutathione-glutaredoxin redox reactions	1- 3- 4- 5- 6- 9
glycine biosynthesis	1- 2- 3- 4- 5- 6- 9
glycolysis/gluconeogenesis	1- 2- 3- 4- 5- 6- 9
glyoxylate cycle	1- 2- 3- 4- 5- 6- 9
hexaprenyl diphosphate biosynthesis	1- 5- 6- 9
histidine biosynthesis	1- 2- 3- 5- 9
homoserine biosynthesis	1- 3- 4- 5- 6- 9
isoleucine biosynthesis	1- 2- 3- 4- 5- 6- 9
leucine biosynthesis	1- 2- 3- 4- 5- 6- 9
lysine biosynthesis	1- 2- 3- 4- 5- 6- 9
methionine salvage pathway	1- 3- 4- 5- 9
mevalonate pathway	1- 2- 3- 4- 5- 6- 9
nonoxidative branch of the pentose phosphate pathway	1- 2- 3- 4- 5- 6- 9
oxidative branch of the pentose phosphate pathway	1- 2- 3- 4- 5- 6- 9
p-aminobenzoate biosynthesis	1- 2- 3- 5- 9
pantothenate and coenzyme A biosynthesis	1- 3- 4- 5- 6- 9
periplasmic NAD degradation	1- 2- 3- 9
phenylalanine biosynthesis	1- 2- 3- 4- 5- 6- 9
phosphatidate biosynthesis	2- 4- 9
proline biosynthesis	1- 2- 3- 4- 5- 6- 9
putrescine biosynthesis	1- 3- 4- 5- 9
pyruvate dehydrogenase	1- 2- 3- 4- 5- 6- 9
S-adenosylmethionine biosynthesis	1- 2- 3- 4- 5- 6- 9
salvage pathways of purines and their nucleosides	1- 2- 3- 9
spermidine and methylthioadenosine biosynthesis	1- 3- 4- 5- 9
sulfate assimilation pathway	1- 2- 3- 4- 5- 6- 9
superpathway of glucose fermentation	1- 3- 4- 5- 6- 9
superpathway of histidine, purine, and pyrimidine biosynthesis	1- 2- 3- 5- 9
TCA cycle, aerobic respiration	1- 3- 4- 5- 9
thioredoxin system	1- 2- 3- 4- 5- 6- 9
tryptophan biosynthesis	1- 2- 3- 5- 6- 9
tryptophan degradation	1- 2- 3- 4- 5- 6- 9

Table 2. Cont.

Metabolic subsystem	Models
tyrosine biosynthesis	1- 2- 3- 4- 5- 6- 9
tyrosine degradation	1- 2- 3- 5- 6
valine biosynthesis	1- 3- 4- 5- 6- 9
valine degradation	1- 2- 3- 4- 5- 6- 9

Models 1–9 refer to the H3 deleted tail model, the H3 substituted tail model, the H4 substituted tail model, the H4 deleted tail model, the H2B deleted tail model, the H2B substituted tail model, the H2A deleted tail model, the H2A substituted tail model and the H3 depletion model, respectively.

behavior [50]. Twelve centrality indices of the undirected enzyme-centric network of yeast were computed by cytoHubba. Results show that the lowly expressed enzymes are in the top rank of centrality indices of each given model. The File S4, lists the lowly expressed enzymes among the 100 top ranked of twelve different centrality indices. Figure 3 shows the distribution of the lowly expressed enzymes in undirected enzyme-centric network of yeast in the H4 substituted tail model as an example. Results show that the properties of yeast metabolic network could change despite its scale-freeness.

3.3. Constraint-based modeling results

3.3.1. FBA and comparison of all reaction fluxes in the nine models. After model construction, to scrutinize the impact of mutated histone tails on the cell growth rate, FBA was carried out for all nine constructed GSMMs while the objective function was a standard biomass function. The biomass function is a hypothetical reaction that experimentally determines and quantifies the specific growth rate of the cell. This reaction reflects the needs of the cell in order to make 1 gr of cellular dry weight.

Figure 4 summarizes the optimal objective values of all constructed models. Results show that the different mutated histone tails have diverse effects on the growth rate of the corresponding models. The H3 depletion model, has the lowest value while the H2A substituted tail model, has no changes in the optimal objective value.

All the computed optimal reaction fluxes of the nine models were compared with the WT model (as it has been described in the method section 2.5). Results as indicated in Figure 5, show that the number of carrying-flux reactions of all models has been decreased in comparison with the WT model, except in the H2B deleted and substituted tail models. The numbers of the negative carrying-flux reactions (according to the direction of reversible flux) have almost no change in comparison with the positive carrying-flux reactions. It means that in the H2B model, despite the increase in the carrying-flux numbers, the optimal objective value decreases. In other words, H2B modifications have a direct effect on the growth rate.

3.3.2. FVA. Biological systems often contain redundancies that contribute to their robustness. FVA is a valuable method to

Table 3. Decreased flux range of yeast metabolic subnetwork in the 9 mutated histone tail models.

Metabolic subsystem	Models
ATPase, cytosolic	3–9
de novo biosynthesis of purine nucleotides	2- 3- 4- 6- 7
de novo biosynthesis of pyrimidine deoxyribonucleotides	4–5
de novo NAD biosynthesis	2–4
fatty acid oxidation pathway	2–6
folate transformations	2
glycerol degradation	2–6
glycine cleavage complex	2
histidine biosynthesis	4
NADH dehydrogenase	3
periplasmic NAD degradation	7
phosphatidylinositol phosphate biosynthesis	4
salvage pathways of adenine, hypoxanthine and their nucleosides	2- 3- 4
salvage pathways of guanine, xanthine and their nucleosides	4
salvage pathways of purines and their nucleosides	4
salvage pathways of pyrimidine ribonucleotides	2- 3- 4
serine biosynthesis	7
TCA cycle, aerobic respiration	2–6

Models 1–9 refer to the H3 deleted tail model, the H3 substituted tail model, the H4 substituted tail model, the H4 deleted tail model, the H2B deleted tail model, the H2B substituted tail model, the H2A deleted tail model, the H2A substituted tail model and the H3 depletion model, respectively.

Table 4. The percentage of the lowly expressed enzymes of nine constructed models in each category of pFBA and Single Gene Deletion.

Reaction in (GPR)	Single gene deletion	essential genes	pFBA optima	ELE	MLE	pFBA no-flux	Blocked
H3 deleted tail	33.33%	33.33%	16.7%	0	50%	0	0
H3 substituted tail	4%	16.8%	11.8%	13.9%	15.9%	5%	36.6%
H4 substituted tail	6.6%	13.3%	26.7%	23.3%	6.7%	0	30%
H4 deleted tail	4.1%	12.5%	16.7%	16.7%	8.3%	0	45.8%
H2B deleted tail	0	20%	40%	0	0	0	40%
H2B substituted tail	0	5%	0	30%	5%	10%	50%
H2A deleted tail	6.3%	19%	4.8%	6.3%	30.1%	4.8%	35%
H2A substituted tail	0	0	50%	0	0	0	50%
H3 repletion	16.6%	66.8%	0	16.6%	16.6%	0	0

Table 5. We set the objective function of the constructed models to the given reactions (acetyl-CoA synthetase and acetyl-CoA carboxylase reactions) and turned the lower bound to zero for maximizing the production and regulation reaction of acetyl-CoA.

	Model 0	Model 1	Model 2	Model 3	Model 4	Model 5	Model 6	Model 7	Model 8	Model 9
Regulation	1.5652	1.5652	0.1318	1.3358	1.3726	1.3643	1.5652	0.5209	1.5652	1.5319
Production	150.8571	25.75	144.4722	150.8571	73.25	144.0963	149.236	0.4	19.8	150.8571

The producing reaction of acetyl-CoA is catalyzed by acetyl-CoA synthetase, which is in turn regulated by acetyl-CoA carboxylase, an enzyme which catalyzes acetyl CoA conversion to malonyl-CoA, the first and rate-limiting reaction in de novo synthesis of fatty acids. The optimal objective values for acetyl-CoA regulation and production calculated by FBA for all models. Models 1–9 refer to the H3 deleted tail model, the H3 substituted tail model, the H4 substituted tail model, the H4 deleted tail model, the H2B deleted tail model, the H2B substituted tail model, the H2A deleted tail model, the H2A substituted tail model and the H3 depletion model, respectively.

examine the redundancy of metabolic network by calculating the full range of the numerical values (maximum and minimum possible fluxes) for each reaction. FVA was carried out for each model and subsequently compared to the WT model. Then, increased and decreased flux range of each reaction within the metabolic subsystems of yeast were determined. Table 2 summarizes the increased flux range of the metabolic subsystems of each model. As results in Table 3 show, the main effects of histone tail modifications are on the increasing of flux range.

3.3.3. pFBA and single gene deletion results. pFBA was used to label all metabolic genes based on their ability to contribute to the optimal growth rate predictions. As results in Table 4 show, the lowly expressed enzymes in each model are important regarding to *SingleGeneDeletion* function and pFBA analysis.

3.3.4. Production and regulation of acetyl-CoA. Table 5, shows the different turnover values of acetyl-CoA according to two reactions, acetyl-CoA synthetase and acetyl-CoA carboxylase reactions, (that are responsible for production and regulation of acetyl-CoA, respectively) in the ten models. All the mutated histone tail models changed the turnover values of this metabolite. Results show that the histone tails play a key role in production and regulation of acetyl-CoA.

Discussion

Cell metabolism is dependent on different factors. Among those, some are related to the external metabolites such as nutrient cultures, while some are internal regulatory factors [51]. It has been shown that metabolites can regulate the chromatin-modifying enzymes activities and dynamical property of the epigenome can modify the gene expression pattern of the cell. These evidences, providing a direct link between metabolism and epigenetics. Scientific approaches that deal with gene expression analysis, which related to the metabolism have several different purposes. While some studies are directly targeting the theoretical foundations, others trigger specific biological questions. In this study, we aim to answer two main theoretical questions:

A) Do the modifications of histone tails (i.e., histone tail deletion or the substitution of amino acid lysine with glutamine in histone tail) affect the whole metabolism of yeast? Moreover, if so, to what rate? In addition, whether the GSMMs are able to explain these changes?

B) Is the nucleocytoplasmic acetyl-CoA, which is the main metabolite pool for acetylation of histones, under the control of histone tail modifications? Moreover, whether the epigenetic modifications can control the main source of acetylation?

To answer the first question, we used GSMM of yeast and nine gene expression profiles of the mutated histone tails. Determination of structural properties of a network and its nature is the first step to analyze the given network [52]. The oldest and the most complicated network in a living cell is metabolic network, which its topological characteristics are known. The metabolic network of yeast shows power-law degree distribution pointing to the fact that the network is robust to random failures and modifications in its structure [53,54]. On the other hand, the essential genes are vital to maintenance cellular life and their deletion will result in lethality or infertility. Although, the main approach for separating these genes from non-essential genes is experiment, but there are some

theoretical approaches for predicting these genes. Network biology, is one of the theoretical approaches, which determines the important nodes in a network by calculating the different centrality parameters. Analyzing the undirected enzyme-centric network of yeast according to twelve centrality indices, shows that some of the lowly expressed enzymes are very important for the network robustness, and suggesting that metabolism could be influenced by them. Concomitantly, pFBA and single gene deletion analysis, as two different theoretical approaches which especially applied to metabolic network, confirmed this expectation. The pFBA analysis categorizes all the reactions in the metabolic network to six clusters. The results of pFBA and single gene deletion show some of the lowly expressed enzymes are crucial for the yeast growth and can affect the metabolic network and metabolism. Therefore, concluded from these results that the connection between histone modifications and metabolism is not implausible. Pioneering works demonstrated that the histone tails or enzymes which are related to the acetylation of histone (e.g. histone acetyl transferase) control the metabolism of yeast, but are not essential for its viability [55]. For the quantification of this crosstalk behavior and gaining more information about how these modifications affect yeast metabolism, FBA has been done. The FBA results (when the objective function was the biomass equation) show considerable changes in the optimal flux values and the growth rate of the constructed models. Although these substantial changes differed in each model, but none of them is lethal. Unlike the growth rate of the H2A substituted tail model remained unchanged, the depletion of the H3 had the highest effect on the growth rate. Subsequently, the deleted and substituted H2B tail models showed the modest changes, while the substituted H3 and the H4 tail models showed more changes. This leads us to the conclusion that lysine in H3 and H4 as the target of histone modifications has a major role in the growth rate of yeast and the different H3 and H4 modifications can have different growth rate. The FBA results confirm the previous experimental results that the histone modifications were not lethal. In 2012 Kim et al., demonstrated that the N-terminal tails of histones reveal functional redundancy in the budding yeast [36]. The results calculated by FVA shows, that the most of the reaction flux has been changed in all nine models and display different metabolic patterns. The common increased range of flux in the subsystem of the mutated H3, H4 and H2B tail models indicated a functional redundancy in regulation of yeast metabolism including amino acid biosynthesis (e.g. glycine, histidine, homoserine, isoleucine, leucine, lysine, methionine, phenylalanine, proline, tryptophan, tyrosine and valine), as well as glycolysis, gluconeogenesis, glyoxylate cycle, pentose phosphate pathway and TCA cycle. By changing the objective functions of the models to a specific reaction, the maximum yields of important metabolites measured. In all models, our data indicates that acetyl-CoA turnover changes in dependence of mutated histone tails, which is potentially indicative of a change in its concentration. Indeed, it shows that the main source of acetylation is under control of epigenetic modifications. Finally, it is broadly accepted that the main sources of epigenetic changes are metabolites, which can be provided by different states of the cell metabolism. On the other hand, epigenetic modifications change the metabolic states of a cell. One of the essential reactions, underlying this change is acetyl-CoA synthetase. Therefore, histone tails have a feedback control on their main acetylation source, as a major target of epigenetic modifications. Thus, we can claim that histone tails are the key players of crosstalk between epigenetics and metabolism.

References

1. Gruning NM, Lehrach H, Ralser M (2010) Regulatory crosstalk of the metabolic network. Trends Biochem Sci 35: 220–227.
2. Mazzio EA, Soliman KF (2012) Basic concepts of epigenetics: impact of environmental signals on gene expression. Epigenetics 7: 119–130.
3. Choi JK (2010) Systems biology and epigenetic gene regulation. IET Syst Biol 4: 289–295.
4. Lu C, Thompson CB (2012) Metabolic regulation of epigenetics. Cell Metab 16: 9–17.
5. Cai L, Sutter BM, Li B, Tu BP (2011) Acetyl-CoA induces cell growth and proliferation by promoting the acetylation of histones at growth genes. Mol Cell 42: 426–437.
6. Colyer HA, Armstrong RN, Mills KI (2012) Microarray for epigenetic changes: gene expression arrays. Methods Mol Biol 863: 319–328.
7. Yun J, Johnson JL, Hanigan CL, Locasale JW (2012) Interactions between epigenetics and metabolism in cancers. Front Oncol 2: 163.
8. Sigurdsson MI, Jamshidi N, Jonsson JJ, Palsson BO (2009) Genome-scale network analysis of imprinted human metabolic genes. Epigenetics 4: 43–46.
9. Donohoe DR, Bultman SJ (2012) Metaboloepigenetics: interrelationships between energy metabolism and epigenetic control of gene expression. J Cell Physiol 227: 3169–3177.
10. Palsson B (2009) Metabolic systems biology. FEBS Lett 583: 3900–3904.
11. Poolman MG, Bonde BK, Gevorgyan A, Patel HH, Fell DA (2006) Challenges to be faced in the reconstruction of metabolic networks from public databases. Syst Biol (Stevenage) 153: 379–384.
12. Reed JL (2012) Shrinking the metabolic solution space using experimental datasets. PLoS Comput Biol 8: e1002662.
13. Herrgard MJ, Lee BS, Portnoy V, Palsson BO (2006) Integrated analysis of regulatory and metabolic networks reveals novel regulatory mechanisms in Saccharomyces cerevisiae. Genome Res 16: 627–635.
14. Covert MW, Palsson BO (2002) Transcriptional regulation in constraints-based metabolic models of Escherichia coli. J Biol Chem 277: 28058–28064.
15. Akesson M, Forster J, Nielsen J (2004) Integration of gene expression data into genome-scale metabolic models. Metab Eng 6: 285–293.
16. Becker SA, Palsson BO (2008) Context-specific metabolic networks are consistent with experiments. PLoS Comput Biol 4: e1000082.
17. Colijn C, Brandes A, Zucker J, Lun DS, Weiner B, et al. (2009) Interpreting expression data with metabolic flux models: predicting Mycobacterium tuberculosis mycolic acid production. PLoS Comput Biol 5: e1000489.
18. Moxley JF, Jewett MC, Antoniewicz MR, Villas-Boas SG, Alper H, et al. (2009) Linking high-resolution metabolic flux phenotypes and transcriptional regulation in yeast modulated by the global regulator Gcn4p. Proc Natl Acad Sci U S A 106: 6477–6482.
19. Jensen PA, Papin JA (2011) Functional integration of a metabolic network model and expression data without arbitrary thresholding. Bioinformatics 27: 541–547.
20. Kim J, Reed JL (2012) RELATCH: relative optimality in metabolic networks explains robust metabolic and regulatory responses to perturbations. Genome Biol 13: R78.
21. Agren R, Bordel S, Mardinoglu A, Pornputtapong N, Nookaew I, et al. (2012) Reconstruction of genome-scale active metabolic networks for 69 human cell types and 16 cancer types using INIT. PLoS Comput Biol 8: e1002518.
22. Wang Y, Eddy JA, Price ND (2012) Reconstruction of genome-scale metabolic models for 126 human tissues using mCADRE. BMC Syst Biol 6: 153.
23. Machado D, Herrgard M (2014) Systematic evaluation of methods for integration of transcriptomic data into constraint-based models of metabolism. PLoS Comput Biol 10: e1003580.
24. Osterlund T, Nookaew I, Nielsen J (2011) Fifteen years of large scale metabolic modeling of yeast: developments and impacts. Biotechnol Adv 30: 979–988.
25. Forster J, Famili I, Fu P, Palsson BO, Nielsen J (2003) Genome-scale reconstruction of the Saccharomyces cerevisiae metabolic network. Genome Res 13: 244–253.
26. Duarte NC, Herrgard MJ, Palsson BO (2004) Reconstruction and validation of Saccharomyces cerevisiae iND750, a fully compartmentalized genome-scale metabolic model. Genome Res 14: 1298–1309.
27. Kuepfer L, Sauer U, Blank LM (2005) Metabolic functions of duplicate genes in Saccharomyces cerevisiae. Genome Res 15: 1421–1430.
28. Nookaew I, Jewett MC, Meechai A, Thammarongtham C, Laoteng K, et al. (2008) The genome-scale metabolic model iIN800 of Saccharomyces cerevisiae and its validation: a scaffold to query lipid metabolism. BMC Syst Biol 2: 71.
29. Mo ML, Palsson BO, Herrgard MJ (2009) Connecting extracellular metabolomic measurements to intracellular flux states in yeast. BMC Syst Biol 3: 37.
30. Zomorrodi AR, Maranas CD (2010) Improving the iMM904 S. cerevisiae metabolic model using essentiality and synthetic lethality data. BMC Syst Biol 4: 178.
31. Heavner BD, Smallbone K, Barker B, Mendes P, Walker LP (2012) Yeast 5 - an expanded reconstruction of the Saccharomyces cerevisiae metabolic network. BMC Syst Biol 6: 55.
32. Herrgard MJ, Swainston N, Dobson P, Dunn WB, Arga KY, et al. (2008) A consensus yeast metabolic network reconstruction obtained from a community approach to systems biology. Nat Biotechnol 26: 1155–1160.
33. Barrett T, Edgar R (2006) Mining microarray data at NCBI's Gene Expression Omnibus (GEO)*. Methods Mol Biol 338: 175–190.
34. Orth JD, Thiele I, Palsson BO (2010) What is flux balance analysis? Nat Biotechnol 28: 245–248.
35. Schellenberger J, Que R, Fleming RM, Thiele I, Orth JD, et al. (2011) Quantitative prediction of cellular metabolism with constraint-based models: the COBRA Toolbox v2.0. Nat Protoc 6: 1290–1307.
36. Kim JA, Hsu JY, Smith MM, Allis CD (2012) Mutagenesis of pairwise combinations of histone amino-terminal tails reveals functional redundancy in budding yeast. Proc Natl Acad Sci U S A 109: 5779–5784.
37. Sabet N, Volo S, Yu C, Madigan JP, Morse RH (2004) Genome-wide analysis of the relationship between transcriptional regulation by Rpd3p and the histone H3 and H4 amino termini in budding yeast. Mol Cell Biol 24: 8823–8833.
38. Nag R, Kyriss M, Smerdon JW, Wyrick JJ, Smerdon MJ (2010) A cassette of N-terminal amino acids of histone H2B are required for efficient cell survival, DNA repair and Swi/Snf binding in UV irradiated yeast. Nucleic Acids Res 38: 1450–1460.
39. Parra MA, Wyrick JJ (2007) Regulation of gene transcription by the histone H2A N-terminal domain. Mol Cell Biol 27: 7641–7648.
40. Gossett AJ, Lieb JD (2012) In vivo effects of histone H3 depletion on nucleosome occupancy and position in Saccharomyces cerevisiae. PLoS Genet 8: e1002771.
41. Floratos A, Smith K, Ji Z, Watkinson J, Califano A (2010) geWorkbench: an open source platform for integrative genomics. Bioinformatics 26: 1779–1780.
42. Lee D, Smallbone K, Dunn WB, Murabito E, Winder CL, et al. (2012) Improving metabolic flux predictions using absolute gene expression data. BMC Syst Biol 6: 73.
43. Blazier AS, Papin JA (2012) Integration of expression data in genome-scale metabolic network reconstructions. Front Physiol 3: 299.
44. Shlomi T, Cabili MN, Herrgard MJ, Palsson BO, Ruppin E (2008) Network-based prediction of human tissue-specific metabolism. Nat Biotechnol 26: 1003–1010.
45. Hyduke DR, Lewis NE, Palsson BO (2013) Analysis of omics data with genome-scale models of metabolism. Mol Biosyst 9: 167–174.
46. Horne AB, Hodgman TC, Spence HD, Dalby AR (2004) Constructing an enzyme-centric view of metabolism. Bioinformatics 20: 2050–2055.
47. Lin CY, Chin CH, Wu HH, Chen SH, Ho CW, et al. (2008) Hubba: hub objects analyzer-a framework of interactome hubs identification for network biology. Nucleic Acids Res 36: W438–443.
48. Segre D, Vitkup D, Church GM (2002) Analysis of optimality in natural and perturbed metabolic networks. Proc Natl Acad Sci U S A 99: 15112–15117.
49. Galdieri L, Vancura A (2012) Acetyl-CoA carboxylase regulates global histone acetylation. J Biol Chem 287: 23865–23876.
50. Ma HW, Zeng AP (2003) The connectivity structure, giant strong component and centrality of metabolic networks. Bioinformatics 19: 1423–1430.
51. Rando OJ, Winston F (2012) Chromatin and transcription in yeast. Genetics 190: 351–387.

Author Contributions

Conceived and designed the experiments: ASY YA AMN. Performed the experiments: ASY YA. Analyzed the data: ASY YA. Contributed reagents/materials/analysis tools: AMN AAS. Wrote the paper: ASY YA AMN.

52. Tang H, Zhong F, Xie H (2012) A quick guide to biomolecular network studies: construction, analysis, applications, and resources. Biochem Biophys Res Commun 424: 7–11.

53. Jeong H, Tombor B, Albert R, Oltvai ZN, Barabasi AL (2000) The large-scale organization of metabolic networks. Nature 407: 651–654.

54. Mahadevan R, Palsson BO (2005) Properties of metabolic networks: structure versus function. Biophys J 88: L07–09.

55. Schuster T, Han M, Grunstein M (1986) Yeast histone H2A and H2B amino termini have interchangeable functions. Cell 45: 445–451.

14

The Transcription Factor Ste12 Mediates the Regulatory Role of the Tmk1 MAP Kinase in Mycoparasitism and Vegetative Hyphal Fusion in the Filamentous Fungus *Trichoderma atroviride*

Sabine Gruber, Susanne Zeilinger*

Research Area Biotechnology and Microbiology, Institute of Chemical Engineering, Vienna University of Technology, Wien, Austria

Abstract

Mycoparasitic species of the fungal genus *Trichoderma* are potent antagonists able to combat plant pathogenic fungi by direct parasitism. An essential step in this mycoparasitic fungus-fungus interaction is the detection of the fungal host followed by activation of molecular weapons in the mycoparasite by host-derived signals. The *Trichoderma atroviride* MAP kinase Tmk1, a homolog of yeast Fus3/Kss1, plays an essential role in regulating the mycoparasitic host attack, aerial hyphae formation and conidiation. However, the transcription factors acting downstream of Tmk1 are hitherto unknown. Here we analyzed the functions of the *T. atroviride* Ste12 transcription factor whose orthologue in yeast is targeted by the Fus3 and Kss1 MAP kinases. Deletion of the *ste12* gene in *T. atroviride* not only resulted in reduced mycoparasitic overgrowth and lysis of host fungi but also led to loss of hyphal avoidance in the colony periphery and a severe reduction in conidial anastomosis tube formation and vegetative hyphal fusion events. The transcription of several orthologues of *Neurospora crassa* hyphal fusion genes was reduced upon *ste12* deletion; however, the Δ*ste12* mutant showed enhanced expression of mycoparasitism-relevant chitinolytic and proteolytic enzymes and of the cell wall integrity MAP kinase Tmk2. Based on the comparative analyses of Δ*ste12* and Δ*tmk1* mutants, an essential role of the Ste12 transcriptional regulator in mediating outcomes of the Tmk1 MAPK pathway such as regulation of the mycoparasitic activity, hyphal fusion and carbon source-dependent vegetative growth is suggested. Aerial hyphae formation and conidiation, in contrast, were found to be independent of Ste12.

Editor: Stefanie Pöggeler, Georg-August-University of Göttingen Institute of Microbiology & Genetics, Germany

Funding: The Vienna Science and Technology Fund WWTF (www.wwtf.at, grant number LS09-036 to SZ) and the Austrian Science Fund FWF (www.fwf.ac.at, grant number V139 to SZ) are acknowledged for funding. The funders had no role in study design, data collection and analysis, decision to publish, or preparation of the manuscript.

Competing Interests: The authors have declared that no competing interests exist.

* Email: szeiling@mail.tuwien.ac.at

Introduction

Mycoparasitic species of the fungal genus *Trichoderma* are potent biocontrol agents and promising substitutes for chemical fungicides as they attack and parasitize plant pathogens, such as *Rhizoctonia spp.*, *Phythium spp.*, *Botrytis cinerea* und *Fusarium spp.* [1]. Mycoparasitic responses are triggered by molecules released from the host fungus and through physical contact accomplished through surface located components (e.g. lectins) [2,3]. As a consequence, *Trichoderma* inhibits or kills the host by parasitizing its hyphae thereby employing hydrolytic enzymes like chitinases, proteases, and glucanases which degrade the host's cell wall. Mycoparasitism further includes shaping of infection structures (coiling response) and the production of antimicrobial secondary metabolites [4]. In the past years, investigation of signaling pathways in the potent mycoparasites *Trichoderma atroviride* and *Trichoderma virens* showed essential roles of conserved signaling routes involving G protein-coupled receptors (GPCRs) and heterotrimeric G proteins, the cAMP pathway and

mitogen-activated protein kinase (MAPK) cascades in regulating vegetative growth, conidiation, and mycoparasitism-associated processes (reviewed in [5,6]).

MAPK cascades are characterized by a three-tiered signaling module comprising a MAPK kinase kinase (MAPKKK), a MAPK kinase (MAPKK) and the MAPK which is hierarchically activated by dual phosphorylation of conserved threonine and tyrosine residues [7]. The proposed mechanism of MAPK signaling comprises the transduction of extracellular and intracellular signals, thereby often regulating transcription factors by MAPK-mediated phosphorylation. Fungal MAPKs are involved in regulating a wide range of processes including cell cycle, stress response and several essential developmental processes such as sporulation, mating, hyphal growth, and pathogenicity [8,9]. In the yeast *Saccharomyces cerevisiae*, mating and filamentous growth are controlled by the Fus3 and Kss1 MAPKs, respectively [10]. Despite their distinct activation mechanisms and signaling output, both MAPKs target the homeodomain transcription factor Ste12, which acts as a central node in both mating and invasive growth and that is under complex regulation by several regulatory proteins and co-factors being tightly controlled by each MAPK. The Fusp/Kss1 MAPK cascade is highly conserved in filamentous fungi which, however, in most cases only posses a single Fus3/Kss1 orthologue [11]. Δmak-2 mutants of the model fungus *Neurospora crassa* showed reduced growth rate, derepressed conidiation, failed to develop protoperithecia, and lacked hyphal fusion – phenotypes which they share with Δpp-1 mutants missing the *ste12* homologue [12]. In the phytopathogenic fungus *Magnaporthe oryzae*, the Fus3/Kss1 homologous MAP kinase Pmk1 is essential for pathogenicity-related processes. Δpmk1 mutants failed to form appressoria and to grow invasively in plants but still recognized hydrophobic surfaces [13]. Studies from several phytopathogenic fungi, including appressorium- and non-appressorium-forming pathogens, necrotrophs and biotrophs, revealed a conserved role of the Pmk1 MAPK pathway for regulating plant infection with respective deletion mutants being affected in pathogenicity-related processes such as appressorium formation, penetration hyphae differentiation, root attachment and the production of plant cell wall-degrading enzymes (reviewed in [9,14]). Concordant with the model of Ste12 being targeted by the Fus3/Kss1 homologous Pmk1-type MAP kinase, *ste12*-deficient mutants of several phytopathognic fungi are either non-pathogenic or suffer from strongly attenuated virulence (reviewed in [11,15]).

Similar to other fungal pathogens, the molecular processes involved in host attack in mycoparasitic fungi are tightly regulated by conserved signaling pathways. In both, *T. atroviride* as well as *T. virens*, the Pmk1 MAPK homologues Tmk1/TmkA (Tvk1) play crucial, albeit species-specific, roles in mycoparasitism [16–18]. *T. virens* Δtvk1/ΔtmkA mutants showed secondary metabolite production similar to the wild-type and unaltered mycoparasitism of *R. solani*, while antagonism against *Sclerotium rolfsii* was reduced [16,18]. In contrast, deletion of *tmk1* in *T. atroviride* resulted in mutants with reduced mycoparasitic activity against *R. solani* and a loss of mycoparasitism of *B. cinerea* although Δtmk1 mutants showed an increased production of antifungal metabolites such as peptaibols and 6-pentyl-α-pyrone [17].

Although recent comparative genomic analyses revealed structural conservation of Fus3/Kss1 MAPK cascade components in taxonomically and biologically diverse fungi [19,20], the available studies showed remarkable functional differences between fungi with phytopathogenic and mycoparasitic lifestyles. While in plant pathogens such as *Fusarium oxysporum* and *Cochliobolus heterostrophus* the expression of extracellular plant-lysing enzymes is positively regulated by the Pmk1-type MAPK [21,22], Tmk1 and Tvk1 repress the production of secreted mycoparasitism-relevant cell wall-lysing chitinases and proteases in the mycoparasites *T. atroviride* and *T. virens* [16,17]. A further dissection of the Fus3/Kss1 MAPK cascade including detailed analyses of factors acting upstream and downstream of the core MAP kinase and discovery of the signals originating from the respective hosts will be necessary for a detailed understanding of this widely conserved signaling pathway in different fungi.

The objectives of this study were to confirm the presence of a functional Ste12 homolog in the mycoparasite *T. atroviride* and to characterize its function as an assumed central component of the mycoparasitism-relevant Tmk1 MAPK signaling pathway. To this end, we deleted the *ste12* gene and comparatively analyzed *T. atroviride* Δste12 and Δtmk1 mutants regarding physiological and differentiation processes. Furthermore, the role of Ste12 in host sensing and mycoparasitism of *T. atroviride* was addressed. Our study provides the first functional characterization of a Ste12-like transcription factor in a fungus exhibiting a mycoparasitic lifestyle

and unveils the Tmk1-Ste12 signaling pathway as key player not only in mycoparasitism but also hyphal avoidance, vegetative hyphal fusion and carbon source-dependent growth of *T. atroviride*.

Materials and Methods

Cultivation conditions

T. atroviride strain P1 (ATCC 74058; teleomorph *Hypocrea atroviridis*), was used in this study. The parental as well as the mutant strains Δste12 and Δtmk1-12 [23] were cultivated at 28°C using a 12 hours light/dark cycle in either rich medium (potato dextrose agar, PDA, or potato dextrose broth, PDB) (BD Dicfo, Franklin Lakes, NJ), or minimal medium (MM, containing [g/l]: $MgSO_4 \cdot 7H_2O$ 1, KH_2PO_4 10, $(NH_4)_2SO_4$ 6, tri-sodium citrate 3, $FeSO_4 \cdot 7H_2O$ 0.005, $ZnSO_4 \cdot 2H_2O$ 0.0014, $CoCl_2 \cdot 6H_2O$ 0.002, $MnSO_4 \cdot 6H_2O$ 0.0017, glucose or glycerol 10). Cultivations in liquid medium were either performed in stationary cultures or shake flask cultures, depending on the respective experiment. For testing hyphal network formation, liquid stationary cultures were inoculated with an agar plug from a sporulating culture and mycelia were harvested from the colony centre and the peripheral hyphal zone as described [24]. For analyzing chitinase gene expression and extracellular endo- and exochitinase activities, *T. atroviride* was inoculated for 20 hours in minimal medium containing 1% glycerol as a carbon source. Mycelia were then harvested by filtration and transferred to media containing 1% N-acetyl-glucosamine (NAG) or 1% colloidal chitin. Mycelia and culture filtrates were harvested after 5, 14, and 24 hours from NAG-induced cultures and after 14, 24, 26, and 48 hours from chitin-induced cultures and stored at −20°C for enzyme assays or at −80°C for RNA extraction.

Plate confrontation assays with *Rhizoctonia solani* and *Botrytis cinerea* as hosts were performed as previously described [3,25]. Pictures were captured from 24 hours until 14 days of growth. For RNA extraction, cultivations were performed on PDA plates covered with a sterile cellophane membrane. *Trichoderma* mycelium was harvested from the confrontation zone (5 mm of the peripheral area) before direct contact between the two fungi (5 mm distance), at direct contact, and after contact (5 mm overgrowth). Self-confrontations between the *Trichoderma* strains tested served as controls. Mycelia of the *T. atroviride* parental and mutant strains were frozen in liquid nitrogen and stored at −80°C.

Enzyme assays

Enzymatic activities of culture supernatants were assayed as previously described [26] using the substrates p-nitrophenyl N-acetyl-β-D-glucosaminide for determination of N-acetyl-glucosaminidase and 4-nitrophenyl-β-D-N,N′,N″-triacetylchitotriose for determination of endochitinase activity. Enzyme activity was measured as U/ml (one unit is defined as the release of 1 μmol of nitrophenol per minute) relative to total intracellular protein assessed by Bradford assay (BioRad) and represented as enzyme activity in U/μg protein.

Molecular techniques and mutant generation

In order to generate *T. atroviride* Δste12 mutant strains, the DelsGate deletion construction methodology [27] was applied. ~1 kb of the up- and downstream flanking non-coding regions of the *ste12* gene were amplified and recombined via phage attachment sites using BP clonase in a "Donor vector" (pDONR) containing the *hph* hygromycin B-phosphotransferase-encoding marker cassette. For amplification of the deletion vector, One Shot Omnimax 2 T1R *Escherichia coli* cells (Invitrogen, Carlsbad, CA)

Table 1. Oligonucleotides used in this study.

gene		Sequence (5′ to 3′)
nag1	FW	TGTCCTACAGCCTCTGCTGCAAAAGTTC
	RV	CATCTCCTCACAGACAAGCGGTGAAAG
prb1	FW	CGCACTGCTTCCTTCACCAACT
	RV	TTTCACTTCATCCTTCGCTCCA
ech42	FW	CGCAACTTCCAGCCTCAGAACC
	RV	TCAATACCATCGAAACCCCAGTCC
sar1	FW	CTCGACAATGCCGGAAAGACCA
	RV	TTGCCAAGGATGACAAAGGGG
act1	FW	GCACGGAATCGCTCGTTG
	RV	TTCTCCACCCCGCCAAGC
ham7	FW	GGCTCTTTACTCTTGCGTCGAC
	RV	CCGCCCAGCCATTGCGAAG
nox1	FW	CTCAAGATTCACACCTACCTCAC
	RV	GCAACAGAGGGACCACAGAAG
hex1	FW	AGGAGTCTTCCTTCATTGCCAAC
	RV	AGAACGAGGACACGGACGC
tmk2	FW	CAGATGCCCACTTCCAATCCTT
	RV	CAAAAGCTATTGATGTATGGTCTGTG
tac6	FW	CGGGACTTATGGTTTGGGCG
	RV	CGAACGGTCCAGATGCGG
ham9	FW	CACAAGGATGCCCAACAGATG
	RV	CCGAGGCATTGGGCTGGAG
tef1	FW	GGTACTGGTGAGTTCGAGGCTG
	RV	GGGCTCAATGGCGTCAATG
ste12	FW	CCTGTGGCCGACTGTTCAAG
	RV	ATTCTTCCTCTTCCTCGGCAG
ste12-C	FW	CACTGCACTGTATTCCGGCTCCC
	RV	CGATGGCAGCGAAGACAATGAG
hph	FW	GCCGATCTTAGCCAGACGAG
	RV	CTGCGGGCGATTTGTGTACG

RT-qPCR analysis

Total RNA was extracted using the peqGOLD TriFast Solution (PeqLab, Erlangen, Germany). Frozen mycelia were homogenized using glass beads by grinding twice for 30 s in a RETSCH MM301 Ball Mill (Retsch, Haan, Germany). Isolated RNA was treated with Deoxyribonuclease I (Fermentas, St. Leon-Rot, Germany), purified with the RNeasy MinElute Cleanup Kit (Qiagen, Hilden, Germany) and reverse transcribed to cDNA using the Revert Aid H Minus First Strand cDNA Synthesis Kit (Fermentas, Vilnius, Lithuania). qPCR was carried out with the Mastercycler ep realplex real-time PCR system (Eppendorf, Hamburg, Germany) using iQ SYBR Green Supermix (Bio-Rad, Hercules, CA).

Relative gene transcript levels were quantified using sar1, act1 or tef1 as reference genes [30] and the expression ratios were calculated according to [31]. For samples derived from chitin- and N-acetyl-glucosamine-induced cultures, gene expression levels were normalized to basal expression levels from cultivations on 1% glycerol. All samples were analyzed in three independent experiments with three replicates in each run.

For expression profiling of genes putatively involved in CAT formation and hyphal fusion, parental and mutant strains were grown in stationary liquid cultures with MM containing 1% glucose, as described in [24] and peripheral and intra hyphal zones were harvested. mRNA levels were quantified and normalized to the corresponding signals of tef1 as reference gene [30]. Statistical analysis was done by relative expression analysis with REST software using the Pair Wise Fixed Reallocation Randomisation Test [32].

Expression of ste12 in T. atroviride during the mycoparasitic interaction with R. solani and during self-confrontation was analyzed by semi-quantitative RT-PCR. tef1 gene expression was used as reference. All primer sequences are listed in Table 1.

Biolog phenotype array analysis

The Biolog FF MicroPlate assay (Biolog Inc., Hayward, CA) which comprises 95 wells with different carbon-containing compounds and one well with water was used to investigate growth rates on pre-filled carbon sources. Conidia were collected from the Trichoderma strains (parental strain, Δtmk1, Δste12) and used as inoculums as described [33]. Inoculated microplates were incubated in darkness at 28°C, and OD750 readings determined after 18, 24, 42, 48, 66, 72, 96 and 168 hours using a microplate reader (Biolog), which measures the turbidity and reflects mycelia production on the tested substrate. All analyses were performed in triplicate. For comparative analysis, values from the 48 hours time point was used as this was within the linear growth phase of T. atroviride on the majority of carbon sources. Values were quantitatively illustrated using the Hierarchical Clustering Explorer 3 (HCE3) [34]. For all analyses the hierarchical clustering algorithm with average linkage and Euclidean distance measure was applied.

Microscopic analyses

Microscopic analyses were mainly performed as described by [35]. Briefly, 500 µl PDA was spread onto glass slides, inoculated by placing T. atroviride and R. solani on opposite sides of the glass slides, and incubated on a moistened filter paper at 28°C in a Petri dish sealed with Parafilm. After 48–72 hours, the fungal hyphae were imaged with an inverted T300 microscope (Nikon, Tokyo, Japan). Images were captured with a Nikon DXM1200F digital camera and digitally processed using Photoshop CS3 (Adobe, San Jose, CA, US). For biomimetic assays, sterile nylon 66 fibers (approximate diameter 14 µm; Nilit, Migdal-Haemek, Israel) were

were used. The resulting ste12 deletion vector pRAM was confirmed by PCR and DNA sequencing. In order to generate stable deletion mutants, the protoplast-based transformation method was applied as previously described [28]. Transformants were selected on PDA containing 200 µg/mL hygromycin B. Mitotically stable transformants were obtained by three rounds of single spore isolation and homologous integration of the deletion cassette was confirmed by PCR and Southern analysis [29] (Figure S1). Loss of ste12 gene expression in the Δste12 mutant was confirmed by RT-qPCR using the parental and the Δtmk1 strains as controls.

For complementation, a 5141-kb fragment bearing the ste12 gene and its 5′ and 3′ regulatory regions was amplified from T. atroviride strain P1 by PCR using primers ste12-C-FW and ste12-C-RV (Table 1) and introduced into the Δste12 mutant by co-transformation with plasmid p3SR2. Transformants were selected on acetamide-containing medium and purified by three rounds of single spore isolation.

placed on the media before inoculation. For analysis in liquid cultures, 1×10^6 spores per ml in 250 μl of MM containing 1% glucose were spread onto the glass slide and pictures were taken at 24 h and 48 h as described above.

Results

Characteristics of the *T. atroviride* Ste12 homologue

The aim of this study was to characterize the role of the Ste12 transcription factor, which is assumed to act as a central component of the mycoparasitism-relevant signaling pathway involving the Tmk1 MAPK in *T. atroviride*.

The *T. atroviride ste12* gene (ID 29631;http://genome.jgi-psf. org/Triat2/Triat2.home.html) consists of an open reading frame of 2142-bp with two C-terminally located introns (60-bp and 63-bp) and is predicted to encode a protein of 672 amino acids. Ste12 displays considerable amino acid identity to the already functionally characterized Ste12 proteins of *Fusarium oxysporum* (ACM80357; 77%), *Magnaporthe oryzae* (*M. grisea*) (AF432913; 74%), *Neurospora crassa* (EAA28575; 65%), and *Aspergillus nidulans* (XP_659894; 60%) [12,36–38]. Phylogenetic analysis

showed Ste12 in a clade together with the orthologues from *Trichoderma virens* (ID 75179, http://genome.jgi-psf.org/TriviGv29_8_2/TriviGv29_8_2.home.html) and *Trichoderma reesei* (ID 36543, http://genome.jgi-psf.org/Trire2/Trire2.home. html), and with *F. oxysporum* Ste12 and *M. oryzae* Mst12 (Fig. 1). Similar to Ste12 proteins from other filamentous fungi, *T. atroviride* Ste12 contains an N-terminally located homeodomain-like (STE) motif (amino acids 54–163) which is presumed to be involved in DNA binding [39] and two distinct C-terminal C_2H_2 zinc finger domains (amino acids 546–582) which are distinguishing features of Ste12-like proteins of filamentous fungi [15].

For analyzing the expression of *ste12*, *T. atroviride* was either grown alone, in confrontation against *R. solani* as host, in self-confrontation, or in the presence of nylon fibers whose diameter resembles that of host hyphae. *ste12* was only moderately expressed when the fungus was grown alone, during self-confrontation, when coming into contact with plain nylon fibers, and at the pre-contact stage of the confrontation with the host (i.e. when *R. solani* was at a distance of 5 mm), whereas mRNA levels increased upon direct host contact (Fig. 1 B). The re-decline of

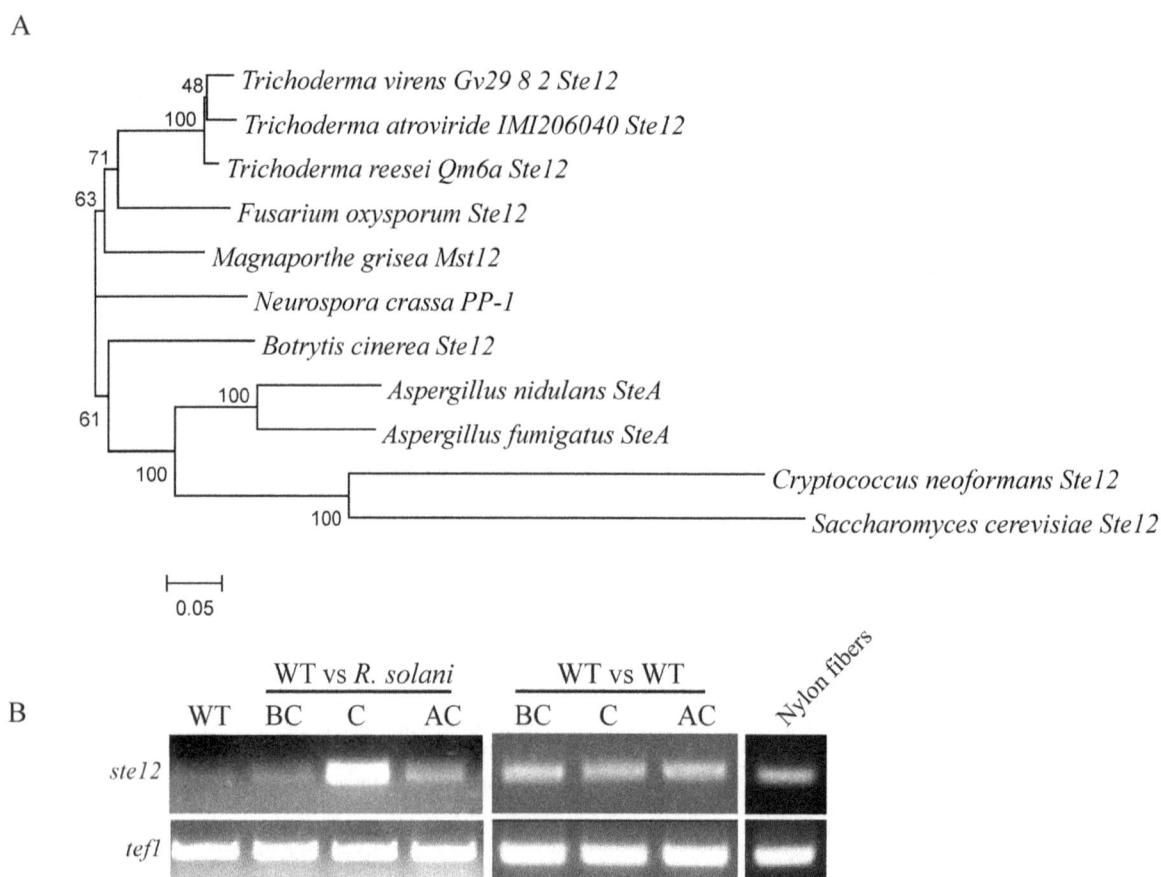

Figure 1. Phylogeny and transcriptional regulation of *T. atroviride ste12* (A) Phylogenetic analysis of Ste12-like proteins from various filamentous fungi. Ste12 orthologues identified in the genomes of the three *Trichoderma* species, *T. atroviride* (ID 29631), *Trichoderma virens* (ID 75179), and *Trichoderma reesei* (ID 36543), and *Fusarium oxysporum* Ste12 (ACM80357), *Magnaporthe oryzae* (*M. grisea*) Mst12 (AF432913), *Neurospora crassa* PP-1 (EAA28575), *Botrytis cinerea* Ste12 (ACJ06644.1), *Aspergillus nidulans* SteA (XP_659894.1), *Aspergillus fumigatus* SteA (EDP51368.1), *Cryptococcus neoformans* Ste12 (XP_776009.1), and *Saccharomyces cerevisiae* Ste12 (CAX80094.1) were aligned using ClustalX and the tree constructed using neighbor-joining algorithm with 1,000 bootstraps. (B) *ste12* transcription during growth of *T. atroviride* alone, in the mycoparasitic interaction with *R. solani* as host, in self-confrontation, and at contact with plain nylon fibers. *Trichoderma* mycelia were harvested from the interaction zone before direct contact (BC), at direct contact (C) and after contact (AC) between the two fungi. Products from RT-PCR reactions with primers targeting *ste12* and *tef1* (loading control) were separated by agarose gel electrophoresis.

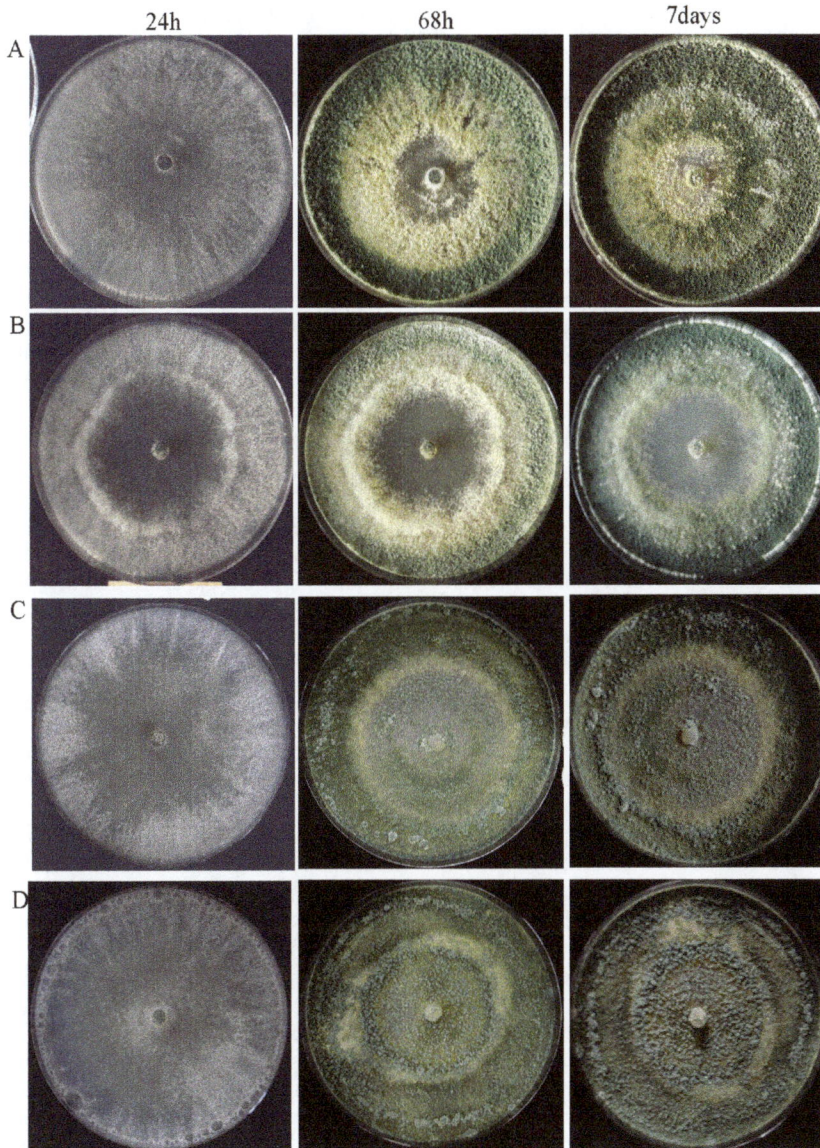

Figure 2. Colony morphology of Δ*ste12* (B) in comparison to the parental strain (A) and the complemented strains *ste12*-C1(C) and *ste12*-C2(D) upon growth on potato dextrose agar at 28°C for up to 7 days.

ste12 mRNA levels at the after contact stage, when *T. atroviride* had overgrown the host by 5 mm, indicates that this host-induced up-regulation of *ste12* expression is only transient and suggests a role of Ste12 in the regulation of mycoparasitism-relevant processes upon direct host contact.

In contrast to *Colletotrichum lindemuthianum* and *Botrytis cinerea*, for which alternative splicing of *ste12* has been described [40,41], only full transcripts (data not shown) were found for *T. atroviride ste12*.

Generation of *T. atroviride* ste12 deletion and complementation mutants

For functional characterization of *T. atroviride ste12*, we transformed the linearized *ste12* deletion vector into *T. atroviride* protoplasts. Although all of the resulting 20 transformants showed hygroymcin B-resistance, PCR- and Southern blot-based screening resulted in only one mutant with homologous integration and

deletion of the *ste12* gene in a mitotically stable manner (Figure S1). Complemented strains were generated by introducing a 5141-kb fragment bearing the *ste12* gene and its 5′ and 3′ regulatory regions into the Δ*ste12* mutant. Two complemented transformants with an ectopically (*ste12*-C1) and homologously (*ste12*-C2) integrated *ste12* gene, respectively, were selected and included in a subset of experiments. The Δ*ste12* and complemented mutants exhibited growth rates similar to the parental strain on solid complete medium (PDA). However, deletion of *ste12* resulted in somewhat altered colony development with a reduced production of aerial hyphae in the colony centre and a delayed concentrical conidial ring formation. Whereas in the parental and both complemented strains the onset of conidiation starts in the middle of the colony, conidial maturation started from the subperipheral zone in Δ*ste12* (Fig. 2). Similar to the parental strain, conidiation in the Δ*ste12* mutant was light-dependent and the number of conidia produced by Δ*ste12* was similar to the parental strain ($1.9\pm0.3 * 10^9$ and $1.2\pm0.2 * 10^9$, respectively).

Ste12 impacts carbon utilization of *T. atroviride*

In order to get additional insights into the phenotypic consequences resulting from *ste12* deletion and to learn more on a putative involvement of Ste12 in the Tmk1 MAPK pathway, we performed comparative nutrient profiling of Δ*ste12* and Δ*tmk1* mutants. The carbon-source utilization profile of *T. atroviride* on 95 different carbon sources has been characterized previously resulting in four clusters [33]. Clusters I (comprising mainly monosaccharides and polyols, but also γ-amino-butyric acid and N-acetyl-D-glucosamine), II (several monosaccharides, some oligosaccharides and arylglucosides), and III (mainly disaccharides and oligosaccharides, arylglucosides and L-amino acids) contain carbon sources allowing fast, moderate and slow growth, respectively, while carbon sources only allowing very poor or no growth at all are contained in cluster IV (containing several L-amino acids, peptides, amines, TCA-intermediates, aliphatic organic acids). Analysis of growth of the Δ*ste12* and Δ*tmk1* mutants on the 95 carbon sources revealed similar carbon utilization profiles for both mutants which, however, significantly differed from the parental strain (Fig. 3). Amongst the cluster I-III carbon sources, D-mannose, D-ribose, dextrin, salicin, amygdalin, L-arabinose, succinic acid, and β-methyl-D-glucoside only allowed reduced growth of both mutants compared to the parental strain. On the other hand, both, Δ*ste12* and Δ*tmk1*, exhibited enhanced growth on 2-keto-D-gluconic acid, glycerol, maltotriose, quinic acid and the amino acids and amino acid derivatives L-glutamic acid, L-asparagine, L-aspartic acid, L-threonine, L-pyroglutamic acid, and L-alanyl-glycine. In addition to this congruent behavior, also differences in the nutritional profiles between Δ*ste12* and Δ*tmk1* mutants were evident. While on i-erythritol, the Δ*tmk1* mutant showed similar growth than the parental strain, the Δ*ste12* mutant showed significantly reduced growth. Vice versa, deletion of *tmk1*, but not *ste12*, resulted in reduced growth on γ-amino-butyric acid which, together with N-acetyl-D-glucosamine, was the best carbon source for the *T. atroviride* parental strain. On the other hand, the Δ*tmk1* mutant exhibited better growth compared to the *ste12* deletion mutant and the parental strain on the monosaccharides D-trehalose, D-xylose, and D-fructose, the disaccharide cellobiose and the tetrasaccharide stachyose, while Δ*ste12* grew better on the polyols D-mannitol and D-arabitol.

Summarizing, vegetative growth and conidiation of *T. atroviride* on rich medium remained largely unaffected by deletion of the *ste12* gene. Nutrient profiling, however, revealed effects of Ste12 and Tmk1 on the utilization of certain carbohydrates and amino acids as carbon sources. Consequently, we conclude that the regulation of growth by Tmk1 is carbon source-dependent and only partially mediated by Ste12.

Ste12 impacts hyphal morphology and negative hyphal autotropism

During normal mycelial growth, hyphal tips are engaged in environmental sensing and usually avoid each other (negative autotropism) allowing the fungus to explore and exploit the available substrate whereas sub-apical hyphal parts generate new branches (reviewed in [42]). Microscopic analyses of Δ*ste12* mycelia from the colony periphery revealed long hyphae with only few branches which aberrantly clustered by growing alongside each other thereby resulting in compact hyphal bundles (Fig. 4 A). This loss of hyphal avoidance was also apparent in Δ*tmk1* mutants but not in the parental strain and the complemented strain *ste12-*

C2 which at the colony periphery formed typical branched hyphae that grew away from their neighbours.

To explore whether the aberrant hyphal aggregation caused by deletion of *ste12* or *tmk1* is due to a de-regulated sensing mechanism, the biomimetic system [43] was used. Frequent attachment to and growth alongside uncoated nylon fibers was observed in Δ*ste12*, similar to the behaviour of Δ*tmk1* mutants but contrasting the parental and the complemented strain (Fig. 4 B). Interestingly, this attachment was accompanied by enhanced branching of the Δ*ste12* mutant's hyphae, a reaction normally displayed by the *T. atroviride* wild-type in response to the presence of a host fungus [35].

We conclude that the role of the Tmk1 MAPK in suppressing attachment to and coiling around own hyphae and foreign hyphal-like structures in the absence of respective cues is mediated by the Ste12 transcription factor.

Ste12 impacts mycoparasitic overgrowth and host lysis

Our previous analyses of Δ*tmk1* mutants revealed reduced mycoparasitic activity and altered host specificity as Δ*tmk1* mutants still could parasitize and at least partially lyse *R. solani*, whereas they completely lost the ability to antagonize *B. cinerea* [17].

In order to assess whether Tmk1 regulates the mycoparasitic activity of *T. atroviride* by employing the Ste12 transcription factor and whether the observed aberrant self-attachment of Δ*ste12* impacts the mycoparasitic interaction with a living host fungus, plate confrontation assays with *R. solani* and *B. cinerea* as fungal hosts were performed. During the early phases of the interaction with *R. solani*, i.e. growth towards the host fungus and establishment of direct contact, the Δ*ste12* mutant behaved similar as the parental strain (Fig. 5 A). After 14 days however, *R. solani* hyphae were completely lysed by the parental and the complemented strain while only incompletely lysed by the Δ*ste12* mutant. A reduction in the mycoparasitic attack and host lysing abilities of the mutant was also evident against *B. cinerea*. While the parental and the complemented strain steadily overgrew the host, *ste12* deletion resulted in a halt shortly after establishment of contact between the two fungi which could hardly be overcome by the Δ*ste12* mutant even under prolonged incubation times (Fig. 5 A).

Imaging of the confrontation zone between *T. atroviride* and *R. solani* revealed typical growth of the parental strain towards the host followed by attachment to and growth alongside host hyphae (Fig. 5 B). Whereas the parental strain grew as well separated hyphae, the Δ*ste12* mutant approached the host primarily in the form of hyphal bundles with only single hyphae attaching to *Rhizoctonia*. Microscopic examination of individual hyphae of the Δ*ste12* mutant during the interaction with either *R. solani* or *B. cinerea* revealed typical mycoparasitism-associated morphological changes, i.e. hyphal attachment to and coiling around the host (Fig. 5 C). From these results we conclude that Ste12 affects mycoparasitic overgrowth and host lysis in a host-specific manner although attachment to and coiling around the host hyphae is unaltered upon *ste12* deletion.

Ste12 affects the expression of cell wall-degrading enzymes

The lysis of the host's cell wall is a key process in mycoparasitism [4]. We therefore tested whether Ste12 mediates the Tmk1-dependent regulation of chitinase gene expression. As reported previously, Δ*tmk1* mutants attained higher *nag1* (N-acetyl-D-glucosamidase I-encoding) and *ech42* (endochitinase 42-encoding) transcript levels and enhanced extracellular N-acetyl-

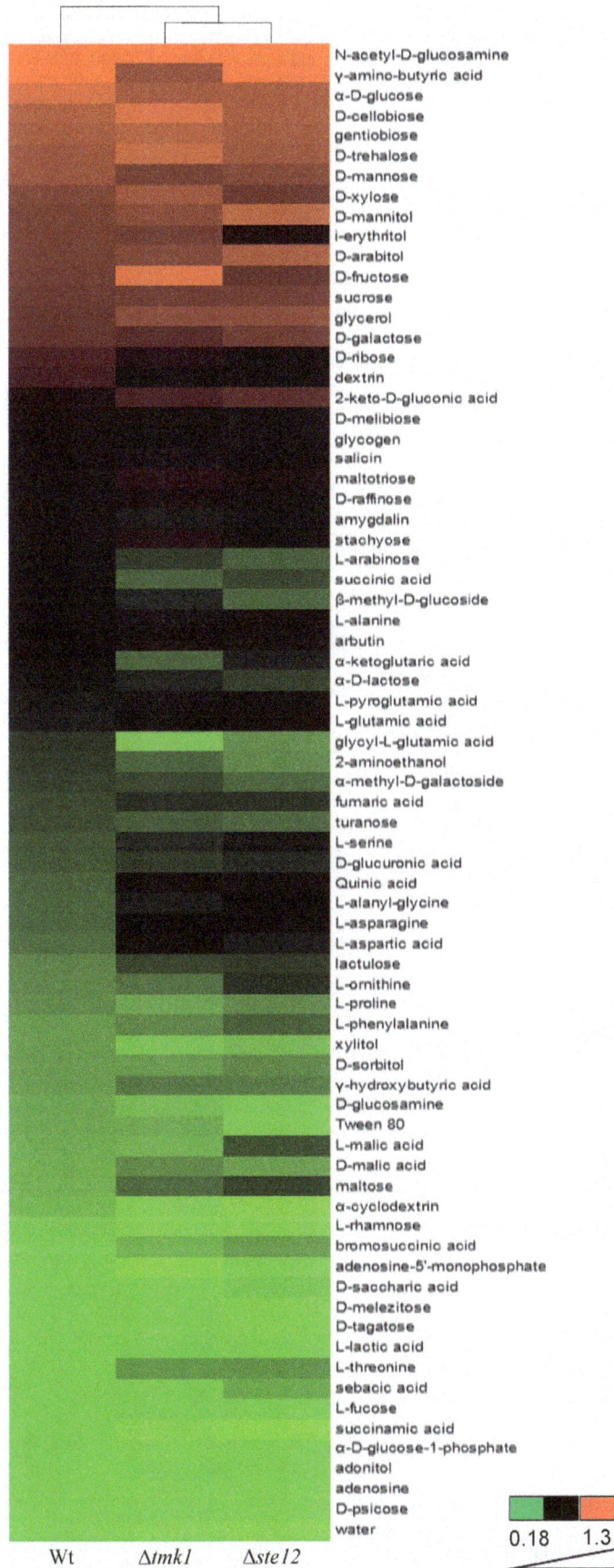

		N-acetyl-D-glucosamine
		γ-amino-butyric acid
		α-D-glucose
		D-cellobiose
		gentiobiose
		D-trehalose
		D-mannose
		D-xylose
		D-mannitol
		i-erythritol
		D-arabitol
		D-fructose
		sucrose
		glycerol
		D-galactose
		D-ribose
		dextrin
		2-keto-D-gluconic acid
		D-melibiose
		glycogen
		salicin
		maltotriose
		D-raffinose
		amygdalin
		stachyose
		L-arabinose
		succinic acid
		β-methyl-D-glucoside
		L-alanine
		arbutin
		α-ketoglutaric acid
		α-D-lactose
		L-pyroglutamic acid
		L-glutamic acid
		glycyl-L-glutamic acid
		2-aminoethanol
		α-methyl-D-galactoside
		fumaric acid
		turanose
		L-serine
		D-glucuronic acid
		Quinic acid
		L-alanyl-glycine
		L-asparagine
		L-aspartic acid
		lactulose
		L-ornithine
		L-proline
		L-phenylalanine
		xylitol
		D-sorbitol
		γ-hydroxybutyric acid
		D-glucosamine
		Tween 80
		L-malic acid
		D-malic acid
		maltose
		α-cyclodextrin
		L-rhamnose
		bromosuccinic acid
		adenosine-5'-monophosphate
		D-saccharic acid
		D-melezitose
		D-tagatose
		L-lactic acid
		L-threonine
		sebacic acid
		L-fucose
		succinamic acid
		α-D-glucose-1-phosphate
		adonitol
		adenosine
		D-psicose
		water

0.18   1.3

Wt   Δtmk1   Δste12

Figure 3. Comparative carbon source utilization profiles of the Δ*ste12* and Δ*tmk1* mutants and the parental strain (WT). Strains were grown on 95 carbon sources (FF-plates) using the Biolog Phenotype Microarray system. Hierarchical clustering results are displayed as coloured mosaics attached to a dendrogram according to growth (OD$_{750}$) after 48 hours of incubation. All data points are the average of three replicates. The threshold cut-off for all conditions was set at 0.225 which corresponded to growth of the parental strain on the water control. Red boxes indicate maximal growth (OD$_{750}$ 0.7–1.3), black boxes medium growth (OD$_{750}$ 0.55–0.7), and green boxes weak growth (OD$_{750}$ 0.2–0.55).

glucosaminidase (NAGase) and endochitinase activities under chitinase-inducing conditions [17].

While secreted NAGase activities in N-acetyl-glucosamine-induced cultures were decreased upon *ste12* deletion at all time points tested, the Δ*ste12* mutant showed elevated extracellular endochitinase activities compared to the parental strain upon induction with colloidal chitin (Fig. 6 A). Further analysis at the transcript level confirmed the enhanced transcription of the *ech42* gene in the Δ*ste12* mutant after cultivation on colloidal chitin for 36 hours and, unexpectedly, also revealed enhanced *nag1* mRNA levels compared to the parental strain upon cultivation in the presence of N-acetyl-glucosamine for 14 and 24 hours (Fig. 6 B). Similar to *ech42*, the *prb1* gene, which encodes a subtilisin-like serine protease whose over-expression has been shown to improve the biocontrol activity of *T. atroviride* [44], can be induced by chitin. *prb1* expression was highest after 36 hours of cultivation in chitin-containing media in both the Δ*ste12* mutant and the parental strain with *prb1* mRNA levels being ~2-fold enhanced in the mutant.

Based on the findings that Ste12 negatively regulates the expression of the cell wall-degrading enzymes tested but positively affects the mycoparasitic activity of *T. atroviride* against *R. solani* and *B. cinerea*, we were interested in analyzing the expression of the mycoparasitism-relevant *ech42*, *nag1* and *prb1* genes in direct confrontation assays. To this end, mycelia from the parental strain and the Δ*ste12* and Δ*tmk1* mutants during direct interaction with *R. solani* were harvested at the early overgrowth stage which corresponded to the stage with the most significant differences between the Δ*ste12* mutant and the parental strain in the mycoparasitism assays (Fig. 5 A). While in the parental strain all three tested genes were significantly induced during overgrowth of *R. solani* compared to the self confrontation control, only *prb1* showed a host-induced expression pattern in the Δ*ste12* mutant with mRNA levels exceeding those of the parental strain by several-fold. Expression of *nag1* was elevated in both, the Δ*ste12* and the Δ*tmk1* mutants, although in a host-independent manner, i.e. also in the self confrontation control. Similarly, *ech42* gene transcription was independent from host-derived signals in both mutants with *ech42* mRNA levels in the Δ*tmk1* mutant significantly exceeding those of the parental strain and of the Δ*ste12* mutant (Fig. 6 C).

These results suggest that Ste12 and Tmk1 negatively regulate the expression of genes important for host cell wall degradation and host lysis in *T. atroviride*.

Ste12 mediates the influence of Tmk1 on CAT formation and hyphal fusion

Vegetative hyphal fusion is important for the development of a mycelial network during colony development in many filamentous fungi. In addition to fusion of hyphae within a mature colony, germlings of *N. crassa* recognize each other shortly after conidial germination and can fuse via specialized structures, the conidial anastomosis tubes (CATs) [45]. In *N. crassa* and *F. oxysporum*, the Fus3/Kss1 homologues Mak-2 and Fmk1, respectively, are required for both CAT and hyphal fusion during vegetative growth [21,46].

Despite the fact that *T. atroviride* Δ*ste12* and Δ*tmk1* mutants lost negative hyphal autotropism in the colony periphery which resulted in the observed hyphal aggregation, we were not able to detect distinct fusions between aggregated hyphae. To further analyze a putative role of Tmk1 and Ste12 in regulating fusion processes, the behavior of Δ*ste12* and Δ*tmk1* germlings was assessed microscopically. Conidial anastomosis tubes as well as fusion bridges between germ tubes were frequently observed in the parental and the complemented strain (Fig. 7). In contrast, CATs could not be detected in the Δ*ste12* and Δ*tmk1* mutants and also fusions between germ tubes were only rarely observed despite frequent contacts between the germlings. It is worth mentioning that Δ*tmk1* conidia showed delayed germ tube formation and extensive aggregation of Δ*tmk1* germlings occurred (Fig. 7). Recent studies in the fungal model *N. crassa* led to the identification of target genes being required for cell fusion which are under control of the Mak-2 MAPK and the Ste12 homolog PP-1 [45]. To further substantiate our above findings of Tmk1 and Ste12 playing key roles in fusion processes in *T. atroviride*, mycelia of Δ*ste12* and Δ*tmk1* mutants and the parental strain were harvested from the internal as well as peripheral zones of the fungal colony. For gene expression analyses, the respective *T. atroviride* orthologues (Ta300768, *tmk2*, *hex1*, Ta302802, Ta294940) of the *N. crassa* fusion genes *ham-7* (encoding a GPI-anchored protein required for activation of the cell wall integrity MAPK MAK-1), *mak-1* (MAPK), *hex-1* (involved in septal plugging), *nox-1* (NADPH oxidase), and *ham-9* (pleckstrin domain protein) were identified in the *T. atroviride* genome database (http://genome.jgi-psf.org/Triat2/Triat2.home.html) by BLAST searches. Moreover, the glycosyl-hydrolase 18 (GH18) subgroup C chitinase-encoding gene *tac6* (Ta348129), which plays a role in hyphal network formation in *T. atroviride* [24], was included in the study.

Of the genes tested, Ta300768/*ham-7*, *hex1*, Ta302802/*nox-1*, *tmk2*, and *tac6* showed a significantly higher transcription in the center than at the peripheral zone of *T. atroviride* colonies. This expression pattern would be indicative of a role of these genes in hyphal fusion which predominantly takes place in the colony center (Figure 8). In accordance with *tmk1* and *ste12* deletion resulting in a loss of cell fusion, mRNA levels of Ta300768/*ham-7*, *hex1*, Ta302802/*nox-1*, and *tac6* were reduced in both the center and the periphery of Δ*ste12* and Δ*tmk1* colonies compared to the parental strain colony center. Ta294940/*ham9* was found to be similarly transcribed throughout the wild-type colony and the Δ*tmk1* peripheral zone but showed heavily reduced expression in Δ*ste12* colonies. A completely different picture was obtained for *tmk2*, which encodes the *T. atroviride* homolog of the cell wall integrity pathway MAPK Slt2 of *S. cerevisiae* and Mak-1 of *N. crassa*. While deletion of *tmk1* was found to result in significantly reduced *tmk2* transcription, Δ*ste12* mutants showed enhanced *tmk2* mRNA levels especially in the peripheral zone of the colony (Figure 8). These results show that expression of the cell wall integrity MAPK Tmk2 in *T. atroviride* is positively regulated by Tmk1 but negatively affected by Ste12 under the conditions tested and suggest that the repressing effect of Ste12 is mediated by upstream components other than Tmk1.

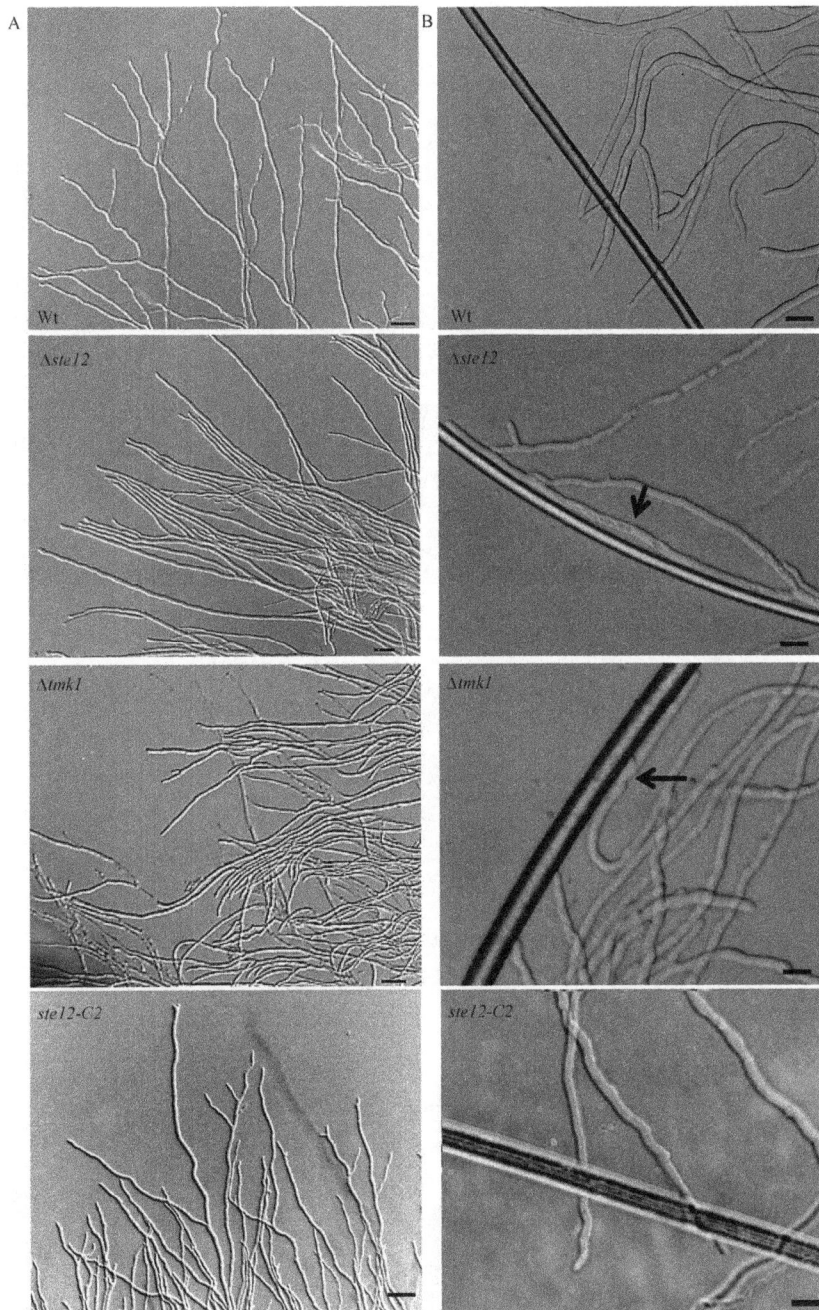

Figure 4. Phenotypes of the Δ*ste12* and Δ*tmk1* mutants compared to the parental strain (WT) and the complemented strain *ste12*-C2 upon growth on potato dextrose agar (PDA). (A) Hyphae of Δ*ste12* and Δ*tmk1* mutants attached and formed hyphal aggregates in the colony periphery whereas the parental and the complemented strain showed hyphal avoidance. (B) Light microscopy of hyphae of the Δ*ste12* and Δ*tmk1* mutants, the parental strain, and the complemented strain *ste12*-C2 upon growth on PDA with plain nylon fibers (approximate diameter 14 μm). Attachment to and growth along the fibers of hyphae of Δ*ste12* and Δ*tmk1* mutants is marked with arrows.

Discussion

The genus *Trichoderma* comprises potent antagonists with many species being able to parasitize and kill other fungi [4]. As a prerequisite for this mycoparasitic lifestyle, *Trichoderma* has to possess appropriate receptors and intracellular signaling pathways for sensing and integrating signals derived from host fungi. Orthologues of yeast Fus3 (mating pathway) and Kss1 (filamentous growth pathway) MAPKs are multifunctional pathogenicity factors

required for virulence of biologically and taxonomically diverse fungi [20]. However, although the pathogenicity MAPK (Pmk) pathway modules have been conserved throughout evolution [47] the output responses are species-specific and it is still not clear how exactly Pmk-type MAPKs regulate fungal virulence. In the foliar plant pathogen *M. oryzae*, the Pmk1 cascade targets the transcription factor Mst12, an ortholog of yeast Ste12. Similar to *pmk1* mutants, *mst12* mutants fail to penetrate the plant surface and are compromised in infectious growth although Mst12, in

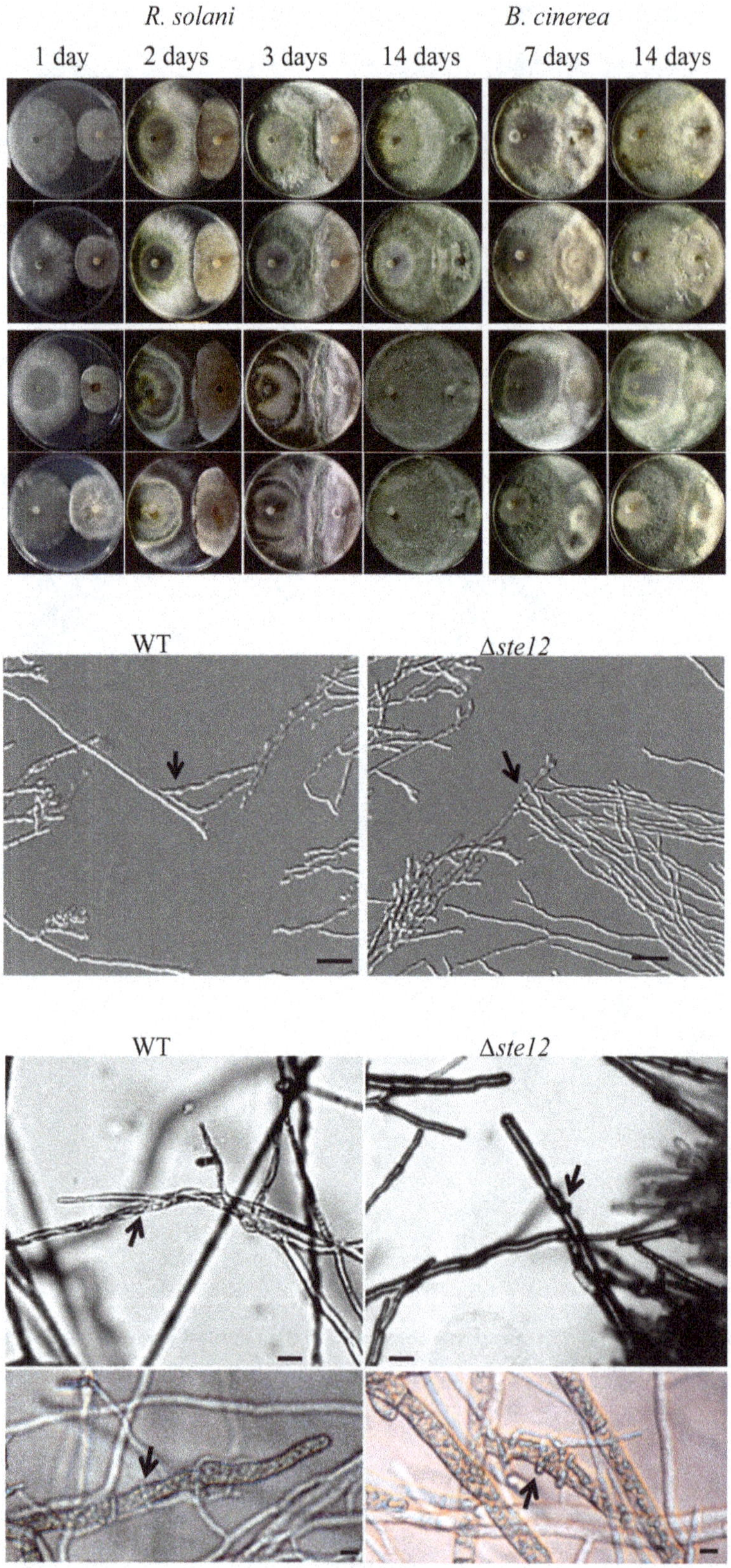

Figure 5. Mycoparasitic activity of the Δ*ste12* mutant against *R. solani* and *B. cinerea* as hosts. (A) Plate confrontation assays of the Δ*ste12* mutant (second panel), the parental strain (upper panel) and the complemented strains *ste12*-C1 (third panel) and *ste12*-C2 (fourth panel) against host fungi. Pictures were taken 1, 2, 3, and 14 days (*R. solani*) and 7 and 14 days (*B. cinerea*) after inoculation of the two fungi on opposite sides of the plate. (B) Microscopic analyses of the confrontation zone between *T. atroviride* (right side) and *R. solani* (left side). The Δ*ste12* mutant approaches the host as aggregated hyphae with only single hyphae attaching to *Rhizoctonia*. Attachments to host hyphae are marked by arrows. The scale bar represents 50 μm. (C) Attachment to and coiling around host hyphae. Despite the inability of the Δ*ste12* mutant to fully overgrow and parasitize *B. cinerea*, the mutant shows the typical mycoparasitism-associated coiling response.

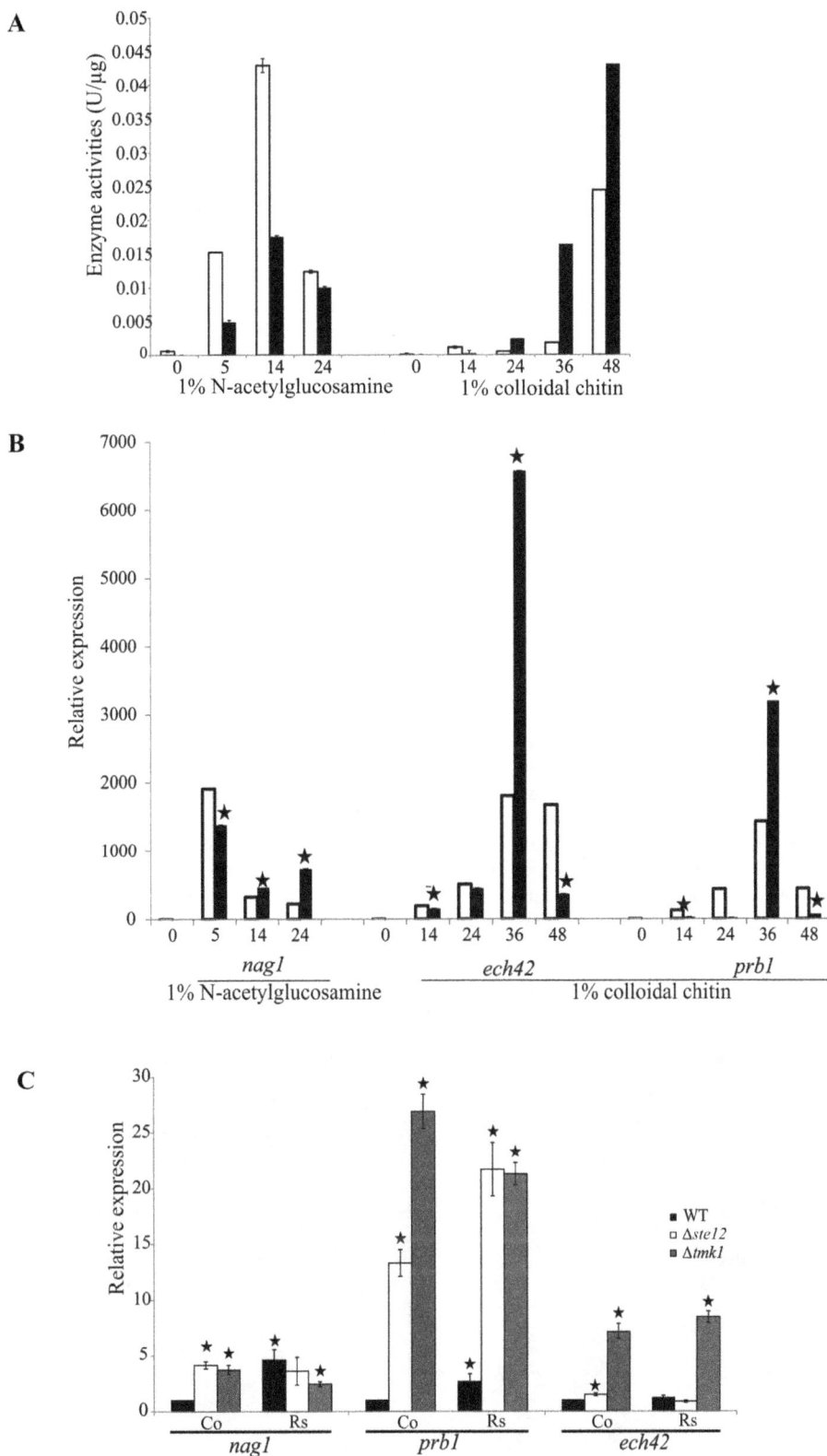

Figure 6. Impact of Ste12 on the expression of mycoparasitism-related cell wall-degrading enzymes. (A) Extracellular N-acetyl-glucosaminidase (NAGase) and endochitinase activities in the Δste12 mutant (black bars) and the parental strain (white bars). After pre-cultivation on 1% glycerol, mycelial biomass was transferred to 1% N-acetyl-glucosamine-containing media for inducing NAGases and to 1% colloidal chitin-containing media for induction of endochitinases. Culture filtrates were harvested at the indicated time points and determined enzyme activities related to intracellular total protein. (B) Relative transcription ratios of the chitinase-encoding nag1 and ech42 genes and the prb1 protease-encoding gene in the Δste12 mutant (black bars) and the parental strain (white bars). RT-qPCR analyses were performed 5, 14, and 24 hours after transfer to N-acetyl-glucosamine (nag1) and 14, 24, 36, and 48 hours after transfer to colloidal chitin (ech42, prb1) using act1 as reference gene. Un-induced

samples of the parental strain harvested after pre-cultivation in glycerol-containing media were arbitrarily assigned the factor 1. Asterisks indicate significantly different (p≤0.05; calculated by REST software) transcription ratios of the mutant compared to the parental strain. (C) Relative transcription ratios of *nag1*, *ech42* and *prb1* in Δ*ste12* (white bars) and Δ*tmk1* (grey bars) mutants and the parental strain (black bars) upon direct confrontation with *R. solani*. Samples were collected from a control (co) where *Trichoderma* was confronted with itself and from the early overgrowth stage in the confrontation with *R. solani* (Rs) and subject to RT-qPCR using *sar1* as reference gene. The control sample of the parental strain was arbitrarily assigned the factor 1 and those samples which show significant differences (p≤0.05; calculated by REST software) to this control are marked with an asterisk. Results shown are means ±SD (n = 3).

contrast to Pmk1, is dispensable for appressorium formation in *M. oryzae* [38,48]. A comparable situation has been reported for non appressorium-forming fungi such as *F. oxysporum* where invasive growth is regulated by the Fmk1 MAPK and the transcription factor Ste12 [19].

Here, we have characterized Ste12 from the mycoparasitic fungus *T. atroviride* by assessing its role in mediating vegetative growth, colony development and mycoparasitism-related functions. The structure of *T. atroviride* Ste12 is similar to Ste12 proteins from other filamentous fungi and contains the typical homeodomain-like STE domain and two C_2H_2 zinc finger motifs. In *S. cerevisiae* Ste12, the homeodomain is required for binding to the regulatory protein Dig2 and the cofactor Tec1, whereas the central and C-terminal regions are involved in homodimerisation and binding to the negative regulator Dig1 and the Mcm1 transcription factor (reviewed by [11,47]). Whereas we found a conserved Mcm1 orthologue in the *T. atroviride* genome (ID 223702), *T. atroviride* does not encode Dig1 and Dig2 like proteins - a situation similar to other filamentous fungi [47]. An orthologue of Tec1, which upon heterodimerisation with Ste12 induces filamentous growth under nutrient limiting conditions in *S. cerevisiae* [10], is encoded in the genome of the filamentous pathogen *A. fumigatus* but could neither be detected in plant pathogenic ascomycetes [47] nor in our study in the mycoparasite *T. atroviride*. The homeodomain of Ste12 has previously been shown to be required for DNA binding in *Cryptococcus neoformans* and *Colletotrichum lindemuthianum* [40,49] and to be essential for all characterized functions of the *N. crassa* Ste12

homolog PP-1 [45]. The C-terminal zinc finger motifs, which are characteristic for Ste12 proteins of filamentous fungi but missing in *Saccharomycotina*, have recently been reported as dispensable in *N. crassa* PP-1, they are, however, together with the homeodomain required for Mst12-mediated virulence in *M. oryzae* [38,50].

ste12 mRNA levels were transiently up-regulated in *T. atroviride* at the stage of direct contact with the host fungus *R. solani*. A similar host-induced regulation of *ste12* expression occurs in *F. oxysporum* where transcription of *fost12* is up-regulated during the early stages of host plant colonization [36].

To substantiate a putative role of Ste12 in the mediation of mycoparasitism-relevant processes and as a functional target of the Tmk1 MAP kinase, the *ste12* gene was deleted in *T. atroviride*. Comparative analyses of the phenotypes of Δ*ste12* and Δ*tmk1* mutants revealed that several of the Tmk1 MAPK outputs are mediated by Ste12 (Fig. 9). The Δ*ste12* and Δ*tmk1* mutants shared defects in hyphal avoidance and anastomosis and showed similar carbon-source utilization profiles and alterations in mycoparasitism-related processes, with the latter, however, being more pronounced upon *tmk1* deletion. Also differences between Δ*ste12* and Δ*tmk1* mutants were found which suggests that additional transcription factors other than Ste12 are targeted by Tmk1. As reported previously, Δ*tmk1* mutants form "flat" colonies with only few aerial hyphae, show reduced hyphal growth rates on PDA and sporulate light-independently [17,23]. In the present study we further found that *tmk1* deletion resulted in a delay in germ tube elongation, similar to what has been reported for *N. crassa mak-2* deletion strains and which indicates that a functional MAPK is

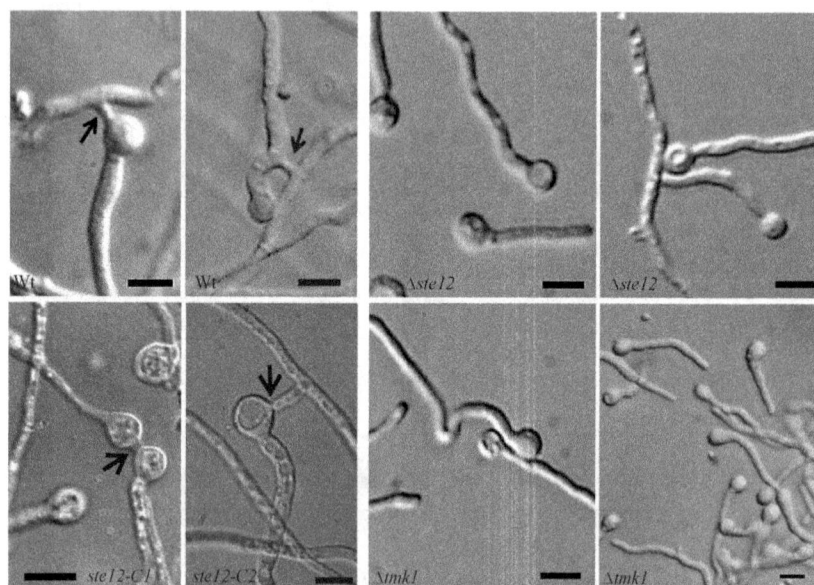

Figure 7. Analysis of conidial anastomosis tube (CAT) formation and hyphal fusion in Δ*ste12* and Δ*tmk1* mutants. Microscopic analyses of germlings of Δ*ste12* and Δ*tmk1* mutants, the parental strain and the complemented strains *ste12*-C1 and *ste12*-C2 16 hours after inoculation of conidia in potato dextrose broth. CATs and fusion bridges between germ tubes of the parental strain are marked with arrows. The scale bar represents 10 μm.

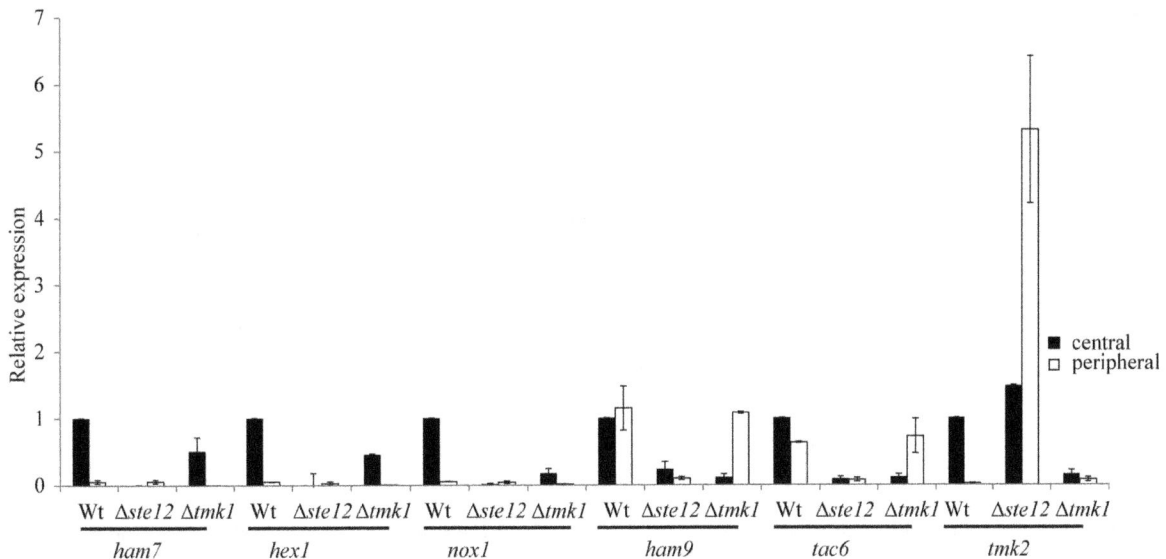

Figure 8. Impact of Ste12 and Tmk1 on the transcription of genes with a putative role in hyphal fusion. Relative transcription ratios of the putative fusion genes (see text) in the colony center and the peripheral zones of colonies of the Δ*ste12* and Δ*tmk1* mutants and the parental strain were determined by RT-qPCR using *tef1* as reference gene. The sample from the central region of parental strain colonies was arbitrarily assigned the factor 1. Results shown are means ±SD (n = 3).

required for optimal apical hyphal extension [46]. In contrast, *T. atroviride* Ste12 is dispensable for hyphal extension and is not involved in mediating the repressing effect of Tmk1 on conidiation in the dark; furthermore, it plays only a minor role in aerial hyphae formation.

Δ*ste12* and Δ*tmk1* mutants showed in large parts overlapping carbon source utilization profiles which, however, significantly differed from that of the parental strain. Regulation of primary metabolic pathways by the Pmk1 MAPK has previously been reported for *M. oryzae* [51,52] and in *C. neoformans*, the mediator protein Ssn8, which in *S. cerevisiae* is involved in carbon utilization, acts downstream of the Cpk1 MAPK and the Ste12 transcription factor [53]. Interestingly, both Δ*ste12* and Δ*tmk1* mutants showed reduced growth on γ-amino-butyric acid (GABA) with a more severe growth reduction resulting from *tmk1* gene deletion. GABA metabolism is required for full pathogenicity in the wheat pathogen *Stagnospora nodorum* [54]. A similar situation may apply to *T. atroviride* where deletion of *ste12* or *tmk1* results in impaired mycoparasitism being more pronounced in Δ*tmk1* than Δ*ste12* mutants.

In the fungal model *N. crassa*, hyphal fusion occurs between germlings during colony establishment and between hyphae in subapical parts of mature colonies. Cell fusion in germinating conidia is associated with the production of specialized fusion structures, the conidial anastomosis tubes (CATs) [55]. *Neurospora* strains carrying deletions of Fus3 MAPK pathway components such as *nrc-1*, *mek-2*, *mak-2* or *pp-1* are defective in cell fusion and are female sterile [45,46,56,57].

Phenotypic analyses of *T. atroviride* Δ*ste12* and Δ*tmk1* mutants revealed a loss of CAT formation and a severe reduction of hyphal fusion events, whereas CATs and fusion bridges were frequently observed in the parental strain. Interestingly, to the best of our knowledge, conidial anastomosis tube formation has never been shown before in *Trichoderma* although there are several reports on inter- and intra-strain anastomosis of vegetative hyphae or protoplast fusion mainly with the aim to obtain new genetic combinations with improved biocontrol activities (e.g. [58], [59]).

Our results suggest that the Tmk1 MAPK pathway regulates CAT formation and hyphal fusion in *T. atroviride* by employing the Ste12 transcription factor. This is similar to the situation in *N. crassa*, but different to *F. oxysporum* where the essential role of the Fmk1 MAPK in vegetative hyphal fusion is not mediated by Ste12 [19].

Formation of an interconnected mycelial network by hyphal fusions is important for communication and translocation of water and nutrients within a filamentous fungus' colony [60]. Loss of or reductions in vegetative hyphal fusion events hence are supposed to result in a decrease in mycelial interconnections and a concomitant reduced nutrient transport from the colony periphery, where fresh medium is available, to the colony center, where nutrients are increasingly exploited. It may be speculated that the altered growth phenotype of *T. atroviride* Δ*ste12* and Δ*tmk1* mutants, i.e. the formation of aggregated hyphal bundles, which typically are formed by fungi on an exhausted substrate, is due to the reduced abilities of the mutants to anastomose.

Transcriptional analysis of conserved *T. atroviride* orthologues being targets of the PP-1 transcription factor in the *N. crassa* cell fusion pathway further supported the fusion defects of the Δ*ste12* mutant. Similar to *N. crassa* PP-1 [45], Ste12 showed an activating role on the expression of putative fusion genes such as Ta300768 (*ham7*), Ta294940 (*ham9*), Ta298536 (*hex1*), and Ta302802 (*nox1*) in *T. atroviride*. However, there were remarkable differences between the two fungi concerning the role of Ste12 in regulating the expression of the cell wall integrity MAPK-encoding gene. Whereas *mak-1* expression is reduced in *N. crassa* Δ*pp-1* strains [45], the *T. atroviride* Δ*ste12* mutant showed elevated *tmk2* mRNA levels. In the peripheral zones of the colonies, *ste12* or *tmk1* deletion resulted in 256-fold and 4-fold enhanced *tmk2* transcript levels, respectively, compared to the parental strain whereas in the colony centre only the Δ*ste12* but not the Δ*tmk1* mutant over-expressed *tmk2*. *T. virens tmkB* mutants, missing the TmkB homologue of the yeast cell wall integrity MAPK Slt2, showed increased sensitivity to cell wall-degrading enzymes and had cell wall integrity defects [61]. It may

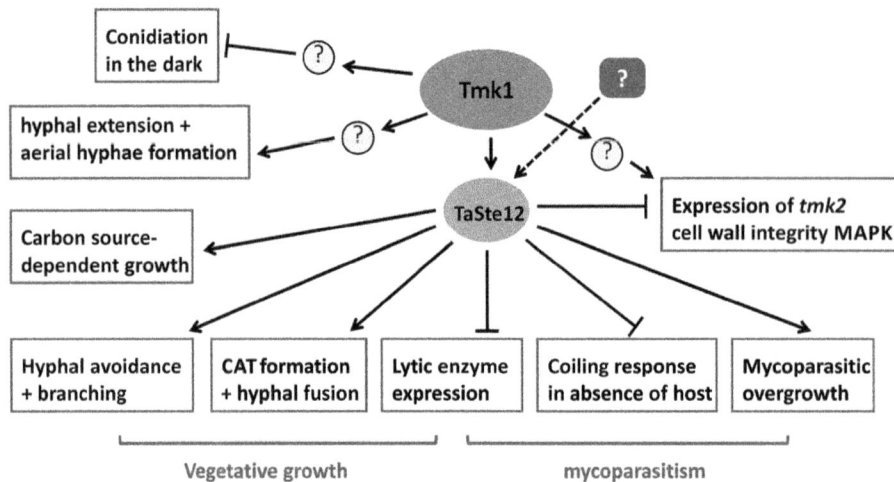

Figure 9. Model of the role of Ste12 as down-stream target of the Tmk1 MAP kinase pathway and as target of additional, still unknown pathways.

be speculated that the elevated transcription of *tmk2* in the periphery of *T. atroviride* Δ*ste12* and Δ*tmk1* mutant colonies is due to enhanced cell wall stress provoked by the observed constitutive up-regulation of chitinase gene expression in the mutants. This would be consistent with previous studies showing that most *T. atroviride* chitinases are not only involved in the degradation of host cell walls during mycoparasitism but are also required for rebuilding and recycling of the fungus' own cell wall during growth [62]. Anyway, whereas in *N. crassa* the cell wall integrity MAPK has been shown to act downstream of the Mak-2/PP-1 pathway during cell fusion [45] this seems not to be the case in *T. atroviride*.

Comparable to hyphal anastomosis, which roughly consists of pre-contact sensing, chemotropism, adhesion, and subsequent cell wall lysis at the fusion site [63], the mycoparasitic fungus-fungus interaction comprises similar steps. The initial interaction between *Trichoderma* and the host fungus is characterized by chemotropic growth of the mycoparasites hyphae towards the host [64] followed by the induction of a set of genes already before direct contact between the two fungi [65,66]. The pre-contact induction of certain genes being involved in host lysis such as the *ech42* endochitinase and the *prb1* protease ([3,67] led to a model in which degradation products of the hosts' cell wall are sensed by respective *Trichoderma* receptors. The resulting activation of intracellular signaling cascades by diffusible host-derived signals and, upon direct contact, lectins on the host surface, finally leads to the full induction of the mycoparasitic response resulting in host attack and lysis [4,5]. In contrast to basidiomycetous fusion parasites [68] and the mycoparasitic interaction between the two zygomycetes *Absidia glauca* and *Parasitella parasitica*, where fusion bridges between the two fungi are formed [69], *Trichoderma* mycoparasites invasively grow inside the host. Ultrastructural studies revealed that coiling hyphae of *Trichoderma* constrict and partially digest the host cell wall at the interaction site followed by penetration and growth of *Trichoderma* invading hyphae inside the host [64]. Nevertheless, our phenotypic characterization of Δ*tmk1* and Δ*ste12* mutants revealed that Tmk1 and Ste12 play important roles in hyphal anastomosis as well as mycoparasitism in *T. atroviride*. The finding that Δ*tmk1* and Δ*ste12* mutants not only showed enhanced attachment to own and foreign hyphae but also to hyphal-like plain nylon fibers suggests a repressing role of the Tmk1 signaling pathway on hyphal attachment in the absence of

respective (host-derived) signals. Taking into consideration that cAMP stimulates coiling in *T. atroviride* [70] and that Δ*tga3* mutants missing the adenylate cyclase-stimulating Gα subunit Tga3 are defective in host recognition and coiling [71], the Tmk1 MAPK cascade and the cAMP pathway seem to have antagonistic roles in regulating the mycoparasitism-relevant coiling response.

In *M. oryzae*, the Mcm1 orthologue MoMcm1 was found to interact with the Mst12 transcription factor. The findings that a *Momcm1 mst12* double mutant formed appressoria even on hydrophilic surfaces whereas this was not the case in *Momcm1* and *mst12* single mutants suggested overlapping functions of MoMcm1 and Mst12 in suppressing appressorium formation under non-conducive conditions [72]. In contrast to *M. oryzae*, the Mcm1 othologue seems to be dispensable for mediating the repressive role of Tmk1 on attachment to and coiling around hyphae and hyphal-like structures in the absence of respective signals in *T. atroviride*.

Deletion of *tmk1* or *ste12* resulted in an aberrant, i.e. host-independent over-expression of the *ech42* and *nag1* chitinases and the *prb1* protease. It is interesting to note that, despite these enhancements, Δ*tmk1* and Δ*ste12* mutants are impaired in both hyphal fusion and mycoparasitic host lysis which suggests a co-regulation of these processes by this signaling pathway and the involvement of additional still uncharacterized genes in *T. atroviride* mycoparasitism. Otherwise, the reduced mycoparasitic abilities of the mutants may result from their failure to build a fully interconnected mycelial network as has previously been predicted to be the cause of the reduced pathogenicity displayed by MAPK mutants of plant pathogens [60].

Detailed analyses of the signals originating from host fungi and their respective *Trichoderma* receptors together with a functional analysis of the downstream targets of conserved signaling cascades will be necessary to fully understand the mycoparasitic fungus-fungus interaction. Comparative genome analysis revealed that transcription factors are expanded in the mycoparasites *T. atroviride* and *T. virens* compared to the saprophyte *T. reesei* [73]. However, there are only few reports on the functional characterization of genes encoding proteins with transcription factor activity from mycoparasitic *Trichoderma* species [74–77]. This study revealed an essential role of the Ste12 transcriptional regulator in mediating the output of the mycoparasitism-relevant Tmk1 MAP kinase pathway (Figure 9) and illustrated the interconnection between hyphal anastomosis and the mycoparasitic activity of *T. atroviride*.

Supporting Information

Figure S1   Genotypic analysis of Δ*ste12* gene deletion and complementation mutants. (A) PCR analysis of the three out of 20 hygromycinB-resistant transformants that showed a stable integration of the *ste12* deletion construct after three rounds of single spore isolation. The primer pair hph-FW and hph-RV (Table 1) amplified a 560-bp fragment of the integrated *hph* gene. (B) Southern hydridization of *Nco*I-digested DNA from parental strain (WT) and the three different putative deletion mutants (D, F, S) with a 2693-bp probe covering 1415-bp of the 5′ non-coding region of the *ste12* gene and 1278-bp of the *hph* selection marker cassette. The parental strain and transformants F and S show a 1597-bp band indicative of the native *ste12* gene, while transformant D lacks this band and instead shows two bands of 2314-bp and 3360-bp confirming transformant D as a *ste12* null mutant resulting from homologous recombination at the *ste12* locus. (C) Confirmation of complementation mutants by PCR using primers ste12-C-FW and ste12-C-RV (Table 1) located 1500-bp 5′ and 3′, respectively, of the *ste12* open reading frame. This primer pair is expected to amplify a 5141-bp fragment in the parental strain (lane 5) and a 6159-bp fragment in the Δ*ste12* mutant (lanes 1 and 4). The amplification of both fragments in complementation mutant *ste12*-C1 (lane 2) confirms ectopic integration of *ste12*, whereas the presence of only the 5141-bp band in complementation mutant *ste12*-C2 (lane 3) is indicative of a rescue of the *ste12* gene at the homologous locus by replacement of the deletion construct. (D) RT-PCR with primers ste12-FW and ste12-RV (Table 1) amplified the expected 320-bp fragment of the *ste12* gene in the parental strain (lane 3), the Δ*tmk1* mutant (lane 2), and the *ste12* complemented strains (lanes 4 and 5) but not in the Δ*ste12* deletion mutant (lane 1).

Acknowledgments

We thank Markus Omann and René Mayer for help with deletion construct design and mutant generation.

Author Contributions

Conceived and designed the experiments: SZ SG. Performed the experiments: SG. Analyzed the data: SZ SG. Contributed reagents/materials/analysis tools: SZ. Wrote the paper: SZ SG.

References

1. Hjeljord L, Tronsmo A (1998) Trichoderma and Gliocladium in Biological Control: An Overview. In: Harman GE, Kubicek CP, editors. Trichoderma and Gliocladium. Vol. 2. Enzymes, Biological Control and Commercial Applications. London: Taylor and Francis Ltd. pp. 131–151.
2. Inbar J, Chet I (1994) A newly isolated lectin from the plant pathogenic fungus Sclerotium rolfsii: purification, characterization and role in mycoparasitism. Microbiology 140: 651–657.
3. Zeilinger S, Galhaup C, Payer K, Woo SL, Mach RL et al. (1999) Chitinase gene expression during mycoparasitic interaction of Trichoderma harzianum with its host. Fungal Genet Biol 26(2):131–140.
4. Druzhinina IS, Seidl-Seiboth V, Herrera-Estrella A, Horwitz BA, Kenerley CM et al. (2011) Trichoderma: the genomics of opportunistic success. Nature Rev Microbiol 9(10):749–759.
5. Zeilinger S, Omann M (2007) Trichoderma biocontrol: signal transduction pathways involved in host sensing and mycoparasitism. Gene Regul Syst Bio 1:227–234.
6. Mukherjee PK, Horwitz BA, Herrera-Estrella A, Schmoll M, Kenerley CM (2013) Trichoderma research in the genome era. Annu Rev Phytopathol 51:105–129.
7. Cargnello M, Roux PP (2011) Activation and function of the MAPKs and their substrates, the MAPK-activated protein kinases. Microbiol Mol Biol Rev 75(1):50–83.
8. Gustin MC, Albertyn J, Alexander M, Davenport K (1998) MAP kinase pathways in the yeast Saccharomyces cerevisiae. Microbiol Mol Biol Rev 62(4):1264–1300.
9. Xu J-R (1997) MAP Kinases in Fungal Pathogens. Fungal Genet Biol 31(3):137–152.
10. Madhani HD, Fink GR (1997) Combinatorial control required for the specificity of yeast MAPK signaling. Science 275(5304):1314–1317.
11. Rispail N, Di Pietro A (2010) The homeodomain transcription factor Ste12: Connecting fungal MAPK signalling to plant pathogenicity. Commun Integr Biol 3(4):327–332.
12. Li D, Bobrowicz P, Wilkinson HH, Ebbole DJ (2005) A mitogen-activated protein kinase pathway essential for mating and contributing to vegetative growth in Neurospora crassa. Genetics 170(3):1091–1104.
13. Xu JR, Hamer JE (1996) MAP kinase and cAMP signaling regulate infection structure formation and pathogenic growth in the rice blast fungus Magnaporthe grisea. Genes Dev 10(21):2696–2706.
14. Zhao X, Xu JR (2007) A highly conserved MAPK-docking site in Mst7 is essential for Pmk1 activation in Magnaporthe grisea. Mol Microbiol 63(3):881–894.
15. Wong Sak Hoi J, Dumas B (2010) Ste12 and Ste12-like proteins, fungal transcription factors regulating development and pathogenicity. Euk Cell 9(4):480–485.
16. Mendoza-Mendoza A, Pozo MJ, Grzegorski D, Martinez P, Garcia JM et al. (2003) Enhanced biocontrol activity of Trichoderma through inactivation of a mitogen-activated protein kinase. Proc Natl Acad Sci U S A 100(26):15965–15970.
17. Reithner B, Schuhmacher R, Stoppacher N, Pucher M, Brunner K, Zeilinger S (2007) Signaling via the Trichoderma atroviride mitogen-activated protein kinase Tmk1 differentially affects mycoparasitism and plant protection. Fungal Genet Biol 44(11):1123–1133.
18. Mukherjee PK, Latha J, Hadar R, Horwitz BA (2003) TmkA, a mitogen-activated protein kinase of Trichoderma virens, is involved in biocontrol properties and repression of conidiation in the dark. Euk Cell 2(3):446–455.
19. Rispail N, Di Pietro A (2009) Fusarium oxysporum Ste12 controls invasive growth and virulence downstream of the Fmk1 MAPK cascade. Mol Plant Microbe Interact 22(7):830–839.
20. Hamel LP, Nicole MC, Duplessis S, Ellis BE (2012) Mitogen-activated protein kinase signaling in plant-interacting fungi: distinct messages from conserved messengers. Plant Cell 24(4):1327–1351.
21. Di Pietro A, Garcia-MacEira FI, Meglecz E, Roncero MI (2001) A MAP kinase of the vascular wilt fungus Fusarium oxysporum is essential for root penetration and pathogenesis. Mol Microbiol 39(5):1140–1152.
22. Lev S, Horwitz BA (2003) A mitogen-activated protein kinase pathway modulates the expression of two cellulase genes in Cochliobolus heterostrophus during plant infection. Plant Cell 15(4):835–844.
23. Zeilinger S (2004) Gene disruption in Trichoderma atroviride via Agrobacterium-mediated transformation. Curr Genet 45(1):54–60.
24. Gruber S, Vaaje-Kolstad G, Matarese F, Lopez-Mondejar R, Kubicek CP, Seidl-Seiboth V (2011) Analysis of subgroup C of fungal chitinases containing chitin-binding and LysM modules in the mycoparasite Trichoderma atroviride. Glycobiology 21(1):122–133.
25. Lorito M, Mach RL, Sposato P, Strauss J, Peterbauer CK, Kubicek CP (1996) Mycoparasitic interaction relieves binding of the Cre1 carbon catabolite repressor protein to promoter sequences of the ech42 (endochitinase-encoding) gene in Trichoderma harzianum. Proc Natl Acad Sci USA 93(25):14868–14872.
26. Harman GE, Hayes C.K., Lorito M., Broadway R.M., Di Pietro A. et al. (1993) Chitinolytic enzymes of Trichoderma harzianum: purification of chitobiosidase and endochitinase. Mol Plant Pathol 83: 313–318.
27. Garcia-Pedrajas MD, Nadal M, Kapa LB, Perlin MH, Andrews DL, Gold SE (2008) DelsGate, a robust and rapid gene deletion construction method. Fungal Genet Biol 45(4):379–388.
28. Peterbauer CK, Brunner K, Mach RL, Kubicek CP (2002) Identification of the N-acetyl-D-glucosamine-inducible element in the promoter of the Trichoderma atroviride nag1 gene encoding N-acetyl-glucosaminidase. Mol Genet Genomics 267(2):162–170.
29. J. Sambrook EFF, Maniatis T (1989) Molecular Cloning: A Laboratory Manual. (2nd ed)Cold Spring Harbor Laboratory Press, New York.
30. Brunner K, Omann M, Pucher ME, Delic M, Lehner SM et al. (2008) Trichoderma G protein-coupled receptors: functional characterisation of a cAMP receptor-like protein from Trichoderma atroviride. Current Genet 54(6):283–299.
31. Pfaffl MW (2001) A new mathematical model for relative quantification in real-time RT-PCR. Nucleic Acids Res 29(9):e45.

32. Pfaffl MW, Horgan GW, Dempfle L (2002) Relative expression software tool (REST) for group-wise comparison and statistical analysis of relative expression results in real-time PCR. Nucleic Acids Res 30(9):e36.

33. Seidl V, Druzhinina IS, Kubicek CP (2006) A screening system for carbon sources enhancing beta-N-acetylglucosaminidase formation in *Hypocrea atroviridis* (*Trichoderma atroviride*). Microbiology 152:2003–2012.

34. Seo J, Shneiderman B (2002) Interactively exploring hierarchical clustering results [gene identification]. Computer 35(7):80–86.

35. Lu Z, Tombolini R, Woo S, Zeilinger S, Lorito M, Jansson JK (2004) In vivo study of *Trichoderma*-pathogen-plant interactions, using constitutive and inducible green fluorescent protein reporter systems. Appl Environ Microbiol 70(5):3073–3081.

36. Asuncion Garcia-Sanchez M, Martin-Rodrigues N, Ramos B, de Vega-Bartol JJ, Perlin MH, Diaz-Minguez JM (2010) Fost12, the *Fusarium oxysporum* homolog of the transcription factor Ste12, is upregulated during plant infection and required for virulence. Fungal Genet Biol 47(3):216–225.

37. Vallim MA, Miller KY, Miller BL (2000) *Aspergillus* SteA (sterile12-like) is a homeodomain-C2/H2-Zn+2 finger transcription factor required for sexual reproduction. Mol Microbiol 36(2):290–301.

38. Park G, Xue C, Zheng L, Lam S, Xu JR (2002) MST12 regulates infectious growth but not appressorium formation in the rice blast fungus *Magnaporthe grisea*. Mol Plant Microbe interact 15(3):183–192.

39. Yuan YL, Fields S (1991) Properties of the DNA-binding domain of the *Saccharomyces cerevisiae* STE12 protein. Mol Cell Biol 11(12):5910–5918.

40. Wong Sak Hoi J, Herbert C, Bacha N, O'Connell R, Lafitte C et al. (2007) Regulation and role of a STE12-like transcription factor from the plant pathogen *Colletotrichum lindemuthianum*. Mol Microbiol 64(1):68–82.

41. Schamber A, Leroch M, Diwo J, Mendgen K, Hahn M (2010) The role of mitogen-activated protein (MAP) kinase signalling components and the Ste12 transcription factor in germination and pathogenicity of *Botrytis cinerea*. Mol Plant Pathol 11(1):105–119.

42. Harris SD (2008) Branching of fungal hyphae: regulation, mechanisms and comparison with other branching systems. Mycologia 100(6):823–832.

43. Dennis C, Webster J (1971) Antagonistic properties of species-groups of *Trichoderma*. Trans Br Mycol Soc 57(1):25–IN23.

44. Flores A, Chet I, Herrera-Estrella A (1997) Improved biocontrol activity of *Trichoderma harzianum* by over-expression of the proteinase-encoding gene *prb1*. Curr Genet 31(1):30–37.

45. Leeder AC, Jonkers W, Li J, Glass NL (2013) Early colony establishment in *Neurospora crassa* requires a MAP kinase regulatory network. Genetics 195(3):883–898.

46. Pandey A, Roca MG, Read ND, Glass NL (2004) Role of a mitogen-activated protein kinase pathway during conidial germination and hyphal fusion in *Neurospora crassa*. Euk Cell 3(2):348–358.

47. Rispail N, Soanes DM, Ant C, Czajkowski R, Grunler A et al. (2009) Comparative genomics of MAP kinase and calcium-calcineurin signalling components in plant and human pathogenic fungi. Fungal Genet Biol 46(4):287–298.

48. Park JY, Jin J, Lee YW, Kang S, Lee YH (2009) Rice blast fungus (*Magnaporthe oryzae*) infects *Arabidopsis* via a mechanism distinct from that required for the infection of rice. Plant Physiol 149(1):474–486.

49. Chang YC, Wright LC, Tscharke RL, Sorrell TC, Wilson CF, Kwon-Chung KJ (2004) Regulatory roles for the homeodomain and C2H2 zinc finger regions of *Cryptococcus neoformans* Ste12. Mol Microbiol 53(5):1385–1396.

50. Park G, Bruno KS, Staiger CJ, Talbot NJ, Xu JR (2004) Independent genetic mechanisms mediate turgor generation and penetration peg formation during plant infection in the rice blast fungus. Mol Microbiol 53(6):1695–1707.

51. Jin Q, Li C, Li Y, Shang J, Li D, Chen B, Dong H (2013) Complexity of roles and regulation of the PMK1-MAPK pathway in mycelium development, conidiation and appressorium formation in *Magnaporthe oryzae*. Gene expression patterns 13(5–6):133–141.

52. Soanes DM, Chakrabarti A, Paszkiewicz KH, Dawe AL, Talbot NJ (2012) Genome-wide transcriptional profiling of appressorium development by the rice blast fungus *Magnaporthe oryzae*. PLoS Pathogens 8(2):e1002514.

53. Wang LI, Lin YS, Liu KH, Jong AY, Shen WC (2011) *Cryptococcus neoformans* mediator protein Ssn8 negatively regulates diverse physiological processes and is required for virulence. PloS One 6(4):e19162.

54. Mead O, Thynne E, Winterberg B, Solomon PS (2013) Characterising the role of GABA and its metabolism in the wheat pathogen *Stagonospora nodorum*. PloS One 8(11):e78368.

55. Roca MG, Arlt J, Jeffree CE, Read ND (2005) Cell biology of conidial anastomosis tubes in *Neurospora crassa*. Euk Cell 4(5):911–919.

56. Fleissner A, Simonin AR, Glass NL (2008) Cell fusion in the filamentous fungus, *Neurospora crassa*. Methods Mol Biol 475:21–38.

57. Fu C, Iyer P, Herkal A, Abdullah J, Stout A, Free SJ (2011) Identification and characterization of genes required for cell-to-cell fusion in *Neurospora crassa*. Euk Cell 10(8):1100–1109.

58. Furlaneto MC, Pizzirani-Kleiner AA (1992) Intraspecific hybridisation of Trichoderma pseudokoningii by anastomosis and by protoplast fusion. FEMS Microbiol Lett 69(2):191–195.

59. Manczinger L, Ferenczy L (1985) Somatic cell fusion of *Trichoderma reesei* resulting in new genetic combinations. Appl Microbiol Biotechnol 22(1):72–76.

60. Glass NL, Rasmussen C, Roca MG, Read ND (2004) Hyphal homing, fusion and mycelial interconnectedness. Trends Microbiol 12(3):135–141.

61. Kumar A, Scher K, Mukherjee M, Pardovitz-Kedmi E et al. (2010) Overlapping and distinct functions of two *Trichoderma virens* MAP kinases in cell-wall integrity, antagonistic properties and repression of conidiation. Biochem Biophys Res Commun 398(4):765–770.

62. Gruber S, Seidl-Seiboth V (2012) Self versus non-self: fungal cell wall degradation in Trichoderma. Microbiology 158(Pt 1):26–34.

63. Glass NL, Jacobson DJ, Shiu PK (2000) The genetics of hyphal fusion and vegetative incompatibility in filamentous ascomycete fungi. Annu Rev Genet 34:165–186.

64. Elad Y, Boyle P, Henis Y (1983) Parasitism of *Trichoderma* spp. on *Rhizoctonia solani* and *Sclerotium rolfsii*: scanning electron microscopy and fluorescence microscopy. Phytopathol 73:85–88.

65. Seidl V, Song L, Lindquist E, Gruber S, Koptchinskiy A et al. (2011) Transcriptomic response of the mycoparasitic fungus *Trichoderma atroviride* to the presence of a fungal prey. BMC Genomics 10:567.

66. Atanasova L, Le Crom S, Gruber S, Coulpier F, Seidl-Seiboth V et al. (2013) Comparative transcriptomics reveals different strategies of *Trichoderma* mycoparasitism. BMC Genomics 14:121.

67. Cortes C, Gutierrez A, Olmedo V, Inbar J, Chet I, Herrera-Estrella A (1998) The expression of genes involved in parasitism by *Trichoderma harzianum* is triggered by a diffusible factor. Mol Gen Genet 260(2–3):218–225.

68. Bauer R, Lutz M, Oberwinkler F (2004) Tuberculina-rusts: a unique basidiomycetous interfungal cellular interaction with horizontal nuclear transfer. Mycologia 96(5):960–967.

69. Kellner M, Burmester A, Wostemeyer A, Wostemeyer J (1993) Transfer of genetic information from the mycoparasite Parasitella parasitica to its host *Absidia glauca*. Current Genet 23(4):334–337.

70. Omero C, Inbar J, Rocha-Ramirez V, Herrera-Estrella A, Chet I, Horwitz BA (1999) G protein activators and cAMP promote mycoparasitic behaviour in *Trichoderma harzianum*. Mycol Res 103(12):1637–1642.

71. Zeilinger S, Reithner B, Scala V, Peissl I, Lorito M, Mach RL (2005) Signal transduction by Tga3, a novel G protein alpha subunit of *Trichoderma atroviride*. Appl Environ Microbiol 71(3):1591–1597.

72. Zhou X, Liu W, Wang C, Xu Q, Wang Y, Ding S, Xu JR (2011) A MADS-box transcription factor MoMcm1 is required for male fertility, microconidium production and virulence in *Magnaporthe oryzae*. Mol Microbiol 80(1):33–53.

73. Kubicek CP, Herrera-Estrella A, Seidl-Seiboth V, Martinez DA, Druzhinina IS et al. (2011) Comparative genome sequence analysis underscores mycoparasitism as the ancestral life style of *Trichoderma*. Genome Biology 12(4):R40.

74. Casas-Flores S, Rios-Momberg M, Bibbins M, Ponce-Noyola P, Herrera-Estrella A (2004) BLR-1 and BLR-2, key regulatory elements of photoconidiation and mycelial growth in *Trichoderma atroviride*. Microbiology 150:3561–3569.

75. Rubio MB, Hermosa R, Reino JL, Collado IG, Monte E (2009) Thctf1 transcription factor of Trichoderma harzianum is involved in 6-pentyl-2H-pyran-2-one production and antifungal activity. Fungal Genet Biol 46(1):17–27.

76. Trushina N, Levin M, Mukherjee PK, Horwitz BA (2013) PacC and pH-dependent transcriptome of the mycotrophic fungus *Trichoderma virens*. BMC Genomics 14:138.

77. Peterbauer CK, Litscher D, Kubicek CP (2002) The *Trichoderma atroviride seb1* (stress response element binding) gene encodes an AGGGG-binding protein which is involved in the response to high osmolarity stress. Mol Genet Genomics 268(2):223–231.

15

Rapid Analysis of Glycolytic and Oxidative Substrate Flux of Cancer Cells in a Microplate

Lisa S. Pike Winer, Min Wu*¤

Seahorse Bioscience Inc., North Billerica, Massachusetts, United States of America

Abstract

Cancer cells exhibit remarkable alterations in cellular metabolism, particularly in their nutrient substrate preference. We have devised several experimental methods that rapidly analyze the metabolic substrate flux in cancer cells: glycolysis and the oxidation of major fuel substrates glucose, glutamine, and fatty acids. Using the XF Extracellular Flux analyzer, these methods measure, in real-time, the oxygen consumption rate (OCR) and extracellular acidification rate (ECAR) of living cells in a microplate as they respond to substrates and metabolic perturbation agents. In proof-of-principle experiments, we analyzed substrate flux and mitochondrial bioenergetics of two human glioblastoma cell lines, SF188s and SF188f, which were derived from the same parental cell line but proliferate at slow and fast rates, respectively. These analyses led to three interesting observations: 1) both cell lines respired effectively with substantial endogenous substrate respiration; 2) SF188f cells underwent a significant shift from glycolytic to oxidative metabolism, along with a high rate of glutamine oxidation relative to SF188s cells; and 3) the mitochondrial proton leak-linked respiration of SF188f cells increased significantly compared to SF188s cells. It is plausible that the proton leak of SF188f cells may play a role in allowing continuous glutamine-fueled anaplerotic TCA cycle flux by partially uncoupling the TCA cycle from oxidative phosphorylation. Taken together, these rapid, sensitive and high-throughput substrate flux analysis methods introduce highly valuable approaches for developing a greater understanding of genetic and epigenetic pathways that regulate cellular metabolism, and the development of therapies that target cancer metabolism.

Editor: Robert W. Sobol, University of Pittsburgh, United States of America

Funding: This work was conducted in and funded by Seahorse Bioscience. LSPW is an employee and MW was an employee at the time of the work in the company. The funder had no role in study design, data collection and analysis, decision to publish, or preparation of the manuscript.

Competing Interests: Lisa S. Pike Winer is an Employee at Seahorse Bioscience, North Billerica, MA, USA. Min Wu was an employee at Seahorse Bioscience at the time of this work, North Billerica, MA, USA.

* Email: mwu13@mgh.harvard.edu

¤ Current address: Center for Human Genetic Research, Massachusetts General Hospital, Boston, Massachusetts, United States of America

Introduction

Cancer cells significantly reprogram their metabolism to drive tumor growth and survival. Otto Warburg first observed that under aerobic conditions, tumors had high rates of glycolysis compared to the surrounding tissue, a phenomenon known as the Warburg effect, or aerobic glycolysis [1]. He postulated that increased glycolysis and impaired mitochondria respiration is the prime cause of cancer [2]. More recently, a large body of evidence indicates that cancer cells undergo metabolic reprogramming, leading to extensive use of and dependence upon glucose or glutamine for their growth and survival [3–9]. This metabolic reprogramming has been shown to be the result of oncogene activation and/or loss of tumor suppressor functions, as well as in response to environmental cues, all of which regulate nutrient substrate uptake and metabolism [10–14]. Depending on the combinations of these factors and a given cellular context, cancer cells can manifest an array of metabolic phenotypes [15], which may impact either treatment selection or response to treatment.

In view of numerous types of genetically and metabolically diverse cancer cells, a rapid, informative, relatively easy-to-perform and higher-throughput substrate flux analysis can facilitate greater understanding of the genetic and epigenetic

pathways that regulate cancer cell metabolism, determining whether there is a finite number of metabolic phenotypes among all type of cancer cells, independent of tissue origin, and discovering agents that target specific metabolic pathways for cancer treatment.

Cells produce ATP via two major energy-producing pathways: glycolysis and oxidative phosphorylation. The glycolytic pathway converts glucose to pyruvate. One fate of the pyruvate is reduction to lactate in the cytosol in an oxygen-independent biochemical reaction resulting in ATP production and net proton production. Protons are pumped out of the cell by various mechanisms to maintain the intracellular pH [16] and the efflux of the protons into the extracellular space or medium surrounding the cells causes extracellular acidification [17–21]. The major nutrient substrates glucose, glutamine, and fatty acids can be completely oxidized to into CO_2 and H_2O via the tricarboxylic acid cycle (TCA cycle) which requires the electron transport chain (ETC) in the mitochondria using oxygen as a terminal electron acceptor, and which is coupled to ATP production by oxidative phosphorylation. The CO_2 produced can be converted to bicarbonate and protons as catalyzed by carbolic anhydrase [16], another source of protons causing medium acidification. In many non-transformed differentiated cells such as neurons, oxidative phosphorylation produces most of the cellular ATP. In contrast, cancer cells rely heavily on glycolysis in addition to oxidative phosphorylation for their ATP production [22]. As well as fueling ATP production, glucose and glutamine are essential carbon sources that provide anabolic precursors, some of which (e.g., citrate and oxaloacetate) are produced through a truncated TCA cycle for the biosynthesis of lipids, nucleic acids and amino acids.

Since living cells do not store ATP, they produce it continuously and on demand, and therefore constantly consume oxygen and fuel substrates. Thus, the demand for ATP in cells (i.e. ADP availability) controls the rate of oxygen consumption. Electrons (energy) stored in nutrient substrates are extracted via the mitochondrial TCA cycle reactions and carried by reduced electron carriers NADH and $FADH_2$ to the ETC. As the electrons flow down the ETC, the energy released is used to pump protons from the matrix into the intramembrane space, forming a transmembrane electrochemical proton gradient across the mitochondrial inner membrane. At the end of the ETC, the electrons are transferred to molecular oxygen, reducing it to water via the terminal cytochrome C oxidase. As protons return to the mitochondrial matrix through the ATP synthase complex, the energy stored in the proton gradient then drives the phosphorylation of ADP to ATP coupling respiration (electron transport) with ATP production. Oxidative phosphorylation, however, is incompletely coupled to respiration. Protons can also re-enter the matrix via proton channels such as uncoupling proteins (UCP), which are located on the inner membrane, bypassing ATP synthase and dissipating the energy gradient without producing ATP in a process known as proton leak. Partially reduced oxygen species (ROS) such as the superoxide anion can be produced at different sites in the ETC depending on conditions [23]; the proton leak has been considered an important cellular mechanism for protecting cells from oxidative damage through lowering ROS produced by the ETC [24].

Given the connection between oxygen consumption and extracellular acidification with nutrient substrate metabolism, increased oxygen consumption is a measure of substrate oxidation when a substrate is added to cells. Likewise, an increase in the rate of extracellular acidification upon glucose addition is a measure of glycolytic flux.

Various traditional experimental methods that analyze substrate metabolism have contributed to the current understanding of cellular metabolism. These include measuring the accumulation of radio-labeled end products such as H_2O and CO_2 metabolized from substrates such as ^3H labeled glucose and ^{14}C labeled fatty acids. Other previously used (and more recently developed) methods to quantify metabolites are stable isotope tracers coupled with mass-spectrometry and NMR analysis, both of which have yielded detailed information on substrate metabolism. These techniques, however, can nevertheless be either labor intensive and cumbersome, and/or relatively inaccessible for many laboratories.

This study had two main objectives. The first was to apply the principles described above to establish a series of rapid and easy methods to analyze glycolytic flux and oxidative substrate flux of cancer cells. This was achieved by measuring cellular oxygen consumption rate and extracellular acidification using the XF Extracellular Flux analyzer [22]. The second was to perform proof-of-principle experiments through applying these substrate flux methods, along with analyzing mitochondrial bioenergetics in human glioblastoma cells SF188s and SF188f to interrogate their metabolic networks.

Materials and Methods

Reagents

Carbonyl cyanide 4-(trifluoromethoxy)phenylhydrazone (FCCP), myxothiazol, antimycin A, rotenone, 2-deoxyglucose, oxamate, aminooxyacetate, glucose, sodium pyruvate and sodium palmitate were obtained from Sigma (St. Louis, MO, USA). L-glutamine was obtained from Invitrogen (Carlsbad, CA). Oligomycin was obtained from EMD (San Diego, CA, USA). Bovine Serum Albumin fraction V (fatty acid ultra-free) was obtained from Roche Diagnostics (Indianapolis, IN, USA). All compounds and medium were prepared according to the manufacturers' instructions unless indicated otherwise.

Cell lines and cell culture

Human glioblastoma SF188 cells were obtained from the University of California at San Francisco Brain Tissue Bank. These cells were originally maintained, as recommended by the originator, in an MEM medium containing 5.5 mM glucose and 2 mM L-glutamine. They were adapted step-wise to DMEM medium containing 25 mM glucose and 6 mM L-glutamine as a model system for the study of glutamine metabolism as reported [6]. As a control, the parental cells were also adapted in parallel to DMEM medium containing 5.5 mM glucose and 2 mM L-glutamine. The former acquired a much more rapid growth rate after 4 weeks culture in the medium and was named SF188f (fast). The latter, however, maintained similar growth rates as those parental cells maintained in MEM, and were named SF188s (slow). To maintain their distinct growth phenotype, SF188s and SF188f cells were always cultured in DMEM containing 5.5 mM glucose and 2 mM L-glutamine and 25 mM glucose and 6 mM L-glutamine, respectively at 37°C in a Forma incubator with 10% CO_2 and 100% humidity at ~80% confluence in 175 cm^2 T-flasks (Corning). Human prostate cancer PC-3 and cervical cancer HeLa cells were purchased from American Type Culture Collection (Manassas, VA, USA), and were maintained in RMPI1640. All cell culture media were supplemented with 10% fetal bovine albumin (FBS, Hyclone, Logan, UT, USA).

XF assay medium

A base medium was used for the assays described in this study directly or supplemented with substrates and cofactors as specified in each of the specific assays (see below) and for each experiment. The base assay medium was prepared as follows. Dulbecco's Modified Eagle's Medium (DMEM) powder (Sigma, catalog number D5030) was the starting material. It contains no glucose, L-glutamine, sodium pyruvate, sodium bicarbonate, or phenol red, and has a low phosphate (see Materials S1 in File S1for details of preparation). In addition, a modified Krebs-Henseleit- bicarbonate buffer (KHB) which contains no bicarbonate and lower phosphate can also be used (Materials S2 in File S1) as an alternative assay medium for fatty acid oxidation. The use of amino acid-free buffer, as opposed to the base medium for other assays is possible but, may result in different experimental outcomes which may require different data interpretations. All the assays described here were performed solely in the base medium with indicated supplementation, with the exception of fatty acid oxidation which was performed in both base medium and KHB, yielding similar results.

Preparation of palmitate-BSA conjugate

Sodium palmitate was solubilized by warming it to 68°C in 150 mM sodium chloride solution. It was then bound to BSA in solution at molar ratio of 6:1. The complete protocol is described in Material S3 in File S1.

Measurement of oxygen consumption rate and extracellular acidification rates

OCR and ECAR measurements were performed using the XF24 or XF96 Extracellular Flux analyzer (Seahorse Bioscience, North Billerica, MA) as described [22]. Briefly, cells were plated into XF24 (V7) or XF96 (V3) polystyrene cell culture plates (Seahorse Bioscience, North Billerica). SF188s cells were seeded at 30,000/well (XF24 plate) or 20,000/well (XF96 plate) and SF188f cells at 20,000/well (XF24 plate) or 15,000/well (XF96 plate), respectively. PC-3 and HeLa cells were plated at 25,000 and 30,000 cells per well, respectively, in XF24 cell culture plates. The cells were incubated for 24 to 28 hours in a humidified 37°C incubator with 10% CO_2 (DMEM medium) or 5% CO_2 (RMPI1640 and MEM medium), respectively. Because the two cell lines proliferate at different rates during the 24–28 hour incubation period, SF188s and SF188f cells were treated with trypsin and then counted to determine the cell number in each well after an assay. These cell counts were used to normalize either OCR or ECAR. Their viabilities, as determined after assays, were nearly indistinguishable regardless of the presence or absence of exogenous substrates or metabolic inhibitors in the assay medium. Prior to performing an assay, growth medium in the wells of an XF cell plate was exchanged with the appropriate assay medium to achieve a minimum of 1:1000 dilution of growth medium. 600 μL (XF24) or 150 μL (XF96) of the assay medium was added to cells for an XF assay. While sensor cartridges were calibrated, cell plates were incubated in a 37°C/non-CO_2 incubator for 60 minutes prior to the start of an assay. All experiments were performed at 37°C. Each measurement cycle consisted of a mixing time of 3 minutes and a data acquisition period of 3 minutes (13 data points) for the XF24, and 2 min and 4 min for the XF96. OCR and ECAR data points refer to the average rates during the measurement cycles. All compounds were prepared at appropriate concentrations in desired assay medium and adjusted to pH 7.4. A volume of 75 μL for XF24 (25 μL for XF96) of compound was added to each injection port. In a typical experiment, 3 baseline measurements were taken prior to the addition of any compound, and 3 response measurements were taken after the addition of each compound. OCR and ECAR were reported as absolute rates (pmoles/min for OCR and mpH/min for ECAR) or normalized against cell counts, or expressed as a percentage of the baseline oxygen consumption. In this study, baseline OCR or ECAR (a technical term) refers to the starting rates prior to the addition of an agent, which can be used for comparisons with those rates after the addition. In contrast, basal OCR or ECAR (a biological term) refers to OCR or ECAR which occur in cells at rest in order to maintain basic cell function. Unless otherwise specified, the third measurement of baseline or after addition of each substrate or compound was used to generate absolute OCR or ECAR values. As well, percentage of baseline OCR values was calculated as OCR at the third measurement after an agent injection divided by the OCR immediately before the injection. Each datum was determined minimally in triplicate.

XF Substrate Flux Assay Conditions

Glycolytic flux and glycolytic capacity. The assay medium consists of the base medium supplemented with 2 mM L-glutamine. Glutamine is required to achieve the maximal glycolysis rate for some, but not all cell lines. The same medium was used to determine glycolytic capacity. The concentration of glucose added to initiate glycolysis and measure glycolytic capacity

was 10 mM, greater than the saturation point under both conditions.

Glucose oxidation. The assay medium was the base medium without any exogenous fuel substrate supplementation. The concentration of glucose added to initiate glucose oxidation was 10 mM, which was determined in preliminary experiments to be above saturation.

Glutamine oxidation. The assay medium was the base medium without any exogenous fuel substrate. The concentration of glutamine added to the cells to initiate glutamine oxidation was 4 mM, which was also pre-determined to be above saturation.

Fatty acid oxidation. The assay medium was the base medium (or KHB) supplemented with 5.5 mM glucose and 50 μM carnitine (required to transport long chain fatty acid into the mitochondria). Fatty acids tested include long chain fatty acid palmitate, medium chain fatty acid octanoate, and short chain fatty acid butyrate. They were titrated for concentrations stimulating maximal OCR response. The working concentration of palmitate conjugated with BSA was 150 μM, and octanoate 1 mM, which were also above saturation.

It is critical that the above assay conditions are strictly adhered to. Any variation in the assay medium composition may result in different interpretations and insights into cellular metabolic network. Saturating substrate concentrations and optimal compound concentrations were determined by performing titration experiments as described in Materials S4 and S5 in File S1. The effect of assay conditions on the interpretation of experimental results will be described elsewhere.

Cell counts

SF188s and SF188f cells were detached with trypsin-EDTA and harvested immediately following an XF assay. The number of cells in each well was determined using a ViCell automated trypan blue counter (Beckman-Coulter, Fullerton, CA), and was used to normalize OCR and ECAR as indicated in the figure legends.

Statistical analysis

The data were presented as mean ± standard deviation, with at least three replicates used for each data point. Unless otherwise indicated, a paired Student's t test was performed for each experimental group to assess the statistical significance against respective controls.

Results

Glycolysis and glycolytic capacity

We have previously shown that glycolysis accounts for ~80% of total ECAR in a number of cancer cells as determined through two methods: a) removing glucose from the assay medium and b) adding glycolytic pathway inhibitors such as hexokinase inhibitor 2-DG and lactate dehydrogenase (LDH) inhibitor oxamate [22]. The remaining 20% of the ECAR can be attributed to other metabolic processes, such as the TCA cycle CO_2 evolution. In order to measure glycolysis using ECAR more accurately and easily, we took the following approach. Glucose is added to cells that are incubated in a glucose-free medium, but supplemented with glutamine (see Materials and Methods). The ECAR increase following the addition of glucose establishes the glycolysis rate. A subsequent addition of a glycolysis inhibitor eliminates the glucose-induced ECAR increase. Any acidification due to other metabolic processes such as the TCA cycle CO_2 release (from any substrate but glucose) is detected as ECAR prior to glucose addition. The OCR response to glucose, monitored concurrently with ECAR,

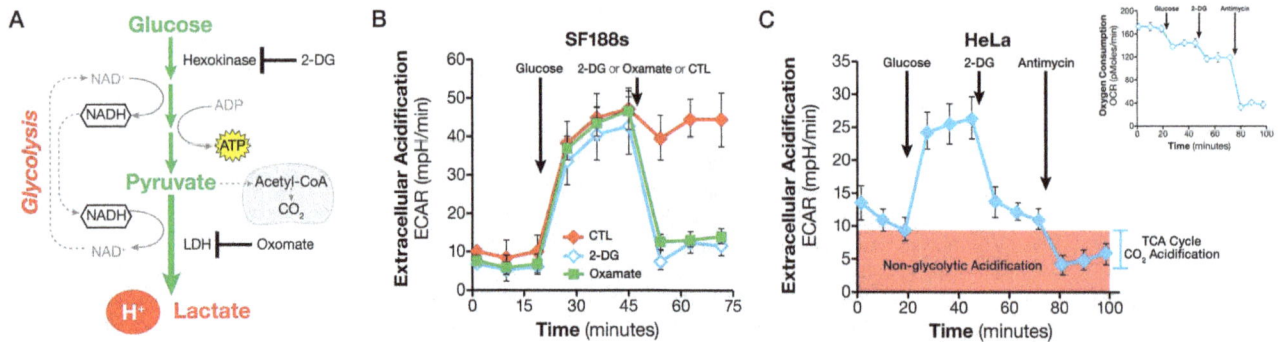

Figure 1. Analyzing Glycolytic flux. A. Schematic illustration of the glycolytic pathway. NADH produced in the cytosol as glucose is converted to pyruvate and is regenerated by LDH in the cytosol. B. Kinetic ECAR response of SF188s cells to glucose (10 mM) and 2-DG (100 mM) or oxamate (100 mM), respectively. SF188s cells were plated at 30,000 cells/well in XF24 V7 cell culture plates 24–28 hours prior to the assays. The assay medium was substrate-free base medium (as described in Material and Methods) supplemented with 2 mM glutamine. The ECAR value was not normalized. A representative experiment out of at least three is shown here. Each data point represents mean $\pm$ SD, n = 4. C. ECAR response of HeLa cells to glucose (10 mM), 2-DG (100 mM) and antimycin (1 µM). Insert: the OCR response in the same experiments showing the Crabtree effect and that glucose did not increase OCR. HeLa cells were plated at 30,000/well in XF24 cell culture plates 24–28 hours prior to the assays. ECAR or OCR values were not normalized. The assay medium was substrate-free base medium (as described in Material and Methods) supplemented with 2 mM glutamine. A representative experiment out of at least three is shown here. Each data point represents mean $\pm$ SD, n = 5.

serves as an indicator of whether glucose is also catabolized through mitochondrial respiration (Figure 1A).

We performed the experiment using SF188s and HeLa cells. As shown in Figure 1B, glucose addition to SF188s cells triggered an instant ECAR increase, 38 ± 4 mpH/min (ECAR of measurement 6 less that of measurement 3), which was subsequently abolished by the addition of glycolysis inhibitors, either 2-DG or oxamate. This experiment indicated that exogenously added glucose was broken down to lactate (because LDH inhibition by oxamate reduces ECAR similarly to 2-DG), causing an ECAR increase and thereby validating our experimental design. Similar results were obtained in HeLa cells (Fig. 1C) which were consistent with our previous study [22], as well as with a number of recent reports showing that the ECAR response parallels that of lactate production [25–27]. Prior to adding glucose to the cells as well as following the addition of 2-DG or oxamate, we again observed a small ECAR, 6 ± 1 mpH/min and 10 ± 2 mpH/min (at measurement 3), respectively in SF188s and HeLa cells (Figure 1B and C). We refer to this small but measurable ECAR as non-glycolytic acidification. The OCR response indicated that the injection of glucose not only failed to trigger an increase, but in fact caused a slight decrease in OCR (Figure 1C), which is similar to the Crabtree Effect, first observed in tumor cells by Crabtree in the 1920s [28].

The two most significant proton sources that can contribute to non-glycolytic acidification are the TCA cycle and breakdown of intracellular glycogen, i.e., glycogenolysis [20]. To determine the contribution of the TCA cycle CO_2 evolution, we used a complex III inhibitor antimycin to stop the electron flow and thus the TCA cycle flux in HeLa cells. Glucose was added first to initiate glycolysis, followed by 2-DG to abolish it, leaving behind non-glycolytic ECAR. The final addition of antimycin stopped the TCA cycle from producing CO_2. Although a residue remained, 4 ± 1.4 mpH/min, antimycin eliminated about half of the non-glycolytic ECAR (Figure 1C), confirming the contribution of TCA cycle CO_2-derived proton to non-glycolytic acidification. We tested whether the antimycin-resistent ECAR was due to glycogenolysis by using CP91149, an inhibitor of glycogen phosphorylase (the first enzyme of glycogen breakdown), but we did not observe any significant effect of CP91149 on non-glycolytic acidification (data not shown), leading us to conclude that the

residual non-glycolytic ECAR had to be accounted for by other metabolic processes, such as decarboxylation reactions catalyzed by glucose-6 phosphate dehydrogenase and/or pyruvate dehydrogenase. Collectively, these results again confirmed that glycolysis accounts for the majority of ECAR observed in cancer cells, and established that TCA-derived CO_2 is a primary contributor of non-glycolytic acidification.

Glycolytic flux determined *in vitro* at ambient oxygen levels reflects basal glycolysis rate. When cells experience loss of mitochondrial ATP production due to inhibition of oxidative phosphorylation, either at low oxygen tension or by oligomycin, they augment their glycolytic flux and make more ATP from glycolytic pathways to maintain cellular ATP homeostasis [22]. We refer to this increased glycolytic flux in response to deficiency in mitochondrial ATP production as glycolytic capacity. Experimentally, we define glycolytic capacity as the glucose-induced ECAR by mitochondrial ATP synthase inhibitor oligomycin.

To determine both glycolytic flux and glycolytic capacity of the same cell population in one experiment, we measured ECAR while consecutively injecting glucose, oligomycin, and 2-DG. As shown in Figure 2A, adding glucose to HeLa cells, as expected, triggered a glycolytic flux of 19 ± 0.9 mpH/min (EACR at measurement 6 less that at measurement 3) in HeLa cells. The subsequent addition of oligomycin caused a further increase in ECAR to 44 ± 3.8 mpH/min (ECAR at measurement 9 less that at measurement 3), indicating an elevated glucose flux toward lactate and revealing the glycolytic capacity of HeLa cells. The final addition of glycolysis inhibitor 2-DG abolished the overall glycolysis (Figure 2A). The calculated glycolytic flux and glycolytic capacity from the glycolysis experiment are shown in Figure 2B.

Two experimental conditions must be met in the above experiment. First, the oligomycin concentration should maximally inhibit respiration. This was achieved by selecting the oligomycin concentration that resulted in maximal OCR inhibition in a titration experiment. For example, 0.5 µM oligomycin was found to be sufficient to achieve maximal inhibition of OCR in HeLa cells (data not shown). Second, it is critical to ensure that the supply of exogenous glucose is saturating, allowing the glycolytic machinery to be the limiting factor. The saturating concentration of glucose for achieving maximal ECAR response was determined in a glucose concentration titration experiment, in which

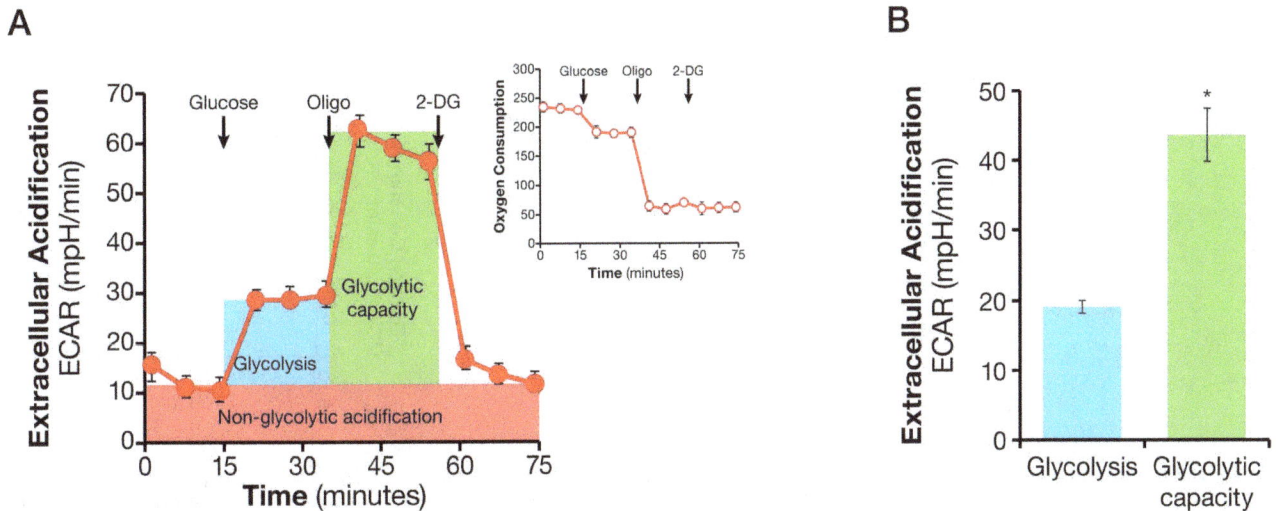

Figure 2. Assaying glycolytic flux and glycolytic capacity. A. Kinetic ECAR response of HeLa cells to glucose (10 mM), oligomycin (2 µM), and 2-DG (100 mM). Insert shows OCR response in response to glucose and oligomycin. HeLa cells were plated at 30,000/well in XF24 V7 cell culture plates 24–28 hours prior to the assays. The assay medium was the substrate-free base medium supplemented with 2 mM glutamine. ECAR or OCR values were not normalized. A representative experiment out of at 5 is shown here. Each data point represents mean ± SD, n = 4. B. Calculated glycolytic flux, glycolytic capacity. Glycolytic flux is the difference between the ECARs of measurement 6 and measurement 3. Likewise, glycolytic capacity describes the difference between the ECAR of measurement 9 and that of measurement 3. * $p < 0.05$.

increasing concentrations of glucose were added to the cells, followed by the addition of control or oligomycin at the concentration that maximally inhibits respiration. In HeLa cells, for instance, we found that ECAR increased continuously until the glucose concentration reached 5 mM, after which there was no further ECAR increase (Figure S1). Similar results were obtained with a dozen transformed and non-transformed cell lines (data not shown). We chose a higher-than-saturating concentration of 10 mM as the standard concentration to determine both glycolytic flux and glycolytic capacity.

Glucose oxidation

Glucose-derived pyruvate can also enter the mitochondria, where it is converted to acetyl CoA by pyruvate dehydrogenase and enters the TCA cycle via citrate synthase (or as oxaloacetate via pyruvate carboxylase [29]. The acetyl moiety is eventually oxidized to CO_2 and H_2O (Figure 3A). The oxygen-consuming process of glucose oxidation first to pyruvate and then to CO_2 and H_2O is referred to here as glucose oxidation.

Experimentally, we used the glucose-induced OCR to measure glucose oxidation. In order to establish the glucose oxidation assay, we chose PC-3 cells which actively oxidize glucose. To determine glucose oxidation, 10 mM glucose was added to the cells in assay medium containing no glucose or glutamine (see Material and Methods). As shown in Figure 3B, the addition of glucose to PC-3 cells caused an immediate increase in OCR, 45±11 pmol/min (OCR at measurement 6 less that at measurement 3), indicating glucose flux into the TCA cycle, and ultimately, the ETC for complete oxidation. This experimental design provided a quantitative measurement of glucose oxidation under this experimental condition. In a variant design, PC-3 cells were exposed to glucose, oligomycin, and FCCP consecutively, with 3 measurements before each compound addition and following each compound addition. FCCP uncouples respiration from oxidative phosphorylation, allowing oxidation of any oxidizable substrate present in the assay medium to occur. As shown in Figure 3C, FCCP stimulated a spike in OCR following glucose addition, but not the control,

further confirming that the biochemical pathways for glucose oxidation were active in PC-3 cells. The OCR response to FCCP confirms and provides a semi-quantitative assessment of cells' ability to oxidize glucose. In this experimental design, cells were pre-incubated in substrate-free base medium for 60 min prior to an assay, so there is a possibility that they may have been stressed and altered their response to glucose addition. However, this seems unlikely, as OCRs obtained from cells pre-incubated in substrate-free medium and given glucose during the assay were indistinguishable from those of cells that had been pre-incubated with glucose supplemented base medium and were not under the same stress.

Glutamine oxidation

As illustrated in Figure 4A, glutamine enters the mitochondria and is converted to CO_2 and H_2O in an oxygen-consuming process, which we refer to here as glutamine oxidation. Glutamine can also be partially oxidized to malate via the TCA cycle, which then exits the mitochondria and is converted to pyruvate by malic enzyme in the cytosol, a process known as glutaminolysis. For simplicity, we refer to both processes as glutamine oxidation.

To measure glutamine oxidation, glutamine was injected into cells in assay medium containing no glutamine (see Material and Methods). We selected SF188f cells which exhibit robust glutamine oxidation to set up the assay. As shown in Figure 4B, the addition of 4 mM glutamine to SF188f cells caused a large increase in OCR, 207±11 pmoles/min (OCR at measurement 6 less that at measurement 3) revealing a high rate of glutamine oxidation. Using the transaminase inhibitor aminooxyacetate (AOA), we determined whether transaminase or glutamate dehydrogenase (GDH) (Figure 4A) was responsible for glutamine oxidation. The addition of AOA following glutamine to SF188f cells abolished glutamine-induced OCR to 20% over the baseline OCR from 60% before the addition of AOA (Figure 4C), suggesting that transaminase-catalyzed α-KG conversion is the major pathway for glutamine's entry into the TCA cycle (Figure 4A). It follows that the remaining AOA-resistant OCR can be attributed to the

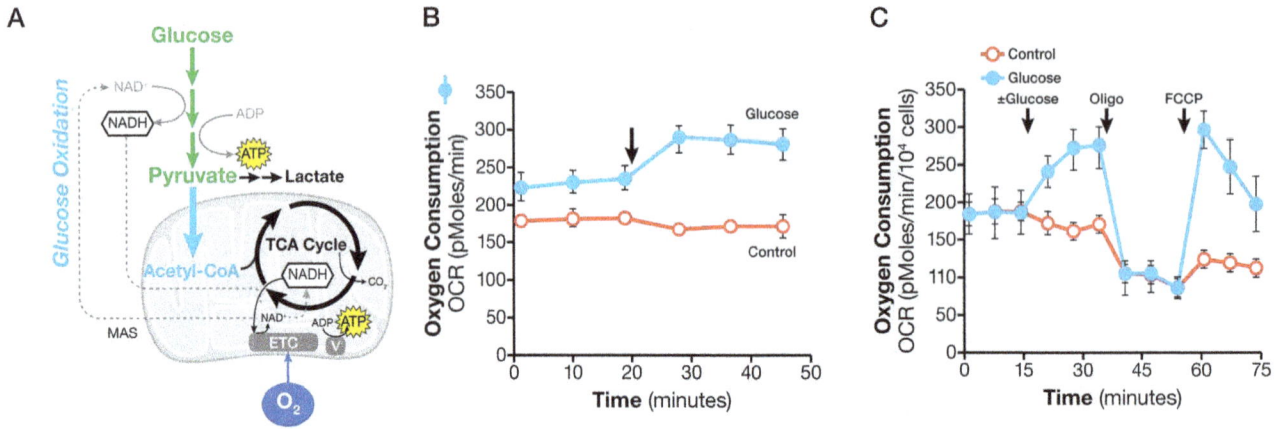

Figure 3. Assaying glucose oxidation. A. Schematic illustration of biochemical pathway for glucose oxidation. The NADH produced in the cytosol as glucose is converted to pyruvate is imported into the mitochondria via the malate-aspartate shuttle and regenerated via the ETC to maintain continuous glucose oxidation. B. Kinetic OCR response of PC-3 cells to glucose (10 mM); C. OCR response to glucose (10 mM), oligomycin (1 μM) and FCCP (0.3 μM). PC-3 cells were plated at 25,000/well in XF24 V7 culture plates. The assay medium was the substrate-free base medium. The OCR values were not normalized. A representative experiment out of three is shown here. Each data point represents mean $\pm$ SD, n = 4.

alternative reaction catalyzed by glutamate dehydrogenase. In an attempt to confirm the GDH-driven glutamine oxidation, we used EGCG, a known GDH inhibitor. Unfortunately, EGCG inhibited OCR regardless of glutamine's presence, suggesting it does not specifically inhibit GDH in this experimental setting, and we were thus unable to confirm GDH-mediated glutamine oxidation. These experiments, however, did allow us to a) measure the glutamine oxidation rate and b) gain insight into the alternate biochemical reactions used for glutamine oxidation. To accurately measure the contribution of transaminase and GDH-to glutamine oxidation, however, additional approaches are necessary. Finally, similar to what observed in glucose oxidation experiment OCRs obtained from cells pre-incubated in substrate-free medium and given glutamine during the assay were indistinguishable from

those of cells that had been pre-incubated with glutamine supplemented base medium.

We noticed that a substantial rate of oxygen consumption, 338$\pm$8 pmoles/min (measurement 3), occurred in the absence of any exogenously added substrate in SF188f cells prior to glutamine addition (Figure 4B). Similarly, PC-3 cells showed considerable OCR, 234$\pm$14 pmoles/min (measurement 3), prior to glucose addition (Figure 3B). This phenomenon is not specific to SF188f or PC-3 cells but is commonly observed, to various extents in a variety of cell lines that we have studied (data not shown). This oxygen consumption in the absence of exogenous substrate is most likely fueled by the oxidation of endogenous substrates; therefore, cancer cells appear to have a significant pool of endogenous

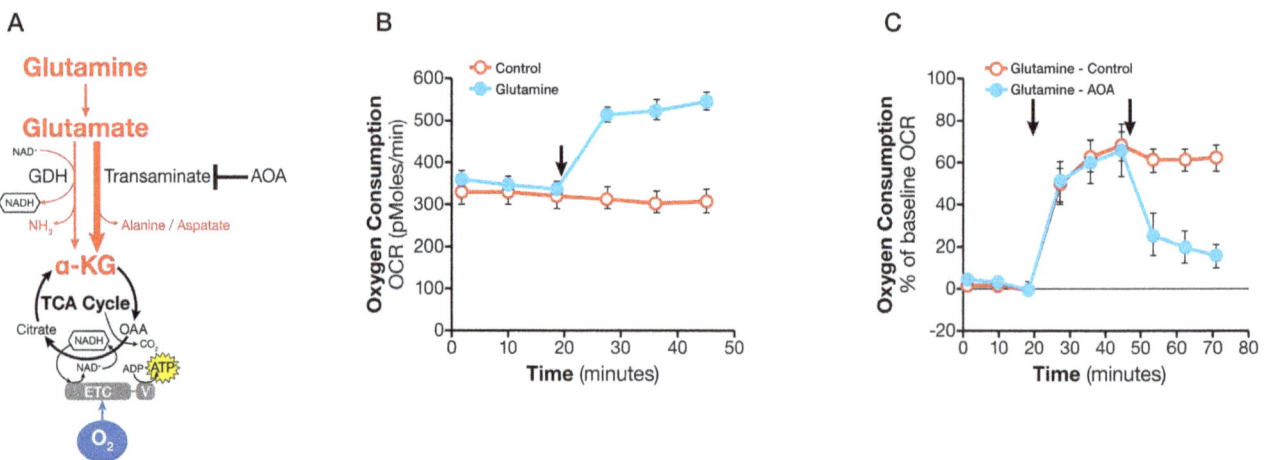

Figure 4. Assaying glutamine oxidation and demonstrating transaminase pathway activity. A. Schematic illustration of biochemical pathway for glutamine oxidation in the mitochondria. B. Kinetic OCR response in SF188f cells to glutamine (4 mM). SF188f cells were plated at 20,000 cells/well in XF24 cell culture plates 24–28 hours prior to the assays. The assay medium was the substrate-free base medium. The OCR value was not normalized. A representative experiment out of three is shown here. Each data point represents mean $\pm$ SD, n = 4. C. OCR response (% of baseline) in SF188f cells to glutamine (4 mM) and AOA (100 μM). Glutamine-induced OCR reached 60% over the baseline (OCR at measurement 6 divided by that at measurement 3) while AOA addition reduced it to 20% (OCR at Measurement 9 divided by measurement 3). SF188f cells were plated at 20,000 cells/well in XF24 V7 cell culture plates 24–28 hours before the assays. The % OCR was plotted using measurement 3 as the baseline. The assay medium was the substrate-free base medium. A representative experiment out of three is shown here. Each data point represents mean $\pm$ SD, n = 4.

oxidizable substrates including fatty acids, amino acids, and glycogen.

Fatty acid oxidation

Fatty acids are an important energy source for meeting a high energy demand and maintaining cellular functions in many cells, such as skeletal and cardio muscle cells. Fatty acid oxidation in cancer cells has also been reported but its relevance remains to be fully understood [30–34]. Exogenous fatty acids are taken up by cells, converted to fatty acyl CoA in the cytosol, then converted to acetyl CoA in the mitochondria via β–oxidation and ultimately broken down to CO_2 and H_2O (Figure 5A). The process of biochemical conversion of fatty acids to CO_2 and H_2O consuming oxygen is referred to here as fatty acid oxidation (FAO).

We used increased OCR following injection of exogenously fatty acids to measure the oxidation of fatty acids of various chain lengths, including long, medium and short chains. SF188f cells, which effectively oxidize fatty acids, were chosen to develop the fatty acid oxidation assay. First, we tested the oxidation of the long-chain fatty acid palmitate. Figure 5B shows that the addition of 150 μ M palmitate [8 Acetyl equivalents) to SF188f cells triggered an immediate increase in OCR of 49±5 pmole/min (measurement 7 less measurement 3). Next, we measured oxidation of the medium-chain fatty acid octanoate. The addition of 1 mM octanoate (4 Acetyl equivalents) also increased OCR by 33±4 pmole/min (Figure 5C). Finally, we tested the oxidization of short-chain fatty acid butyrate (2 Acetyl equivalents). The addition of butyrate induced a small but consistent OCR increase (Figure S2). In short, we found that fatty acids of all chains lengths were oxidized by SF188 cells.

A Shift to oxidative metabolism fueled by glutamine oxidation in rapidly proliferating SF188f glioblastoma cells

Having established the methods to analyze glycolysis and oxidation of exogenous substrates, we performed proof-of-principle experiments in SF188s and SF188f cells (see Material and Methods). These two cell lines were derived from the same parental cell line SF188 (which harbor c-*MYC* amplification) [35], but proliferated at very different rates, with the former much

slower than the latter (Figure 6A). The fast-growing behavior of SF188 cells occurred only after, and not within, the initial four weeks' culture in DMEM containing 25 mM glucose and 6 mM glutamine, whereas SF188 cells cultured in DMEM containing 5.5 mM glucose and 2 mM glutamine maintained the same growth rate as the parental cells, suggesting an intrinsic change in SF188f cells' growth program as they acquired the fast growing behavior. We were curious as to whether there were any metabolic alterations associated with SF188f cells' fast growing behavior. Upon examining their basal OCR and ECAR in assay medium containing 25 mM glucose, 6 mM glutamine and 1 mM pyruvate, we found that SF188f cells displayed a lower ECAR and a higher OCR compared with SF188s cells (after adjusting for the number of cells being measured) (Figure 6B). This suggested that the fast growing SF188f glioblastoma cells adopted a more oxidative metabolism (perhaps reprogramming their metabolic network via certain epigenetic events), shifting away from glycolytic metabolism.

Having determined the basal OCR and ECAR, we investigated glycolytic and oxidative substrate flux in these cells. First, we examined the glycolytic arm of metabolism. The normalized basal glycolytic flux was much lower in SF188f cells, at 5.6±0.8 mpH/10^4 cells, compared with SF188s cells, at 14.4±0.8 mpH/10^4 cells (Figure7A). Interestingly, oligomycin stimulated a large ECAR increase over basal glycolysis in SF188f cells (Figure 7A), but failed to evoke a significant increase in SF188s cells. Thus, under basal conditions, glycolysis in SF188s cells occurs at full capacity while SF188f cells possess a substantial unused glycolytic capacity. The respective glycolysis flux and glycolytic capacity of the pair are shown in Figure 7B.

Second, we investigated the oxidation of glutamine and fatty acids in the cells. Glutamine oxidation occurred in both, but at a much higher rate in SF188f cells (21±0.3 pmol/min/10^4 cells) than in SF188s cells (4.6±0.6 pmol/min/10^4 cells) (Figure 8A). To distinguish the responsible biochemical pathways, we used transaminase inhibitor AOA. As shown in Figure 8A, AOA largely abolished glutamine-stimulated OCR in SF188f cells, but had little effect on SF188s cells. These results suggested that the transaminase pathway in the rapidly dividing SF188f cells was not only activated but became the primary pathway in the oxidation of glutamine. In contrast, the GDH pathway, as inferred from AOA-

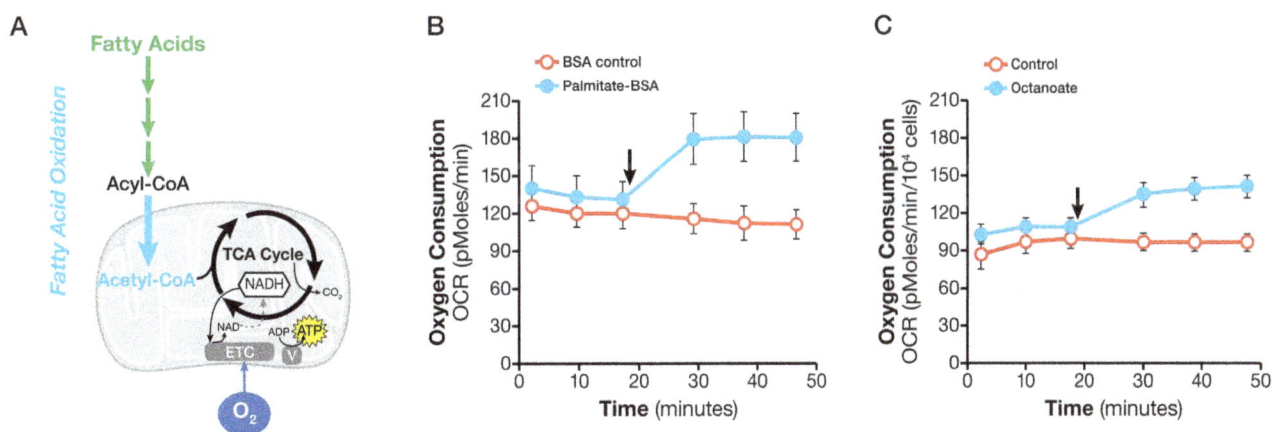

Figure 5. Determining fatty acid oxidation. A. Schematic illustration of biochemical pathway of fatty acid oxidation. B. Kinetic OCR response of SF188f cells to palmitate (150 μM). C. Kinetic OCR response in SF188f cells to octanoate (1 mM). SF188f cells were plated at 15,000 cells/well in XF96 V3 cell culture plates 24–28 hours prior to the assays. Assay medium was the substrate-free base medium supplemented with 5.5 mM glucose and 50 μM carnitine. The OCR value was not normalized. A representative experiment out of four is shown here. Each data point represents mean ± SD, n = 6.

A

B

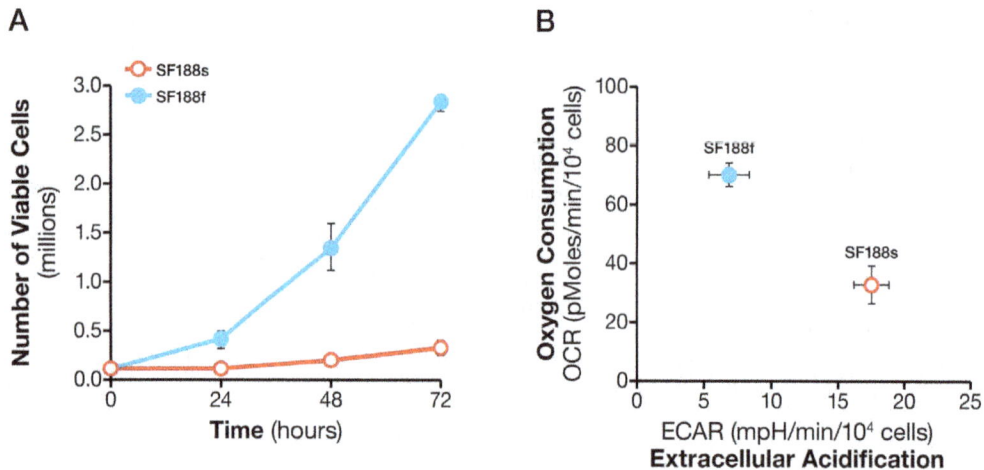

Figure 6. High proliferation rates of SF188f cells and associated shift in their basal OCR and ECAR compared with SF188s cells. A. Number of SF188s and SF188f cells at 24, 48 and 72 hours in their respective culture medium. B. Basal OCR and ECAR of SF188s and SF188f cells. SF188s and SF188f cells were plated at 30,000 and 20,000 cells/well, respectively, in XF24 V7 cell culture plates 24–28 hours prior to the assays. Upon completion of an assay, cells were treated with trypsin and counted for the purpose of normalization; Assay medium was the substrate-free base medium supplemented with 25 mM glucose, 6 mM Glutamine and 1 mM pyruvate. A representative experiment out of three is shown here. The OCR and ECAR values were normalized to pmoles/min/10^4 cells or mpH/min/10^4cells. Each data point represents mean ± SD, n = 3.

resistant fraction of glutamine –evoked OCR remained the sole pathway carrying out glutamine oxidation in the slowly-dividing SF188s cells. Similarly, palmitate and octanoate oxidation took place in SF188f cells, but not in the slower-growing SF188s cells (Figure 8B), suggesting SF188f cells had acquired the ability to take up and oxidize both long and medium-chain fatty acids. Collectively, these results indicated that glutamine and fatty acid oxidation was activated as SF188 cells adapted to a nutrient-excessive environment.

Glucose oxidation of SF188s and SF188f cells was also examined. Glucose addition did not stimulate any increase in OCR in either cell line (data not shown) suggesting neither oxidizes glucose.

Having established differential oxidation of glutamine and fatty acids in SF188s and SF188f cells, we proceeded to investigate the

mitochondrial bioenergetics machinery [36,37] by consecutively adding oligomycin, FCCP, and complex III inhibitor myxothiazol in the presence of saturating amounts of a full set of energy substrates (25 mM glucose, 6 mM glutamine and 1 mM pyruvate for both cell lines). As shown in Figure 9, the mitochondrial bioenergetic state of SF188f differed strikingly from that of SF188s. First of all, in addition to a higher basal OCR, the mitochondrial respiratory capacity (determined by FCCP-stimulated OCR, Figure 9A)) was much higher while the fraction of basal OCR contributing to ATP-coupled respiration (revealed by oligomycin-sensitive OCR) in SF188f cells was greatly attenuated (Figure 9B). Intriguingly, proton leak (the difference between oligomycin-resistant but myxothiazol-sensitive OCR) in SF188f was markedly increased to 48% of the baseline OCR compared with 17% in SF188s cells (Figure 9B). These results suggested that basal

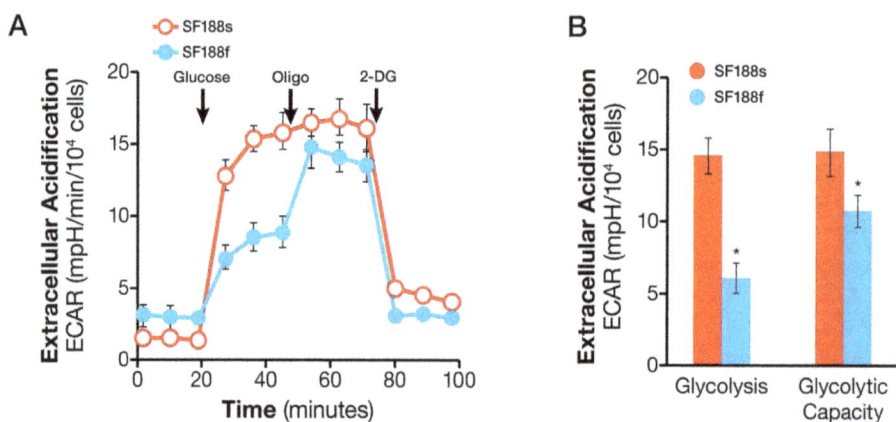

A

B

Figure 7. Lowered basal glycolytic flux but acquired glycolytic capacity in SF188f cells compared with SF188s cells. A. ECAR response of SF188s and SF188f cells to glucose (10 mM), oligomycin (1 μM), and 2-DG (100 mM). SF188s and SF188f cells were plated at 30,000 and 20,000 cells/well, respectively, in XF24 V7 cell culture plates 24–28 hours prior to the assays. The assay medium was the substrate-free base medium supplemented with 2 mM glutamine. Upon completion of an assay, cells were treated with trypsin and counted for the purpose of normalization. ECAR values were normalized to mpH/10^4 cells. A representative experiment out of three is shown here. Each data point represents mean ± SD, n = 4. B. Calculated glycolytic flux and glycolytic capacity of SF188s and SF188f cells normalized to mpH/min/10,000 cells. * p<0.05.

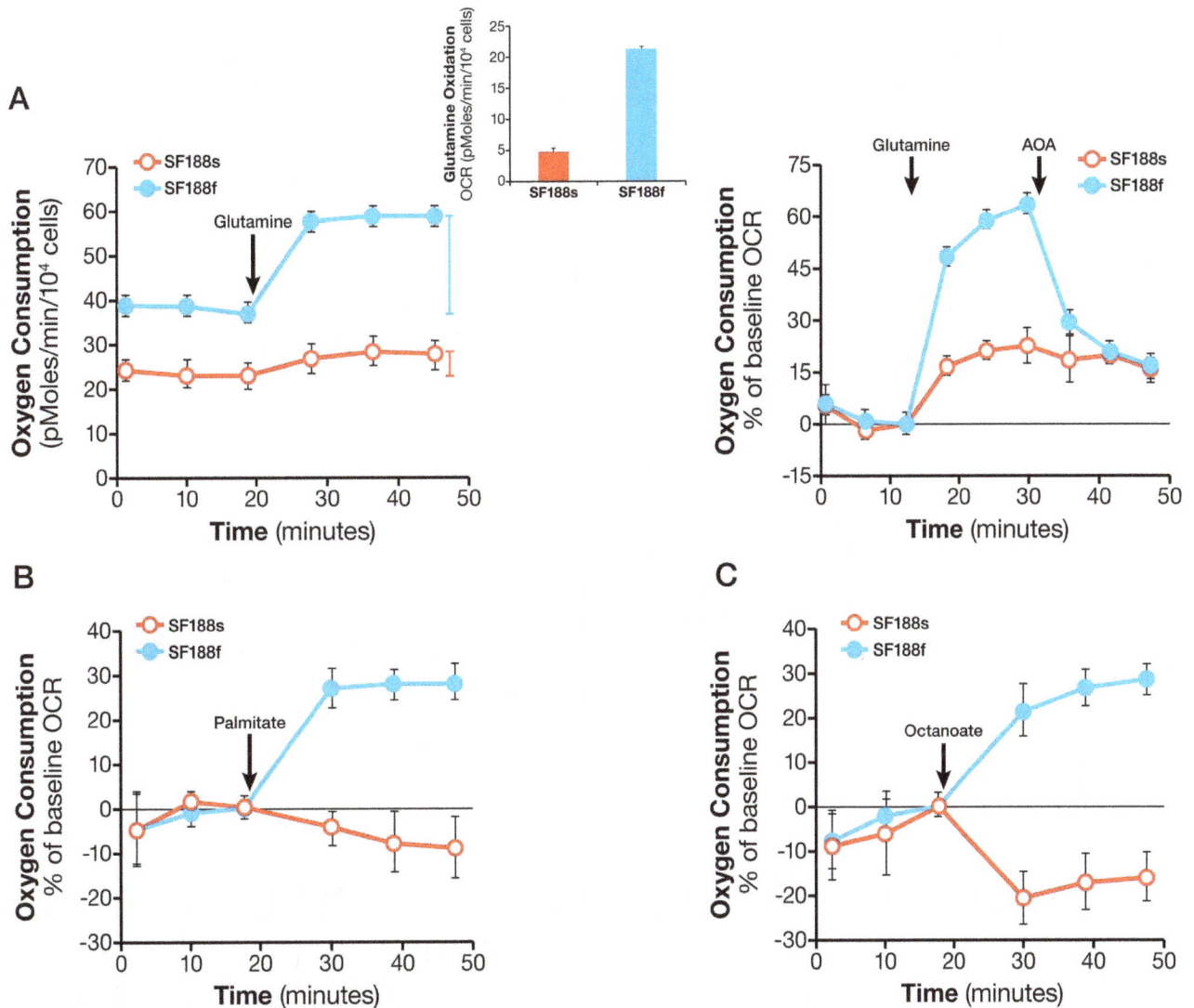

Figure 8. Enhanced glutamine oxidation and activation of fatty acid oxidation in SF188f cells. A. Kinetic OCR response of SF188s and SF188f cells to glutamine (4 mM) (left panel). Insert: calculated glutamine oxidation rate of SF188s and SF188f cells. OCR response (% of baseline) to glutamine and AOA (100 μM) (right panel). SF188s and SF188f cells were plated at 30,000 and 20,000 cells/well, respectively, in XF24 V7 cell culture plates 24–28 hours prior to the assays. The assay medium was the substrate-free base medium. Upon completion of an assay, cells were treated with trypsin and counted for the purpose of normalization. The OCR values were normalized to pmoles/min/10^4 cells. A representative experiment out of three is shown here. Each data point represents mean ± SD, n = 4. B and C. OCR response of SF188s and SF188f cells to palmitate-BSA (150 mM) (B) and octanoate (1 mM) (C). SF188s and SF188f cells were plated at 20,000 and 15,000 cells/well, respectively, in XF96 V3 cell culture plates 24–28 hours prior to the assays. The assay medium was the substrate-free base medium supplemented with 5.5 mM glucose and 50 μM carnitine. Fatty acid oxidation was expressed as % OCR plotted using measurement 3 as the baseline. A representative experiment out of three is shown here. Each data point represents mean ± SD, n = 6.

respiration in SF188f cells was largely uncoupled from phosphorylation of ADP to ATP.

Discussion

In this study, we presented the concept, design, and development of several methods to analyze the glycolytic flux and oxidative flux of glucose, glutamine, and fatty acids in cancer cells. In the proof-of- principle experiments using human glioblastoma SF188s and SF188f cells, we provided initial evidence suggesting a mechanism that may allow continuous anaplerotic glutamine flux by increasing mitochondrial proton leak partially uncoupling the

TCA cycle from oxidative phosphorylation in the fast proliferating SF188f cells.

Glycolysis and glycolytic capacity

Extracellular acidification, or ECAR, is comprised of glycolytic and non-glycolytic acidification. The former is the major contributor to total ECAR in cancer cells studied to date, while the latter accounts for significant albeit small fraction, which is mainly due to TCA cycle CO_2 production (Figure 1). Our experimental design makes it possible to directly identify and calculate glycolytic and non-glycolytic ECAR (Figure 1 and 2). Glucose-derived CO_2 obviously can contribute to glycolytic ECAR. This can be ruled out by examining the OCR response

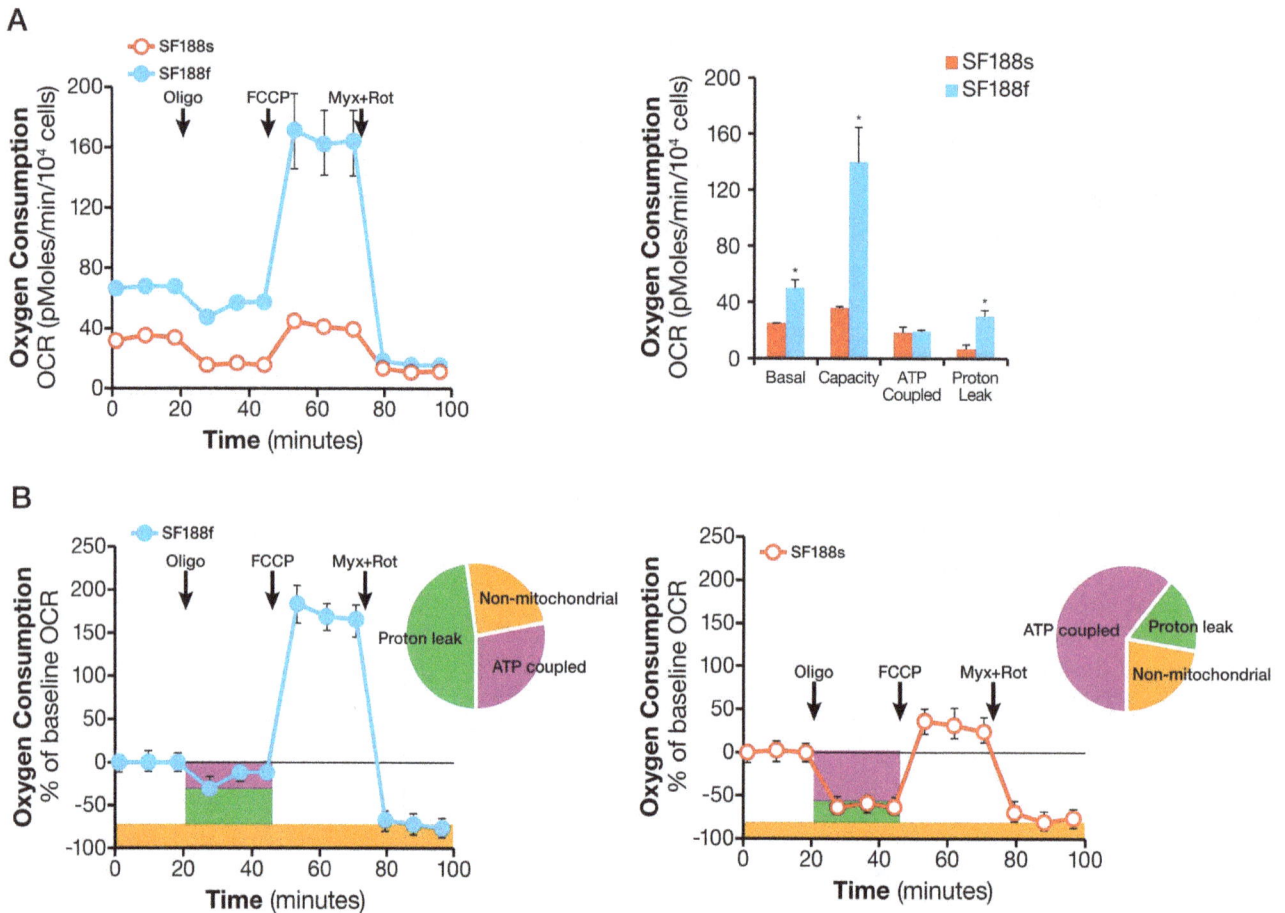

Figure 9. Markedly reprogrammed mitochondrial bioenergetic machinery of SF188s and SF188f cells. A. Kinetic OCR response of SF188s and SF188f cells to oligomycin (1 and 2 μM respectively) to determine ATP coupled respiration, FCCP (0.3 μM) to establish maximal respiratory capacity, and myxothiazol (1 μM) and rotenone (1 μM) cocktail) to define mitochondrial respiration (left). Calculated ATP-coupled respiration, proton leak-linked respiration, maximal mitochondrial respiratory capacity (right). SF188s and SF188f cells were plated at 30,000 and 20,000 cells/well, respectively, in XF24 V7 cell culture plates 24–28 hours prior to the assays. The assay medium was the substrate-free base medium supplemented with 25 mM glucose, 6 mM glutamine and 1 mM pyruvate. Upon completion of an assay, cells were treated with trypsin and counted for the purpose of normalization. OCR values were normalized to pmoles/10^4 cells. A representative experiment out of three is shown here. Each data point represents mean ± SD, n = 3. *p<0.05. B. Kinetic OCR response (% of baseline, baseline = measurement 3) of SF188f (right) and SF188s cell (left) with distribution of ATP-coupled respiration (% of oligomycin-sensitive at measurement 4) and proton leak-linked respiration (oligomycin-resistant at measurement 4 but myxothiazol-sensitive mitochondrial OCR at measurement 12) and non-mitochondria respiration (myxothiazol-resistant OCR at measurement 12).

to glucose in the same experiment. If the addition of glucose does not increase OCR of the cells, or inhibits OCR (the Crabtree effect) as we observed in HeLa cells (Figure 1C and 2A), then there is no glucose-derived CO_2 in the assay. In fact, most of the cancer cells we studied exhibit the Crabtree effect (data not shown). Thus in this experimental setting, glucose-induced ECAR increase is an accurate measurement of glycolytic flux. In any case where glucose increases the OCR, additional experiments such as the addition of a carbonic anhydrase inhibitor will help ascertain whether glucose-derived CO_2 is converted to protons at all, and if so, how significant a contribution to ECAR it might be.

The glycolytic capacity of cells under basal conditions upon which cells can draw in the face of increased energy demand, for example, a condition imposed by the loss of mitochondrial ATP, allows cells to maintain energy homeostasis and cellular function. By extension, it can serve as a mechanism by which cancer cells adapt to and survive under hypoxic conditions [8]. A recent study provides evidence in favor of this notion [38]. These authors found

that prostate cancer cells PC-3 displayed a significant glycolytic capacity in addition to higher glycolytic flux compared with normal prostate epithelial cells. However, our observation that the glycolytic capacity of glioblastoma SF188s and SF188f cultured under different nutrient condition were either nonexistent or substantial (Figure 7) suggest a more complex story which requires further investigation. The glycolytic flux and glycolytic capacity assay offers a very useful tool for those future studies.

At least three mechanisms can explain glycolytic capacity. In the first, cells can simply augment glucose uptake and/or the activity of glycolytic enzymes. For example, glucose transporters have been shown to relocalize rapidly from the cytoplasm to the cell surface when under energy stress, resulting in an immediate increase in glucose uptake [39–41]. In the second mechanism, glucose-derived pyruvate can be redirected away from the mitochondrial TCA cycle to lactate production, when oxidative phosphorylation is suppressed by oligomycin. In the third mechanism, glucose carbons can be shunted away from the production of metabolic intermediates for biosynthesis (including ribose, lipids, and serine synthesis) [13,42] and toward lactate production. The first two mechanisms most likely account for the increased glycolytic flux (i.e., glycolytic capacity), but exactly which mechanism(s) are responsible for glycolytic capacity may depend on cellular context of the cells under study.

Oxidative Substrate Flux

To produce ATP via oxidative phosphorylation, all nutrient substrates taken up by cells are ultimately oxidized via the TCA cycle and the ETC. In order to measure substrate oxidation, we applied this theory using the availability of an exogenous substrate (glucose, glutamine and fatty acids) at above saturation concentration to control oxygen consumption (i.e., electron transport). This principle is generally applicable to any metabolic substrate such as pyruvate, ketone bodies, and succinate (unpublished observations). Although being rapid and with higher throughput, these methods evidently do not yield as detailed information as metabolomics does. The overall metabolic insights gained, however, can either guide and/or be complementary to tracer-based metabolomics analysis to quantitatively follow substrate carbon flux through specific metabolic pathways [43].

Two technical points warrant further discussion. First, by selecting an above saturation substrate concentration in the substrate flux assays, the OCR or ECAR values are not governed by the amount of substrate available, but rather by the cells' ability to catabolize each substrate in meeting the demand of cell proliferation and maintaining basic cell functions. This experimental design makes it possible to evaluate substrate metabolism across many cell lines, regardless of the substrate concentrations in their growth media. For example, the growth media for SF188s and SF188f contained 2 and 6 mM glutamine, respectively; however, 4 mM (above saturation) glutamine, was used to compare glutamine oxidation rate of both (Figure 8A). Second, substrate oxidation rates as determined by the acute substrate addition described here are approximations of those in culture conditions (in which serum is present. They do, however, reveal the cells' ability to oxidize the substrates – in other words, the existence of the metabolic programs – providing clues for further investigation using OCR/ECAR and/or other technologies, such as tracing with stable isotope labeled substrates. For example, the effect of serum or growth factors on substrate oxidation can be ascertained by using variations upon the original assays, either by including them in the assay medium or by adding them acutely following substrate addition. Therefore, the substrate oxidation methods presented here can serve as the foundation upon which additional studies can be devised to gain further insights into the dynamic cellular metabolic network.

In this study, we not only determined basal respiration but also identified the type of exogenous nutrient substrates cells are able to oxidize and the rates at which they can be oxidized under the experimental conditions. Our results show that SF188f cells possess the ability to oxidize glutamine at a higher rate compared with SF188s cells. As shown in Figure 8A, OCR increases by ~20 pmoles/min/10^4 cells from ~40 to ~60 pmoles/min/10^4 cell following glutamine addition. The glutamine oxidation results obtained here are consistent with previously reported high glutamine utilization in SF188 cells employing classical glutamine consumption method [6]. Interestingly, unlike glutamine and fatty acids, glucose oxidation remained unaltered in SF188f cells. These results again highlight cancer cells' substrate preferences and their dynamic shift in response to environmental alterations, which can be readily revealed by these methods.

In addition to respiring on exogenously supplied substrates, we observed substantial respiration with endogenous substrates in cancer cells; for example, SF188f and PC-3 cells (Figure 3 and 4). As we reported previously, fatty acids are one such type of endogenous substrate [30]. However, oxidation of endogenous substrates is a poorly understood process, as well as its physiological role of in normal or cancer cells. Recent studies showed that macroautophagy and chaperone-mediated autophagy are required to maintain energy metabolism for tumor growth and survival [43,44]. Linking the observed respiration with endogenous substrates, we speculate that autophagy may, at least in part, provide endogenous substrates in the form of amino acids and fatty acids. The methods established in this study can measure respiration of both exogenous and endogenous substrates, thus affording an excellent opportunity for conducting such studies.

A potential role of glutamine oxidation along with proton leak in SF188f cells

The high rate of glutamine oxidation observed in SF188f cells was also accompanied by an unusually large proton leak. 48% of the cells' basal respiration was found to be diverted to proton leak, with a smaller fraction of 28% devoted to ATP production (Figure 9). We suggest that this unusually large proton leak, along with high levels of glutamine oxidation, may allow on-demand glutamine anaplerotic flux to meet the biosynthetic demand required for biomass duplication in rapidly dividing SF188f cells. Specifically, proton leak can partially uncouple the TCA cycle flux from mitochondrial ATP production (reducing ATP yield) and thus bypass the control of oxidative phosphorylation exerted on the anaplerotic TCA cycle flux. A similar result which supports our hypothesis has been recently reported in H-RAS transformed mouse embryonic fibroblasts (MEF) [45]. In the H-RAS expressing MEF, an overexpression of kinase suppressor of ras 1 (KSR1), a scaffold protein in the ERK signaling pathway, Leads to significantly increased glutamine oxidation as well as proton leak compared with the KSR1 deficient MEF. Similarly, these metabolic alterations in the H-RAS MEF were also accompanied by faster cell proliferation. Our results provide the initial evidence suggesting a potential role for proton leak in modulating anabolic precursor production in rapidly proliferation SF188f cells. Of course, much more broad and in-depth studies are required to substantiate this possibility.

Conclusion

By measuring OCR and ECAR responses of living cells to the addition of substrates along with metabolic perturbation agents, the methods presented here have cleared a path for easy and systematic analyses of metabolic substrate flux in wide range of cancer cell lines (Figure 10). These microplate-based substrate flux methods use only small amounts of biological materials, making it possible to analyze large numbers of natural, genetically altered, or compound-treated cancer or non-cancer cell samples in a short period of time. The approach is highly valuable, particularly when integrated with genomic, proteomic, and metabolomics technologies with the purpose of understanding of genetic, epigenetic, and signaling pathways that regulate cancer cell metabolism and cancer cells' metabolic vulnerabilities. Given the growing appreciation of metabolic shifts and nutrient substrate switches in many other diseases, including cardiovascular, neurological and immune ailments, these new methods are also very useful for investigating pathobiology in a wide variety of disease models.

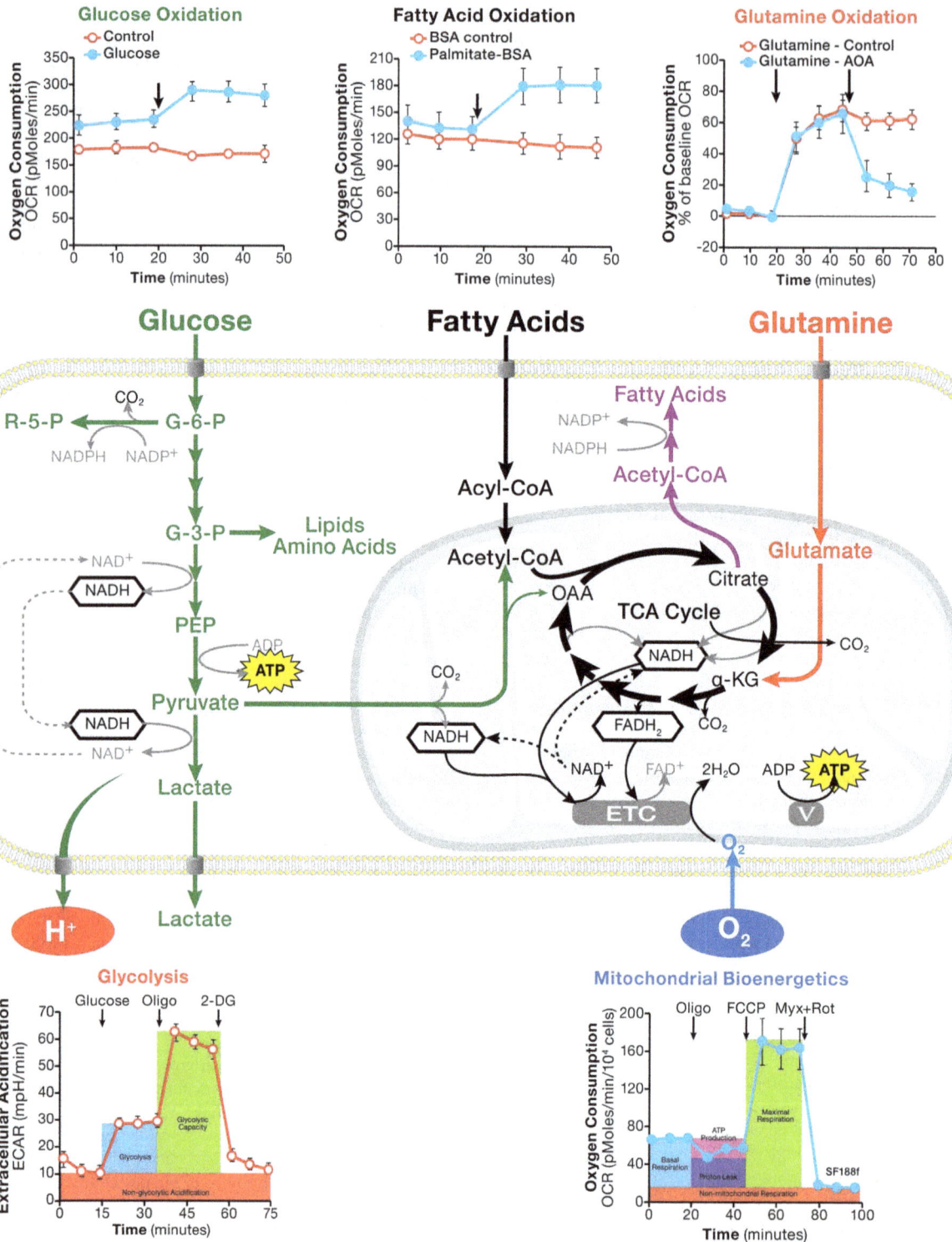

Figure 10. Schematic illustration of cellular metabolism pathways along with assays of glucose, glutamine, and fatty acid oxidation, glycolytic flux and mitochondrial bioenergetics.

Supporting Information

Figure S1 Determining saturating glucose concentration at which maximal ECAR response were achieved under both basal condition and inhibition of oxidative phosphorylation by oligomycin. A. Glucose at concentra-

tions of 1, 5, 10 and 25 mM were injected sequentially to PC-3 cells. PC-3 cells were plated at 25,000/well in XF24 V7 culture plates. The assay medium was the base medium supplemented with 2 mM L-glutamine. The OCR values were not normalized. A representative experiment out of three is shown here. Each data point represents mean $\pm$ SD, n = 4. B. Oligomycin (1 μM) was added to PC3 cells in the same assay medium as in A followed by three sequential injections of glucose at concentrations of 5, 10, and 25 mM as indicated.

Figure S2   SF188f cells are able to oxidation short chain fatty acid bytyrate. A. OCR response of SF188s and SF188f cells to butyrate (0.3 mM). SF188s and SF188f cells were plated at 20,000 and 15,000 cells/well, respectively, in XF96 V3 cell culture plates 24–28 hours prior to the assays. The assay medium was the substrate-free base medium supplemented with 5.5 mM glucose and 50 μM carnitine. Fatty acid oxidation was expressed as % OCR and plotted using measurement 3 as the baseline. A representative experiment out of three is shown here. Each data point represents mean $\pm$ SD, n = 6.

References

Acknowledgments

We would like to thank Dr. Andrew Lane (University of Louisville, Louisville, KY) for the insightful discussions and critical comments; Dr. Justin Cross (Memorial Sloan-Kettering Cancer Center, New York, NY) for valuable comments; Daniel Fitzpatrick (Seahorse Bioscience) for the artwork; and members of Seahorse Bioscience for discussions and support during the course of the work.

Author Contributions

Conceived and designed the experiments: MW. Performed the experiments: LSPW. Analyzed the data: LSPW MW. Wrote the paper: LSPW MW.

1. Warburg O, Posener K, Negelein E (1924) Über den Stoffwechsel der Carcinomzelle. Biochem Zeitschr 152: 35.
2. Warburg O (1956) On the origin of cancer cells. Science 123: 309–314.
3. Baggetto LG (1992) Deviant energetic metabolism of glycolytic cancer cells. Biochimie 74: 959–974.
4. Medina MA (2001) Glutamine and cancer. The Journal of nutrition 131: 2539S–2542S; discussion 2550S–2531S.
5. DeBerardinis RJ, Mancuso A, Daikhin E, Nissim I, Yudkoff M, et al. (2007) Beyond aerobic glycolysis: transformed cells can engage in glutamine metabolism that exceeds the requirement for protein and nucleotide synthesis. Proceedings of the National Academy of Sciences of the United States of America 104: 19345–19350.
6. Wise DR, DeBerardinis RJ, Mancuso A, Sayed N, Zhang XY, et al. (2008) Myc regulates a transcriptional program that stimulates mitochondrial glutaminolysis and leads to glutamine addiction. Proceedings of the National Academy of Sciences of the United States of America 105: 18782–18787.
7. Gao P, Tchernyshyov I, Chang TC, Lee YS, Kita K, et al. (2009) c-Myc suppression of miR-23a/b enhances mitochondrial glutaminase expression and glutamine metabolism. Nature 458: 762–765.
8. Gatenby RA, Gillies RJ (2004) Why do cancers have high aerobic glycolysis? Nature reviews Cancer 4: 891–899.
9. Pedersen PL (2007) Warburg, me and Hexokinase 2: Multiple discoveries of key molecular events underlying one of cancers' most common phenotypes, the "Warburg Effect", i.e., elevated glycolysis in the presence of oxygen. Journal of bioenergetics and biomembranes 39: 211–222.
10. Vander Heiden MG, Cantley LC, Thompson CB (2009) Understanding the Warburg effect: the metabolic requirements of cell proliferation. Science 324: 1029–1033.
11. Koppenol WH, Bounds PL, Dang CV (2011) Otto Warburg's contributions to current concepts of cancer metabolism. Nature reviews Cancer 11: 325–337.
12. DeBerardinis RJ, Lum JJ, Hatzivassiliou G, Thompson CB (2008) The biology of cancer: metabolic reprogramming fuels cell growth and proliferation. Cell metabolism 7: 11–20.
13. Deberardinis RJ, Sayed N, Ditsworth D, Thompson CB (2008) Brick by brick: metabolism and tumor cell growth. Current opinion in genetics & development 18: 54–61.
14. Ward PS, Thompson CB (2012) Metabolic reprogramming: a cancer hallmark even warburg did not anticipate. Cancer cell 21: 297–308.
15. Yuneva MO, Fan TW, Allen TD, Higashi RM, Ferraris DV, et al. (2012) The metabolic profile of tumors depends on both the responsible genetic lesion and tissue type. Cell metabolism 15: 157–170.
16. Casey JR, Grinstein S, Orlowski J (2010) Sensors and regulators of intracellular pH. Nature reviews Molecular cell biology 11: 50–61.
17. Hochachka PW, Mommsen TP (1983) Protons and anaerobiosis. Science 219: 1391–1397.
18. Lane AN, Fan TW, Higashi RM (2009) Metabolic acidosis and the importance of balanced equations. Metabolomics 5: 3.
19. Kemp G (2005) Lactate accumulation, proton buffering, and pH change in ischemically exercising muscle. American journal of physiology Regulatory, integrative and comparative physiology 289: R895–901; author reply R904–910.
20. Ipata PB, Francesco (2012) Glycogen as a fuel: metabolic interaction between glycogen and ATP catabolism in oxygen-independent muscle contraction. Metabolomics 8.
21. Brooks GA (2010) What does glycolysis make and why is it important? Journal of applied physiology 108: 1450–1451.
22. Wu M, Neilson A, Swift AL, Moran R, Tamagnine J, et al. (2007) Multiparameter metabolic analysis reveals a close link between attenuated mitochondrial bioenergetic function and enhanced glycolysis dependency in human tumor cells. Am J Physiol Cell Physiol 292: C125–136.
23. Turrens JF (2003) Mitochondrial formation of reactive oxygen species. The Journal of physiology 552: 335–344.
24. Brookes PS (2005) Mitochondrial H(+) leak and ROS generation: an odd couple. Free radical biology & medicine 38: 12–23.
25. Xie H, Valera VA, Merino MJ, Amato AM, Signoretti S, et al. (2009) LDH-A inhibition, a therapeutic strategy for treatment of hereditary leiomyomatosis and renal cell cancer. Mol Cancer Ther 8: 626–635.
26. Chan DA, Sutphin PD, Nguyen P, Turcotte S, Lai EW, et al. (2011) Targeting GLUT1 and the Warburg Effect in Renal Cell Carcinoma by Chemical Synthetic Lethality. Sci Transl Med 3: 94ra70.
27. Zhang J, Nuebel E, Wisidagama DR, Setoguchi K, Hong JS, et al. (2012) Measuring energy metabolism in cultured cells, including human pluripotent stem cells and differentiated cells. Nat Protoc 7: 1068–1085.
28. Crabtree HG (1929) Observations on the carbohydrate metabolism of tumours. The Biochemical journal 23: 536–545.
29. Fan TW, Lane AN, Higashi RM, Farag MA, Gao H, et al. (2009) Altered regulation of metabolic pathways in human lung cancer discerned by (13)C stable isotope-resolved metabolomics (SIRM). Molecular cancer 8: 41.
30. Pike LS, Smift AL, Croteau NJ, Ferrick DA, Wu M (2010) Inhibition of fatty acid oxidation by etomoxir impairs NADPH production and increases reactive oxygen species resulting in ATP depletion and cell death in human glioblastoma cells. Biochim Biophys Acta.
31. Samudio I, Kurinna S, Ruvolo P, Korchin B, Kantarjian H, et al. (2008) Inhibition of mitochondrial metabolism by methyl-2-cyano-3,12-dioxooleana-1,9-diene-28-oate induces apoptotic or autophagic cell death in chronic myeloid leukemia cells. Molecular cancer therapeutics 7: 1130–1139.
32. Carracedo A, Cantley LC, Pandolfi PP (2013) Cancer metabolism: fatty acid oxidation in the limelight. Nature reviews Cancer 13: 227–232.
33. Caro P, Kishan AU, Norberg E, Stanley IA, Chapuy B, et al. (2012) Metabolic signatures uncover distinct targets in molecular subsets of diffuse large B cell lymphoma. Cancer cell 22: 547–560.
34. Zaugg K, Yao Y, Reilly PT, Kannan K, Kiarash R, et al. (2011) Carnitine palmitoyltransferase 1C promotes cell survival and tumor growth under conditions of metabolic stress. Genes & development 25: 1041–1051.
35. Trent J, Meltzer P, Rosenblum M, Harsh G, Kinzler K, et al. (1986) Evidence for rearrangement, amplification, and expression of c-myc in a human glioblastoma. Proceedings of the National Academy of Sciences of the United States of America 83: 470–473.
36. Choi SW, Gerencser AA, Nicholls DG (2009) Bioenergetic analysis of isolated cerebrocortical nerve terminals on a microgram scale: spare respiratory capacity and stochastic mitochondrial failure. J Neurochem 109: 1179–1191.
37. Hill BG, Dranka BP, Zou L, Chatham JC, Darley-Usmar VM (2009) Importance of the bioenergetic reserve capacity in response to cardiomyocyte stress induced by 4-hydroxynonenal. Biochem J 424: 99–107.
38. Ibrahim-Hashim A, Wojtkowiak JW, Ribeiro MdLC, Estrella V, Bailey KM, et al. (2011) Free Base Lysine Increases Survival and Reduces Metastasis in Prostate Cancer Model. J Cancer Sci Ther S1.
39. Rivenzon-Segal D, Boldin-Adamsky S, Seger D, Seger R, Degani H (2003) Glycolysis and glucose transporter 1 as markers of response to hormonal therapy in breast cancer. International journal of cancer Journal international du cancer 107: 177–182.
40. Artemov D, Bhujwalla ZM, Pilatus U, Glickson JD (1998) Two-compartment model for determination of glycolytic rates of solid tumors by in vivo 13C NMR spectroscopy. NMR in biomedicine 11: 395–404.
41. Mathupala SP, Heese C, Pedersen PL (1997) Glucose catabolism in cancer cells. The type II hexokinase promoter contains functionally active response elements

for the tumor suppressor p53. The Journal of biological chemistry 272: 22776–22780.

42. Ye J, Mancuso A, Tong X, Ward PS, Fan J, et al. (2012) Pyruvate kinase M2 promotes de novo serine synthesis to sustain mTORC1 activity and cell proliferation. Proceedings of the National Academy of Sciences of the United States of America 109: 6904–6909.

43. Guo JY, Chen HY, Mathew R, Fan J, Strohecker AM, et al. (2011) Activated Ras requires autophagy to maintain oxidative metabolism and tumorigenesis. Genes Dev 25: 460–470.

44. Singh R, Kaushik S, Wang Y, Xiang Y, Novak I, et al. (2009) Autophagy regulates lipid metabolism. Nature 458: 1131–1135.

45. Swift AL, Chen D, Pike LS, Fisher K, Winer M, et al. (2011) KSR1, a molecular scaffold of the Ras-Raf-MEK-ERK kinase cascade that is necessary for H-RasV12-induced oncogenic transformation, is also required for uptake/consumption of glucose and glutamine in mouse embryonic fibroblasts. F1000 Posters 2: 52 (poster)

16

Irisin Promotes Human Umbilical Vein Endothelial Cell Proliferation through the ERK Signaling Pathway and Partly Suppresses High Glucose-Induced Apoptosis

Haibo Song[1,2,4], Fei Wu[1,2], Yuan Zhang[1,2], Yuzhu Zhang[1,2], Fang Wang[1], Miao Jiang[1], Zhongde Wang[4], Mingxiang Zhang[2], Shiwu Li[1], Lijun Yang[1], Xing Li Wang[2], Taixing Cui[2,3]*, Dongqi Tang[1,2]*

1 Center for Stem Cell & Regenerative Medicine, The Second Hospital of Shandong University, Jinan, P.R.China, 2 Shandong University Qilu Hospital Research Center for Cell Therapy, Key Laboratory of Cardiovascular Remodeling and Function Research, Qilu Hospital of Shandong University, Jinan, P.R.China, 3 Department of Cell Biology and Anatomy, University of South Carolina, Columbia, South Carolina, United States of America, 4 Center for Reproductive Medicine, Zibo Maternal and Child health hospital, Zibo, P.R.China

Abstract

Irisin is a newly discovered myokine that links exercise with metabolic homeostasis. It is involved in modest weight loss and improves glucose intolerance. However, the direct effects and mechanisms of irisin on vascular endothelial cells (ECs) are not fully understood. In the current study, we demonstrated that irisin promoted Human Umbilical Vein Endothelial Cell (HUVEC) proliferation. It was further demonstrated that this pro-proliferation effect was mediated by irisin-induced activation of extracellular signal–related kinase (ERK) signaling pathways. Inhibition of ERK signaling with U0126 decreased the pro-proliferation effect of irisin on HUVECs. It was also demonstrated that irisin reduced high glucose-induced apoptosis by up-regulating Bcl-2 expression and down-regulating Bax, Caspase-9 and Caspase-3 expression. In summary, these results suggested that irisin plays a novel role in sustaining endothelial homeostasis by promoting HUVEC proliferation via the ERK signaling pathway and protects the cell from high glucose-induced apoptosis by regulating Bcl-2,Bax and Caspase expression.

Editor: Shang-Zhong Xu, University of Hull, United Kingdom

Funding: This work was supported by the Shandong University National Qianren Scholar Fund, and the Taishan Scholar Fund. The funders had no role in study design, data collection and analysis, decision to publish, or preparation of the manuscript.

Competing Interests: The authors have declared that no competing interests exist.

* Email: tangdq@sdu.edu.cn (DT); Taixing.Cui@uscmed.sc.edu (TC)

Introduction

Many vascular diseases are caused by endothelial cell (EC) injury and dysfunction, which occurs in chronic metabolic diseases such as metabolic syndrome and type II diabetes mellitus [1,2]. In many chronic metabolic diseases, vascular endothelial integrity is affected by EC proliferation and apoptosis, which assures blood vessel function [3]. Therefore, restoration of injured EC via regulating endothelial cell proliferation and apoptosis may have very important significance. Thus, extensive efforts were made to find more metabolic related factors that can promote endothelial cell proliferation and avoid their death, but the outcomes were not encouraging.

The benefits of exercise in metabolic and cardiovascular disease prevention and progression have been well documented [4]. Irisin is a newly discovered myokine that links exercise with increased energy expenditure to produce fundamental exercise-based health benefits [5]. Irisin is released from skeletal muscles and is increased with exercise when the fibronectin type III domain containing 5 (Fndc5) is proteolyzed. Irisin is highly conserved across species [5]. Irisin has been proposed to be a bridge between exercise and metabolic homeostasis and to be involved in modest weight loss and improved glucose intolerance in mice [5]. Recent studies

discovered that type 2 diabetic patients displayed significantly lower levels of circulating irisin compared with non-diabetic control subjects [6,7]. Circulating irisin levels were decreased in patients with chronic kidney disease (CKD) and were independently associated with high-density lipoprotein cholesterol levels [8]. Intriguingly, a new study demonstrated that pharmacological irisin concentrations promote mouse H19-7 HN cell proliferation via the STAT3 signaling pathway [9]. This finding suggests that irisin may have a pro-proliferation effect in addition to its role in regulating metabolic homeostasis. However, no previous studies have evaluated whether irisin may directly regulate human EC.

In this study, we treated Human Umbilical Vein Endothelial Cells (HUVECs) with human recombinant irisin (r-irisin), which was expressed and purified in our laboratory to detect its direct effects on HUVEC proliferation and apoptosis [10]. The possible signaling mechanisms by which irisin exerts its effects were also characterized. These studies demonstrated for the first time that irisin can promote HUVEC proliferation via extracellular signal–related kinase (ERK) pathway activation. Irisin can also reduce high glucose-induced apoptosis by up-regulating Bcl-2 and down-regulating Bax, Caspase-9 and Caspase-3 expression.

A

B

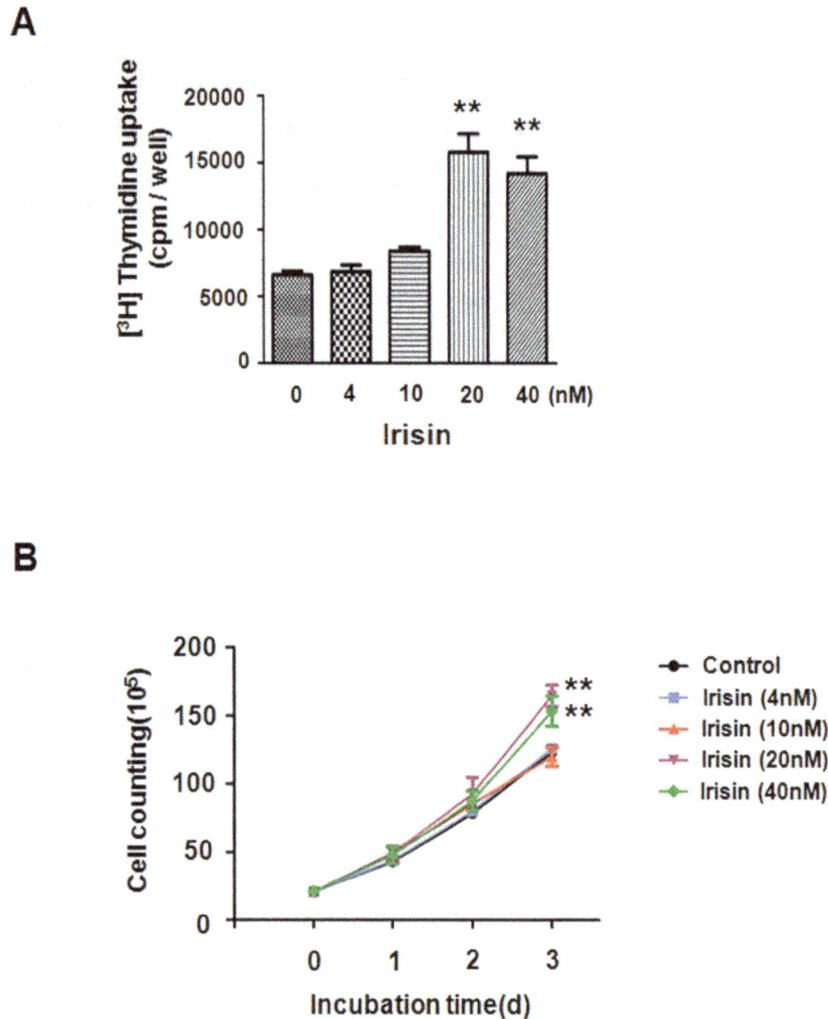

Figure 1. Irisin promoted HUVEC proliferation. (A) [³H] thymidine uptake in HUVECs that were cultured in M199 and treated with or without irisin at the indicated concentrations for 40 h. The data were expressed as the mean ± SE of three independent experiments,**p<0.01 vs. the untreated group. (B) Growth curves of HUVECs. HUVECs treated with or without irisin were constructed by plotting cell numbers that were counted using a hemocytometer over three days of incubation. ** p<0.01 vs. untreated, the data were expressed as the mean±SE of three independent experiments.

Materials and Methods

Expression and Purification of Human Irisin

The expression and purification of human irisin were performed as previously described [10]. Briefly, the cDNA (360 bp) of human irisin was designed and synthesized by Life Technologies. The synthesized human irisin cDNA was cloned into EcoR1/Xba1 sites of the pPICZaA plasmid. A linearized pPICZaA-irisin plasmid was use to transforme the P. pastorisX-33 according to the kit manual (PichiaEasycomp Transformation Kit; Invitrogen). The induction of protein expression and culture of yeast were performed as previously described [10]. Then the r-irisin protein in the supernatant was purified and used in our study.

Primary culture of Human Umbilical Vein Endothelial Cells (HUVECs)

Human umbilical vein endothelial cells(HUVECs) were isolated from human umbilical cords using 300 units/ml collagenaseII(-Sigma-Aldrich, St. Louis, MO) and cultured in Medium 199

(Invitrogen) with 10% (v/v) fetal bovine serum (FBS, Invitrogen) and conditioned supplement (10 ng/mL EGF and 10 ng/mL bFGF,Peprotech) at 37°C in a 5% CO_2 and 95% air atmosphere. HUVECs were used at passages 3–6 in all of the experiments.

The study protocol conformed to the ethical guidelines of the 1975 Declaration of Helsinki with the approval of the Institutional Medical Ethics Committee of Qilu Hospital, Shandong University. All of the donors provided written informed consent.

[3H] Thymidine uptake

HUVECs (P6) were cultured in M199 medium with 10% FBS along with EGF(10 ng/mL, Peprotech) and bFGF(10 ng/mL, Peprotech) at 50% confluence. The cells were cultured in serum-free M199 for 24 hours and treated with or without irisin for 40 hours. [3H] thymidine (final concentration 1 uCi/ml) was added to the media during the last 24 hours of culture. After washing with ice-cold PBS two times, the cells were precipitated with ice-cold 5% trichloroacetic acid (TCA, Sigma-Aldrich, St. Louis, MO) for 4 hours and washed with ice-cold 5% TCA two

A

Control   Irisin (20nM)

(400x)

80 μm

(200x)

80 μm

B

Ki67 positive cells in HUVECs (%)

Control
Irisin(20nM)

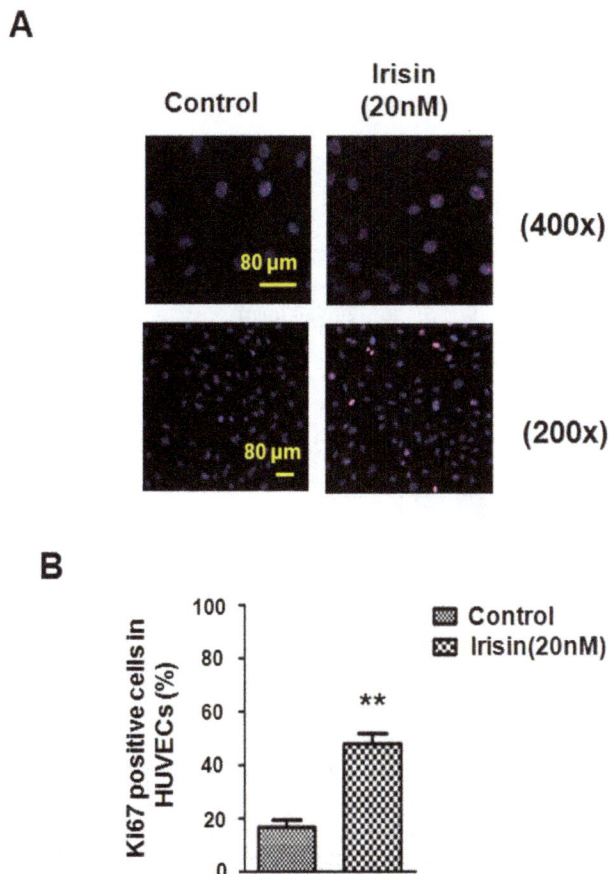

Figure 2. Ki67 staining in HUVECs. (A) Ki67 staining in HUVECs that had been treated with PBS (control) or irisin (20 nM) for 24 h. Images were taken with a confocal microscope. (B) Proliferation was assessed by the ratio of the average number of Ki67-positive cells to total cells in 10 random high-magnification fields. The data were expressed as the mean ± SE of three independent experiments,**p<0.01 vs. the untreated group.

times followed by two additional washes with ice-cold PBS. Then, they were lysed with 0.2 ml 0.5 M NaOH for 30 minutes at 37°C. DNA synthesis was measured by [3H] thymidine uptake.

Cell Counting

HUVECs were planted in a 10-cm diameter culture dish in the presence or absence of 20 nM Irisin. After digestion with trypsin every 24 h, the cells were subjected to cell size calculation using a counting slide (BIO-RAD) according to the manufacturer's instructions. The data were analyzed to investigate the influence of irisin on HUVEC proliferation. All of the experiments were performed in triplicate.

Immunofluorescent staining

HUVECs were grown on glass cover slips and fixed with 4% paraformaldehyde and permeabilized with 0.5% Triton X-100. Then, the cells were placed in 5% normal goat serum for 1 hour. They were then incubated with rabbit anti-Ki67 (1:500, Cat. No. ab15580, Abcam) antibodies overnight at 4°C. The cells were visualized with Alexa Fluor 640 conjugated goat anti-rabbit IgG (Invitrogen) for 1 hour at room temperature. The cells were incubated in 4′, 6-diamidino -2-phenylindole (DAPI)/PBS

(1:5000, cat. no. D9542, Sigma-Aldrich, St. Louis, MO) for 3 minutes at room temperature and washed 3 times in PBS for 5 minutes per wash. Finally, the cells were photographed using a Nikon eclipse Ti and UltraVIEW:emoji:VOX confo- cal microscope, and images were analyzed using Volocity software (PerkinElmer).

Western blot

Total cell protein concentrations were determined using the BCA protein assay kit (Pierce, Rockford, IL, USA). Equal amounts of protein from cell lysates were loaded in 12% sodium dodecyl sulfate-polyacrylamide gels. After electrophoresis, proteins were transferred to polyvinylidene fluoride membranes, blocked with 5% fat-free milk at room temperature for 1 h, and incubated with the indicated primary antibodies overnight at 4°C at 1:1000 dilution (rabbit anti-ERK1/2, anti-phospho-ERK1/2, anti-Akt, anti-phospho-Akt, anti-P38 MAPK, anti-phospho-P38 MAPK anti-phospho-GSK-3β, anti-GSK-3β,, anti-Bcl-2, anti-Bax, anti-Bad, anti-Caspase-9, anti-Caspase-3 antibodies [Cell Signaling Technology, Inc.]). Then, the membranes were washed for 15 min three times with tris-buffered saline with Tween 20 and incubated with HRP-conjugated secondary antibodies for 1 h at room temperature. After washing again for 15 min three times with Tris-buffered saline with Tween 20, immune complexes were detected with enhanced chemiluminescence reagents, and the blots were quantified by densitometric analysis using the Alpha Imager 2200.

Apoptosis Analysis

HUVEC apoptosis was determined with the annexin V-FITC/ propidium iodide assay. HUVECs were seeded into 6-well plates at 1×10^6/well in M199 containing 10% FBS, 10 ng/ml bFGF and 10 ng/ml EGF. After starvation in serum free media for 24 h, the cells were treated with or without 20 nM Irisin and 30 mM glucose for 24 h. They were then harvested, washed and re-suspended in PBS. Apoptotic cells were determined with an Annexin V-FITC apoptosis detection kit (BD Biosciences, USA) according to the manufacturer's protocol. Briefly, the cells were washed and subsequently incubated for 15 minutes at room temperature in the dark in 100 μl 1× binding buffer containing 5 μl Annexin V-FITC and 10 μl PI. Apoptosis data were determined by the BD accuriC6 flow cytometer and processed using FlowJo(FlowJo, Ashland, OR, USA) software.

Statistics

The data are expressed as the mean±standard deviation (SD). All of the experiments were repeated at least three times. Comparisons among values for all groups were performed by one-way analysis of variance (ANOVA). Holm's t-test was used for analysis of differences between different groups. Differences were considered to be significant at P<0.05.

Results

Effect of irisin on HUVEC proliferation

To investigate the role of irisin on HUVEC proliferation, [3H] thymidine uptake was measured as described in the materials and methods. It was observed that the increase of [3H] thymidine uptake induced by irisin (20 nM) was 2.4 times higher than the control group in serum-free conditions (Fig. 1A). To assess the actual cell number changes, the effects of irisin on HUVEC proliferation was detected by direct cell counting. The result demonstrated that irisin over a range of concentrations (20 and 40 nM) can significantly accelerate HUVEC proliferation

A

Irisin(20nM)

Con   5   10   20   30   (min)

P-ERK1/2

ERK1/2

P-Akt

Akt

P-p38

p38

B

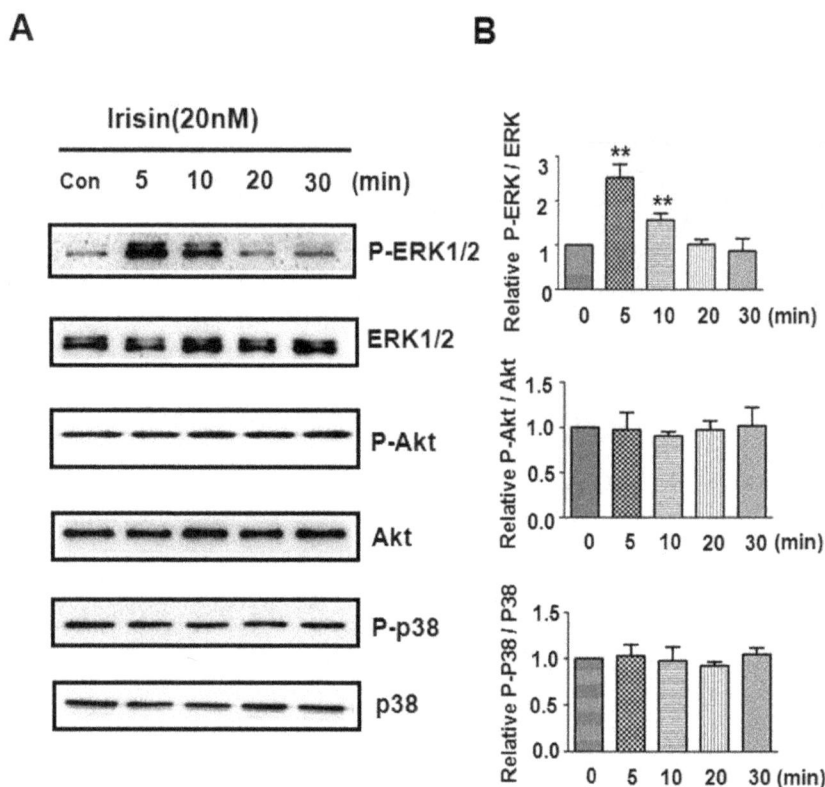

Figure 3. Irisin mediates proliferation via the ERK pathway. HUVECs were treated with or without irisin (20 nM) at indicated time points. (A) Phospho- and total ERK, p38 and AKT levels in cell lysates were analyzed by Western blot. (B) Densitometric analysis of the related bands was expressed as the relative optical band density, which was corrected using respective total proteins as a loading controls and normalized against the untreated control. The data were expressed as the mean ± SE of three independent experiments,**p<0.01 vs. untreated.

(Fig. 1B). Considering that the maximum effect appeared when the irisin concentration was 20 nM, 20 nM Irisin was chosen for the following experiments. To further reveal the effect of irisin on HUVEC proliferation, anti-Ki67 immunofluorescent staining was used. Ki67 is a nuclear protein which is expressed in proliferating cells; thus, it may be essential for maintaining cell proliferation [11]. The results demonstrated that there were more Ki67-expressing cells in the irisin stimulation group than the control group ($48.0 \pm 4.0\%$ and $16.6 \pm 2.9\%$, respectively) (Fig. 2A), and the difference was statistically significant ($P < 0.01$) (Fig. 2B).These data taken together suggested that irisin effectively promoted HUVEC proliferation in serum-free medium.

Irisin Mediates HUVEC Proliferation through the ERK Signaling Pathway

To gain further insights into the relationship between irisin and HUVEC proliferation, the protein levels of ERK, p38 MAPK, and AKT signaling molecules, which are critically involved in cell proliferation, were investigated. As demonstrated in Figure 3, after irisin treatment of HUVECs, phosphorylated ERK (P-ERK) levels were significantly increased at 5 minutes, peaked between 5 and 10 minutes, and decreased at 20 minutes (Fig. 3A). The increase was statistically significant, as quantified by densitometry (Fig. 3B). However, treating HUVECs with irisin had no effect on the level of phospho-p38 and AKT, indicating that the p38 and AKT signaling pathways are not involved in irisin-mediated cell proliferation (Fig. 3A and B).

To further examine that irsin enhances HUVEC proliferation through the ERK signaling pathway, U0126 (ERK inhibitor) was used to block ERK activation. Western blot results demonstrated that the irisin-induced ERK phosphorylation was significantly suppressed, while there was no reduction in the amount of total ERK protein (Fig. 4A and B). The irisin-stimulated HUVEC proliferation was evaluated by measuring Ki67 staining in each group (Fig. 4C and D) and [³H] thymidine uptake (Fig. 4E). The data demonstrated that the r-irsin-induced increase of [³H] thymidine uptake was significantly reduced by U0126 treatment. Similar results were observed using immunofluorescent staining, which also demonstrated that blocking ERK activation mitigated irisin-mediated proliferation.

Irisin protects HUVECs from high glucose-induced apoptosis

To determine whether irisin has a direct effect on HUVEC apoptosis, the cells were incubated with high glucose and irisin for 24 hours. As demonstrated in the flow cytometry results, irisin effectively attenuated high glucose-induced apoptosis in HUVECs at a dose of 20 nM (Fig. 5A and B). The percentage of cells undergoing apoptotic cell death decreased from $33.8 \pm 3.2\%$ in the high glucose group to $22.0 \pm 2.4\%$ after being exposed to 20 nM irisin for 24 hours.

Irisin down-regulates Bax, Caspase-9, Caspase-3 and up-regulates Bcl-2 expression in HUVECs on high glucose conditions

To further investigate the potential mechanism of irisin on high glucose-induced HUVEC apoptosis, the impact of irisin on

A

B

C

D

E

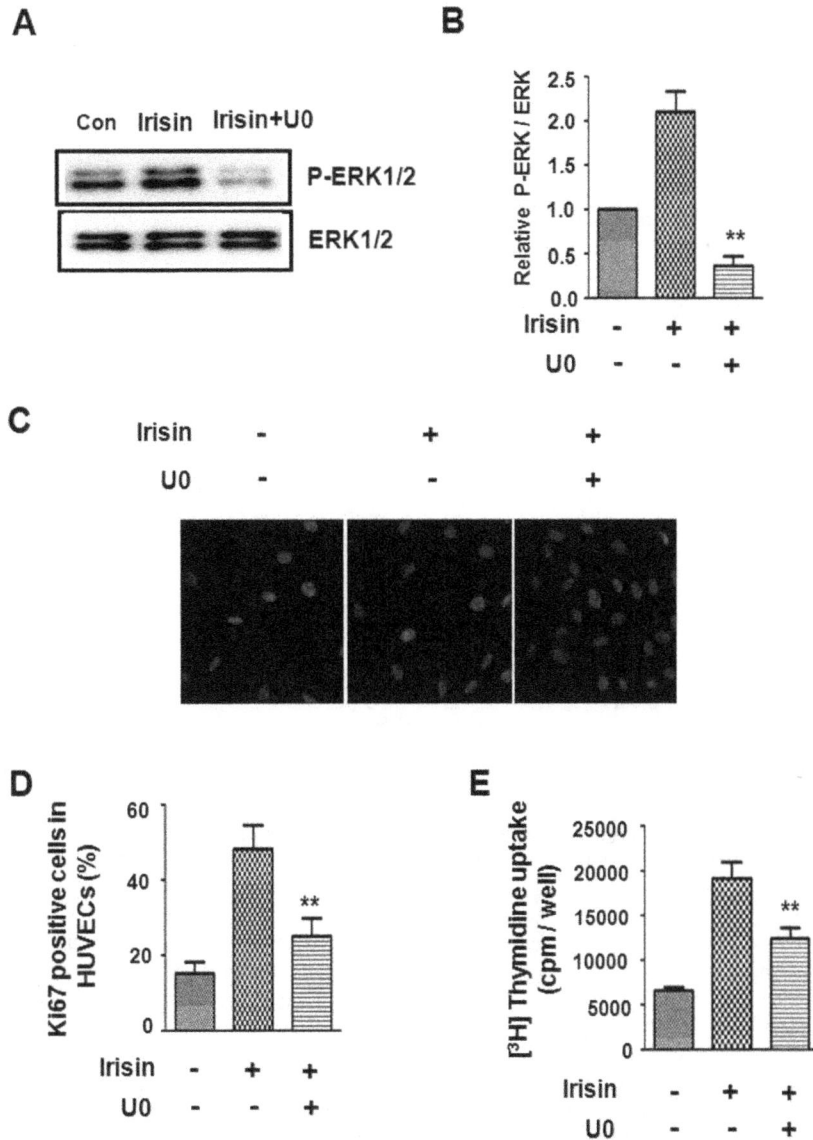

Figure 4. ERK inhibitor attenuated the irisin-induced HUVEC proliferation. HUVECs were pretreated with the ERK inhibitor U0126 for 30 min followed by irisin treatment. (A) Western blots analyzed phosphorylated and total ERK proteinexpression. (B) Densitometric analysis of the related bands was expressed as the relative optical band density, which was corrected using respective total proteins as a loading control and normalized against the untreated control. The data were expressed as the mean ± SE of three independent experiments,**p<0.01 vs. untreated group. The effect of the ERK inhibitor on irisin-induced HUVECproliferation was analyzed by Ki67 staining (C, D) and [^{3}H] thymidine uptake (E). The data were expressed as the mean ± SE of three independent experiments, **p<0.01 vs. the irisin-treated group.

expression of Bcl-2, Bax, Bad, GSK-3β,Caspase-9 and Caspase-3, several key apoptosis regulator proteins, was examined [12,13]. The Western blot results indicated that following the treatment with 20 nM irisin, expression of protein Bax, Caspase-9 and Caspase-3 decreased and the anti-apoptotic protein Bcl-2 increased. The expression of GSK-3β and Bad were not influenced. (Fig. 6A and B).

Discussion

Irisin, which was discovered in 2012 by Bostrom and colleagues, is a cleaved and secreted fragment of FNDC5. Irisin is released by skeletal muscle and is increased with exercise, can act on different tissues and functions as a muscle-derived energy expenditure signal by promoting brown adipocyte thermogenesis in WAT both

invitro and *in vivo* [14]. Furthermore, a recent study in the hippocampus demonstrated that endurance exercise increased the expression of irisin which lead to a significant increase in BDNF gene expression [15]. And another study showed the potential role of irisin on bone metabolism via modulating osteoblast differentiation [16]. Over-expression of irisin by intravenous injection of adenoviral vectors significantly increased total body energy expenditure, reduced body weight, and improved metabolic parameters such as insulin sensitivity in mice that were fed a high fat diet [5]. Moreover, serum irisin levels are lower in type 2 diabetes mellitus patients than in controls with normal glucose tolerance [6,7]. Irisin concentrations in breast milk and plasma are also lower in lactating women with gestational diabetes mellitus than in non-lactating or healthy lactating women [17]. Because of

A

B

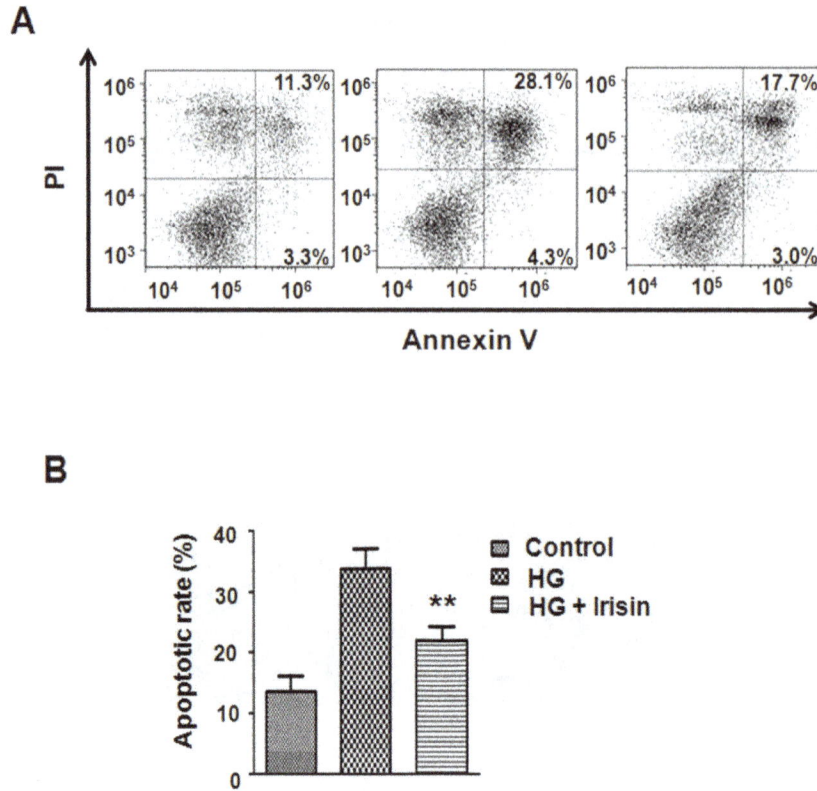

Figure 5. Irisin rescued HUVECs from high glucose-induced apoptosis. (A) Annexin V binding and propidium iodide (PI) staining was analyzed by FlowJo. (B)The bar graph represents the results of three independent experiments, **p<0.01 vs. the high glucose-treated group.

these data, irisin has been proposed as a meaningful therapeutic target for diseases caused by inactivity or chronic caloric excess such as diabetes and obesity. Meanwhile, some controversial results and conclusions against the beneficial roles about irisin have recently been reported [18–20]. Among of these reports, a study found a positive correlation between circulating irisin levels and body weight, BMI, fat mass [18]. But another study have drawn an opposite conclusion [19]. The discrepant results raise serious concerns about its application prospect. So further research need to be made.

Diabetes and hyperglycemia are intimately involved with endothelial dysfunction and markedly increase the risk of all forms of cardiovascular complications including critical limb ischemia and foot ulcers [21,22]. However, besides an interaction between physical activity and metabolic homeostasis, no previous studies have evaluated whether irisin may directly regulate endothelial cell proliferation and apoptosis. Recently, our laboratory successfully established an efficient system for the expression and purification of human recombinant irisin protein in *Pichiapastoris* [10]. In the current study, it was demonstrated that irisin can effectively promote HUVEC proliferation by activating the ERK signaling pathway, while p38 MAPK and Akt appeared to be uninvolved. Moreover, in addition to serving as a potent proliferation factor, our data revealed that irisin also inhibited high glucose-induced HUVEC apoptosis. Because regulating endothelial cell proliferation and apoptosis can significantly affect vascular endothelial integrity and sustain endothelium homeostasis [3], the effect of irisin on regulating HUVEC proliferation and apoptosis of may elucidate both the prevention and treatment of various vascular diseases, especially for some

metabolism-related vascular diseases, because of its potential therapeutic role in metabolism.

It is widely believed that ERK and PI3K/Akt signaling promotes cell proliferation by affecting DNA synthesis [23,24]. Our previous study reported that the ERK pathway can be activated by irisin to modulate the irisin-induced emergence of brown adipocytes. In this study, our data demonstrate that irisin significantly increased ERK phosphorylation, while the levels of P-p38 and P-Akt were not different. Furthermore, because treatment with the ERK inhibitor U0126 inhibited cell proliferation, it was concluded that irisin-induced HUVEC proliferation is dependent on ERK signaling pathway activation. However, the pro-proliferative effect of irisin in HUVECs was only partially abolished, implying that there were alternative signaling pathways contributing to irisin-induced HUVEC proliferation.

In general, apoptosis is regulated by pro-apoptotic and anti-apoptotic proteins and is executed through caspases [25]. Bcl-2 protein family members play an essential role in regulating and executing many cell apoptotic pathways. Examples include Bcl-2, Bax and Bad, which are members of the Bcl-2 family that function as inhibitors and activators of apoptosis, respectively [12,13]. The present study demonstrated that irisin attenuated high glucose-induced HUVEC apoptosis. The flow cytometry result was accompanied by increased Bcl-2 expression and decreased Bax expression. It is well known that Akt is a potent mediator of cell proliferation and cell survival. Previous study demonstrated that Akt promoted cell survival and block apoptosis via its downstream targets, including GSK-3, Bad, caspase-9 [26]. Akt inhibits the conformational change and translocation to mitochondria of Bax protein, so as to prevent the caspase-3 activation [27]. The present

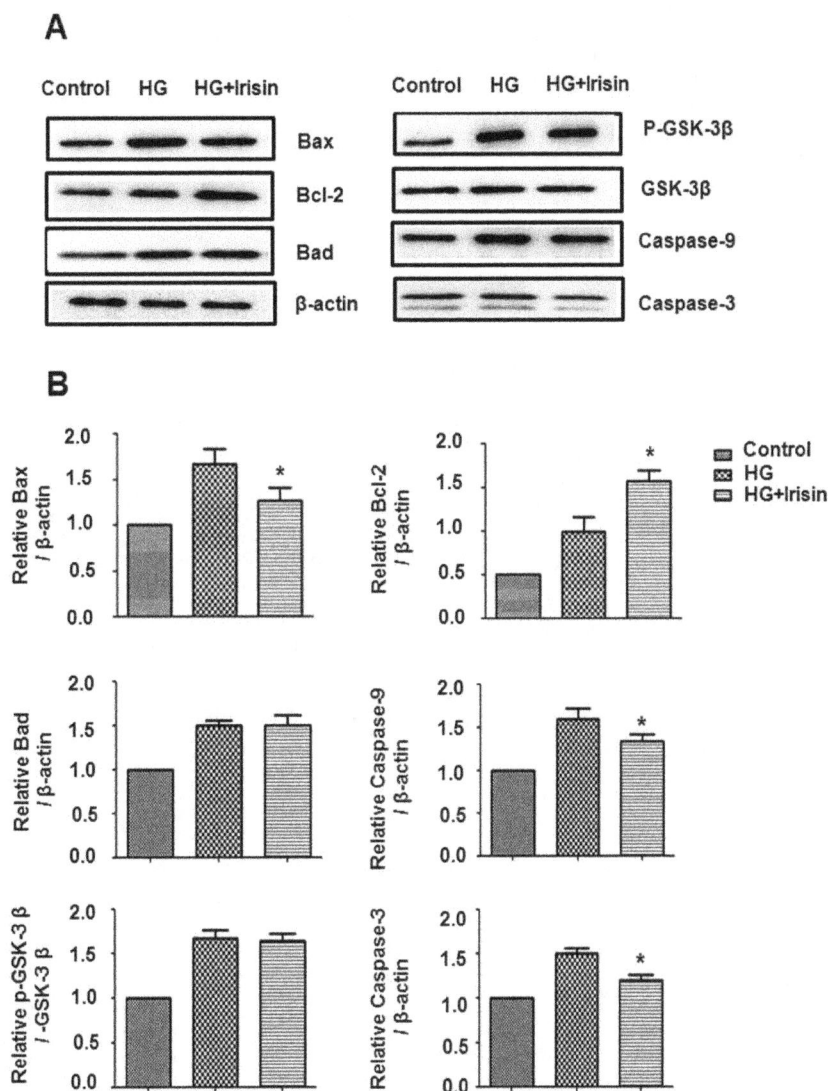

Figure 6. Irisin mediates Bax,Bcl-2,GSK-3β, Caspase-9 and Caspase-3 protein levels in HUVECs. HUVECs were cultured with or without irisin (20 nM) for 24 h. Bax, Bcl-2, Bad, GSK-3β, Caspase-9 and Caspase-3 in cell lysates were analyzed by Western blot. (B) Densitometric analysis of the related bands was expressed as the relative optical band density and was corrected using respective total proteins as loading controls and normalized against the untreated control. The data were expressed as the mean ± SE of three independent experiments, *p<0.05 vs. the high glucose-treated group.

study found that irisin did not affect the expression of GSK-3β and Bad, the downstream targets of Akt. But the expression of caspase-9 and caspase-3 were downregulated when treated with irisin. These results imply that irisin may mediated the HUVEC survival via Akt-independent pathway and its precise mechanism need to be further study.

Conclusions

We presented data demonstrating that recombinant irisin promoted HUVEC proliferation and activated the ERK signaling pathway. Irisin also protected cells from high glucose-induced apoptosis by up-regulating Bcl-2 levels together with down-

regulating Bax,Caspase levels. Although we studied several signaling pathways that are primary targets of cell proliferation and apoptosis, there could still be other signaling pathways that are involved. Because the actions of irisin may be different between *in vitro* and *in vivo*, future work is needed to determine the physiological effects of irisin in mice and humans.

Author Contributions

Conceived and designed the experiments: DT. Performed the experiments: HS FW Yuan Zhang Yuzhu Zhang F. Wang MJ. Analyzed the data: HS F. Wu DT ZW MZ. Contributed reagents/materials/analysis tools: XLW SL LY DT TC. Wrote the paper: HS F. Wu DT.

References

1. Mendizábal Y, Llorens S, Nava E (2013) Hypertension in metabolic syndrome: vascular pathophysiology. International journal of hypertension 2013.

2. Reaven GM (1988) Role of insulin resistance in human disease. Diabetes 37: 1595–1607.

3. Triggle CR, Samuel SM, Ravishankar S, Marei I, Arunachalam G, et al. (2012) The endothelium: influencing vascular smooth muscle in many ways. Canadian journal of physiology and pharmacology 90: 713–738.

4. Strasser B (2013) Physical activity in obesity and metabolic syndrome. Annals of the New York Academy of Sciences 1281: 141–159.

5. Boström P, Wu J, Jedrychowski MP, Korde A, Ye L, et al. (2012) A PGC1-[agr]-dependent myokine that drives brown-fat-like development of white fat and thermogenesis. Nature 481: 463–468.

6. Liu J-J, Wong MD, Toy WC, Tan CS, Liu S, et al. (2013) Lower circulating irisin is associated with type 2 diabetes mellitus. Journal of Diabetes and its Complications 27: 365–369.

7. Choi Y-K, Kim M-K, Bae KH, Seo H, Jeong J-Y, et al. (2013) Serum irisin levels in new-onset type 2 diabetes. Diabetes research and clinical practice 100: 96–101.

8. Wen M-S, Wang C-Y, Lin S-L, Hung K-C (2013) Decrease in irisin in patients with chronic kidney disease. PloS one 8: e64025.

9. Moon H-S, Dincer F, Mantzoros CS (2013) Pharmacological concentrations of irisin increase cell proliferation without influencing markers of neurite outgrowth and synaptogenesis in mouse H19-7 hippocampal cell lines. Metabolism 62: 1131–1136.

10. Zhang Y, Li R, Meng Y, Li S, Donelan W, et al. (2014) Irisin stimulates browning of white adipocytes through mitogen-activated protein kinase p38 MAP kinase and ERK MAP kinase signaling. Diabetes 63: 514–525.

11. Toi M, Saji S, Masuda N, Kuroi K, Sato N, et al. (2011) Ki67 index changes, pathological response and clinical benefits in primary breast cancer patients treated with 24 weeks of aromatase inhibition. Cancer science 102: 858–865.

12. Reed JC (1995) Bcl-2: prevention of apoptosis as a mechanism of drug resistance. Hematology/oncology clinics of North America 9: 451–473.

13. Brady HJ, Gil-Gómez G (1998) Molecules in focus Bax. The pro-apoptotic Bcl-2 family member, Bax. The international journal of biochemistry & cell biology 30: 647–650.

14. Novelle MG, Contreras C, Romero-Picó A, López M, Diéguez C (2013) Irisin, Two Years Later. International journal of endocrinology 2013.

15. Christiane D, Wrann JPW, Salogiannnis J (2013) Exercise induces hippocampal BDNF through a PGC-1α/FNDC5 pathway. Cell Metabolism 18: 649–659.

16. Colaianni G, Cuscito C, Mongelli T, Oranger A, Mori G, et al. (2014) Irisin Enhances Osteoblast DifferentiationIn Vitro. International Journal of Endocrinology 2014.

17. Aydin S, Kuloglu T, Aydin S (2013) Copeptin, adropin and irisin concentrations in breast milk and plasma of healthy women and those with gestational diabetes mellitus. Peptides 47: 66–70.

18. Pardo M, Crujeiras AB, Amil M, Aguera Z, Jiménez-Murcia S, et al. (2014) Association of irisin with fat mass, resting energy expenditure, and daily activity in conditions of extreme body mass index. Int J Endocrinol 2014.

19. Sanchis-Gomar F, Alis R, Pareja-Galeano H, Sola E, Victor VM, et al. (2014) Circulating irisin levels are not correlated with BMI, age, and other biological parameters in obese and diabetic patients. Endocrine 46: 674–677.

20. Sanchis-Gomar F, Alis R, Pareja-Galeano H, Romagnoli M, Perez-Quilis C (2014) Inconsistency in Circulating Irisin Levels: What is Really Happening? Horm Metab Res 46.

21. Hamilton SJ, Chew GT, Watts GF (2007) Therapeutic regulation of endothelial dysfunction in type 2 diabetes mellitus. Diabetes and Vascular Disease Research 4: 89–102.

22. Brem H, Tomic-Canic M (2007) Cellular and molecular basis of wound healing in diabetes. Journal of Clinical Investigation 117: 1219–1222.

23. Kisielewska J, Philipova R, Huang J-Y, Whitaker M (2009) MAP kinase dependent cyclinE/cdk2 activity promotes DNA replication in early sea urchin embryos. Developmental biology 334: 383–394.

24. Wang Y, Shenouda S, Baranwal S, Rathinam R, Jain P, et al. (2011) Integrin subunits alpha5 and alpha6 regulate cell cycle by modulating the chk1 and Rb/E2F pathways to affect breast cancer metastasis. Mol Cancer 10: 84.

25. Zhang Y, Zhuang Z, Meng Q, Jiao Y, Xu J, et al. (2014) Polydatin inhibits growth of lung cancer cells by inducing apoptosis and causing cell cycle arrest. Oncology letters 7: 295–301.

26. Datta SR, Brunet A, Greenberg ME (1999) Cellular survival: a play in three Akts. Genes Dev 13: 2905–2927.

27. Yamaguchi H, Wang HG (2001) The protein kinase PKB/Akt regulates cell survival and apoptosis by inhibiting Bax conformational change. Oncogene 20: 7779–7786.

17

Perforin Competent CD8 T Cells Are Sufficient to Cause Immune-Mediated Blood-Brain Barrier Disruption

Holly L. Johnson[1,2,3], Robin C. Willenbring[2,4], Fang Jin[2], Whitney A. Manhart[2], Stephanie J. LaFrance[2], Istvan Pirko[1], Aaron J. Johnson[1,2]*

1 Department of Neurology, Mayo Clinic, Rochester, Minnesota, United States of America, 2 Department of Immunology, Mayo Clinic, Rochester, Minnesota, United States of America, 3 Neurobiology of Disease Graduate Program, Mayo Graduate School, Rochester, Minnesota, United States of America, 4 Virology and Gene Therapy Graduate Program, Mayo Graduate School, Rochester, Minnesota, United States of America

Abstract

Numerous neurological disorders are characterized by central nervous system (CNS) vascular permeability. However, the underlying contribution of inflammatory-derived factors leading to pathology associated with blood-brain barrier (BBB) disruption remains poorly understood. In order to address this, we developed an inducible model of BBB disruption using a variation of the Theiler's murine encephalomyelitis virus (TMEV) model of multiple sclerosis. This peptide induced fatal syndrome (PIFS) model is initiated by virus-specific CD8 T cells and results in severe CNS vascular permeability and death in the C57BL/6 mouse strain. While perforin is required for BBB disruption, the cellular source of perforin has remained unidentified. In addition to CD8 T cells, various innate immune cells also express perforin and therefore could also contribute to BBB disruption. To investigate this, we isolated the CD8 T cell as the sole perforin-expressing cell type in the PIFS model through adoptive transfer techniques. We determined that C57BL/6 perforin$^{-/-}$ mice reconstituted with perforin competent CD8 T cells and induced to undergo PIFS exhibited: 1) heightened CNS vascular permeability, 2) increased astrocyte activation as measured by GFAP expression, and 3) loss of linear organization of BBB tight junction proteins claudin-5 and occludin in areas of CNS vascular permeability when compared to mock-treated controls. These results are consistent with the characteristics associated with PIFS in perforin competent mice. Therefore, CD8 T cells are sufficient as a sole perforin-expressing cell type to cause BBB disruption in the PIFS model.

Editor: Roberto Furlan, San Raffaele Scientific Institute, Italy

Funding: This work is supported by the National Institutes of Health grant R01 NS060881 and Mayo Graduate School. The funders had no role in study design, data collection and analysis, decision to publish, or preparation of the manuscript.

* Email: Johnson.Aaron2@mayo.edu

Introduction

Numerous devastating neurological disorders, including multiple sclerosis, acute hemorrhagic leukoencephalitis (AHLE), dengue hemorrhagic fever, stroke, glioblastoma multiforme, epilepsy, HIV dementia, and cerebral malaria, are characterized by blood-brain barrier (BBB) disruption [1,2,3,4,5,6,7,8,9,10,11]. Although immune cells have the capacity to initiate CNS vascular permeability, there is relatively little known about how inflammation promotes BBB disruption due to a lack of suitable model systems. This currently undermines the development of therapeutic strategies to ameliorate pathology associated with these disorders. In order to define the mechanisms of BBB disruption, our lab has developed an inducible model of CNS vascular permeability using a variation of the Theiler's murine encephalomyelitis virus (TMEV) model commonly used to study multiple sclerosis [12,13,14,15]. C57BL/6 mice respond to TMEV infection by mounting an antiviral CD8 T cell response that is highly focused on the immunodominant TMEV peptide, VP2$_{121-130}$, presented in the context of the D^b class I molecule [16,17]. However, injection of this immunodominant peptide 7 days post-TMEV infection results in increased astrocyte activation, alteration of BBB tight junctions, severe CNS vascular permeability, and morbidity

within 48 hours. This peptide induced fatal syndrome (PIFS) is dependent on virus-specific CD8 T cells and perforin expression [12,18].

Perforin is a pore forming protein that plays an important role in controlling viral infections and tumors [19]. Perforin has also been shown to play a critical role in an inducible mouse model of seizures, as mice deficient in perforin displayed reduced BBB disruption [6]. When analyzing the effector functions of CD8 T cells in our PIFS model system, we found that perforin, but not Fas ligand, was required for pathology associated with PIFS to develop. In these experiments, we determined C57BL/6 perforin$^{-/-}$ mice are resistant to PIFS and are devoid of CNS vascular permeability as measured by magnetic resonance imaging (MRI) analysis and leakage of FITC-albumin into the CNS parenchyma. Astrocyte activation, as measured by glial fibrillary acidic protein (GFAP) expression, was also found to be dependent on perforin expression in the PIFS model. Events indicative of BBB disruption are dependent on perforin expression [18]. However, the cellular source of perforin required for promoting BBB disruption is unknown. In addition to CD8 T cells, natural killer (NK) cells and γδ T cells express perforin and have been shown to use perforin-mediated cytotoxicity during viral infections [20,21,22]. Neutrophils have also recently been shown to express perforin to regulate immune responses in allergic contact dermatitis [23]. Therefore, while we have previously demonstrated that both CD8 T cells and perforin are critical factors causing BBB disruption, it remained unknown the extent other perforin-expressing immune cell types assisted in the development of PIFS.

Since PIFS is initiated by class I-restricted virus antigen, we hypothesized that CD8 T cells directly use perforin to cause BBB disruption independent of other immune cell types. We tested this hypothesis using adoptive transfer techniques to isolate the CD8 T cell as the sole perforin-expressing cell type in the PIFS model. After reconstituting perforin$^{-/-}$ mice with perforin competent CD8 T cells, mice were intracranially infected with TMEV and administered either PIFS-inducing VP2$_{121-130}$ peptide or mock E7 peptide 7 days post-infection. Mice were then evaluated for activation of astrocytes, disruption of the tight junction organization, and CNS vascular permeability in order to determine whether perforin competent CD8 T cells alone are sufficient to cause BBB disruption.

Materials and Methods

Animals and ethics statement

C57BL/6 perforin deficient male mice (strain #002407) were obtained from Jackson laboratories (Bar Harbor, ME) at 5 weeks of age. C57BL/6-Tg(UBC-GFP) (strain #004353) male mice were bred in-house at Mayo Clinic. C57BL/6 Ly5.1 male mice were provided by Dr. Adam Schrum, Mayo Clinic. All experiments were approved by the Institutional Animal Care and Use Committee of Mayo Clinic (protocol number A57913).

Adoptive transfer

Spleens of GFP+ or Ly5.1+ mice were removed and strained through a nylon mesh 100 μm filter. CD8+ cells were purified from the resulting lymphocyte population using MACS LS cell purification columns (Miltenyi Biotec, Auburn, CA) according to the manufacturer's protocol. Positive cell sorting was used to obtain GFP+ CD8+ cells and negative cell sorting was used to obtain Ly5.1+ CD8+ cells. Both methods resulted in ~90% purity as determined by flow cytometric analysis (data not shown). C57BL/6 perforin$^{-/-}$ mice were irradiated with 400 rads of irradiation, given one day to recover, and then intravenously injected with 10^7 Ly5.1+ CD8+ or GFP+ CD8+ splenocytes (Figure 1A).

Induction of CNS vascular permeability using the PIFS model

CNS vascular permeability was induced using the PIFS model as previously described [12,18,24]. Briefly, all mice were intracranially infected with 2 x 10^6 PFU Daniel's strain of TMEV. Seven days post-TMEV infection, during the peak of CD8 T cell expansion in the brain, mice were intravenously administered 0.1 mg VP2$_{121-130}$ (FHAGSLLVFM) peptide or mock control E7 (RAHYNIVTF) peptide (Genescript) (Figure 1A).

Injection of rhodamine-dextran or FITC-albumin to assess permeability

Mice were given an intravenous injection of 10 mg rhodamine B isothiocyanate-dextran (Sigma #R9379) or 10 mg FITC-albumin (Sigma #A9771) 23 hours after VP2$_{121-130}$ or E7 peptide administration. Brains were harvested at 24 hours.

Flow Cytometry

Brain-infiltrating lymphocytes were isolated from the left hemisphere through collagenase digestion and a percoll gradient as previously described [25]. Isolated brain lymphocytes were then incubated with anti-CD45 PerCP (BD Biosciences, 557235) and anti-CD8 PE-Cy7 (BD Biosciences, 552877) for 40 minutes. Cells were washed twice with FACS buffer and then fixed in 4% paraformaldehyde and 1X PBS. Samples were analyzed on a BD LSRII flow cytometer (BD Biosciences) using FACS Diva software (BD Biosciences).

Rhodamine-dextran permeability assay

Mice were intravenously injected with 10 mg rhodamine B isothiocyanate-dextran (Sigma #R9379) 23 hours after VP2$_{121-130}$ or E7 peptide administration. Brains were harvested one hour later and frozen on aluminum foil on dry ice. The right hemisphere was homogenized with radioimmunoprecipitation assay (RIPA) buffer [10 mmol/L Tris, 140 mmol/L NaCl, 1% Triton X-100, 1% Na dexycholate, 0.1% SDS and protease inhibitor cocktail (Pierce #78410) pH 7.5] and centrifuged for 10 minutes at 10,000 rpm. Protein concentration was assessed using the BCA protein assay (Pierce #23223) and then samples were normalized for protein content. Homogenates were read on a fluorescent plate reader at 540 nm excitation and 625 nm emission to detect rhodamine-dextran leakage into the brain. Data were collected using SpectraMax software (Molecular Devices).

Confocal Microscopy

Fresh frozen coronal slices from mice injected with FITC-albumin, were cut on a cryostat and placed onto positively charged slides. Slides were washed twice with PBS and then fixed in 3% paraformaldehyde for 15 minutes. After the fixation step, slides were rinsed three times in PBS and then incubated for 1 hour in 5% normal goat serum + 0.5% Igepal CA-630 (Sigma I3021) in PBS. To determine the percentage of CNS-infiltrating Ly5.1+ CD8 T cells, slides were incubated with the following primary antibodies at a concentration of 1:100 overnight at room temperature: anti-mouse CD45.1 PE (eBioscience, 12-0453-82) and rat anti-mouse CD8 (AbD Serotec, MCA2694). The following secondary antibodies were used at a concentration of 1:250 for 1 hour at room temperature: AlexaFluor 532 goat anti-mouse IgG (Invitrogen, A-11002) and AlexaFluor 647 goat anti-rat IgG (Invitrogen, A21247). To visualize the tight junction organization, slides were incubated with the following primary antibodies at a concentration of 1:100 overnight at room temperature: mouse anti-claudin-5 (Invitrogen, 35–2500) and rabbit anti-occludin (Invitrogen, 71–1500). The following secondary antibodies were then added at a concentration of 1:250 for 1 hour at room temperature: AlexaFluor 532 goat anti-mouse IgG (Invitrogen, A-11002) and AlexaFluor647 goat anti-rabbit IgG (Invitrogen, A-21244). To assess astrocyte activation, slides were incubated with mouse anti-GFAP-Cy3 (Sigma, 106K4775) at a concentration of 1:500 overnight at room temperature. All slides were rinsed 5 times in PBS before and after incubation with secondary antibodies. This was followed by Hoechst staining at a concentration of 1:500 in PBS for 5 minutes. Slides were then rinsed 5 times in PBS, dried, and covered with Vectashield medium (Vector lab, H-1000). Images were acquired using a Leica (Germany) DM 2500 confocal microscope equipped with a 63x oil immersion objective (numerical aperture 1.30). All images were collected at room temperature using Type F immersion liquid (Leica Microsystems) and analyzed using LAS AF stimulator AF

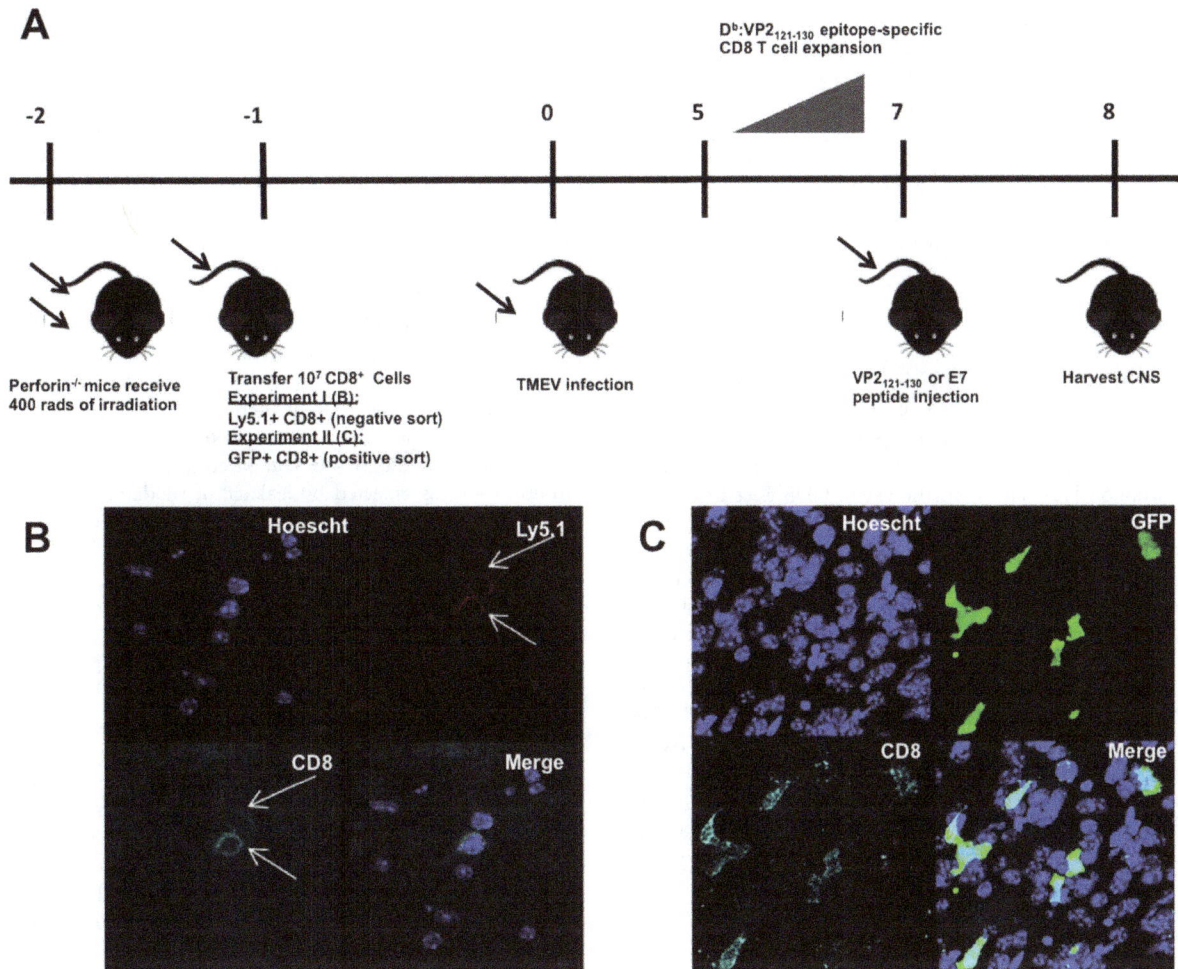

Figure 1. Adoptive transfer of CD8+ cells successfully migrate to brain post induction of PIFS. (A) Schematic illustrating the adoptive transfer of sorted perforin competent CD8+ cells into perforin$^{-/-}$ mice followed by induction of PIFS. C57BL/6 perforin$^{-/-}$ mice were irradiated with 400 rads of irradiation and then intravenously injected with 10^7 GFP+ CD8+ splenocytes (n = 10) or 10^7 Ly5.1+ CD8+ splenocytes (n = 10). Mice were intracranially infected with TMEV on the following day. PIFS-inducing VP2$_{121-130}$ peptide or mock E7 control peptide were intravenously administered on day 7, during the peak of CD8 T cell expansion. MRI analysis was performed on the following day to visualize the extent of CNS vascular permeability and then the CNS was harvested for additional assays. Both negative (Experiment I) and positive (Experiment II) sort experiments yielded high purity of CD8+ T cell transfer. (B) Representative confocal microscopic images illustrating co-localization of Ly5.1 and CD8 protein (Experiment I). CNS-infiltrating Ly5.1+ cells colocalized with CD8+ cells 85.7% of the time as measured by confocal analysis. (C) Confocal microscopy showing of representative brain tissue slice showing successful transfer of GFP+ CD8+ T cells (Experiment II). Purity of transfer was analyzed via flow cytometry, 98.0%+− 0.5% of cells that were CD8 positive and GFP positive.

6000 acquisition software. All images shown in manuscript are at 63x magnification.

MRI Acquisition

A Bruker Avance II 7 Tesla vertical bore small animal MRI system (Bruker Biospin) was used for image acquisition to evaluate CNS vascular permeability as previously described [9,26]. Briefly, inhalation anesthesia was induced and maintained using 3–4% isoflurane. An MRI compatible vital sign monitoring system (Model 1030, SA Instruments, Inc., Stony Brook NY) was used to monitor respiratory rate during acquisition. Gadolinium was intraperitoneally administered to mice using weight based dosing of 100 mg/kg. After a standard delay of 15 minutes, a volume acquisition T1-weighted spin echo sequence was used (TR-150 ms, TE = 8 ms, FOV: 32 mm x 19.2 mm x 19.2 mm, Matrix: 160 x 96 x 96, number of averages = 1) to obtain T1-weighted images.

Image Analysis

3D volumetric analysis was performed as previously described [25,27,28,29]. Briefly, Analyze 11.0 software (Biomedical Imaging Resource, Mayo Clinic) was used to quantify the 3D volume of CNS vascular permeability. The 3D volume extractor tool was used to extract brains from gadolinium-enhanced T1-weighted images. The 3D ROI tool was then used to define areas of gadolinium leakage using semi-automated methods. The volume of these areas, which corresponds to the volume of gadolinium leakage, was calculated using the 3D sampling tool. 3D object rendering using the volume rendering tool was performed to visualize the identified volumes of contrast enhancement. All 3D volumes were standardized to the mean of control (mice receiving E7 peptide on day 7) within an experiment and plotted as individual mice, shown as Ratio of Gadolinium Volume.

Statistical Analysis

Mean and standard error values were calculated using SigmaStat software (SYSTAT Software Inc). GraphPad Prism Software was used to construct graphs with standard error bars. A Student's t-test was performed using SigmaStat to evaluate CNS vascular permeability, as detected by leakage of rhodamine-dextran into the CNS, and FITC intensity from confocal microscopic images. A Welch test was used to evaluate 3D volumetric analysis of vascular leak visible on T1-weighted MRI scans.

Results

Adoptively transferred perforin competent CD8 T cells induce CNS vascular permeability

Model and adoptive transfer purity. CD8 T cells that recognize the immunodominant D^b:$VP2_{121-130}$ epitope initiate BBB disruption [12]. However, the extent CD8 T cell expressed perforin contributes to CNS vascular permeability remains unknown. Therefore, we developed an adoptive transfer approach to isolate the CD8 T cell as the sole perforin-expressing cell type (Figure 1A). CD8+ cells were purified from the spleens of Ly5.1+ perforin+/+ mice using negative cell sort (Experiment I), or GFP+ perforin$^{+/+}$ mice using positive cell sorting (Experiment II) resulting in ~90% purity as determined by flow cytometric analysis (data not shown). Recipient C57BL/6 perforin$^{-/-}$ mice were irradiated with 400 rads of irradiation, given one day to recover, and then intravenously injected with 10^7 sorted CD8+ splenocytes. The following day mice were intracranially infected with TMEV. PIFS-inducing $VP2_{121-130}$ peptide or mock E7 control peptide were intravenously administered on day 7, during the peak of CD8 T cell expansion (Figure 1). Gadolinium-enhanced T1-weighted MRI was performed on the following day to visualize CNS vascular permeability. Immediately following MRI, the CNS was harvested to be further analyzed by confocal microscopy, flow cytometry and CNS vascular permeability was measured by fluorescent molecule diffusion across the BBB.

To determine the percentage of CNS-infiltrating Ly5.1+ CD8 T cells post-induction of PIFS, tissue slices were stained with antibody to Ly5.1 and CD8 and observed under a confocal microscope. Comparable numbers of CNS-infiltrating Ly5.1+ CD8 T cells were observed in both $VP2_{121-130}$ peptide (n = 6) and E7 peptide-treated (n = 4) groups, with the average being 85.7% (Figure 1B, data not shown). Flow cytometric analysis was used to determine the percentage of CNS-infiltrating GFP+ CD8 T cells post-induction of PIFS. Isolated CNS-infiltrating lymphocytes were stained with antibody to CD45 and CD8. The CD45^{hi} population was then gated and analyzed for the percentage of GFP+ CD8 T cells. Comparable numbers of CNS-infiltrating CD8+ cells as a percentage of GFP+ cells were observed in both $VP2_{121-130}$ peptide (n = 6) and E7 peptide-treated (n = 4) groups, with the average being 98.0% ± 0.5% (data not shown). Furthermore, tissue slices were stained with antibody to CD8 and GFP and observed using confocal microscopy and defined to be high purity (Figure 1C).

CNS permeability measured by small animal MRI and fluorescent molecule diffusion. In order to quantitatively evaluate the full volume of CNS vascular leak in animals undergoing PIFS, where CD8 T cells were the sole perforin-expressing cell type, we performed volumetric analysis of 3D gadolinium-enhanced T1-weighted MRI scans using Analyze 11.0 software developed by Mayo Clinic's Biomedical Imaging Resource [30,31]. Quantification of the 3D volume of gadolinium leakage from vasculature revealed that $VP2_{121-130}$ peptide-treated

mice reconstituted with perforin competent CD8 T cells had increased CNS vascular permeability (Figure 2 B,D) when compared to mock E7 peptide-treated controls (Figure 2 A,C) (p<0.05, Figure 2E). CD8+ cells were selected using a negative (Figure 2 A,B) or positive (Figure 2 C,D) sort method. Importantly, an additional $VP2_{121-130}$ peptide-treated mouse, not included in this data set, was moribund and unable to be scanned. Post-mortem analysis revealed that this mouse had the highest transfer of CD8 T cells (data not shown).

Another group of animals was injected with rhodamine-dextran 23 hours post-administration of mock E7 peptide or $VP2_{121-130}$ peptide to induce PIFS. Rhodamine-dextran was allowed to circulate for one hour. Brains were then harvested and the right hemisphere was used to assess CNS vascular permeability. In concordance with the T1-weighted MRI analysis, a significant increase in vascular permeability was observed in the $VP2_{121-130}$ peptide-treated group (n = 6) compared to the E7 peptide-treated group (n = 4), as detected by leakage of rhodamine-dextran into the CNS using a fluorescent plate reader (A_{540}) (p<0.05) (Figure 2F).

Increased astrocyte GFAP expression in perforin$^{-/-}$ mice reconstituted with perforin competent CD8 T cells

Activation of astrocytes, as measured by GFAP expression, co-localizes with CNS vascular permeability in C57BL/6 mice induced to undergo PIFS [18]. To determine the extent astrocytes were activated in perforin$^{-/-}$ mice reconstituted with perforin competent CD8 T cells, we stained for astrocyte expression of GFAP and performed confocal microscopy. We found that mice treated with $VP2_{121-130}$ peptide displayed an increase in astrocyte size consistent with astrogliosis (Figure 3B), when compared to E7 peptide-treated mice (Figure 3A). We also observed colocalization of increased astrocyte activation with CNS vascular permeability. This is consistent with previous studies in which PIFS was evaluated in perforin competent wild-type C57BL/6 mice [18]. In this study, in accordance with previous studies, there is a clear increase in astrocyte activation in areas with vascular leak as compared to unaffected tissue (Figure 3B).

Disorganized tight junction proteins colocalize with increased CNS vascular permeability in perforin$^{-/-}$ mice reconstituted with perforin competent CD8 T cells

The increased vascular permeability in perforin$^{-/-}$ mice reconstituted with perforin competent CD8 T cells prompted additional analysis of focal areas of CNS vascular permeability and the state of the BBB tight junction proteins. We determined that mock E7 peptide-treated C57BL/6 perforin$^{-/-}$ mice reconstituted with perforin competent CD8 T cells display preservation of vascular integrity, as shown by lack of FITC-albumin leakage into the brain parenchyma from the brain vasculature, and linear organization of BBB tight junction proteins claudin-5 and occludin (Figure 4A). However, C57BL/6 perforin$^{-/-}$ mice reconstituted with perforin competent CD8 T cells and administered $VP2_{121-130}$ peptide to induce PIFS presented with a loss of linear organization of tight junction proteins claudin-5 and occluding in the microvessels. Disorganization of BBB tight junctions colocalizes with FITC-albumin leakage from brain vasculature (Figure 4B). FITC intensity from confocal microscopic images was then quantified using ImageJ software in brain tissue slices. Two random representative images from each animal were analyzed. FITC intensity values were normalized through subtraction of background values. We determined using this method that $VP2_{121-130}$ peptide-treated mice (n = 6) displayed a significant

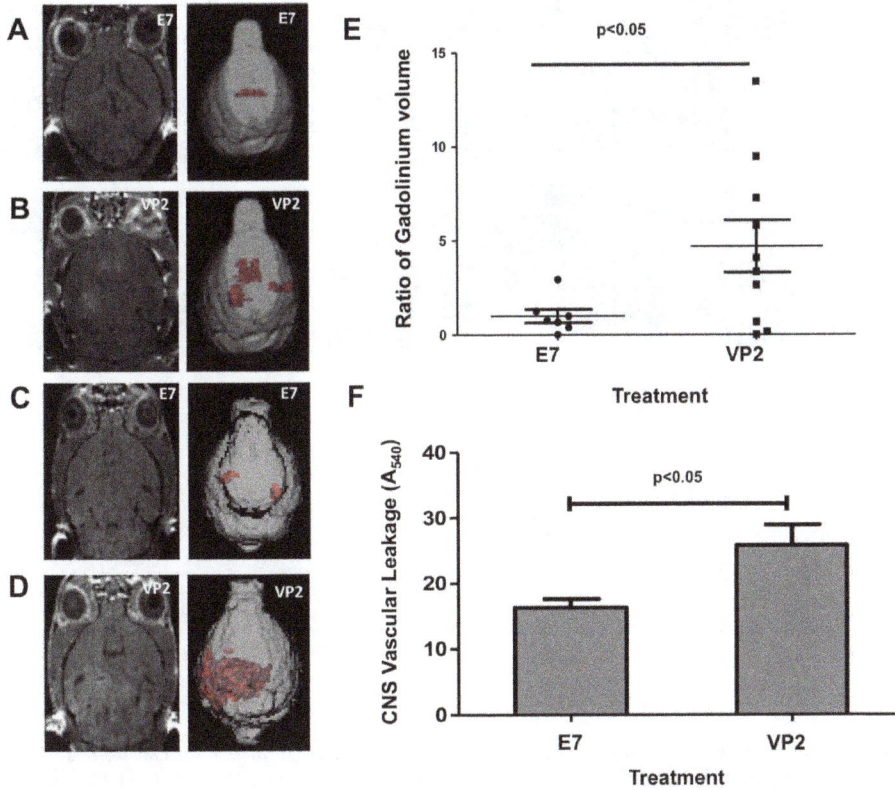

Figure 2. CD8 T cells as sole perforin wielding source are sufficient to induce CNS vascular permeability. The extent of CNS vascular permeability is illustrated by raw image and 3D transparency rendering of gadolinium-enhancing areas from T1-weighted MRI scans in mice administered (A,C) mock E7 control peptide or (B,D) $VP2_{121-130}$ peptide. MRI images are separated to show both methods of CD8+ T cell selection (A,B) negative sort or (C,D) positive sort to isolate and transfer CD8+ cells. (E) Quantification of the 3D volume of vascular leakage from all mice (both negative and positive sort) revealed that $VP2_{121-130}$ peptide-treated mice having CD8 T cells as the sole perforin-expressing cell type have a significantly more amount of gadolinium leakage (p<0.05) (n = **5,9** per group). (F) A significant increase in CNS vascular permeability detectable by intravenously injected rhodamine dextran accumulation in brain homogenates was observed in $VP2_{121-130}$ peptide-treated animals (n = 6) compared to mock E7 peptide-treated animals (n = 4) (p<0.05). Results are depicted as mean ± SEM.

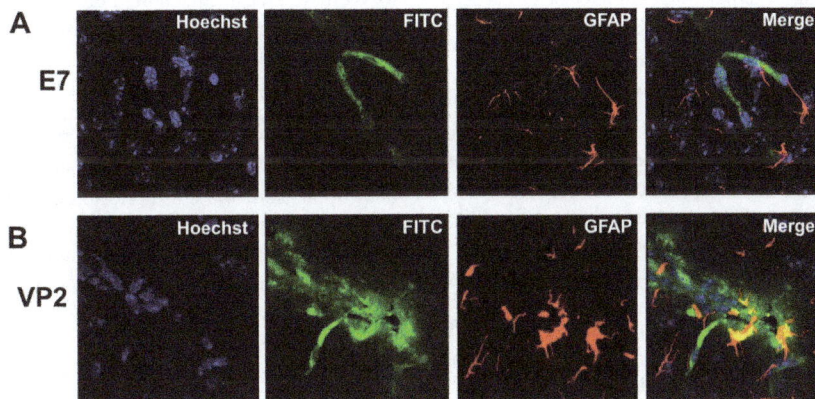

Figure 3. Increased astrocyte activation colocalizes with CNS vascular permeability post-induction of PIFS in perforin$^{-/-}$ mice reconstituted with perforin competent CD8 T cells. Representative confocal microscopic images illustrating astrocyte expression of GFAP in (A) E7 peptide-treated (n = 4) and (B) $VP2_{121-130}$ peptide-treated (n = 6) perforin$^{-/-}$ mice reconstituted with perforin competent CD8 T cells. Mice treated with (B) $VP2_{121-130}$ peptide displayed increased astrocyte activation, as measured by heightened GFAP expression, in areas of CNS vascular permeability when compared to (A) mock E7 peptide-treated negative control mice. CNS permeability is determined by leakage of intravenously administered FITC-albumin.

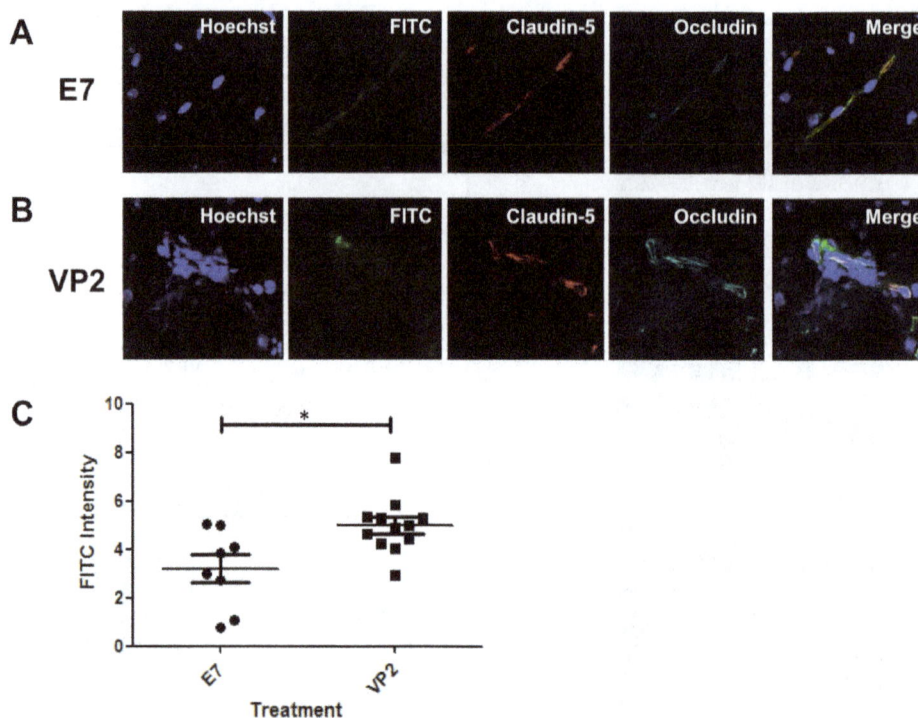

Figure 4. Disorganization of tight junction proteins and increased CNS vascular permeability can be induced in perforin$^{-/-}$ mice reconstituted with perforin competent CD8 T cells. Confocal microscopic images from a representative (A) mock E7 peptide-treated C57BL/6 perforin$^{-/-}$ mouse reconstituted with perforin competent CD8 T cells depict a preservation of vascular integrity and intact tight junction proteins claudin-5 and occludin. (B) VP2$_{121-130}$ peptide-treated C57BL/6 perforin$^{-/-}$ mice reconstituted with perforin competent CD8 T cells display degradation of tight junction proteins claudin-5 and occludin in areas of CNS vascular permeability as seen through FITC-albumin leakage. (C) FITC intensity in brain tissue from two representative confocal microscopic images of each animal was quantified using ImageJ software. VP2$_{121-130}$ peptide-treated mice (n = 6) displayed a significant higher intensity of FITC-albumin in brain parenchyma when compared to E7 peptide-treated mice (n = 4) (p<0.05).

higher intensity of FITC-albumin when compared to E7 peptide-treated mice (n = 4) (p<0.05) (Figure 4C). The colocalization of disorganized tight junction proteins with CNS vascular permeability in this experiment is consistent with previous studies performed in perforin competent wild-type C57BL/6 mice [18,25]. Vascular leak and disorganization of BBB tight junction proteins is evident with transfer of perforin competent CD8 T cells into a perforin-/- mouse and induced to undergo PIFS.

Discussion

Investigating the underlying molecular mechanisms leading to pathology associated with BBB disruption is of critical importance for the development of therapeutic approaches to treat diseases characterized by CNS vascular permeability. Using the PIFS model system, our lab has previously demonstrated critical roles for CD8 T cells and perforin in promoting activation of astrocytes, loss of linear organization of the tight junction, and extensive CNS vascular permeability. We have also shown that inhibiting the functions of these critical players improves pathology and survival [12,18]. Furthermore, we have demonstrated that other molecular players implicated in causing BBB disruption, such as GR-1+ neutrophils, CD4 T cells, TNF-α, IFN-γ, LTβR, and IL-1, do not contribute to lethality in CD8 T cell-initiated BBB disruption [16,25]. However, whether CD8 T cells are the cellular source of perforin required for promoting BBB disruption was not known.

In addition to CD8 T cells, NK cells, γδ T cells, and neutrophils also use perforin as an effector mechanism. [20,21,22,23,32,

33,34]. Mice deficient in NK cells displayed a higher degree of inflammation and demyelination after induction of experimental autoimmune encephalomyelitis (EAE) [32,33,34]. Depletion of NK cells has also been shown to result in more severe encephalitis in mice post-TMEV infection [21]. Additionally, mice deficient in γδ T cells exhibited increased viral load and were more susceptible to West Nile Virus infection. These mice also displayed decreased levels of intracellular perforin in splenocytes post-infection with West Nile Virus and had a lower cytotoxicity compared to wild-type mice [22]. Furthermore, neutrophils have been shown to use perforin to regulate CD8 T cell infiltration into the skin to mediate allergic contact dermatitis [23]. This raises the question as to whether other molecular players such as NK cells, γδ T cells, and neutrophils contribute to perforin-dependent CD8 T cell-mediated BBB disruption during neuroinflammation.

In this study, we tested our hypothesis that CD8 T cells employ perforin to cause BBB disruption independent of other immune cell types. We developed an adoptive transfer approach to isolate perforin competent CD8 T cells and determine if they are sufficient to cause activation of astrocytes, disorganization of BBB tight junction proteins, and CNS vascular permeability. Using this adoptive transfer approach, we found that perforin$^{-/-}$ mice reconstituted with perforin competent CD8 T cells and induced to undergo PIFS exhibited a significant increase in CNS vascular permeability, measured by fluorescent molecule diffusion across the BBB, compared to mock E7 peptide-treated controls. This prompted a more comprehensive analysis using small animal MRI and confocal microscopy. Quantification of the 3D volume of gadolinium leakage from vasculature visible on T1-weighted MRI scans illustrated that PIFS-inducing VP2$_{121-130}$ peptide-treated mice reconstituted with CD8 T cells as the sole perforin-expressing cell type displayed significant increase in CNS vascular permeability when compared to mock E7 peptide-treated controls. This analysis showed that perforin competent CD8 T cells are sufficient to induce CNS vascular permeability.

We then used confocal microscopy to analyze BBB disruption in more detail by examining astrocyte activation, the organization of BBB tight junction proteins claudin-5 and occludin, and focal areas of vascular permeability. This analysis demonstrated that perforin competent CD8 T cells are sufficient to cause increased astrocyte activation and disorganization of the tight junctions in areas of CNS vascular permeability.

Studies using the EAE model have also put forward a role for astrocytes in causing disorganization of BBB tight junction proteins and ensuing CNS vascular permeability. These studies have proposed a mechanism in which CD4 T cells induce astrocytes to release vascular endothelial growth factor (VEGF), resulting in BBB disruption [35]. However, in contrast to EAE, CD4 T cells do not contribute to lethality in the PIFS model [36]. Furthermore, while we have also demonstrated a critical role for VEGF in promoting BBB disruption, we found that neurons were the major source of VEGF expression during CNS vascular permeability [24]. However, since we have observed increased astrocyte activation colocalizing with CNS vascular permeability, there is still a potential role for astrocytes in the PIFS model. It is possible that increased astrocyte activation causes a permeable state or is a consequence of BBB disruption. This will be a topic of further investigation.

This study has expanded our knowledge on the mechanism of BBB disruption by illustrating that perforin competent CD8 T cells are sufficient to cause traits associated with immune-mediated BBB disruption. It remains necessary to investigate the functional mechanism used by perforin to cause BBB disruption. In experimental cerebral malaria, granzyme B has been defined as a critical factor in promoting fatal cerebral pathology [37]. The extent granzyme B contributes to BBB disruption in PIFS remains to be defined. Furthermore, the complete mechanism of how perforin enables delivery of granzymes to cause apoptosis of target cells also remains to be defined. It was originally thought that perforin created pores on target cells to allow entry of granzymes, resulting in cell death [38]. Recently it has been suggested that the perforin pore actively, and selectively, passes cationic molecules, such as granzyme B, more efficiently that anionic and neutral molecules [39]. Alternative models propose that granzymes are endocytosed independent of perforin. This is followed by perforin acting on the endosomal membrane to form pores, causing the release of granzymes, which in turn promotes apoptosis of target cells [40,41]. A recent proposed mechanism involves a combination of these two pathways. In this third model, perforin forms pores in the plasma membrane of the target cell. This results in an influx of Ca^{2+} that triggers repair of the plasma membrane. Granzymes and perforin are then endocytosed together into large endosomes, and perforin acts on the endosomal membrane to release granzymes into the cytosol. However, the mechanism of how perforin causes release of granzymes from the endosome remains unclear [42].

Perforin is commonly defined as providing cytolytic activity against target cells [38]. Recently new, non-classical, mechanisms have been put forward that perforin can function in non-cytolytic pathways. Perforin has been identified, along with granzyme, as having a noncytolytic role in controlling the reactivation of HSV-1 neuronal infections without inducing cytotoxicity [43]. Additionally, in mouse model of obesity related insulin resistance and visceral adipose tissue (VAT), a lack of perforin showed reduced insulin sensitivity and changed the inflammation status within the VAT lesions, suggesting an immunoregulatory role for perforin in this disease state. Importantly, in this study, the formation of crown like structures, indicative of adipocyte death, was not changed in mice lacking perforin, further emphasizing a non-cytolytic role of perforin [44]. Other noncytolytic roles for perforin extend to contributing to CD8 T cell activation during *arenavirus* infection and regulating antigen presentation of dendritic cells (DCs) [45,46]. Previous work, in our lab, shows caspase-3 cleavage is not present until after BBB disruption has already occurred [18]. Given these new roles for perforin and lack of evidence for apoptosis, it is possible that during BBB disruption, perforin is acting in a non-classical mechanism. However, the exact mechanism by which perforin wielding CD8 T cells can induce BBB disruption is yet to be defined.

The contribution of molecules delivered by perforin beyond granzyme B also needs to be evaluated. For example, orphan granzymes have recently been proposed to play a role in pathogen clearance and cell-mediated death [47]. Nevertheless, based on the findings put forward in this manuscript, investigating mechanisms of perforin-dependent cytotoxicity are important to the development of novel therapeutic strategies to ameliorate pathology associated with BBB disruption in several devastating neurological disorders. The demonstration that CD8 T cells can serve as a sole source of perforin to induce BBB disruption will greatly aid in this process.

Acknowledgments

We would like to thank Dr. Adam Schrum (Mayo Clinic, Rochester, MN) for providing us with Ly5.1 congenic mice and Michael Hansen (Mayo Clinic, Rochester, MN) for technical assistance.

Author Contributions

Conceived and designed the experiments: HLJ RCW IP AJJ. Performed the experiments: HLJ RCW WAM FJ SJL. Analyzed the data: HLJ RCW WAM IP AJJ. Wrote the paper: HLJ RCW AJJ.

References

1. Brown H, Hien TT, Day N, Mai NT, Chuong LV, et al. (1999) Evidence of blood-brain barrier dysfunction in human cerebral malaria. Neuropathology and applied neurobiology 25: 331–340.

2. Eugenin EA, Osiecki K, Lopez L, Goldstein H, Calderon TM, et al. (2006) CCL2/monocyte chemoattractant protein-1 mediates enhanced transmigration of human immunodeficiency virus (HIV)-infected leukocytes across the blood-brain barrier: a potential mechanism of HIV-CNS invasion and NeuroAIDS. The Journal of neuroscience : the official journal of the Society for Neuroscience 26: 1098–1106.

3. Huber JD, Egleton RD, Davis TP (2001) Molecular physiology and pathophysiology of tight junctions in the blood-brain barrier. Trends in neurosciences 24: 719–725.

4. Lacerda-Queiroz N, Rodrigues DH, Vilela MC, Rachid MA, Soriani FM, et al. (2012) Platelet-activating factor receptor is essential for the development of experimental cerebral malaria. The American journal of pathology 180: 246–255.

5. Marchi N, Granata T, Ghosh C, Janigro D (2012) Blood-brain barrier dysfunction and epilepsy: pathophysiologic role and therapeutic approaches. Epilepsia 53: 1877–1886.

6. Marchi N, Johnson AJ, Puvenna V, Johnson HL, Tierney W, et al. (2011) Modulation of peripheral cytotoxic cells and ictogenesis in a model of seizures. Epilepsia 52: 1627–1634.

7. Medana IM, Turner GD (2006) Human cerebral malaria and the blood-brain barrier. International journal for parasitology 36: 555–568.

8. Minagar A, Alexander JS (2003) Blood-brain barrier disruption in multiple sclerosis. Multiple sclerosis 9: 540–549.

9. Pirko I, Suidan GL, Rodriguez M, Johnson AJ (2008) Acute hemorrhagic demyelination in a murine model of multiple sclerosis. Journal of neuroinflammation 5: 31.

10. Schneider SW, Ludwig T, Tatenhorst L, Braune S, Oberleithner H, et al. (2004) Glioblastoma cells release factors that disrupt blood-brain barrier features. Acta Neuropathologica 107: 272–276.

11. Talavera D, Castillo AM, Dominguez MC, Gutierrez AE, Meza I (2004) IL8 release, tight junction and cytoskeleton dynamic reorganization conducive to permeability increase are induced by dengue virus infection of microvascular endothelial monolayers. The Journal of general virology 85: 1801–1813.

12. Johnson AJ, Mendez-Fernandez Y, Moyer AM, Sloma CR, Pirko I, et al. (2005) Antigen-specific CD8+ T cells mediate a peptide-induced fatal syndrome. Journal of immunology 174: 6854–6862.

13. McDole J, Johnson AJ, Pirko I (2006) The role of CD8+ T-cells in lesion formation and axonal dysfunction in multiple sclerosis. Neurological research 28: 256–261.

14. Rodriguez M, Oleszak E, Leibowitz J (1987) Theiler's murine encephalomyelitis: a model of demyelination and persistence of virus. Critical reviews in immunology 7: 325–365.

15. Johnson HL, Jin F, Pirko I, Johnson AJ (2013) Theiler's murine encephalomyelitis virus as an experimental model system to study the mechanism of blood-brain barrier disruption. Journal of neurovirology.

16. Johnson AJ, Njenga MK, Hansen MJ, Kuhns ST, Chen LP, et al. (1999) Prevalent class I-restricted T-cell response to the Theiler's virus epitope D-b : VP2(121-130) in the absence of endogenous CD4 help, tumor necrosis factor alpha, gamma interferon, perforin, or costimulation through CD28. Journal of Virology 73: 3702–3708.

17. Johnson AJ, Upshaw J, Pavelko KD, Rodriguez M, Pease LR (2001) Preservation of motor function by inhibition of CD8+ virus peptide-specific T cells in Theiler's virus infection. FASEB journal : official publication of the Federation of American Societies for Experimental Biology 15: 2760–2762.

18. Suidan GL, McDole JR, Chen Y, Pirko I, Johnson AJ (2008) Induction of blood brain barrier tight junction protein alterations by CD8 T cells. PloS one 3: e3037.

19. Janigro D (2012) Are you in or out? Leukocyte, ion, and neurotransmitter permeability across the epileptic blood-brain barrier. Epilepsia 53: 26–34.

20. Bukowski JF, Woda BA, Habu S, Okumura K, Welsh RM (1983) Natural killer cell depletion enhances virus synthesis and virus-induced hepatitis in vivo. Journal of immunology 131: 1531–1538.

21. Paya CV, Patick AK, Leibson PJ, Rodriguez M (1989) Role of natural killer cells as immune effectors in encephalitis and demyelination induced by Theiler's virus. Journal of immunology 143: 95–102.

22. Wang T, Scully E, Yin Z, Kim JH, Wang S, et al. (2003) IFN-gamma-producing gamma delta T cells help control murine West Nile virus infection. Journal of immunology 171: 2524–2531.

23. Kish DD, Gorbachev AV, Parameswaran N, Gupta N, Fairchild RL (2012) Neutrophil expression of Fas ligand and perforin directs effector CD8 T cell infiltration into antigen-challenged skin. Journal of immunology 189: 2191–2202.

24. Suidan GL, Dickerson JW, Chen Y, McDole JR, Tripathi P, et al. (2010) CD8 T cell-initiated vascular endothelial growth factor expression promotes central nervous system vascular permeability under neuroinflammatory conditions. Journal of immunology 184: 1031–1040.

25. Johnson HL, Chen Y, Jin F, Hanson LM, Gamez JD, et al. (2012) CD8 T Cell-Initiated Blood-Brain Barrier Disruption Is Independent of Neutrophil Support. Journal of immunology 189: 1937–1945.

26. Denic A, Macura SI, Mishra P, Gamez JD, Rodriguez M, et al. (2011) MRI in rodent models of brain disorders. Neurotherapeutics : the journal of the American Society for Experimental NeuroTherapeutics 8: 3–18.

27. Pirko I, Gamez J, Johnson AJ, Macura SI, Rodriguez M (2004) Dynamics of MRI lesion development in an animal model of viral-induced acute progressive CNS demyelination. NeuroImage 21: 576–582.

28. Pirko I, Johnson AJ, Chen Y, Lindquist DM, Lohrey AK, et al. (2011) Brain atrophy correlates with functional outcome in a murine model of multiple sclerosis. NeuroImage 54: 802–806.

29. Pirko I, Nolan TK, Holland SK, Johnson AJ (2008) Multiple sclerosis: pathogenesis and MR imaging features of T1 hypointensities in a [corrected] murine model. Radiology 246: 790–795.

30. Robb RA (1999) 3-D visualization in biomedical applications. Annual review of biomedical engineering 1: 377–399.

31. Robb RA (2001) The Biomedical Imaging Resource at Mayo Clinic. Ieee Transactions on Medical Imaging 20: 854–867.

32. Hao J, Liu R, Piao W, Zhou Q, Vollmer TL, et al. (2010) Central nervous system (CNS)-resident natural killer cells suppress Th17 responses and CNS autoimmune pathology. The Journal of experimental medicine 207: 1907–1921.

33. Xu W, Fazekas G, Hara H, Tabira T (2005) Mechanism of natural killer (NK) cell regulatory role in experimental autoimmune encephalomyelitis. Journal of neuroimmunology 163: 24–30.

34. Zhang B, Yamamura T, Kondo T, Fujiwara M, Tabira T (1997) Regulation of experimental autoimmune encephalomyelitis by natural killer (NK) cells. The Journal of experimental medicine 186: 1677–1687.

35. Argaw AT, Gurfein BT, Zhang Y, Zameer A, John GR (2009) VEGF-mediated disruption of endothelial CLN-5 promotes blood-brain barrier breakdown. Proceedings of the National Academy of Sciences of the United States of America 106: 1977–1982.

36. Johnson AJ, Njenga MK, Hansen MJ, Kuhns ST, Chen L, et al. (1999) Prevalent class I-restricted T-cell response to the Theiler's virus epitope Db:VP2121-130 in the absence of endogenous CD4 help, tumor necrosis factor alpha, gamma interferon, perforin, or costimulation through CD28. Journal of virology 73: 3702–3708.

37. Haque A, Best SE, Unosson K, Amante FH, de Labastida F, et al. (2011) Granzyme B expression by CD8+ T cells is required for the development of experimental cerebral malaria. Journal of immunology 186: 6148–6156.

38. Tschopp J, Masson D, Stanley KK (1986) Structural/functional similarity between proteins involved in complement- and cytotoxic T-lymphocyte-mediated cytolysis. Nature 322: 831–834.

39. Stewart SE, Kondos SC, Matthews AY, D'Angelo ME, Dunstone MA, et al. (2014) The perforin pore facilitates the delivery of cationic cargos. The Journal of biological chemistry 289: 9172–9181.

40. Froelich CJ, Orth K, Turbov J, Seth P, Gottlieb R, et al. (1996) New paradigm for lymphocyte granule-mediated cytotoxicity. Target cells bind and internalize granzyme B, but an endosomolytic agent is necessary for cytosolic delivery and subsequent apoptosis. The Journal of biological chemistry 271: 29073–29079.

41. Metkar SS, Wang B, Aguilar-Santelises M, Raja SM, Uhlin-Hansen L, et al. (2002) Cytotoxic cell granule-mediated apoptosis: perforin delivers granzyme B-serglycin complexes into target cells without plasma membrane pore formation. Immunity 16: 417–428.

42. Pipkin ME, Lieberman J (2007) Delivering the kiss of death: progress on understanding how perforin works. Current opinion in immunology 19: 301–308.

43. Knickelbein JE, Khanna KM, Yee MB, Baty CJ, Kinchington PR, et al. (2008) Noncytotoxic lytic granule-mediated CD8+ T cell inhibition of HSV-1 reactivation from neuronal latency. Science 322: 268–271.

44. Revelo XS, Tsai S, Lei H, Luck H, Ghazarian M, et al. (2014) Perforin is a Novel Immune Regulator of Obesity Related Insulin Resistance. Diabetes.

45. Lykens JE, Terrell CE, Zoller EE, Risma K, Jordan MB (2011) Perforin is a critical physiologic regulator of T-cell activation. Blood 118: 618–626.

46. Terrell CE, Jordan MB (2013) Perforin deficiency impairs a critical immunoregulatory loop involving murine CD8(+) T cells and dendritic cells. Blood 121: 5184–5191.

47. Grossman WJ, Revell PA, Lu ZH, Johnson H, Bredemeyer AJ, et al. (2003) The orphan granzymes of humans and mice. Current opinion in immunology 15: 544–552.

Permissions

The contributors of this book come from diverse backgrounds, making this book a truly international effort. This book will bring forth new frontiers with its revolutionizing research information and detailed analysis of the nascent developments around the world.

We would like to thank all the contributing authors for lending their expertise to make the book truly unique. They have played a crucial role in the development of this book. Without their invaluable contributions this book wouldn't have been possible. They have made vital efforts to compile up to date information on the varied aspects of this subject to make this book a valuable addition to the collection of many professionals and students.

This book was conceptualized with the vision of imparting up-to-date information and advanced data in this field. To ensure the same, a matchless editorial board was set up. Every individual on the board went through rigorous rounds of assessment to prove their worth. After which they invested a large part of their time researching and compiling the most relevant data for our readers.

The editorial board has been involved in producing this book since its inception. They have spent rigorous hours researching and exploring the diverse topics which have resulted in the successful publishing of this book. They have passed on their knowledge of decades through this book. To expedite this challenging task, the publisher supported the team at every step. A small team of assistant editors was also appointed to further simplify the editing procedure and attain best results for the readers.

Apart from the editorial board, the designing team has also invested a significant amount of their time in understanding the subject and creating the most relevant covers. They scrutinized every image to scout for the most suitable representation of the subject and create an appropriate cover for the book.

The publishing team has been an ardent support to the editorial, designing and production team. Their endless efforts to recruit the best for this project, has resulted in the accomplishment of this book. They are a veteran in the field of academics and their pool of knowledge is as vast as their experience in printing. Their expertise and guidance has proved useful at every step. Their uncompromising quality standards have made this book an exceptional effort. Their encouragement from time to time has been an inspiration for everyone.

The publisher and the editorial board hope that this book will prove to be a valuable piece of knowledge for researchers, students, practitioners and scholars across the globe.

List of Contributors

Vesna Tosic and Edward J. Roy
Department of Molecular and Integrative Physiology, University of Illinois at Urbana-Champaign, Urbana, Illinois, United States of America

Diana L. Thomas
Neuroscience Program, University of Illinois at Urbana-Champaign, Urbana, Illinois, United States of America

David M. Kranz
Department of Biochemistry, University of Illinois at Urbana-Champaign, Urbana, Illinois, United States of America

Jia Liu and Grant McFadden
Department of Molecular Genetics and Microbiology, University of Florida, Gainesville, Florida, United States of America

Joanna L. Shisler
Department of Microbiology, University of Illinois at Urbana-Champaign, Urbana, Illinois, United States of America

Amy L. MacNeill
Department of Pathobiology at College of Veterinary Medicine, University of Illinois at Urbana Champaign, Urbana, Illinois, United States of America

Xin Zhou, Zenghong Xu and Yan Shui
Key Laboratory of Freshwater Fisheries and Germplasm Resources Utilization, Ministry of Agriculture, Freshwater Fisheries Research Center, Chinese Academy of Fishery Sciences, Wuxi, China

Huaishun Shen, Yacheng Hu and Yuanchao Ma
Key Laboratory of Freshwater Fisheries and Germplasm Resources Utilization, Ministry of Agriculture, Freshwater Fisheries Research Center, Chinese Academy of Fishery Sciences, Wuxi, China Wuxi Fisheries College, Nanjing Agricultural University, Nanjing, China

Chunyan Li, Peng Xu and Xiaowen Sun
The Center for Applied Aquatic Genomics, Chinese Academy of Fishery Sciences, Beijing, China

Youngshang Pak
Department of Chemistry and Institute of Functional Materials, Pusan National University, Busan, Republic of Korea
Laboratory of Molecular Biology, National Cancer Institute, National Institutes of Health, Bethesda, Maryland, United States of America

Ira Pastan, Robert J. Kreitman and Byungkook Lee
Laboratory of Molecular Biology, National Cancer Institute, National Institutes of Health, Bethesda, Maryland, United States of America

Thomas Lars Andresen
Dept. of Micro- and Nanotechnology, Center for Nanomedicine and Theranostics, Technical University of Denmark, Kgs. Lyngby, Denmark

Sofie Trier
Dept. of Micro- and Nanotechnology, Center for Nanomedicine and Theranostics, Technical University of Denmark, Kgs. Lyngby, Denmark
Diabetes Research Unit,Novo Nordisk, Maaloev, Denmark

Lars Linderoth, Simon Bjerregaard and Ulrik Lytt Rahbek
Diabetes Research Unit,Novo Nordisk, Maaloev, Denmark

Enida Gjoni, Loredana Brioschi, Alessandra Cinque, Paola Giussani, Laura Riboni and Paola Viani
Department of Medical Biotechnology and Translational Medicine, Universita` di Milano, LITA Segrate, Milano, Italy

Nicolas Coant, Christophe Magnan and Hervé Le Stunff
Unité Biologie Fonctionnelle et Adaptative –UMR CNRS 8251, Université PARIS- DIDEROT (7), Paris, France

M. Nurul Islam and Carl K. -Y. Ng
School of Biology and Environmental Science and UCD Earth Institute, University College Dublin, Belfield, Ireland

Claudia Verderio
Department of Medical Biotechnology and Translational Medicine, CNR Institute of Neuroscience, Universitá di Milano, Milano, Italy

Reem R. Al Olaby and Hassan M. Azzazy
Department of Chemistry, The American University in Cairo, New Cairo, Egypt

Laurence Cocquerel, Laure Saas and Jean Dubuisson
Center for Infection and Immunity of Lille, CNRS-UMR8204/Inserm-U1019, Pasteur Institute of Lille, University of Lille North of France, Lille, France

Adam Zemla
Pathogen Bioinformatics, Lawrence Livermore National Laboratory, Livermore, CA, United States of America

Jost Vielmetter
Protein Expression Center, Beckman Institute, California Institute of Technology, Pasadena, CA, United States of America

Joseph Marcotrigiano and Abdul Ghafoor Khan
Department of Chemistry and Chemical Biology, Rutgers University, Piscataway, NJ, United States of America

Felipe Vences Catalan and Shoshana Levy
Department of Medicine, Stanford University Medical Center, Stanford, CA, United States of America

Alexander L. Perryman
Department of Medicine, Division of Infectious Diseases, Center for Emerging & Re-emerging Pathogens, Rutgers University-New Jersey Medical School, Newark, NJ, United States of America

Joel S. Freundlich
Department of Medicine, Division of Infectious Diseases, Center for Emerging & Re-emerging Pathogens, Rutgers University-New Jersey Medical School, Newark, NJ, United States of America
Department of Pharmacology and Physiology, Rutgers University-New Jersey Medical School, Newark, NJ, United States of America

Stefano Forli
Department of Integrative Structural and Computational Biology, The Scripps Research Institute, La Jolla, CA, United States of America

Rod Balhorn
Department of Applied Science, University of California Davis, Davis, CA, United States of America

Lequn Zhao, Liang Qu, Jing Zhou, Zhengda Sun, Hao Zou, James D. Marks and Yu Zhou
Department of Anesthesia and Perioperative Care, University of California San Francisco, San Francisco General Hospital, San Francisco, California, United States of America

Yunn-Yi Chen
Departments of Pathology & Laboratory Medicine, University of California San Francisco, San Francisco, California, United States of America

Md. Gulam Musawwir Khan, Jean-François Jacques, Jude Beaudoin and Simon Labbé
Département de Biochimie, Faculté de Médecine et des Sciences de la Santé , Université de Sherbrooke, Sherbrooke, Québec, Canada

Premraj Rajkumar, William H. Aisenberg, Omar W. Acres, Ryan J. Protzko and Jennifer L. Pluznick
Department of Physiology, Johns Hopkins University School of Medicine, Baltimore, Maryland, United States of America

Jing Wang, Guangdong Hu, Zhi Lin, Lei He, Lei Xu and Yanming Zhang
College of Veterinary Medicine, Northwest A&F University, Yangling, Shaanxi, China

Jennifer L. Chinnici, Ci Fu, Lauren M. Caccamise, Jason W. Arnold and Stephen J. Free
Department of Biological Sciences, SUNY University at Buffalo, Buffalo, New York, United States of America

Hong-Po Dong, Kai-Xuan Huang, Hua-Long Wang, Song-Hui Lu, Jing-Yi Cen and Yue-Lei Dong
Research Center for Harmful Algae and Marine Biology, Key Laboratory of Eutrophication and Red Tide Prevention of Guangdong Higher Education Institutes, Jinan University, Guangzhou, China

Ali Salehzadeh-Yazdi, Yazdan Asgari, Ali Akbar Saboury and Ali Masoudi-Nejad
Laboratory of Systems Biology and Bioinformatics (LBB), Institute of Biochemistry and Biophysics, University of Tehran, Tehran, Iran

Sabine Gruber and Susanne Zeilinger
Research Area Biotechnology and Microbiology, Institute of Chemical Engineering, Vienna University of Technology, Wien, Austria

Lisa S. Pike Winer and Min Wu
Seahorse Bioscience Inc., North Billerica, Massachusetts, United States of America

Fang Wang, Miao Jiang, Shiwu Li and Lijun Yang
Center for Stem Cell & Regenerative Medicine, The Second Hospital of Shandong University, Jinan, P.R.China

Fei Wu, Yuan Zhang, Yuzhu Zhang and Dongqi Tang
Center for Stem Cell & Regenerative Medicine, The Second Hospital of Shandong University, Jinan, P.R.China
Shandong University Qilu Hospital Research Center for Cell Therapy, Key Laboratory of Cardiovascular Remodeling and Function Research, Qilu Hospital of Shandong University, Jinan, P.R.China

Haibo Song
Center for Stem Cell & Regenerative Medicine, The Second Hospital of Shandong University, Jinan, P.R.China
Shandong University Qilu Hospital Research Center for Cell Therapy, Key Laboratory of Cardiovascular Remodeling and Function Research, Qilu Hospital of Shandong University, Jinan, P.R.China
Center for Reproductive Medicine, Zibo Maternal and Child health hospital, Zibo, P.R.China

Mingxiang Zhang and Xing Li Wang
Shandong University Qilu Hospital Research Center for Cell Therapy, Key Laboratory of Cardiovascular Remodeling and Function Research, Qilu Hospital of Shandong University, Jinan, P.R.China

Taixing Cui
Shandong University Qilu Hospital Research Center for Cell Therapy, Key Laboratory of Cardiovascular Remodeling and Function Research, Qilu Hospital of Shandong University, Jinan, P.R.China
Department of Cell Biology and Anatomy, University of South Carolina, Columbia, South Carolina, United States of America

Zhongde Wang
Center for Reproductive Medicine, Zibo Maternal and Child health hospital, Zibo, P.R.China

Istvan Pirko
Department of Neurology, Mayo Clinic, Rochester, Minnesota, United States of America

Aaron J. Johnson
Department of Neurology, Mayo Clinic, Rochester, Minnesota, United States of America
Department of Immunology, Mayo Clinic, Rochester, Minnesota, United States of America

Holly L. Johnson
Department of Neurology, Mayo Clinic, Rochester, Minnesota, United States of America
Department of Immunology, Mayo Clinic, Rochester, Minnesota, United States of America
Neurobiology of Disease Graduate Program, Mayo Graduate School, Rochester, Minnesota, United States of America

Fang Jin, Whitney A. Manhart and Stephanie J. LaFrance
Department of Immunology, Mayo Clinic, Rochester, Minnesota, United States of America

Robin C. Willenbring
Department of Immunology, Mayo Clinic, Rochester, Minnesota, United States of America
Virology and Gene Therapy Graduate Program, Mayo Graduate School, Rochester, Minnesota, United States of America

Index

www.ingramcontent.com/pod-product-compliance
Lightning Source LLC
Chambersburg PA
CBHW061244190326
41458CB00011B/3572